Ecology & Evolution in the Tropics

Ecology & Evolution

The University of Chicago Press   •   *Chicago & London*

in the Tropics

A Herpetological Perspective

Edited by Maureen A. Donnelly,
Brian I. Crother, Craig Guyer, Marvalee H. Wake,
& Mary E. White

Maureen A. Donnelly is associate professor of biology at Florida International University. **Brian I. Crother** is professor of biological sciences at Southeastern Louisiana University and the editor of Caribbean Reptiles. **Craig Guyer** is professor of biology at Auburn University. **Marvalee H. Wake** is professor emerita of integrative biology at the University of California, Berkeley, and the editor of The Origin and Evolution of Larval Forms and Hyman's Comparative Vertebrate Anatomy, published by the University of Chicago Press. **Mary E. White** is professor of biological sciences at Southeastern Louisiana University.

The University of Chicago Press, Chicago 60637
The University of Chicago Press, Ltd., London
© 2005 by The University of Chicago
All rights reserved. Published 2005
Printed in the United States of America

14 13 12 11 10 09 08 07 06 05   1 2 3 4 5

ISBN: 0-226-15657-5 (cloth)
ISBN: 0-226-15658-3 (paper)

Library of Congress Cataloging-in-Publication Data

Ecology and evolution in the tropics : a herpetological perspective / edited by Maureen A. Donnelly . . . [et al.].
    p.   cm.
Includes bibliographical references and index.
ISBN 0-226-15657-5 (cloth : alk. paper) — ISBN 0-226-15658-3 (pbk. : alk. paper)
1. Amphibians—Ecology—Tropics.   2. Reptiles—Ecology—Tropics.
3. Amphibians—Evolution—Tropics.   4. Reptiles—Evolution—Tropics.
I. Donnelly, Maureen A., 1954–
QL664.6.E36 2005
597.9′1734′0913—dc22

                                                            2004014642

This book is printed on acid-free paper.

To the memory of James Edward DeWeese,
David Joseph Morafka, & Joseph Bruno Slowinski

Contents

Part II: Ecology, Biogeography, and Faunal Studies

Foreword

Of Heroes and Other Beasts

Historians of science would probably say that in Costa Rican herpetology there are three landmarks: E. D. Cope, E. H. Taylor, and J. M. Savage. But those of us who have followed and analyzed the evolution of the local natural history in the last fifty years know that, in reality, Costa Rican herpetology is divided into two eras: before and after Savage.

Throughout a lifetime of fieldwork and rigorous curatorial work in academia, including the parturition and nurturing of many distinguished professionals of today, Jay Savage has been a role model, a mentor, a friend, and an engine in generating new perspectives in tropical biology. He can justly be acknowledged as a modern hero of tropical studies, and I was delighted to find that very feeling in Harry Greene and Roy McDiarmid's chapter in this volume, "Wallace and Savage: Heroes, Theories, and Venomous Snake Mimicry" (chap. 9).

Even in his official retirement, Jay has been a motivating force behind this outstanding collection of essays, which brings together more than two dozen authors in the eighteen rich chapters that follow. All of these authors share the same goal: to push our knowledge and understanding of tropical herpetology further, from the Neotropics to the confines of Papua New Guinea, from a philosophical and ontological analysis of taxonomy to long-term frog monitoring for climate prediction, to the state-of-the-art DNA sequencing and phylogenesis of most groups of tropical amphibians and reptiles. That is Savage, the unexcelled mentor, at his best.

After his monumental opus, *The Amphibians and Reptiles of Costa Rica* (2002), this low-profile, humble man with the textbook diastema and his perennial jovial disposition, precipitates a deluxe array of authors who celebrate his accomplishments by giving us, and several generations to come, the blue-print to what is known of the tropical herpetofauna and to what would be desirable to learn in the near future. That is Savage, the engine, at his best.

A perfunctory look at Jay Savage's curriculum vitae would be enough to inspire many pages for a presentation of this volume, but in Jay's case it is his whole life and exemplary dedication to his discipline that speak volumes. That is Jay once more, in his role model mode, inspiring us to imitate and achieve. A difficult feat to follow. I need not write any more. Here it is. Our testimony of deepest appreciation for the sage and for the person.

Luis D. Gómez

Organization for Tropical Studies
National Academy of Sciences, Costa Rica

Preface

In 2000, Jay Mathers Savage submitted to the University of Chicago Press a book-length manuscript describing the amphibians and reptiles of Costa Rica (*The Amphibians and Reptiles of Costa Rica: A Herpetofauna between Two Continents, between Two Seas,* 2002). This project comprised a lifetime of study, and such an achievement warranted a celebration. To that end, the five editors of this volume organized a symposium, which was presented at the 2000 joint meeting of the American Society of Ichthyologists and Herpetologists, Society for the Study of Amphibians and Reptiles, and Herpetologists' League in La Paz, Baja California, Mexico. Because of Jay's long interest in the tropical regions of Latin America, it was fitting that our celebration occurred in tropical Mexico. Because Jay advocated for long-term study of organismal diversity, it was fitting that the symposium included speakers who shared Jay's interest in this aspect of biology.

Most of the papers presented during the symposium are included in this volume. The essays herein focus on tropical amphibians and reptiles, and most of the authors developed their interests in tropical organisms as a consequence of their interactions with Jay. The chapters in the volume fall into two broad categories that are not mutually exclusive: evolution and ecology. Some of the chapters that deal with biogeography are in part 1, whereas others are combined with the ecology set in part 2 to provide balance of presentation. Jay's influence on the field of tropical herpetology has been impressive, and these chapters are a glimpse of that influence. The diversity of topics covered reflects Jay's style as a mentor of herpetologists. He is open to and supportive of any avenue of scientific query, and this has led to his having colleagues with a great diversity of interests. The specific topics in the volume range from taxonomic practices to faunal surveys to working with native peoples in conservation projects. Yet the chapters all have a common theme: the study of amphibians and reptiles. Jay wore several research hats during his scientific career, but they were related to his herpetological research, and this is reflected in the diversity of essays in this volume. Some may argue that taxon-oriented research programs are of secondary importance to question-oriented programs. Jay's research program is unabashedly taxon-oriented in approach, as are those of the contributors to this volume. The power in such a program is that it breeds diversity, arguably a highly desirable quality in our quest to understand the world. This volume clearly depicts scientific diversity derived from studies on reptiles and amphibians.

Several groups and individuals provided key support in convening the symposium and publishing its results. We extend our thanks to the American Society of Ichthyologists and Herpetologists, the Herpetologists' League, and the Society for the Study of Amphibians and Reptiles for providing funds to support the participation of graduate students (D. P. Bickford, S.-H. Chen, J. Lament, J. McKnight, K. Nicholson). We thank Drs. Carlos Villavicencio Garayzar and Richard Vogt for their help with the logistics of the symposium, and Raquel Bernáldez King, senior conference planner, for her help with arrangements for the banquet that marked an end to our symposium celebrations.

We thank the contributors for their patience and their attention to deadlines and requests for changes. All of you made this volume possible. We also thank those scientists who presented papers during the symposium (Jasper Lament, Jenna McKnight, and Dave Morafka) for their participation in 2000.

The manuscript of each chapter was reviewed by two or more peers, and to the following colleagues we extend thanks for careful editorial comments: Teresa Avila-Pires, Edmund D. Brodie Jr., Edmund D. Brodie III, Marguerite Butler, Janalee C. Caldwell, Jonathan Campbell, Paul Chippendale, C. Jay Cole, David Cundall, Kevin de Queiroz, Rafael de Sá, Darrel R. Frost, Claude Gascon, Michael Ghiselin, Lou Guillette, Diana Hews, Richard Highton, Marinus S. Hoogmoed, Lynne Houck, Arlyne Johnson, Scott Keogh, Julian C. Lee, Andrew L. Mack, William Magnusson, Stewart Marsh, Jim McGuire, Joseph Mendelson III, Molly Morris, Christopher Murphy, Steven Poe, Tod Reeder, Robert Reynolds, Olivier Rieppel, Javier Rodriguez, Alex Rooney, H. Bradley Shaffer, Jack Sites, Joseph B. Slowinski, Linda Trueb, Graham G. Watkins, James I. Watling, Larry David Wilson, Debra Wright, and Wolfgang Wüster. The first draft of the entire volume was reviewed by Edmund D. Brodie Jr. and two anonymous reviewers. We thank everyone for their comments, which helped improve the essays.

We thank Kathryn Gohl for her careful editing of the entire volume. It was a pleasure to work with Kathy. Jennifer Howard and Christine Schwab at the University of Chicago Press provided invaluable assistance, especially during the final push to complete the project.

Last we thank Christie Henry for encouraging us to assemble this collection of essays into a book. Although at the outset we never considered the possibility that these papers would be published, they serve as a powerful statement to the influence Jay M. Savage has had on modern herpetology.

Part I
Evolution and Biogeography

Evolution refers to changes in organisms through time, and this broad theme is the topic of the first nine chapters in this collection. Most textbook discussions of evolution focus on the process by which changes in gene frequencies lead to altered characteristics of organisms, generally in response to changing environments. However, evolutionary studies also include examinations of the patterns observed among organisms and the inferences that can be drawn about historical relationships from such studies of pattern. Both general approaches to evolutionary biology are included in this section of the volume, but the primary focus is on examining patterns and exploring the many implications of such studies. The goals in packaging these contributions together are to explore important new examples of evolutionary patterns derived from studies of tropical reptiles and amphibians, to explore new uses for such patterns, and to provoke new examinations of the mechanisms that explain their stunning diversity.

In chapter 1, Arnold Kluge proposes a system of taxonomy to replace that of Linnaeus, whose system has been in use for 250 years. Kluge discusses modernization of taxonomy by linking evolutionary theory with phylogenetic hypotheses. He describes the role science plays in systematization, provides description of universals and particulars, and compares definition from intension with definition by ostention. Classification, according to Kluge, is the orderly arrangement of a set of particulars. All of the historical entities in Kluge's phylogenetic taxonomy (phylospecies) can be systematized because they are naturally related to one another. Phylogenesis is the process of phylospecies origination, and the phylospecies is the smallest historical individual. Kluge points out that explanatory power is maximized deductively in phylogenetic systematics, especially when one uses all available data. In Kluge's phylogenetic systematics, a diagnosis is a definition from extension (i.e., a description of particulars). Some may argue that shifting taxonomies are unstable, and in some cases such shifts generate irritation when scientists must learn new names. But Kluge asks what is so great about stability in taxonomy when it is wrong. He examines the Linnaean classification system and descent classification (as proposed by Kevin de Queiroz and colleagues). Kluge's system results in a rankless hierarchy of names. Each phylospecies is identified by a type-specimen, and each higher taxon includes at least two phylospecies. Names are determined by ostension and typification, following the phylospecies code. Kluge demonstrates this new system by using eublepharid geckos as an example.

Marvalee Wake, Gabriela Parra-Olea, and Judy Sheen, in chapter 2, describe the phylogenetic relationships among Central American genera of caecilians, an odd radiation of fossorial, limbless amphibians that is restricted to tropical habitats. The key evolutionary point raised in this chapter is that phylogenetic studies, even preliminary ones, can shape our understanding of key features of secretive organisms. On the basis of morphological and fossil evidence, the genera *Dermophis* and *Gymnopis* are discovered to be Laurasian representatives, and all other Recent caecilians are derived from Gondwanan ancestors. Novel molecular data (16S ribosomal and cytochrome *b* sequences) are beginning to provide additional information on relationships of these elusive amphibians. But as is frequently the case with such ancient groups, the basal relationships of caecilians are not well supported by any data. However, high distance values for molecular data suggest that the Central American caecilians are old and well-defined lineages. Two clades correlate with Taylor's Caeciliinae and Dermophiinae, but the relationships of *Gymnopis* and *Dermophis* are complex, as are the relationships among these taxa and the South American genera *Microcaecilia* and *Siphonops*. Wake and colleagues argue that additional sampling of taxa is critical to understanding the relationships among these but that current data aid in interpretation of patterns of evolution of reproductive modes within the group.

Chapter 3 summarizes information gathered by David Wake on the diversity of Costa Rican salamanders. Wake's curiosity about salamanders has stimulated a tremendous amount of research on these amphibians, and in this chapter he uses phylogenetic and distributional information to infer how these animals radiated to fill their current niches. The Neotropical clade of salamanders is species rich and includes one-half of all salamander species, most of which are found in the smallest part of the Neotropics. This monophyletic group of salamanders has direct development, acute eyes, an accurate mechanism for protruding the tongue, large cells and genomes, and tail autotomy. In Costa Rica, 45 to 48 species of salamanders currently are known, a doubling of the number indicated by the earlier works of the influential herpetologist Edward Taylor. The Costa Rican animals belong to three genera: *Nototriton* (7 species in two clades), *Oedipina* (14 species in two subgenera), and *Bolitoglossa* with 23 species. Wake describes the general ecological patterns of Costa Rican salamanders and notes that although diversity is high across elevational transects, few species occur together at a single site. This fauna uses unusual microhabitats (e.g., moss mats) not used elsewhere in Middle America. From these observations, he argues that a distinctive group of salamanders was restricted to the Talamancan region of Central America for much of the Tertiary and that this region has not been invaded by lineages found to the north in Nuclear Central America. Wake concludes with prospects for future study in the face of amphibian declines that have swept through Costa Rica in recent years.

In chapter 4, the focus of the evolutionary theme changes to one of systematic problems associated with single species. In this chapter, Ronald Heyer, Rafael de Sá, and Sarah Muller examine the enigmatic frog *Leptodactylus silvanimbus,* a species found only in cloud forests of Honduras. Heyer and colleagues tested three hypotheses regarding the relationships of the species: (1) that *L. silvanimbus* is most closely related to *L. melanonotus;* (2) that *L. silvanimbus* is a member of the *L. melanonotus* species group but is not most closely related to *L. melanonotus;* and (3) that *L. silvanimbus* is a relict in Honduras and has no close relationships with any Middle American species. Separate analyses of both 812 molecular characters and 51 nonmolecular characters support the third hypothesis. This study is a classic example of the process by which evolutionary biologists investigate species that appear to represent distributional disjunctions. It also serves as a counterexample to the procedure advocated by Kluge for dealing with multiple data sets for the same group.

Shyh-Hwang Chen, in chapter 5, examines the utility of chromosome data for determining historical relationships of species within the *rhodopis* and *diastema* groups of the species-rich genus *Eleutherodactylus.* Chen builds on promising work by the late Jim DeWeese, who first noted that morphological patterns of chromosomes appeared to convey phylogenetic information within Costa Rican *Eleutherodactylus.* Chen describes the novel chromosomal data for six species of *Eleutherodactylus* and adds these to reexaminations of other taxa. These data suggest that the *E. rhodopis* group is monophyletic, but the content of his group is not equivalent to previous attempts to delimit the group; the data also suggest that there are two clades within the *E. rhodopis* group.

In chapter 6, Sharon Emerson takes a mechanistic approach to the study of evolution by exploring the physiological basis of sexual dimorphism in frogs. She relates patterns of dimorphism to the organization–activation theory of sexual differentiation, a theory that hypothesizes that sex steroids act early in development to organize secondary sexual characteristics and at sexual maturity to activate their expression. Emerson points out that sexually dimorphic morphologies can arise through diverse physiological processes and have different growth characteristics. She then hypothesizes that the ease with which features produced by hypertrophy of cells and the activational effects of hormones are gained or lost is inversely related to the ease with which features produced by hyperplasy and the organizational effects of hormones are gained or lost. Emerson suggests that studies of androgens in free-ranging frogs will allow for the types of tests requisite to the examination of hypotheses.

The last three chapters in part 1 discuss aspects of snake evolution and explore various aspects of patterns associated with species-rich radiations of organisms. Two of the chapters examine snake phylogeny, but from opposite ends of the phylogenetic tree. In chapter 7, Mary White, Maria Kelly-Smith, and Brian Crother examine the origin of snakes and the basal radiations within this group as inferred from morphological and molecular data. These authors explore problems with the use and correct interpretation of molecular data and the relationship of these data to morphological data. Additionally, they describe persistent problems with understanding the early evolution of large clades, like that of snakes, in which key divergences might be supported by very few recoverable characters. Joseph Slowinski and Robin Lawson, in chapter 8, examine the relationships of elapids, the group to which the late Joe Slowinski devoted much of his professional life. This contribution represents an example of how molecular data can elucidate relationships of large but relatively recent clades. This large group of venomous snakes (nearly 300 species in approximately 60 genera) remains poorly known, and its relationships to other snakes are poorly understood. Slowinski and Lawson used two different types of DNA (mitochondrial and nuclear) to understand elapid phylogeny and found a well-supported link between elapids, atractaspidids, *Homoroselaps,* and other colubrid genera. However, their data also demonstrate the care with which molecular data should be treated, because the c-*mos* and mitochondrial DNA did not always provide consistent information.

In the final chapter of part 1, Harry Greene and Roy McDiarmid honor Alfred Russel Wallace and Jay Mathers Savage for their contributions to our understanding of venomous snake mimicry and indicate how recent evolutionary information about snakes can shed new light on old ideas about hypotheses of interactions of species. Greene and McDiarmid update Pough's review of mimicry and describe additional mimicry systems in venomous snakes as well as unsolved puzzles. They also describe the possibility that invertebrates serve as noxious models for snakes. They conclude from this review that four macro-evolutionary patterns define a Savage-Wallace effect within mimicry systems. These patterns include the following: (1) mimicry is likely among related organisms that share a common body plan; (2) mimicry spanning distantly related organisms depends on taxa sharing a simplified body form; (3) snake mimicry is widespread because venomous snakes can injure or kill predators; and (4) the origin of noxious attributes can increase diversity within a clade because the models protect the mimics. This chapter serves as an example to revitalize the role of natural history in synthesizing information from evolutionary, ecological, and behavioral biology.

1

Taxonomy in Theory and Practice, with Arguments for a New Phylogenetic System of Taxonomy

Arnold G. Kluge

Theory without fact is fantasy, but facts without theory is chaos.
CHARLES OTIS WHITMAN

Taxonomy as it is currently practiced is nearly 250 years old, and it is only now undergoing its first major revolution (Pennisi 1996). There is widespread agreement that categorical ranks must go, and with this change alone there is repudiation of what has been to date the venerable Linnaean approach to classification (Linnaeus 1758; see review in Ereshefsky 2001) as well as the rejection of the well-entrenched International Codes of Nomenclature (ICBN 1994; ICNB 1992; ICZN 1999). Making explicit connections to evolutionary theory and phylogenetic hypotheses is also considered important to the modernization of taxonomy. The issues underlying these attacks on the Linnaean tradition are especially complex because they involve the whole of science, from the philosophical to the pragmatic (Ereshefsky 2001). My attempt to understand the theory and practice of taxonomy, including the current revolution, is founded largely on the distinction between classification and systematization (Griffiths 1974; see also Griffiths 1973, 1976). Most of the discussion to follow explores the role science plays in systematization, and with that understanding I propose and exemplify a new phylogenetic system of taxonomy.

Definitions and Distinctions

There are several different kinds of definitions, which unfortunately are conflated in ordinary discourse, and whose record of use in science is not much better. Particularly important terms in distinguishing the classification of Linnaeus from the systematization of G. C. D. Griffiths are *extension, intension,* and *ostension* (table 1.1). Intension is used to define universals. Ostension and extension are used to define particulars. Extensional definitions, however, may look like intensional definitions, thus opening the door for confusion such that particulars seem to be defined in universal terms. Also potentially confusing are those things that may be judged abstract (table 1.1), such as the Equator and the North and South Poles. Suffice it to say that these kinds of particulars are not so abstract as to keep us from knowing when we have crossed the Equator or have arrived at either of the Poles.

Intension (connotation) is the principle according to which a class, or set, of things is picked out; it is the condition a thing must satisfy to be precisely described by the predicate (Blackburn 1994). Intension involves a defining rule whereby members are included, or excluded, from the set. The conditions are the properties common to all those things (and only them) that are referred to by a word (Angeles 1992). Definition from intension is a form of prescription, because the members of a set are prescribed as necessarily having certain qualities. Prediction requires intensional definition, and that kind of definition is important in the identification of properties that may constitute a generality, such as a universal law of nature (Kluge 2003b).

Ostension is definition by reference, by enumeration or pointing at particulars (by showing the object to which a name is given). This kind of definition is used in identifying configurational and contingent relations, as in the case of a particular taxon and its historical relationships.

Definition from extension is a form of ostension, and it is also used to define particulars. Extensional definitions appear to differ from intensional ones only semantically, but this is not really the case. Consider that the denotation of a predicate is the class of things (extensions) that a term (intension) picks out. For example, the extension of the universal "red" is the class of red things (particulars) (Blackburn 1994). Extension can also be used to evaluate the characteristics of the members of a demarcated class, in order to formulate a defining "rule" (a descriptive generalization) written in the form of an intension.

Classification is the orderly arrangement of a set of particulars that instantiate some intensionally defined, abstract, generality—a universal, or what some would call a natural kind (tables 1.1, 1.2). For example, the Periodic Table is a classification, one of material essentialism, in which an element's place in the table is defined intensionally according to the inherent property of atomic number—the number of protons possessed by the element. Thus, gold and

Table 1.1 The different roles particulars and universals play in science

Particulars: Concrete objects (composite wholes, entities, individuals, parts, things)	Universals: Abstract objects (classes, concepts, functions, generalities, kinds, natural kinds, numbers, propositions, sets)
1. A particular is "defined" ostensively (by reference, pointing, or enumeration) or extensionally (denotatively),	1. A universal is defined intensionally (connotatively),
2. and with such imprecision a particular can be wrongly identified.	2. and with such necessary and sufficient qualification a universal cannot be wrongly identified.
3. A particular is always spatially restricted.	3. A universal is always spatially unrestricted.
4. A particular is always temporally restricted.	4. A universal is usually temporally unrestricted.
5. A particular is a member of a universal *or* a part of some other particular.	5. A universal has members.
6. Those parts *of parts* exhibiting replicator and interactor qualities are referred to as contemporary individuals, whereas those that are the effects of inheritance, but not effectors, are historical individuals.	6. The members of a universal are denied historical connections, because they are spatiotemporally unrestricted.
7. The parts of an individual need not be, and frequently are not, similar.	7. Members are identical to the extent they share exactly the attributes that define the class name.
8. Postdictive explanation is description of individuals, which requires ostensive or extensional definition.	8. Predictive explanation is achieved through abstract generalization, which requires intensional definition.
9. A particular (including a contemporary or a historical individual) is temporary and changeable (mutable).	9. A universal is permanent and unchangeable (immutable).
10. An individual has a "fuzzy" boundary (physically speaking, as in the case of an organism's origin and death, or the origin and extinction of a historical individual).	10. A universal has a precise boundary (conceptually speaking, given the necessary and sufficient qualifications for membership).
11. The names of historical individuals are proper nouns.	11. The names of universals are predicates (assumed attributes).
12. Particulars are rarely considered philosophically and scientifically important.	12. Universals are usually considered philosophically and scientifically important.

Table 1.2   Distinguishing taxonomic classifications and systematization

	Linnaean classification (Linnaeus 1758)	Annotated Linnaean classification (Wiley 1981b)	Descent classification (de Queiroz and Gauthier 1992)	Set theory classification (Papavero et al. 2001)	Phylogenetic system
Higher categorical ranks (family, etc.)	yes	yes	no	no	no
Each higher taxon is a proper noun	yes	yes	yes	no	yes
Genus categorical rank	yes	yes	yes	yes (praenomen)	no
Genus group name italicized	yes	yes	no	yes	no
Species categorical rank	yes	yes	yes	yes	no
Sequencing convention[a]	no	yes	yes	no	no
Polytomous relationships convention[b]	no	yes (sedis mutabilis)	no	yes (F-genos)	yes (sedis mutabilis)
Unresolved relationships convention[c]	no	yes (insertae sedis)	no	yes	no
Plesion convention[d]	no	yes	no	no	no
Stem/crown group convention[e]	no	no	yes	no	no
Metaspecies convention (*)[f]	no	no	yes	no	no
Redundant names	yes	no	no	no	no
Taxonomic names defined	intensionally	intensionally (PSD)[g]	intensionally (PSD)[g]	intensionally (PSD)[g]	ostensively (PSR)[g]

Table 1.2   (continued)

	Linnaean classification (Linnaeus 1758)	Annotated Linnaean classification (Wiley 1981b)	Descent classification (de Queiroz and Gauthier 1992)	Set theory classification (Papavero et al. 2001)	Phylogenetic system
Primary emphasis	stability	stability	stability	stability	descriptive efficiency
Species concept(s) defined	usually qualitatively	causally	qualitatively or causally	qualitatively	causally
Monophyletic taxa	not emphasized	where possible	at all levels, except species	at all levels, including species	at all levels[h]

Note: PSD, phylogenetic system of definition; PSR, phylogenetic system of reference.

[a] See convention 3 of annotated Linnaean classification for further details (Wiley 1981b, 209).

[b] See convention 4 of annotated Linnaean classification for further details (Wiley 1981b, 211).

[c] See convention 5 of annotated Linnaean classification for further details (Wiley 1981b, 212).

[d] See convention 7 of annotated Linnaean classification for further details (Wiley 1981b, 219).

[e] See de Queiroz and Gauthier 1992, fig. 7.

[f] See Donoghue 1985, 177; see also de Queiroz and Donoghue 1988.

[g] Härlin 1998.

[h] Phylospecies, like all higher taxa in the phylogenetic system of taxonomy, are necessarily monophyletic, because of their historical individuality, whereas "species" terminal taxa of the other taxonomic approaches are not necessarily monophyletic but may become so in time.

lead are classified (ordered) in relation to each other, and to other elements in the table, according to their atomic numbers, 79 and 82, respectively. To obtain gold from lead does nothing to the intensionally defined categories of either "goldness" or "leadness." There still exists a place in the Periodic Table for atomic numbers 79 and 82. Atomic theory is the major premise underlying the Periodic Table, with the table successfully predicting elements new to science. Because predictiveness is considered a condition of natural laws, atomic theory is spoken of in law-like terms (table 1.1).

Laws of nature, and their accompanying theories, are subject to critical testing and revision, and even rejection. For example, in formulating the original version of the Periodic Table in 1871, Mendeleyev defined the classification in terms of atomic weight. At that time, gold and lead were intensionally defined as having weights 197.2 and 207.2. In 1942, the definition of lead was changed to 207.21. In spite of these kinds of redeterminations of atomic weights, some of the elements in Mendeleyev's Periodic Table were still required by their properties to be put in positions out of order of atomic weight. The problem was resolved when the structure of the atom became better understood, whereupon the Periodic Table was defined according to the theory of atomic number. Webb et al.'s (2001) recent findings that call into question the constancy of the speed of light further illustrate that no scientific theory is immune to testing, even one of nature's most well-known laws.

All of the historical entities in the new phylogenetic taxonomy proposed herein, those that I call phylospecies, and the more inclusive parts of history to which they belong, can be systematized, because they are naturally related to one another (see the following section). Particular instances of those kinds of historical entities are not matters of intensional definition, because they are not *prescribed* in terms of inherent qualities, that is, properties necessary and sufficient for membership in some group. This is unlike such abstractions as the categorical ranks of the traditional Linnaean and annotated classifications, which can only be instantiated by qualitatively defined things, things analogous to the instances of the elements in the Periodic Table. In no sense can there be laws of history (Popper 1957). History is not predictable (e.g., contra Schwenk 1994; Griffiths 1999), even though the retrodiction of trends in phylogenetic history is undeniable. History simply is, and its retrodiction is potential only in the marks it may leave that are not subject to an information-destroying process (Sober 1988, 3−5).

The parts of phylogenetic history, such as species and the more inclusive groups to which they belong, must be testable and subject to revision for the system to be considered scientific (Kluge 1999, 2003a). This involves the use of discovery operations that are capable of exposing false hypotheses and providing novel, testable explanations. Those operations must be ideographic (historical, retrodictive), given the nature of the system. They cannot be nomothetic (universal, predictive) (Grant 2002).

From the Philosophical to the Pragmatic

INDIVIDUATING HISTORICAL ENTITIES

The importance of individuation in taxonomy is clear. As Quine (1958, 1960) convincingly argued, there can be "no entity without identity" if language is to express proper scientific facts. The individuation of historical entities is straightforward because they are naturally related to one to another by virtue of inheritance (Darwin 1859, chap. 13; Ghiselin 1966; Hull 1978). Indeed, those entities are self-defining because of the causal relation of inheritance, and it is that self-definition that provides the basis for their objectivity. It is then the causal relation of inheritance that constitutes the basis for the *system* of taxonomy (table 1.1). The system is genealogical, that which represents the natural, hierarchical arrangement of species.

A PHYLOGENETIC SPECIES CONCEPT

A rigorous scientific research program requires that an abstract, theoretical definition of a class of particulars (table 1.1) logically precede the formulation and testing of the discovery operations designed to detect those things (Grant 2002). For example, Kluge (1990, 423) defined the concept *species* as "the smallest historical individual within which there is a parental pattern of ancestry and descent." This definition follows from Darwin's (1859) theory of evolution, "descent, with modification," from those conditions pertaining to the inheritance of historical individuals, and from the prevailing notion that species are the effects of evolution, not effectors (Lidén 1990). Note that this definition of species does not presuppose any discovery operation, nor that the entities are of any particular kind, bisexual or asexual, and extant or extinct.

Kluge (1990, 422; see also Hull 1980; Sober 1991; de Queiroz et al. 1995) also drew attention to the distinction between historical and contemporary individuals (table 1.1). Contemporary individuals, such as a population or a deme, are less inclusive than historical individuals, and although all kinds of contemporary individuals are historical, insofar as they are extended in time, their unity is the result of the cohesive and integrative processes of their replicator-continuum (Lidén 1990). In contrast, historical individuals are united by common history, not by their current interactions.

The most inclusive entity in a historical system is earth-bound life, whereas the least inclusive entity is the smallest historical individual, the phylogenetic species, or more simply what I call phylospecies. Although the concept of phylospecies, as defined above, does not necessarily exclude any scientific discovery operation, the identification of phylospecies, as well as the natural groups of which they are a part, will quite likely involve phylogenetic systematic analysis (Hennig 1966). In that regard, the identification of phylospecies is consistent with the scientific goals of the phylogenetic system of taxonomy proposed in this chapter. The phylospecies concept is then not the same as the operational phylogenetic species concept that requires *fixed* character differences (e.g., sensu Eldredge and Cracraft 1980; Cracraft 1983; see also Vrana and Wheeler 1992; for review, see Frost and Kluge 1994).

Asexual phylospecies can be discovered just like bisexuals, according to phylogenetic systematic analysis (Kluge 1990). Asexuals may as well be judged good phylospecies in theory (Kluge 1990). They are historically connected and exhibit inherited traits, just as bisexuals do. Although asexuals do not engage, at least regularly, in those processes that lead to the cohesiveness of bisexual populations, such as gene flow and genetic homeostasis, nonetheless they may be rendered "cohesive" by developmental canalization as well as by natural selection acting on the similar behavioral and ecological traits that they have inherited. Thus, the phylospecies concept is unique in that it covers the extant as well as the extinct, and bisexuals as well as asexuals (Ereshefsky 2001).

Having explicated phylospecies, we can consider certain fundamentals relating to their origin. Arguing by analogy, we cannot liken the origin of phylospecies to reproduction, gestation, and birth. These are the kinds of processes that are identified with the origin of contemporary individuals and can only apply metaphorically to the singular model of species (Lidén 1990, 184). Rather, the origin of phylospecies is like mitosis or schizogony (replication by fission) (Frost and Kluge 1994). If we assume the latter analogy, the process of phylospecies origination should be termed phylogenesis (Hennig 1965, 97). Phylogenesis is then the patterns and processes relating to the origin of the smallest historical individuals (e.g., see Cracraft 1992; Hovenkamp 1997), as distinct from speciation, which involves the replicator-continua of populations and genes. Arguably, phylogenesis in asexuals is also one of fission. Lidén (1990, 184) makes especially clear the nature of those studies that focus on historical individuals: "Macroevolution . . . can be defined as differential [phylogenesis] and extinction (sorting) of gene pool continua, which in turn of course leads to change in diversity at higher (taxa) levels, but as integrating processes are lacking, a 'hierarchical theory of evolution' . . . above the level of replicators is theoretically empty. . . . If the clades manifested as *Aves* and *Anemone nemorosa* become extinct, they do so as a sum of replicators. We do not need theories for that!"

I do not object to using the terms *species* and *speciation*, providing they are nominal, or taken to mean phylospecies and phylogenesis, respectively. Phylospecies should not be equated to the Linnaean categorical rank of species, or to any one of the familiar qualitative conceptualizations of species, such as those definitions having to do with interbreeding (Mayr 1970, 12; Paterson 1985) or ecological (Van Valen 1976) processes. An exclusive use of phylospecies and phylogenesis, rather than species and speciation, clearly sets systematization apart from classification and the population/process thinking of the neo-Darwinian synthesis, the latter not being concerned with historical individuality but with the replicator-continua of populations and genes. As Lidén (1990; see also Gardiner 1952) has emphasized, the study of contemporary individuals is legitimate in its own right, but it must not be confused with necessarily unique particulars, like phylospecies, and the study of their origin and extinction.

MONOPHYLY

In a taxonomic system defined in terms of inheritance, each taxon is a historical individual; it is a spatiotemporally restricted, as well as necessarily unique, part of phylogeny (Kluge 1990, 1997, 1998a, 1999; Siddall and Kluge 1997). Therefore, each historical individual *is* monophyletic; "A 'thing' cannot be paraphyletic" (Lidén 1990, 183; contra de Queiroz and Donoghue 1988). The monophyly of taxa inclusive of two or more phylospecies follows Hennig's (1966) definition, that is, an ancestor and all of the phylospecies hypothesized to have evolved there from. The ancestor, or common ancestral phylospecies, "is *identical* with *all* of the species that have arisen from it" (Hennig 1966, 71; my emphasis); it is not one of the parts of the more inclusive taxon being defined ostensively. Neither does a common ancestral phylospecies depend on an operationalism for its existence. It is irrelevant, for example, that few, if any, of the common ancestral species that have been hypothesized have survived all of the empirical tests demanded by phylogenetic systematic analysis. Moreover, that common ancestral species are verified all the time with so-called missing links is irrelevant, because those induced speculations involve special knowledge, such as plausible adaptive scenarios, or overall similarity. These are not scientific tests, because there is no attempt to disconfirm competing hypotheses of relative recency of common ancestry (Kluge 2003b).

In the case of phylospecies, monophyly concerns only a spatiotemporally restricted, necessarily unique point of origin. In the parlance of phylogenetic systematics, the monophyly of phylospecies does not include sister or common ancestral lineages. This interpretation of monophyly is substantially different from that of evolutionary systematists, who define monophyly in terms of process, for example, as the "natural outcome of the evolutionary process" of descent (Donoghue and Cantino 1988, 107), the occurrence of "long-term extrinsic [zoogeographic] barriers to gene flow" (Avise et al. 1987, 517), or "the stage . . . at which point neither species is 'paraphyletic' with respect to the other" (Goldstein and DeSalle 2000, 371). As an example of what is wrong with process as a definition, paraphyletic species can *become* monophyletic if the excluded part were to *go* extinct. Consider further that in the case of the biological species concept, all paraphyletic sets of interbreeding populations have the potential to *become* monophyletic.

To equate historical individuality and monophyly at all levels of taxonomy, including that of phylospecies, has other significant consequences. On the decidedly positive side, all taxa can be evaluated on the same scale, that of historical individuality, with all phylospecies being the same unit of measure, that is, the smallest historical individual. Also, this interpretation of monophyly calls into question how it has been used in other historical contexts, such as the history of nucleotides, genes, cells, tissues, organs, organisms, populations, and demes, that is, the replicator-continua models of contemporary individuals. Whereas the historical individual interpretation of monophyly does not prescribe any discovery operation, which is as it should be for conceptual definitions, process kinds of monophyly are defined in terms of inherent qualities, and for which a particular kind of discovery operation is presupposed.

TO FIX A TAXONOMIC NAME

Taxonomic names are fixed by definition or reference (Härlin 1998). In a phylogenetic system of definition, a name can be fixed intensionally in either of three ways: (1) by having certain evolved traits (e.g., Tetrapoda as specified by having arms and legs); (2) by including two or more species (e.g., Vertebrata as specified by *Petromyzon marinus* and *Squalus acanthias*); or (3) by including one or more species and excluding one or more other species (e.g., Vertebrata as specified by including *Petromyzon marinus* but excluding *Drepanaspis gemuendensis*). These are the apomorphy-based, node-based, and stem-based definitions of de Queiroz and Gauthier (1990). Cantino and de Queiroz's (2000) Principle of Reference is a phylogenetic system of definition, not one of reference (sensu Härlin 1998; M. Härlin, pers. comm.). In sharp contrast, a phylogenetic system of reference is situational, where a name is fixed in reference to a particular historical individual. This is an ostensive definition, which leads us to the discovery of those entities that are to be defined. This kind of ostension is not to be confused with the ostensively defined type approach of Ghiselin (1974) and Hull (1978).

TAXONOMY AS SCIENCE

Darwin (1859) set forth three logically connected principles of evolution: "descent, with modification, due to natural selection." As discussed in detail elsewhere (Kluge 1997, 1999, 2001a), assuming the first two assumptions in a deductive model of explanation is sufficient to identify and explain the ideographic system of phylospecies relationships. Further, it is well known that phylogenetic systematics (sensu Hennig 1966) provides scientific operations sufficient to delimit that history, where the most parsimonious hypothesis identified is the least disconfirmed (Kluge 2003a).

This view of science excludes induction, and in doing so excludes the need for explanation provided by natural kinds and essentialism of the relational (Griffiths 1999) or the causal homeostatic mechanism kind (Boyd 1999). There should be no enthusiasm for natural kinds and essentialism in phylogenetic inference, because all epistemological needs are met by deduction (Kluge 2003a). Further, in common ancestry as relational essentialism, "species are necessarily related because they are related by common ancestry," and that is tautological. The essentialism of a causal homeostatic mechanism begs the question as to "what makes a homeostatic mechanism a homeostatic mechanism for a particular kind" (Ereshefsky 2001, 107). Yet more fundamental, what is homeostatic cannot evolve, by definition.

In general, explanatory power is the ability of a hypothesis to explain the evidence (Salmon 1966, 46). This has nothing necessarily to do with a statistical assessment of when sufficient data have been acquired, or with the plausibility or verisimilitude of the resulting hypothesis (Watkins 1984). In phylogenetic systematics, parsimonious hypotheses have long been sought for their power to explain hypothesized shared-evolved character states (synapomorphies) in terms of inheritance (homology), and it was Farris (1979, 1980, 1983, 1989, 2000) who argued most convincingly that maximizing that power in the phylogenetic system of the optimal hypothesis (sensu Hennig 1966) means minimizing instances of independent evolution (homoplasy) as ad hoc hypotheses (Popper 1959, 145). I have argued elsewhere (Kluge 1997, 1998a, 1998b, 1999, 2001a, 2002, 2003a) that explanatory power is maximized deductively in phylogenetic systematics, not inductively. It is maximized in the form of the least-refuted, most parsimonious, phylogenetic hypothesis, which is the most highly corroborated and severely tested proposition. Popper (1959, 145) was clear when it came to this deductive aspect of science—it is "the principle of parsimony . . . that restrains us from indulgence in . . . auxiliary hypotheses," the least-falsified hypothesis being the most parsimonious (see also pp. 272–73). Severity of test requires critical evidence (Kluge 1997, 2001a, 2002; Farris et al. 2001), and the tests must be sophisticated if they are to lead to novel explanations, that is, explanations beyond what are contained in the data (Kluge 1998a, 2003a).

Parsimony in either of these senses does not constitute a model relating to the simplicity of nature, whereas differentially weighting characters in a parsimony analysis is a model, because counterfactual process assumptions are made, in addition to the background knowledge of "descent, with modification" (Kluge 1998a). Effectively, explanatory power is maximized with parsimony at all levels of discoverable biodiversity, including that of phylospecies, with phylogenetic systematic methods and unmodeled (unprobabilified) data (Kluge 1997, 2003a). The operational phylogenetic species concept defined in terms of fixed character difference does not provide that guarantee.

Hybridogenesis may be hypothesized when a reticulate pattern of relationships describes the data more parsimoniously than does the most parsimonious dichotomous phylogenetic hypothesis (Nelson 1983; see also McDade 1992). In this sense, the identification of hybrid phylospecies may further maximize explanatory power and thereby increase the descriptive efficiency of the corresponding taxonomy (see below). Such hypotheses must be subject to test.

When it was first introduced to phylogenetic inference, the concept of total evidence was argued deductively (Kluge 1989a; see also Farris 1983), in terms of maximizing the explanatory power of the optimal, least ad hoc, hypothesis. It can also be understood inductively, where any partitioned analysis of the available evidence does not maximize severity of test, and consequently neither does it maximize explanatory power or degree of corroboration (Kluge 1997, 2001a, 2002, 2003a).

Phylogenetic hypotheses are expected to change as they become more severely tested with critical evidence (Kluge 2001a, 2002) and as technology reveals new sources of variation that may constitute additional evidence. As hypotheses change as a result of the weight of this "new" evidence, the nomenclature of taxonomy must follow if it is to be of any use in communicating the updated scientific knowledge. There is but one history, for which, in the case of the phylogenetic system of taxonomy outlined later in the chapter, an ostensively defined monophyletic nomenclature is a perfect match.

OBJECTIVITY IN TAXONOMY

Diagnosability is one of the better-argued operationalisms in taxonomy, because of its obvious relationship to descriptive efficiency (Farris 1979, 1980, 2000). When taxonomy is based on the monophyletic parts of the most parsimonious hypothesis of phylogenetic relationships, descriptive efficiency (descriptive power) is maximized; that is to say, synapomorphies most efficiently diagnose (characterize) the named groups of taxa. Effectively, for the set of data used in deducing the phylogenetic hypothesis of relationships, descriptive efficiency is maximized on the hierarchy of names that has the smallest number of diagnostic entries (fewest steps). Although diagnosability itself cannot be justified in terms of an evolutionary epistemology, it is nonetheless consistent with the results of phylogenetic systematic analysis that is covered by such an epistemology (Hull 1964, 3–4; Wiley 1981a; Kluge 1999). In this regard, it is worth emphasizing again that the characters used to deduce monophyletic groups are contingent (Kluge 1997, 1999), not logically necessary to the origin of any historical individual or to the corresponding taxonomy (Ghiselin 1995; Härlin 1998; Härlin and Sundberg 1998), whereas those who claim induction of phylogenetic pattern and process presuppose that parts of history are necessarily accompanied by character evidence of a certain kind and quality (Kluge 2001a, 2002). That supposition is especially clear in modeled assumptions and weighted data (Siddall and Kluge 1997).

In phylogenetic systematics, a diagnosis amounts to a descriptive kind of definition (more precisely, a postscriptive definition), that is, a definition from extension (Frost and Kluge 1994). Diagnosis is used as a descriptor of particulars that cannot be defined prescriptively. Although diagnosis has been interpreted as a kind of extrapolation (Farris 1979), that cannot mean such characterizations are predictive of other as-yet-unstudied characters (contra Goloboff 1993; Griffiths 1999; Schuh 2000, 52; for additional comments, see Kluge 1998a).

Although the phylogenetic systematist understands the relationship between diagnosis and ostension to be an enumeration, or a list, of the observed features of a self-defining individual, which have an indefinite number of characteristics (Popper 1957, 77), other kinds of systematists, such as evolutionary systematists and pattern cladists, take diagnosis to mean an intensional definition, a finite list of properties necessary and sufficient for membership. That intensional definition attributes a material essence to taxa, an attribution that is not just neutral to evolutionary theory but one that is antagonistic to that theory. Although the actual content of these two kinds of diagnoses may be indistinguishable, important theoretical and pragmatic consequences obtain, such as how characters are individuated in the first place, and how they may be re-analyzed subsequently, and whether or not character weighting routines and ambiguously optimized synapomorphies should be employed in the analysis of phylogenetic relationships (Kluge 1998a, 1999, 2003b). Moreover, diagnosis by intensional definition requires that all members of the group so defined must exhibit the necessary and sufficient properties of the group. Strictly speaking, with that kind of definition, snakes could not be classified as members of Tetrapoda. One might argue that these kinds of exceptions can be accommodated when they are known by broadening the definition of the inclusive group (e.g., Tetrapoda). However, two problems accompany this strategy: (1) claiming to "know" something is a risky business, at least when it comes to history; and (2) changing an intensional definition only leads to a different intensional definition. In other words, there are real and important downsides in phylogenetic inference to conflating ostension and intension.

Priority and typification are widely misunderstood as being unscientific. Indeed, both are fundamental scientific concerns. Priority is a social necessity, the giving of credit to the first discoverer. Credit of that kind is what drives scientists forward and therefore increases understanding in science (T. Grant, pers. comm.). Typification is concerned with repeatability, and hence testability and objectivity.

To actually fix a taxonomic name requires a set of rules or guidelines. For example, in Linnaean classification, the rule is one of priority—officially, the oldest name given to a type-specimen(s) representative of a taxon is the senior synonym of that taxon. As noted earlier, ostension is definition by enumeration or pointing at particulars, a kind of definition used in characterizing configurational and contingent relations, as in the case of phylospecies and their historical relationships. In this system of reference, the rule involves a form of ostension, which references a unique part of history. To avoid referencing a particular most recent common ancestor (see section titled "Monophyly") and putting forward a circular definition in which both ancestor and descendants are specified, it is recommended that the phylogenetic definition of a higher taxonomic name be phrased as "the least inclusive monophyletic group comprising these two or more historical individuals (A, B, . . .)" (Schander and Thollesson 1995). A phylospecies' name is also fixed ostensively, but in reference to a name-bearing type-specimen(s).

Nomenclatural typification cannot be ignored for scientific reasons, because the parts of a hypothesized taxon can change as the result of future study (contra de Queiroz and Gauthier 1994, 28). A name-bearing type-specimen(s) provides an objective standard of reference by which the application of the phylospecies and more inclusive taxonomic names can be traced, no matter how the content of a taxon changes (Lidén and Oxelman 1996; Forey 2001, 89). The

Oxford English Dictionary would be much shorter than it is if the English language were governed by rules as strict as those pertaining to the phylogenetic definition of names and typification.

STABILITY IN TAXONOMY

Stability in taxonomy simply means that the names of taxa do not change over time, as when, for example, different relationships are identified and new entities are named, and seeking ways to minimize that change has long been the principal concern of taxonomy (e.g., see ICBN 1994; ICNB 1992; ICZN 1999). That stability continues to be an important issue in taxonomy is evident in the fact that the need to increase stability is one of the arguments for overthrowing Linnaean classification (Cantino and de Queiroz 2000). Because there is no basis for stability in science, arguments for advocating it and seeking its increase in taxonomy must be reexamined (contra Hillis 1987; Siddall 2002, 96).

In keeping with the most basic tenets of science, phylogenetic hypotheses are subject to testing and revision, and the corresponding system of taxonomic names must match all changes if the nomenclature is to be judged scientifically useful. As pointed out earlier in the chapter, even laws of nature, and their corresponding intensionally defined classifications, such as the Periodic Table, are critically examined and revised. However, it is irrational to fix taxonomic names by intensional definition, because the ideographic hypotheses to which the names refer are not matters of intensional definition (see, however, Cantino and de Queiroz 2000). A monophyletic group is hypothesized as a historical part of the most parsimonious phylogenetic hypothesis. A monophyletic group can only be defined ostensively, and relatively imprecisely at that, because its boundaries do not exhibit the precision of intensional definition (table 1.1). When a historical individual originates, and when it goes extinct, can be no more precisely determined than when an organism originates and dies. These boundaries are necessarily fuzzy, because of the myriad of processes that occur in replicator-continua, such as populations and genes (de Queiroz et al. 1995, 660).

As discussed previously, one still might argue that stability can be achieved by a phylogenetic system of definition (Cantino and de Queiroz 2000), but none of the three kinds of specification considered above controls for the content of a monophyletic group (e.g., see Forey 2001, 85–86). The parts of monophyletic groups are not fixed by intensional definition, they too being hypotheses subject to testing and revision, and so those corresponding names cannot be stabilized.

Some argue that taxonomic nomenclature is like an ordinary language, that is, taxonomic names *must* be stabilized if systematists are to communicate their results effectively and efficiently to nonsystematists, including the public at large (J. M. Carpenter, pers. comm.; Nixon and Carpenter 2000). I believe there are two interrelated issues that negate this position. To begin with, ordinary languages do in fact change, in syntax and etymologically, and some do so in relatively short periods of time. Indeed, languages are judged inherently unstable, because there is rarely if ever a compelling basis for absolutely fixing grammar and the meaning of the vast majority of the words used in ordinary discourse (e.g., again see the *Oxford English Dictionary*). As Popper (1976, 28) succinctly put it, the quest for universality, "in words or concepts or meanings, is a wild-goose chase." In science, the focus is on testing "statements or propositions or theories" (p. 21).

In summary, stability may be an ideal goal of scientific discourse, but that goal is rendered mute in the absence of being able to prove the objective reality of any particular hypothesis to which a name is attributed. Systematists, other scientists, and the general public must learn to accept the instability that comes from choosing a different hypothesis of taxonomic relationships when it is demanded by weight of evidence. This demand is no greater than changing the definition of the Periodic Table from atomic weight to atomic number and having to relearn the order of some of the elements. Many traditional systematists misinterpret rules of nomenclature, such as priority and typification, as stabilizing the language of systematics. Any set of rules that cannot be judged scientific must be subordinate to the changing hypotheses of phylogenetic systematists, and only then if they are heuristic, that is, if they point to testable propositions. Further, it is impossible for a system of taxonomy to be as precise as a classification of universals, because of the particular nature and biological complexity of taxa.

CODES OF NOMENCLATURE

All of the codes of nomenclature governing Linnaean classification are carefully crafted legalisms. The rules are set forth precisely in one or the other of the international codes (ICBN 1994; ICNB 1992; ICZN 1999), with the General Assembly of the International Union of Biological Science periodically updating them. The international commissions responsible for the codes, and for arbitrating particular cases, only assume responsibility for nomenclature. More specifically, each code provides a set of rules that seek, over the group for which they have authority, maximum universality and continuity in the scientific names of classes of things, and groupings thereof. The International Codes of Nomenclature are primarily concerned with fixing a reference; they do not contain a criterion for extensionally defining taxa. For example, ICZN (1999, 17), under "Criteria of Availability" (chap. 4), article 13.1, only requires that a name be "accompanied by a description or definition that states in words characters that are purported to differentiate the taxon," or "accompanied by a bibliographic reference to such a published statement," or "proposed expressly as a new replacement name (nomen novum) for an available name." In other words, their focus is "stability in nomenclature." Although the published conventions pertaining to the annotated Linnaean (Wiley 1981b) and rankless set theory classifications (Papavero et al. 2001, 45–82), and the draft PhyloCode relating to the descent classification (Cantino and de Queiroz 2000; Forey 2001), have yet to achieve such formal international status, they are nonetheless offering rules for doing taxonomy.

All of these codes are fiats, and in that respect they may act as an impediment to science, where there is a demand for severely testing and overturning hypotheses of relationships, regardless of nomenclatural stability. Such legalisms only hamper scientific study and its communication, particularly when the rules have the appearance of an externalist force (Hull 1988, 1), as they so obviously do in the case of the PhyloCode (e.g., see Cantino, Bryant, et al. 1999; Donoghue 2001, 756–57). No organization of taxonomists, no matter how well intentioned, or even if democratically elected, should have the authority to legislate conventions that run counter to the nature of science, as do the rules for categorical ranks (Cogger 1987) and the basis for defining taxonomic names (e.g., the Principle of Reference in Cantino and de Queiroz 2000). Mere convention cannot be the final word in science. Moreover, the very fact that each of the international codes focuses on a different *kind* of life—animals, bacteria, or plants—underscores their lack of concern for natural entities.

In the spirit of the scientific ideal of the open society (Magee 1973), I believe that taxonomic guidelines must be allowed to change according to the consequences of scientific discourse, conjecture, and refutation. As such, rules should only be heuristic devices, declarations used to initiate a round of critical discussion, which ideally is one of indefinitely many. There is no reason that taxonomic conventions cannot be judged scientifically, assuming of course that taxonomy itself is a scientific enterprise. The distinction between classification and systematization would seem to be the most obvious place to start the debate in the case of the phylogenetic system of taxonomy described in the following section, because it is fundamental to that enterprise (tables 1.1, 1.2). Obviously debatable guidelines accompanying that system of taxonomy are the reference to historical individuals, including the monophyly of phylospecies and the more inclusive groups of which they are a part.

Classification Approaches to Taxonomy

LINNAEAN CLASSIFICATION

Linnaean classification is a hierarchical taxonomy whose inclusive levels correspond to intensionally defined categorical ranks, the original ranks being variety < species < genus < order < class < kingdom (Linnaeus 1758). The ranks of family (< order) and phylum (< kingdom) were added later. The mandatory ranks in today's zoological nomenclature are species < genus < family < order < class < phylum < kingdom. Ranks may be added by using standardized prefixes (supra-, sub-, and infra-). Also, there are nomenclatural conventions that require proper names at a particular rank to carry unique endings. In zoological nomenclature, these include the following: superfamily, -oidea (e.g., Hominoidea); family, -idae (e.g., Hominidae); and subfamily, -inae (e.g., Homininae). Linnaean classification is also notably "binomial," that is, each species group name must be accompanied by a genus group name. Additional defining features of Linnaean classification are summarized in table 1.2.

It is argued that one merely learns the categorical ranks of the taxonomic names (or the unique endings attached to the names) and the hierarchy of those names is thereby revealed. But revealed in what sense? For example, if revealed relative to a particular phylogenetic hypothesis, then one has to have additional conventions, such as the convention of monophyly. Moreover, if the phylogenetic hypothesis and the hierarchy of names are givens, then categorical ranks are redundant when they accompany the taxonomic names.

Taxa of any categorical rank in Linnaean classification do not have to be monophyletic or diagnosable; they can be nonevolutionary and without an empirical basis for their delimitation. It is also telling that the names at the rank of genus and above are *expressly* considered to be plural rather than singular, which is logically inconsistent with the necessarily unique nature of the ideographic things to which names refer in a taxonomic system. To promote the idea that higher taxa are plurals is to emphasize that they are universals (not historical particulars) and that membership in one is a matter of intensional definition. Further, it has been demonstrated that the codes' recommended use of categorical ranks actually leads to nomenclatural instability. For example, lumping taxonomic names can leave groups without names, because the conventions state that those other names at the same categorical rank must become synonyms, and splitting, recognizing more taxonomic names at a categorical rank, can leave the larger group without a name. Moreover, the use of categorical ranks actually leads to nomenclatural instability when avoiding paraphyletic taxa, because lumping or splitting is involved (e.g., see de Queiroz and Gauthier 1994).

Ranks did not originate in a tradition of classification that was scientific, certainly not by current standards, nor was it evolutionary. Linnaeus assumed creationism and transcendental essentialism. Like other naturalists of his day, Linnaeus employed Aristotle's principle of downward classification by logical subdivision. The classes recognized, including species, were of fixed, unchanging types—with membership in a taxon determined by the essential properties of that taxon. Being a devout Christian, Linnaeus attributed the eternal immutability of those sets to the hand of God. Originally, the higher ranks of order, class, and kingdom merely expressed relative subordination (rank) in classification and were otherwise arbitrary. The essential nature of the category of genus was considered especially useful for organizing the increasingly large numbers of species being recognized during the eighteenth century. Although these reasons may have been well received during Linnaeus's time, they only stigmatize today's evolutionary systematics when categorical ranks are used.

Darwin (1859, 422) attempted to save Linnaean ranks by giving them an evolutionary definition: "the natural system is genealogical in its arrangement, like a pedigree; but the degrees of modification which the different groups have undergone, have to be expressed by ranking them under different so-called genera, sub-families, families, sections, orders, and classes." Although the definitions of categorical ranks, including that of species, can be even further refined in contemporary terms of science and evolutionary theory (e.g., see the annotated Linnaean classification of Wiley 1981b), they only become more redundant with an indented list of names.

That kind of redundancy aside, it still might be argued that categorical ranks are a useful index to *general* and *exclusive* kinds of information, under circumstances in which it is inconvenient or unnecessary to refer to the phylogenetic hypothesis that maximizes the explanatory power of the available evidence (e.g., see Nixon and Carpenter 2000). I disagree. General information pertains to the set (whole), or most of its members (parts). Further, an exclusive kind of information is the "not *a*" condition in a logical argument concerning sets. As such, set exclusivity is nominal, that is, existing only in name or form (it is the information that is not real or actual). Moreover, in whatever way this justification for categorical ranks is made, it is a relative one, involving having to learn the particular ranks for the group in question, in addition to the phylogenetic hypothesis/indented list taxonomy, relative to the usefulness of the general or nominal information communicated. Still further, the problem with general phylogenetic information is that it can be no greater than what can be described in detail at any particular level of generality. Also, general (set) information in phylogenetic inference does not take account of the distinction between plesiomorphy and apomorphy (Farris 1976b). Even some combination of general and nominal information is hardly a compelling reason for retaining categorical ranks. Just ask any student of comparative biology who wants the most efficient way of obtaining detailed information about real-world objects and events. The phylogenetic system of taxonomy described in the following section provides just that way, with the clarity of synapomorphic description and diagnosis, and it does so with maximum efficiency.

An additional problem with categorical ranks is that different taxa of the same rank may be falsely considered equivalent in quality or kind of divergence on that basis. Even if all kinds of categorical ranks were required to be monophyletic, the ranks would still represent arbitrary "degrees of modification which the different groups have undergone" (Darwin 1859), because rates of evolution vary among the parts of species history. If taxonomy is to relate to phylogenetic history, and if it is to be used objectively in the study of evolutionary pattern and process, then there is no necessary basis for comparing different taxa of the same rank. Indeed, those kinds of comparisons have proven to be misleading (e.g., Smith and Patterson 1988). That the abolition of ranks can make it difficult for "people who compile 'diversity indices' based on generic counts or family counts" (Forey 2001, 89) might just as well be argued to be a good thing, given the potential for misinformation in such indices when ranks are used.

I agree with those who argue that the Linnaean classification must be abandoned (e.g., see the review in Ereshefsky 2001); however, my primary reasons for doing so appear to be unique. I reject Linnaean classification because it is inconsistent with, if not an impediment to, the objectives of science. I also agree with those who point out that taxonomy must be based on the hierarchy of phylogeny and that Linnaean classification is problematic because it requires having to deal with redundant names associated with categorical rank. I disagree with those who argue against ranks, because they promote instability (e.g., de Queiroz and Gauthier 1994). Rather, I see no scientific reason to embrace stability.

ANNOTATED LINNAEAN CLASSIFICATION

Wiley (1981b) annotated the Linnaean approach to taxonomy with a set of conventions. For example, one of the principal rules declares that taxonomy is consistent to some degree with a phylogenetic hypothesis (i.e., they are not independent of one another). Although possibly viewed as an improvement over the traditional Linnaean approach to classification, these annotations are insufficient to render the Linnaean type of classification one of systematization.

As knowledge of diversity has increased, it has become increasingly difficult to rank taxa according to category. Wiley's (1981b) solution to this problem is to adopt a sequencing convention—that three or more names at the same level of indentation are to be interpreted as successively more-included taxa. Although this convention may be efficient of ranks (see, however, Ereshefsky 2001), asymmetrical, comb-like hypotheses will necessarily receive fewer names than symmetrically arranged propositions, and this seems an arbitrary basis on which to reduce the number of ranks employed. If the goal of taxonomy is to represent consistently and comprehensively the results of phylogenetic systematic analysis, then it follows that each level of diversification must be named. If the history is extremely bushy, then that is what taxonomy should be capable of representing. If it is a perfectly dichotomous pattern, symmetrical or asymmetrical, then taxonomy should have that potential as well. As stated by de Queiroz and Gauthier (1992, 457), "given that the primary task is to represent phylogeny—and acknowledging that there are already more taxon names than anyone can remember—then naming clades seems preferable to leaving them unnamed" (see also Platnick 1977; Ax 1987; Willmann 1987, 1989). These and other defining features of the annotated Linnaean classification are summarized in table 1.2.

Ereshefsky's (2001, 268–74) recent review of Linnaean classification covers several other major difficulties with Wiley's (1981b) approach. Ereshefsky's summary, while rather general, captures the sense in which the annotations fail to significantly improve on Linnaean classification: "First, its rules of nomenclature are those of the traditional Linnaean system. So all of the problems associated with traditional nomenclature are carried over to the annotated system. Second, the annotated system adopts the vacuous and misleading categories of the traditional system. And third, the annotated system fails to achieve two of its principal aims: the preservation of the Linnaean system, and the reconciliation of that system with cladism. In short, the annotated system fails to alleviate past problems of the Linnaean system while at the same time creating its own problems."

DESCENT CLASSIFICATION

The current effort in English-language journals to exorcise Linnaean classification from taxonomy is being led for the most part by Kevin de Queiroz (1987, 1988, 1992, 1994, 1995a, 1995b, 1997a, 1997b, 1998, 1999; Rowe 1987, 1988; Estes et al. 1988; Gauthier, Estes, et al. 1988; de Queiroz and Donoghue 1988, 1990a, 1990b; Cannatella and de Queiroz 1989; de Queiroz and Gauthier 1990, 1992, 1994, 1995; O'Hara 1992, 1993; Rowe and Gauthier 1992; Bryant 1994, 1996, 1997; Schwenk 1994; Sundberg and Pleijel 1994; Donoghue 1995; Schander and Thollesson 1995; Lee 1996a, 1996b, 1998a, 1999; Wyss and Meng 1996; Cantino et al. 1997; Crane and Kenrick 1997; Kron 1997; Cantino 1998; Eriksson et al. 1998; Hibbett and Donoghue 1998; Schander 1998a, 1998b; Cantino, Bryant, et al. 1999; Cantino, Wagstaff, et al. 1999; Ereshefsky 2001). Even though de Queiroz and his associates do not completely agree on these matters (e.g., see Cantino, Wagstaff, et al. 1999; Ereshefsky 2001), they are of a like mind as to the revolutionary nature of their enterprise, and they pursue their goal on a variety of fronts, metaphysical and scientific, theoretical and pragmatic, and sociological and political (Nixon and Carpenter 2000). It is for these reasons that I refer to their collective effort as "K. de Queiroz et al." Summarizing, I find that K. de Queiroz et al.'s goal is to formulate a rankless binominal hierarchy of names without Linnaean categorical ranks (except, perhaps, for the rank of species) by only assuming the "descent" clause of the Darwinian theory of evolution. In pursuing that objective, they downplay the importance of assigning a type-specimen(s) and type-species, as well as the extensional definition of taxonomic names. These and other defining features of the descent classification are summarized in table 1.2. (*Binominal* is the clear choice when it is used in reference to paired taxonomic names. *Nominal* pertains to a noun or word group that functions as a noun. There is no word *nomial,* and *binomial* is ambiguous because it pertains either to two terms, e.g., mathematics, or two names, e.g., taxonomy.)

K. de Queiroz et al. have labeled their approach in various ways, usually as "phylogenetic" or "evolutionary," and they often refer to it as a "system." Because it is not a complete system, that is, its references to evolution and phylogenetics are theoretically incomplete and empirically empty, I have chosen to call their approach a descent classification. Also, that the original proponents of the descent classification, de Queiroz and Gauthier, are changing their minds on certain arguments cannot be denied, and that is no more evident than in the changing face of typification. For example, in the draft copy of the Phylogenetic Code of Biological Nomenclature (PhyloCode; Cantino and de Queiroz 2000, 4), which is intended to conventionalize the descent classification with precisely stated legalisms, "specifiers . . . are used to specify the clade to which the name applies," where "[s]pecies, specimens, and synapomorphies cited within these definitions are called specifiers." In other words, typification now appears to be acceptable, providing it is referred to as specification (see also Forey 2001, 89)!

K. de Queiroz et al. argue that by assuming evolution and eliminating all reference to ranks, types, and diagnostic characters, the descent classification is made "explicit, universal, and stable." To be explicit simply means that the association between a name and a taxon is clear. Universality in the descent classification does not mean that the association between a name and a taxon is unrestricted in space and time (table 1.1). Rather, universality has a more colloquial meaning, like a uniformity of understanding among authors as to an association (de Queiroz and Gauthier 1994, 28). Stability, as noted earlier, means that the names of taxa should not change.

K. de Queiroz et al. (e.g., de Queiroz 1992, 299–300) reject the idea that ostension can "provide satisfactory definitions of taxon names," because ostension does not provide definitions "in terms of necessary and sufficient properties." K. de Queiroz et al.'s solution to this problem is to assume just "descent" as necessary and sufficient for intensionally defining taxonomic names. As Cantino and de Queiroz (2000, 12; my emphasis) state in the descent classification's Principle of Reference: "The primary purpose of taxon names is to provide a means of referring to taxa, *as opposed to* indicating their characters, relationships, or membership." Thus, all biologists might be expected to endorse K. de Queiroz et al.'s approach to taxonomy. Descent classification has the appearance of being universal and historically meaningful at the same time (see table 1.1)!

These advantages are more apparent than real. To begin with, K. de Queiroz et al.'s intention to use "descent" in defining names does not make their approach more, or less, consistent with evolution than any other monophyletic system of nomenclature. Ironically, K. de Queiroz et al.'s basis for naming things is more universal than they think, because it would apply even if evolutionary theory were to be proven false and organisms formed nested sets for some reason other than inheritance (D. Frost, pers. comm.). Industry provides many such examples, with automobiles being one of the most obvious.

Further, the descent classification cannot claim that a particular taxonomic name is defined by a reference to common ancestry because *descent* is just the property of the class of all monophyletic groups (Ghiselin 1995, 1997; see also Mahner and Bunge 1997, 58; Siddall and Kluge 1997, 316; Härlin and Sundberg 1998); *descent* is *not* the defining condition of any particular group or corresponding name (contra de Queiroz 1992, 305). Thus, the descent classification is deficient with respect to its intention to give evolutionary meaning to the names of particular groups in its rankless list. A basis for hierarchy, such as provided by a node, stem, or apomorphy definition, lies outside the argument for the descent classification. The fact that *descent* only references the class of all monophyletic groups, not any particular group deserving a name, means that the K. de Queiroz et al. approach to classification is some unspecified kind of intensionally defined common ancestry taxonomy. Paradoxically, it is both nomothetic and ideographic!

As Mahner and Bunge (1997, 267) observed, the sense of "descent, with modification" is "the emergence of qualitative novelty," descentism alone (sensu the "descent" of de Queiroz and Gauthier 1992) being empty of empirical meaning. No effect is possible with just the premise of descent. A premise of trait evolution, the "with modification" conditional, must be stated if synapomorphies are to be explained as homologous and therefore judged as evidence of common ancestry (Kluge 1999). It is in this regard that the descent classification must be judged as some unscientific, untestable kind of metaphysics.

If these were not problems enough, de Queiroz (1992, 309; see also de Queiroz and Gauthier 1990) admits that the "descent" definition of taxa need not be restricted to the names of monophyletic taxa, because that definition applies as well to the name of a paraphyletic taxon, an incomplete assemblage of lineages descending from a common ancestor. Yet another argument is required to logically justify the descent classification—otherwise, the "descent" definition does not apply exclusively to the names of monophyletic taxa. In the absence of such a rationalization, the descent classification must again be judged deficient. Something more than "descent" is required.

Phylogenetic systematists firmly believe there is only one history. Its parts, such as a particular phylospecies, or a monophyletic group of such entities, are not a matter of intensional definition but can only be identified, referenced, pointed to, enumerated, or diagnosed (Kluge 2002). Thus, a phylogenetic definition from intension requires setting aside the criterion that the named thing is a class or set of organisms (Härlin 1998). K. de Queiroz's (1992, 305) solution to this dilemma has been to either blur or deny the empiricists' long-recognized distinction (p. 307) between universals and particulars (Locke 1700, 330; table 1.1; also see review in Ghiselin 1997). In doing so, de Queiroz effectively denies the differences between natural kinds (laws of nature) and historical individuals (necessarily unique particulars) (table 1.1; see also Hull 1974).

K. de Queiroz and Gauthier (1992, 452) go further in attempting to avoid the conclusion that phylogenetic inference involves something more than an intensionally defined descent hypothesis, by arguing that "organisms are parts of taxa not because they possess certain characters, but because of their particular phylogenetic relationships," which they followed with the assertion that "adopting [their descent classification] nomenclature requires neither agreement about, nor a detailed knowledge of, phylogenetic relationships" (de Queiroz and Gauthier 1994, 29–30). As such, descent classification can have nothing to do with the empirical world, where synapomorphies can be explained as homologues, where shared derived character states may be explained in terms of their inheritance (Farris 1983).

According to K. de Queiroz et al., there is stability when the association between a name and a taxon is consistent over time (Härlin 1998). When that association changes, as it can when categorically ranked taxa are lumped or split, or paraphyletic taxa are eliminated (de Queiroz and Gauthier 1990, figs. 2–6; 1994, boxes 2, 3), there is nomenclatural instability, and this is one of the bases on which K. de Queiroz et al. most strongly argue that the traditional Linnaean classification, which requires categorical ranks, should be rejected. Sereno (1999) takes their argument one step further, claiming that choosing among methods of classification (node, stem, or apomorphy) should then be based on which one stabilizes the taxonomic content of a taxon more than another in the face of local changes of relationships. However, as discussed earlier, stability cannot be justified scientifically.

In their quest for stability as well as universality, K. de Queiroz et al. have also sought standardization by attributing two names to each higher taxon. For each higher taxon there is a name for the crown clade and a name for the more inclusive clade of which the crown is a part, that is, the crown clade plus its stem taxa. The more widely used name is attributed to the crown clade, the less-well-known name to the stem clade, the latter followed by the former at any higher level in the hierarchy (e.g., Anthracosauria—Amniota; de Queiroz and Gauthier 1992, fig. 7). An apparent problem with the crown/stem clade convention, not to mention the idea that the "stem-group concept is partly typological and

not purely genealogical" (Hennig 1981, 3), is that the stem taxa will be dominated by extinct taxa because they are expected to have the less-well-known
names (e.g., Myopterygii—Vertebrata; de Queiroz and Gauthier 1992, fig. 7).
Thus, the names of extinct taxa in the descent classification may not contribute
as much to understanding the patterns and processes underlying biodiversity as
do the names of extant taxa. This convention seems wrong-headed, because organisms, as bearers of potential evidence for explaining biodiversity, must be
allowed to speak for themselves on a case-by-case basis. Sometimes the living
organism may be more critical to testing phylogeny; sometimes it is the fossil
(Gauthier, Kluge, et al. 1988). Conventions that promote bias in science must be
rejected, and that includes the crown/stem clade convention.

Also problematic is the apomorphy approach that is used to intensionally
define a name. As already noted, to define Tetrapoda as having arms and legs is
necessarily to exclude apodous vertebrates, such as caecilians and most snakes.

Last, Cantino, Bryant, et al. (1999; see also Pleijel and Rouse 2000) admit that
they disagree about whether the categorical rank of species should be a part of
the descent classification. This is significant because it indicates K. de Queiroz
et al.'s approach to classification is not sufficiently coherent to convince even
its most ardent supporters to eliminate all abstract levels of rank. If only the categorical rank of species is retained, while all other ranks used in Linnaean
classification are eliminated, then the descent classification must be judged
arbitrary.

SET THEORY CLASSIFICATION

Papavero and his colleagues, Llorente-Bousquets and Abe, have also been consistently critical of Linnaean classification since the early 1990s. Curiously, K. de
Queiroz et al. (see also Ereshefsky 2001) have given little if any notice to this extensive literature in Spanish, although Papavero et al. have often cited the efforts of their counterparts to the north. In any case, Papavero et al.'s recently
published review in English (2001 and references therein) leaves no doubt that
the revolution to overthrow Linnaean classification is international in scope.
Although Papavero et al. also advocate a rankless hierarchy of names, like K. de
Queiroz et al., they assume a phylogenetic systematic hypothesis up front,
which is unlike the scheme of K. de Queiroz et al. Papavero et al.'s approach to
naming taxa is also different in that it uses conventions from set theory, and it
is on that basis that I label it a rankless kind of set theory classification. The
defining features of their classification are summarized in table 1.2.

The set theory classification, like the descent classification (table 1.2), can be
criticized because taxa are defined intensionally, according to set theory. The
context is idiographic, not nomothetic. A criticism peculiar to set theory classification is the fact that the reference capability ascribed to apomorphy is nom-

inal and therefore empirically empty. Like the other classifications, stability is sought through absolute priority of names, which runs counter to the nature of science. Peculiar to set theory classification, higher taxonomic epithets are strings of species or generic names, with the higher the taxon the longer the string, and that makes diversity increasingly difficult to reference and especially difficult to communicate when its parts are discussed orally.

A New Phylogenetic System of Taxonomy

ARGUMENTS

If "[b]iological nomenclature is the basic language by which scientists communicate about the diversity of living things" (de Queiroz and Gauthier 1994, 27), then there must be a concern for the scientific nature of what is being communicated. This is the major premise of the new system of phylogenetic taxonomy set forth in this section, where choice of a historical hypothesis and corresponding taxonomy are based on maximizing degree of corroboration and descriptive efficiency, respectively. Thus, the new system is expressly phylogenetic and scientific.

There are several additional advantages to such a system of taxonomy, as revealed coincidentally by Ereshefsky (2001). To begin with, the species concept is not pluralistic. There can be no advantage to having more than one concept, such as having one based on interbreeding (Mayr 1970; Paterson 1985) and another based on ecology (Van Valen 1976), when those classifications ultimately require knowledge of inheritance and consequently a phylogenetic system of relationships. The monistic historical approach I have identified references all taxa, extant and extinct, bisexual and asexual, according to the causal relation of inheritance, rather than any of the inherent qualities of wholeness. Inheritance is considered sufficient in the phylogenetic system of taxonomy, because it alone provides a hierarchical arrangement of all historical individuals. This is not the "relational essentialistic" interpretation that P. E. Griffiths (1999, 210) has attributed to inheritance. Of course, qualities such as gene flow, genetic homeostasis, and natural selection may be relevant to understanding the cohesiveness and integration of extant bisexual phylospecies, but they play no role in the definition of those species, or any of the entities of which they are a part.

Further, the phylospecies concept may be judged superior to species concepts that attempt to cover the possibility of future reticulation, such as the evolutionary species concept of Mayr (1970; see also Paterson 1985), the ecological concept of Van Valen (1976), and the phylogenetic concept of Mishler and Donoghue (1982; see also Mishler and Brandon 1987; and including that of Frost and Kluge 1994). Having to wait for an answer is not part of the phylogenetic system. The focus of taxonomy should be on referencing past lineage systems, whereas systematics more generally is concerned with scientific discovery of the relationships of those individuals and tracing the history of their traits. None of these endeavors concerned with the causal relation of inheritance is affected by future events. To concentrate on real things, not abstract generalities, where testability does not include some time in the indefinite future, is an advantage. Of course, future events such as hybridization *might* affect our understanding of phylogenetic history, but there is no incentive to waiting unless we can be sure that it *will* have such an effect.

Further, the phylogenetic system of reference used to assign a name to a taxon is considered scientific for two reasons. To begin with, each phylospecies is identified with a type-specimen(s), and each higher taxon with two or more type-phylospecies, and those types are selected from the least disconfirmed phylogenetic hypothesis on which the monophyletic groups are identified. It is this hypothesis that exhibits the highest degree of corroboration, because it is the result of a simultaneous analysis of all of the relevant available data (total evidence). A type-specimen(s) exhibits the diagnostic features distinguishing the phylospecies from its sister lineage, and type-phylospecies exhibit those diagnostic features that distinguishes a higher taxon from its sister. Unlike the rankless set theory classification of Papavero et al. (2001; see also the group inclusion method of Farris 1973), these are not nominal apomorphies but hypotheses of character transformation (sensu Hennig 1966, 89). A type-specimen(s) must be chosen carefully, however, because it can only represent one stage in the organism's life history. A semaphoront, or character bearer, is defined in terms of a certain theoretically infinitely small time-span of its life, during which it is considered unchangeable (Hennig 1966, 65). Thus, all taxonomic names are determined objectively and scientifically in the same manner, by ostension and typification.

Stability can be achieved in three ways in science, by fiat, tautology, or forever failing to refute highly testable hypotheses, despite our best efforts. Only the latter is scientific, because it is the product, even if the unintended consequence, of good science. I argue that the phylogenetic system of taxonomy outlined here is an inherently rational approach to increasing stability, because stability will necessarily increase in time, as a function of the testability and reciprocally clarifying nature of the scientific enterprise on which it depends (Popper 1959; Hennig 1966; Kluge 1991, 1998a, 2003b). Indeed, phylogenetic systematic hypotheses seem certain to become severely tested, with increasingly critical evidence, as data matrices grow large in terms of characters, specimens, and taxa. More severely tested phylogenetic hypotheses provide increased opportunities for testing incongruent characters and for eliminating those kinds of errors from the data matrix (Kluge 1999). Stability comes coincidentally with the removal of error. This claim cannot be met by any of the other approaches to taxonomy (table 1.2), the traditional Linnaean and annotated Linnaean classifications employing ranks, or the rankless descent and set theory classifications.

To maximize nomenclatural stability, we must pay special attention to documenting the diagnostic characteristics of the designated type-specimen(s) and type-phylospecies. Further, these arguments for stability must not be confused with seeking the highest degree of corroboration and documenting the levels achieved with indices such as Bremer support, the bootstrap, and the jackknife (e.g., see Forey 2001, 90). Admittedly, other approaches such as the traditional Linnaean classification may provide greater nomenclatural stability in the short run because fiat is more efficient. However, this is a meaningless kind of stability as far as knowledge of phylogeny and its explanatory power are concerned.

GUIDELINES

The guidelines to follow are different from the published and unpublished codes of nomenclature that govern classification (table 1.2), because these declarations are to be evaluated on the same scientific grounds as the phylogenetic system of taxonomy for which they are intended (see the earlier section titled "Codes of Nomenclature"). These guidelines are merely talking points.

1. A taxon is a named group of organisms.
2. Each name, including that of phylospecies, is defined ostensively, in reference to a monophyletic entity (a particular historical individual; a clade) on the most parsimonious, total evidence, hypothesis of phylogenetic relationships. Each name is then accompanied by a maximally descriptively efficient diagnosis. That diagnosis excludes ambiguously optimized synapomorphies, as well as those that appeal to special knowledge, such as assumed by ACTRAN and DELTRAN (contra Forey 2001, 92). There is no basis for redundant names.
3. This system of taxonomy consists of a hierarchically arranged list of names. The names of subordinate taxa within each monophyletic group are indented to reflect their causal relationships of inheritance (Farris 1976a). Any particular name, including that of a phylospecies, is thereby systematized. The indented list of names is simply learned. The indented parts of a lengthy list of names (Wiley 1981b, 203–4), such as might appear on successive pages of a published paper or PowerPoint slide show, are to be connected by any form of notation, providing it is explicitly stated.
4. Phylospecies is the smallest taxon. Those pattern classes referred to as subspecies, or varieties, in the Linnaean classification are not referenced (Kluge 1990; Frost and Kluge 1994). The species categorical rank is not employed, nor is the concept species in any of its familiar forms, although the term *species* can be used in a nominal sense.

5. Phylospecies names are uninominal (e.g., see Michener 1963), with each name standing alone, like those of all other taxa.

6. Only phylospecies names are italicized. Phylospecies are set apart from higher taxa in this regard because there is no less inclusive, purely histori-cal entity within them. It signals that within these smallest historical indi-viduals, there are only contemporary individuals such as populations and individual organisms. It is in this sense that phylospecies is unique (contra Lidén 1990, 185).

7. Each phylospecies name requires the designation of a type-specimen(s) at the time the taxon is formally referenced in a scientific, peer-reviewed, publication. The type-specimen(s) exemplifies the diagnostic features that serve to delimit the phylospecies from its sister lineage (Moore 1998), and unlike the rule of priority applied in Linnaean classification, this is the basis in evidence that is used to decide synonymy.

8. Two or more type-phylospecies (A, B, . . .) are designated for each higher taxon (X) at the time the higher taxon is formally referenced in a scientific, peer-reviewed publication. That designation is to be phrased as "the mono-phyletic group X comprising A and B." Type-phylospecies exemplify the diagnostic features that serve to diagnose the higher taxon, as historically different from its sister lineage.

9. All names are treated as proper nouns. Each monophyletic group is re-ferred to in the singular and without the article *the,* as in "the Eublepha-ridae." All previously published names that express plural inflectional endings should be changed to the singular (e.g., Eublepharini becomes Eublepharinus). Ease of pronunciation is the only criterion for forming names. Names with categorical-rank suffixes do not indicate rank when they are used in the phylogenetic system of taxonomy.

10. All names, including those of phylospecies, are capitalized, whereas the corresponding vernacular names are not.

11. Names with the same spelling cannot refer to different monophyletic groups, that is, homonymy is unacceptable. Homonymy may be judged a significant problem when the current binominal Linnaean system is con-verted to the uninominal phylogenetic system of taxonomy. One solution is to require that each phylospecies name be accompanied by the name of the group that includes its sister phylospecies. In this arrangement, the phylospecies would be the *praenomen* (e.g., see table 1.6).

12. Shutter quotes surround the name of an unresolved group in which there are equally highly corroborated hypotheses of sister-group relationships among the three or more named entities included in the assemblage (Wiley 1981b). The metataxon convention of Donoghue (1985, 177; see also de Queiroz and Donoghue 1988; Estes et al. 1988; Gauthier, Estes, et al. 1988) is not used (for arguments, see Kluge 1989b).

13. Names are synonyms when they designate the same monophyletic group. The name whose type-specimen(s) is consistent with the monophyletic group's diagnosis is the senior synonym. When type-specimens specifying different names are equally consistent, it is the oldest name that references the monophyletic group.

14. The name of a hybrid phylospecies follows one of its parental taxa, and the hybrid origin is indicated by the names of the parental phylospecies placed in parentheses beside the hybrid's name (Wiley 1981b).

15. These guidelines are adhered to when names formulated under the traditional Linnaean, annotated Linnaean, descent, or set theory classifications are converted to the phylogenetic system of taxonomy.

APPLICATIONS

The task of moving the phylogenetic system of taxonomy from theory to practice is daunting. Linnaean classification has been widely used for nearly 250 years, and there are now several million taxonomic names to be converted to any other taxonomy. In addition, phylogenetic systematics, with its basis in testability and explanatory power (Kluge 1999, 2001a, 2002, 2003a), has yet to completely replace the practice of evolutionary systematics, where verification is sought and ad hoc assumptions are used to avoid contradictory data and results (sensu Mayr and Ashlock 1991). Indeed, it might be argued that the currently popular maximum likelihood and Bayesian analysis of gene sequence data are just explicit forms of evolutionary systematics, the inductive estimation of phylogeny, in which model assumptions probabilify the character evidence necessary to establish the truth of historical hypotheses (Kluge 1998b). Converting existing names could be accomplished simply by fiat, whereas changing the way systematists think and practice science is the truly difficult task in applying the phylogenetic system of taxonomy. Examples from gekkotan systematics are used to explore these two issues.

Squamata is a well-corroborated taxon, of which Gekkota is a large part (Gauthier, Estes, et al. 1988). Gekkota (geckos in the vernacular) consists of both fully tetrapodous and nearly limbless forms. Although the sister-group relationship of the extinct "bavarisauroids" to Gekkota exhibits a high degree of corroboration (Kluge 1987, 39, 1997), the relationship of that monophyletic group to other extant squamates continues to be tested and vigorously debated (e.g., compare the results of Lee 1998b, fig. 2, with those of Harris et al. 1999, fig. 1).

There are more than 1,000 currently recognized extant "species" of geckos (Kluge 2001b), new historical lineages continue to be delimited, and most systematic revisions identify additional junior synonyms. A more difficult problem than the sheer number of "species" is the fact that students of gekkotan sys-

Table 1.3   Higher taxonomic names of geckos

Gekkones Gmelin 1788, 1067	Nyctisauria Gray 1845, 142
Geckoides Oppel 1811, 22	Gecconidae Cope 1871, 236
Geckotiens Cuvier 1817, 44	Eublepharidae Boulenger 1883, 308
Ascalabotae Merrem 1820, 39	Geckonidae Boulenger 1884, 119
Geckotidae Gray 1825, 198	Pygopodidae Boulenger 1884, 119
Geckones Spix 1825, 15	Uroplatidae Boulenger 1884, 119
Ascalabotoidea Fitzinger 1826, 13	Geckonomorpha Fürbringer 1900, 13
platyglossae Wagler 1830, 141	Pygopodomorpha Fürbringer 1900, 13
Gekkonidae Bonaparte 1831, 65	Uroplatimorpha Fürbringer 1900, 16
Aprasiadae Gray 1841, 428	Ophiopsisepidae Jensen 1901 (1900), 325
Lialisidae Gray 1841, 427	Diplodactylinae Underwood 1954, 477
Pygopidae Gray 1841, 427	Sphaerodactylidae Underwood 1954, 476
Pygopoda Fitzinger 1843, 23	Carphodactylini Kluge 1967, 1021
Stenodactyli Fitzinger 1843, 89	Pletholaxini Kluge 1976, 69
Ptyodactyli Fitzinger 1843, 93	Teratoscincinae Kluge 1987, 40
Platydactyli Fitzinger 1843, 97	Aeluroscalabotinae Grismer 1988, 414
Hemidactyli Fitzinger 1843, 103	

Note: References mentioned in the table are not available in the reference list.

tematics rarely cite a species concept. Without details about such a concept, and corresponding discovery operation, what has been named and what has been placed in synonymy cannot be evaluated scientifically.

The currently recognized higher gekkotan taxa are rarely better documented than are the "species" of geckos. Few of the sister-group relationships are the result of anything resembling a phylogenetic systematic analysis, and there is little reason to accept the higher taxa as hypothesized parts of history (e.g., see the review in Kluge and Nussbaum 1995). The extent of the problem facing gecko systematists in converting to the phylogenetic system of taxonomy can be estimated from the 107 generally recognized Linnaean genus-group names (Kluge 2001b). Given just the number of currently identified extant "species," more than 400 additional names are required to define the relationships among pairs of extant "species." Moreover, most of those names would be new to science, because few epithets are currently available in synonymy (table 1.3; Kluge 2001b). Relationships among some groups of "species" are more obviously in need of phylogenetic systematic analysis than are others. Indeed, some of the most diverse gekkotan taxa have yet to be analyzed with phylogenetic systematic methods, such as *Cyrtodactylus, Hemidactylus,* and *Sphaerodactylus,* with 66, 79, and 93 "species," respectively. Thus, gekkotans are an excellent group with which to illustrate the costs of applying the phylogenetic system of taxonomy. A large number of "species" are recognized without the benefit of a coherent species concept, Linnaean classification is well entrenched, and for the most part phylogenetic relationships have not been tested scientifically (sensu

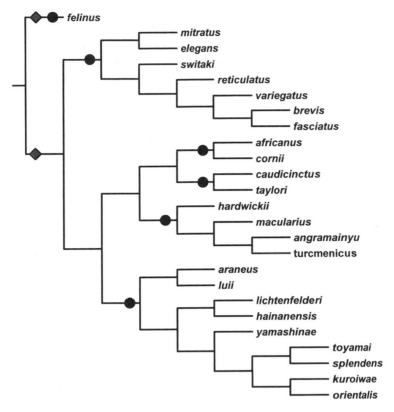

Figure 1.1   Phylogenetic systematic relationships among 25 of the 26 currently recognized extant "species" of eublepharid gekkotans (after Grismer 1983, 1988; Grismer et al. 1994, 1999). Only *fuscus* has been omitted. The small filled circles and diamonds refer to the monophyletic Linnaean taxa recognized in table 1.4, genus and family ranks, respectively.

Kluge 1997, 2003a). Conversion to the phylogenetic system of taxonomy must be a function of phylogenetic systematic reanalysis at all levels, and these kinds of studies will be extensive in the case of gecko taxonomy.

Although the vast majority of geckos require considerable phylogenetic systematic research in pursuit of a scientifically stable phylogenetic system of taxonomy, there are notable exceptions, such as eublepharids. Largely through Grismer's phylogenetic systematic studies (e.g., Grismer 1983, 1988; Grismer et al. 1994, 1999) there now exists a well-corroborated phylogenetic hypothesis for 25 of the currently recognized 26 extant "species" of eublepharids (fig. 1.1). The only two groups resolved in Ota et al.'s (1999, fig. 5) molecular hypothesis of eublepharid relationships are consistent with those assemblages identified by Grismer—(*africanus, caudicinctus*) and (*splendens* (*kuroiwae, orientalis*)). In addition, a corresponding Linnaean classification has been proposed for them

Table 1.4   Linnaean taxonomy of eublepharid gekkotans

Family Eublepharidae
  Subfamily Aeluroscalabotinae
    Genus *Aeluroscalabotes*
      Species *felinus*
  Subfamily Eublepharinae
    Genus *Coleonyx*
      Species *mitratus*
      Species *elegans*
      Species *switaki*
      Species *reticulatus*
      Species *variegatus*
      Species *brevis*
      Species *fasciatus*
    Genus *Goniurosaurus*
      Species *araneus*
      Species *luii*
      Species *kuroiwae*
      Species *yamashinae*
      Species *orientalis*
      Species *toyamai*
      Species *splendens*
      Species *hainanensis*
      Species *lichtenfelderi*
    Genus *Eublepharis*
      Species *hardwickii*
      Species *macularius*
      Species *turcmenicus*
      Species *angramainyu*
    Genus *Hemitheconyx*
      Species *caudicinctus*
      Species *taylori*
    Genus *Holodactylus*
      Species *africanus*
      Species *cornii*

Source: After Grismer 1983, 1988; Grismer et al. 1994, 1999.

(table 1.4), which includes six names of Linnaean categorical rank genus and two names of subfamily rank. Although those eight higher taxonomic names are monophyletic, they are not sufficient to comprehensively describe the 24 sister-lineage relationships (cf. fig. 1.1 and table 1.4). A phylogenetic system of taxonomy for the 24 higher taxonomic names corresponding to the accepted phylogenetic hypothesis (fig. 1.1) is set forth in tables 1.5 and 1.6, expanded and

Table 1.5 Phylogenetic taxonomy of eublepharid gekkotans (expanded)

Eublepharidae
  Aeluroscalabotes
    Felinus
  Eublepharinae
    new name A
      Coleonyx
        Mitratus
        Elegans
      Anarbylus
        Switaki
        new name B
          Reticulatus
          new name C
            Variegatus
            new name D
              Brevis
              Fasciatus
    Eublepharini
      new name E
      new name F
        Araneus
        Luii
        new name F′
        new name G
          Hainanensis
          Lichtenfelderi
        new name G′
          Yamashinae
        new name H
          Goniurosaurus
            Kuroiwae
            Orientalis
            new name I
              Toyamai
              Splendens
     new name E′
      new name J
        Hardwicki
      Eublepharis
        Macularius
        new name K
          Turcmenicus
          Angramainyu

(continued)

Table 1.5   (continued)

new name J'
Hemitheconyx
Caudicinctus
Taylori
Holodactylus
Africanus
Cornii

condensed versions, respectively. An expanded phylogenetic system of taxonomy refers to the perfectly dichotomous hierarchy of names. The phylospecies name precedes the name of the proximate clade of which it is a part, on the same line, in the condensed version of the phylogenetic system of taxonomy. The latter, however, is not an example of traditional Linnaean binominal nomenclature, because the succeeding name is without reference to rank (e.g., genus), and the phylospecies is the praenomen of the pair. None of the names published originally with plural inflectional endings has been changed to the singular.

Although the phylogenetic hypothesis describes the taxonomic hierarchy exactly, 16 new names must be ostensively defined. Nonetheless, eublepharids provide an example in which the practical and scientific costs and benefits of a monophyletic Linnaean classification and the phylogenetic system of taxonomy are evident: (1) the genus and subfamily categorical ranks are redundant with the hierarchy of names; (2) the current Linnaean hierarchy of names does not completely describe the phylogenetic hypothesis, and the corresponding taxonomy does not maximize descriptive efficiency; and (3) the phylogenetic system of taxonomy requires 16 new higher taxonomic names to comprehensively describe the phylogenetic hypothesis, where the corresponding monophyletic taxonomy maximizes descriptive efficiency.

The higher classification of geckos provides insight into how difficult it will be to completely eliminate the practice of evolutionary systematics. In the case of gecko–pygopod relationships, there is an example of phylogenetic systematic analysis being subverted by appealing to "special, ad hoc, or prior knowledge"—gambits typical of evolutionary systematics. The basic problem is that even when the most highly corroborated, least-refuted, phylogenetic systematic hypotheses have been published, some systematists continue to choose other, less parsimonious hypotheses. This issue seems to be relatively more common in a group like that of the geckos, which exhibit distinctly different modes of form, fully limbed and nearly completely limbless. In choosing less parsimonious hypotheses, researchers often allow special knowledge to dictate what is inferred to be a gecko and what synapomorphies are explainable as

Table 1.6 Phylogenetic taxonomy of eublepharid gekkotans (condensed)

Eublepharidae
  Felinus Aeluroscalabotes
  Eublepharinae
    new name A
      Coleonyx
        Mitratus
        Elegans
      Switaki Anarbylus
        Reticulatus **new name B**
          Variegatus **new name C**
          new name D
            Brevis
            Fasciatus
  Eublepharini
    new name E
      new name F
        Araneus
        Luii
      new name F′
        new name G
          Hainanensis
          Lichtenfelderi
        Yamashinae **new name G′**
        new name H
          Goniurosaurus
            Kuroiwae
            Orientalis
          new name I
            Toyamai
            Splendens
    new name E′
      Hardwicki **new name J**
        Macularius Eublepharis
        new name K
          Turcmenicus
          Angramainyu
    new name J′
      Hemitheconyx
        Caudicinctus
        Taylori
      Holodactylus
        Africanus
        Cornii

homologues. Maximizing explanatory power is not the goal of evolutionary systematics, and gecko taxonomy cannot be claimed to be most descriptively efficient when it is based on a less parsimonious hypothesis of relationships.

McDowell and Bogert (1954; Underwood 1957; see fig. 1.2A) were the first to refer the largely limbless pygopods to Gekkota, all of which are otherwise fully limbed. A phylogenetic systematic analysis of all the available evidence was then published by Kluge (1987; see also Kluge 1976a, 1976b), which indicated that pygopods were not only gekkotans but also sister to diplodactyline geckos. That conclusion was based on 27 synapomorphies diagnostic of Gekkota, 4 delimiting gekkonoids within Gekkota, and one feature characteristic of the pygopod + diplodactyline part of gekkonoids (fig. 1.2B). In spite of this highly corroborated hypothesis, Estes et al. (1988, 204–6; my emphases) responded that "the evidence for pygopod–diplodactyline relationships, however, is contradicted by *other* characters, and for this reason we retain for the present the traditional arrangement [fig. 1.2A] that has eublepharines and diplodactylines as parts of Gekkonidae, and pygopodids as a separate taxon (for further discussion see comments on Gekkonidae and Pygopodidae)." Under their discussion of Gekkonidae, Estes et al. reported that "if [Kluge's 1987] hypothesis that pygopodids are nested within what we call gekkonids is accepted, changes in the level at which *some* of the above characters exist as synapomorphies will need to be made." Without being specific, Estes et al. also spoke of "the rather *extensive* character conflict" and their finding of "*more* derived characters supporting gekkonid monophyly." Under their discussion of Pygopodidae, they turned

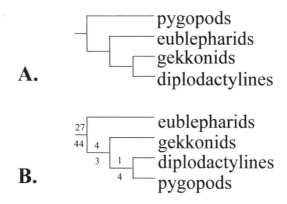

Figure 1.2   (A) Pygopods as the sister group to all other gekkotans (after McDowell and Bogert 1954). (B) Pygopods as the sister group to diplodactyline gekkotans. The number of synapomorphies identified by Kluge (1987) in support of a monophyletic group is listed above the line; those numbers below the line refer to the characters summarized herein (appendix 1.1).

to the evidence for Kluge's (1987) claim, admitting that "a *number* of [their] characters support nesting of pygopodids within gekkonids (characters 35-1, 38-1, 145-0; possibly also 21-0, 45-1, 65-2; in addition, clutch with two eggs, and presence of spectacle)" (35-1, supratemporal lost; 38-1, vomer fused; 145-0, stapedial artery passes through, or anterior to, stapes; 21-0, parietals fused; 45-1, ectopterygoid contacts palatine; 65-2, splenial absent). However, they went on to say, "we have treated the two taxa as separate because there are *other* characters in which pygopodids appear to retain a primitive condition not seen in gekkonids (see diagnosis and comments for Gekkonidae)." Estes et al. did not discuss those other characters, nor did they document the supposed extensive character conflict in Kluge's characters. Thus, although the study of Estes et al. had the appearance of being phylogenetic systematic in so far as it used a parsimony computer program, it was strongly influenced by ad hoc propositions of character choice and rejection. It is an example of the verificationist nature of evolutionary systematics.

To let this situation stand would be a serious impediment to applying the phylogenetic system of taxonomy to gekkotans (e.g., see Pough et al. 1998), because the explanatory power of the competing hypotheses would appear to be in doubt and the higher classification unresolved. It was for this reason that the evidence for pygopod–gecko relationships was reviewed once again (Kluge and Shea, manuscript in preparation). The details relating to the quality and quantity of that reexamined and expanded body of data are summarized in appendix 1.1. The sister-group relationship of geckos and pygopods must continue to be considered strongly corroborated, given the large number of novelties they share (fig. 1.2B). Conservatively, there are 44 such synapomorphies, and there is no reason to believe that this is all of the character evidence. Then there is the additional evidence indicating that pygopods are part of gekkonoid history (fig. 1.2B), and that which points to a pygopod–diplodactyline sister-group relationship (also see appendix 1.1), three and four synapomorphies, respectively. Thus, the gekkotan phylogenetic hypothesis, which includes pygopods, is even more highly corroborated than before (fig. 1.2B). Moreover, Donnellan et al.'s (1999, fig. 1A) recent phylogenetic systematic analysis of 12S rRNA and c-*mos* gene sequence data, albeit weighted for the effects of transitions to transversions, does not disconfirm the hypothesis on the basis of the expanded gekkotan data set (fig. 1.2B; appendix 1.1). Thus, the pygopod + diplodactyline hypothesis maximizes the explanatory power of all the available evidence, to which the phylogenetic system of taxonomy should be applied.

Donnellan et al. (1999, fig. 1B) also performed a maximum likelihood analysis of the 12S rRNA and c-*mos* gene sequence data, assuming a HKY-85 two-parameter model for unequal base frequencies, which resulted in the same branching pattern, except for the gekkonids *Phyllodactylus marmoratus* and *Gehyra variegata* being sister to all other gekkotans (see also Ota et al. 1999, fig. 5). This conclusion is disconfirmed by their parsimony analysis of those same data, as well as by the phylogenetic systematic analysis of the morphological data summarized in this chapter (fig. 1.2B; appendix 1.1). Obviously, nothing was gained by Donnellan et al.'s evolutionary systematic approach, in which the impressive technicalities of maximum likelihood were employed. Future tests of the pygopod + diplodactyline hypothesis must focus on character reanalysis and the careful description of new character hypotheses, not on speculation, that is, model assumptions, which go beyond what is sufficient to make an inference and which cannot be tested empirically (Kluge 2001a, 2002). Those who embrace the likelihoodist, inductionist approach to phylogenetic inference rather than the deductionist framework of testability (Kluge 1997) can take solace in the fact that a maximum likelihood tree is the same as a most parsimonious phylogenetic hypothesis (Farris 1999); only the common mechanism assumption of homogeneity of evolution need be discarded, each character being allowed to have its own suite of branch lengths (Tuffley and Steel 1997).

Summary

The new phylogenetic system of taxonomy described in this chapter is intended to replace the traditional Linnaean approach to taxonomy (table 1.2). The new system is a uninominal hierarchy of proper nouns without categorical ranks; it is simply an indented list of names. Each name in the list, including that of phylospecies, is ostensively defined, in reference to a monophyletic group on the most parsimonious total evidence phylogenetic hypothesis. The phylogenetic system of taxonomy is therefore considered scientifically determined, because that is the least disconfirmed hypothesis of phylogeny, as well as the proposition that maximizes the explanatory power of all of the available evidence. It also follows that the phylogenetic system of taxonomy is maximally descriptively efficient of the evidence for phylogenetic relationships, because the hypothesis of phylogeny and the corresponding hierarchy of names are logically consistent. A type-specimen(s) is designated for each phylospecies, and two or more type-phylospecies are designated for each higher taxon, because those types provide the objective basis for relating names to taxa when the content of groups changes as the result of testing with new and critical evidence. The accompanying types may also be considered scientifically determined when they are chosen for the diagnostic features of the particular taxa they reference. The new phylogenetic system of taxonomy must be judged evolutionary as well, because the Darwinian conditionals of "descent, with modification" constitute the major premises in the inference of the most parsimonious phylogenetic hypothesis. In addition, stability of nomenclature in the phylogenetic system of taxonomy is expected as a by-product of the reciprocally clarifying nature of phylogenetic systematics.

A major issue in applying the phylogenetic system of taxonomy is the consistent use of phylogenetic systematic analysis at all levels of historical individuality, including the smallest, the one I call phylospecies. Recent phylogenetic systematic studies of eublepharid gekkotans underscore the advantage of ostensively defining all monophyletic groups. The relationship of pygopods to other gekkotans documents the importance of explaining the available evidence, assuming just "descent, with modification" as background knowledge. Ad hocisms and model assumptions of evolutionary systematics unnecessarily reduce the severity of the tests (Kluge 1997). Moreover, likelihoodists can no longer claim an advantage in their pursuit of Truth, or even verisimilitude, given the potential for their methodology to be statistically inconsistent (Farris 1999; Steel and Penny 2000). Theory and axiomatization by themselves are not sufficient, as the rankless descent and set theory classifications teach us.

All other forms of taxonomy that were reviewed—the traditional and annotated Linnaean and the rankless descent and set theory approaches—are obviously flawed because they are classifications (table 1.2). All suffer from their treatment of taxa as class concepts (table 1.1), as matters of intensional definition. Although attempts to replace Linnaean classification with a rankless taxonomy may be applauded, they are otherwise problematic. For example, the descent classification of K. de Queiroz and his supporters must be rejected, because the Darwinian conditional of "descent" is not a sufficient assumption on which to base the name of any particular taxon. Further, the descent classification must be judged an unscientific kind of metaphysics because the assumption of "descent" is empirically empty. Moreover, the assumption of "descent" by itself provides no mechanism for increasing nomenclatural stability, and one is left having to legislate rules of nomenclature. Also, in rejecting nomenclatural typification, such as designating type-specimens, there is then no basis in descent classification for tracking the parts of taxa that change as a function of new evidence.

The rankless set theory classification (table 1.2) of Papavero et al. (2001 and references therein) is also rejected because in that classification, taxa are defined intensionally, according to set theory. Also, the reference capability ascribed to apomorphy is nominal and therefore empirically empty, and like descent classification, stability is sought through absolute priority of names. Last, higher taxonomic epithets in the set theory classification are strings of species or generic names, the higher the taxon the longer the string, and that makes diversity increasingly difficult to reference.

Appendix 1.1   Synapomorphies as an Extensional Definition of Taxa

Those synapomorphies diagnostic of Gekkota (fig. 1.2B) are as follows: 1, paired egg-teeth present; 2, premaxilla and vomer largely, if not entirely, separated by maxilla; 3, jugal reduced in size, postorbital arch incomplete; 4, lacrimal absent; 5, ontogenetic fusion of frontals; 6, parietal foramen absent; 7, supratemporal arch absent; 8, quadrate abuts paroccipital process; 9, internal process on extracolumella absent; 10, bipartite occipital condyle present; 11, pterygoid teeth absent; 12, *crista cranii* of frontal completely surround olfactory lobe of brain and fuse below it; 13, epipterygoid abuts directly against ventral surface of alar process of prootic rather than lying lateral or anterior to it; 14, *crista prootica* extends forward onto basipterygoid process, thereby forming bony canal for lateral head vein (possibly reversed in some gekkonids); 15, medial aperture within *recessus scalae tympani* divided into two foramina; 16, adult dentary completely surrounds Meckel's canal—the place of ontogenetic fusion not evident along lingual surface of dentary; 17, splenial does not extend anteriorly beyond tooth row midpoint; 18, hyoid cornu present; 19, notochordal canal usually persists throughout life; 20, autotomy septa in caudal vertebrae located posterior to a single pair of transverse processes; 21, postcloacal bones present; 22, carpal intermedium absent; 23, sublingual glands diffusely scattered across floor of mouth; 24, very thin, flat foretongue, with broad and untapered tip; 25, short, flat-topped peg-like papillae on foretongue; 26, postcloacal sacs present; 27, stapedial (facial) artery passes anterior to stapes; 28, *m. levator anguli oris* weakly differentiated; 29, *m. pseudotemporalis* inserts on parietal posterior to epipterygoid; 30, quadrate aponeurosis of *m. adductor mandibulae externus* absent; 31, bodenaponeurosis forms tendinous gutter on coronoid bone; 32, *m. rectus abdominis lateralis* absent; 33, extracolumella muscle present; 34, meatal closure muscle present; 35, multiple longitudinal bundles of *m. hyoglossus;* 36, spindle body (thickening of tectorial membrane near origin from cochlear limbic lip) present; 37, segmented pattern of hair cells occurs along auditory papilla; 38, cochlear duct and basilar membrane elongate; 39, cochlear limbus present; 40, periventricular tectal lamina present; 41, tectum synoticum absent; 42, a graded series of acrocentric chromosomes present—distinct break between macro- and microchromosomes absent; 43, eye-licking behavior present; 44, clutch consists of two or one eggs.

Those synapomorphies diagnostic of Gekkonoidea (fig. 1.2B) are as follows: 1, parietals paired; 2, supratemporal absent; 3, spectacle present.

Those synapomorphies diagnostic of Pygopodidae (fig. 1.2B) are as follows: 1, stapedial foramen absent; 2, sacral pleurapophyses absent; 3, pupil more or less straight; 4, O-shaped meatal closure muscle present.

Acknowledgments

The students in my University of Michigan courses, especially those in Chordate Anatomy and Phylogeny (Biology 252) and Principles of Phylogenetic Systematics (Biology 491), have for more than twenty years been "exposed to the benefits" of the phylogenetic system of taxonomy described in this chapter. Of particular note over the years has been the wonderment of the newest students to biology as to "why any one would choose the traditional, Linnaean, form of classification over that of the scientific phylogenetic system." The phylogenetic system of taxonomy described herein was first peer-reviewed in a seminar at the Department of Biology, University of Miami (December 1986), and in a symposium sponsored by the International Gecko Society, Palm Springs, Calif. (June 1995). The criticisms offered by David Hillis, Julian Lee, C. Richard Robins, Jay M. Savage, Aaron Bauer, and Lee Grismer, subsequent to those presentations, helped shape the ideas presented here. Glenn Shea was responsible in large part for summarizing the new evidence for pygopod relationships. Darrel Frost, Taran Grant, Mikael Härlin, Maureen Kearney, Roberto Keller, and David Lahti provided more criticisms of a preliminary draft of this chapter than I care to reveal. I am grateful to Roberto for providing me with his English translation of Papavero et al.'s 1992 publication and for sharing his enthusiasm for natural kinds and essentialism. I am especially indebted to Taran for insisting that I reorganize and expand the original manuscript and for pointing out my logical inconsistencies. I remain responsible for all those cases where I have not taken the reviewers' good advice. This chapter was completed at the Cladistics Institute, Harbor Springs, Michigan.

2

Biogeography and Molecular Phylogeny of Certain New World Caecilians

Marvalee H. Wake, Gabriela Parra-Olea,
and Judy P.-Y. Sheen

We begin our discussion of the relationships of New World caecilians, with emphasis on those of Mexico and Central America, by quoting the illustrious E. H. Taylor (1973, 227): "Since this paper went to press a paper on the partial revision of the genera *Dermophis* and *Gymnopis* has appeared by Savage and Wake [1972]. This is Savage's second attempt at a revision of these genera, the first appearing in *Trans. Kansas Acad. Sci.* [1953]. One should look forward to his next several revisions."

Savage's third attempt (and Wake's second) resulted in a careful examination of Central American caecilians, based largely on substantial new material that had accumulated since their 1972 paper. A search for new characters resulted in the resurrection of a number of Taylor taxa, but with non-overlapping identifying characters (Savage and Wake 2001). In addition, molecular systematic studies of Latin American caecilians (as well as other caecilian taxa) have been under way by Wake and her colleagues. We now review the current knowledge of Central American caecilians and present new information on the molecular systematics of several Central and South American taxa. We speculate on the implications of the data and the analyses in terms of the relationships of the taxa studied, higher-order caecilian relationships, and predictions about a more complete understanding of the biology of caecilians.

Materials and Methods

The specimens examined for morphological analysis, the characters noted, and the methods used in data recording and analysis are summarized by Savage and Wake (1972, 2001).

The mtDNA study used the following sampling design. We obtained partial sequences of 16S and cytochrome *b* (cyt *b*) genes for the following taxa: *Oscaecilia ochrocephala, Caecilia volcani, Typhlonectes natans, Gymnopis multiplicata, Dermophis mexicanus* (two samples, one from Guatemala and one from Costa Rica), *D. parviceps, Siphonops annulatus, Microcaecilia* sp. nov. 1 and 2, *Ichthyophis bannanicus,* and *Boulengerula taitana* (see table 2.1). In addition, we used published 16S sequences for *T. natans* (therefore sampling two different individuals probably representing different populations), *Caecilia* sp., and *Epicrionops* sp. (Hedges et al. 1993).

The DNA sequencing protocol included extracting genomic DNA from small quantities of frozen tissue by using Qiagen DNeasy tissue kits. We sequenced 544 base pairs (bp) of the large 16S subunit ribosomal mtDNA gene and 390 bp of the cyt *b* gene. These genes were selected in order to recover maximum phylogenetic information both at the terminal nodes and at the base of the tree. Amplification was done by means of polymerase chain reaction (PCR) (Saiki et al. 1988), using the primers MVZ16 and MVZ25 (Moritz et al. 1992) for

Table 2.1 Species and their localities

Species	Museum no.	Locality
Ichthyophis bannanicus	MVZ 226265	Tam Dao, Vinh Phu Province, Vietnam
Typhlonectes natans	MVZ 500992	Baranquilla, Colombia
Oscaecilia ochrocephala	MVZ FC 13876	Gamboa, Panama Province, Panama
Caecilia volcani	MVZ 231242	Fortuna, Chiriqui Province, Panama
Dermophis mexicanus	MVZ FC 13178	Finca Los Andes, Volcan Atitlan, Depto. Suchitepequez, Guatemala
Dermophis mexicanus	MVZ FC 14021	Costa Rica
Dermophis parviceps	MVZ FC 14084	Costa Rica
Gymnopis multiplicata	MVZ FC 13567	Green Turtle Station, Tortuguero, Limón Province, Costa Rica
Microcaecilia sp. nov. 1	IWK 0128	Kurupkari, Guyana
Microcaecilia sp. nov. 2	MAD 496	Cowfly Camp, Iwokrama Forest, Guyana
Siphonops annulatus	MVZ 162588	Huampami, Depto. Amazonas, Peru
Boulengerula taitana	MVZ 179505	Wundanyi, Taita Hills, Kenya

Note: Specimens of species used in the molecular phylogenetic analysis are listed by museum number and locality.

cyt *b* and the primers 16Sar and 16Sbr (Palumbi et al. 1991) for 16S. The PCRs consisted of 38 cycles, with a denaturing temperature of 92°C for 1 min, annealing at 48°C for 1 min, and extension at 72°C for 1 min in a Techne PHC-1 thermocycler. The PCRs were run in a total volume of 25 μl, using 0.5 pmol of each primer. Double-strand templates were cleaned using a QIAquick PCR purification kit. We used 5.5 μl of double strand as the template for cycle-sequencing reactions in a 10-μl total volume with the Perkin-Elmer Ready Reaction Kit to incorporate dye-labeled dideoxy terminators. Thermal cycling was performed using standard conditions. Cycle-sequencing products were purified using ethanol precipitation and separated on a 6% polyacrylamide gel, using an ABI 377 DNA sequencer (Applied Biosystems).

Sequence analyses included partial sequences of 16S (544 bp) and cyt *b* (390 bp), which were read from both strands and aligned by eye with each other and with the outgroup in the program Sequence Navigator, version 1.0.1 (Applied Biosystems). For 16S, 87 bp were excluded from the analyses because of ambiguity of alignment. Pairwise comparisons of observed proportional sequence divergence (p-distance) and corrected sequence divergence (Kimura two-parameter; Kimura 1980), and of number of transitions and transversions, were obtained using the program PAUP*4.0b2a (Swofford 1998). Corrected sequence divergences were estimated using the Kimura two-parameter distance (K2p) in order to correct for multiple hits. To test for the possibility that some types of nucleotide substitutions had become saturated, we plotted p-distance (y) versus corrected (K2p) estimates of proportional sequence divergence (x) for first, second, and third codon positions, and for transitions and transversions separately (not shown).

Phylogenetic inference was based primarily on maximum parsimony (MP) analyses (Swofford 1998). MP phylogenies were estimated using the heuristic search strategies for each tree-building methodology. We used 10 repeated randomized input order of taxa for all MP analyses to minimize the effect of entry sequence on the topology of the resulting cladograms. MP analyses were conducted without the steepest descent option, and with accelerated character transformation (ACCTRAN) optimization, tree bisection-reconnection (TBR) branch swapping, save minimal trees (MULPARS), and zero-length branches collapsed to yield polytomies. We used nonparametric bootstrapping (100 pseudoreplicates, 50% majority rule) to assess the stability of internal branches in cladograms. Nonparametric bootstrap values generally provide a conservative measure of the probability that a recovered group represents a true clade (Felsenstein and Kishino 1993; Felsenstein 1985). Transversion to transitions weights 1, 4, and 10 were used for the analysis in order to determine whether resolution was thereby increased, especially at the base of the tree (Moritz et al. 1992). Each base position was treated as an unordered character with four alternative states.

Maximum likelihood (ML) analysis (Felsenstein 1981, 1993) was used with the heuristic algorithm. We randomly selected as the starting tree one of the trees found during the MP searches. Using empirical nucleotide frequencies and five rate categories, we fixed the probabilities of the six possible nucleotide transformations, the proportion of invariable sites, and the gamma "shape" parameter of the gamma distribution of rate heterogeneity across nucleotide positions (Yang 1996) to the empirical values calculated from the starting tree in a search for a better ML tree (a tree with a higher log-likelihood value) under the general time-reversible model of nucleotide substitution (Yang 1994a; Gu et al. 1995; Swofford et al. 1996).

For all phylogenetic analyses, a sequence of a bolitoglossine salamander of the genus *Batrachoseps* was used as the outgroup. The data sets were analyzed separately for each gene as well as in combination.

Current Systematic Status and Biogeographic History

SYSTEMATIC STATUS

The current status of the systematics of Central American caecilians—based on work by Savage and Wake (1972), Taylor (1973), and Nussbaum (1988), and the descriptions of *Caecilia volcani* (Taylor 1969) and *Oscaecilia osae* (Lahanas and Savage 1992)—is that three species are recognized in the genus *Dermophis: mexicanus*, which ranges from southern Mexico to Panama; *oaxacae*, in central and southwestern Mexico; and *parviceps*, from Costa Rica to northern Colombia. As recognized by Savage and Wake (1972), *D. mexicanus* includes four and *D. parviceps* three of Taylor's (1968) ten species in the genus. *Gymnopis* currently includes two species, *multiplicata* and *syntrema*. *Caecilia* and *Oscaecilia*, genera that are wide-ranging in South America, reach central and eastern Panama (*Oscaecilia ochrocephala* and *O. elongata; Caecilia nigricans, C. isthmica*, and *C. leucocephala*); one species in western Panama reaches to the Costa Rican border (*Caecilia volcani*), and one is restricted to southwestern Costa Rica (*Oscaecilia osae*).

Given the accumulation of new material representing Central and South American taxa, and given a considerable new literature on the geology and biogeographic history of the complex Central and northern South American regions, Savage and Wake (2001) developed new ideas about the kinds of morphological characters that are appropriate to distinguish caecilian taxa, and they took a new look at the question of the systematics and biogeography of the northern New World caecilians.

Savage and Wake (2001) recognized seven species of *Dermophis* (*mexicanus, oaxacae, gracilior, costaricense, glandulosus, parviceps*, and *occidentalis*), two species of *Gymnopis* (*syntrema* and *multiplicata*), three of *Oscaecilia* (*osae, elongata*, and *ochrocephala*), and four of *Caecilia* (*nigricans, volcani, isthmica,* and *leucocephala*). *Dermophis mexicanus*–like caecilians occur from central Mexico south into central Nicaragua; they are large and robust, and are characterized by annular grooves bordered by dark lines that are especially evident ventrally and by relatively large numbers of both primary and secondary annuli. *Dermophis mexicanus* and *D. oaxacae* are members of that group. The populations of Costa Rican and Panamanian caecilians with numerous secondary folds are smaller than members of the *D. mexicanus* group, and their annuli, though numerous, are not marked with dark coloration. Two populations, one on the Atlantic versant and one on the Pacific, differ substantially in annular counts. Taylor (1955) called the Atlantic form *D. costaricense* and the Pacific one *D. gracilior.* It is possible that more extensive collecting (difficult with changing land use in Costa Rica) might reveal intermediate forms, but currently the two are allopatric and differ in annular counts, so we recognize both taxa as species. *Dermophis glandulosus* is a moderate-sized form, not a small one allied with *D. parviceps,* as Taylor (1955) suggested on the basis of the single juvenile that was the type for his species designation (Taylor's intuition was sound). Savage and Wake (2001) therefore recognized *D. glandulosus* because it has fewer secondary annuli than does *gracilior* and because of the two species' sympatry in the San Vito area of central Costa Rica. Taylor's *D. balboai* is conspecific with *D. glandulosus,* so Savage and Wake did not recognize it. *D. glandulosus* is the only species of *Dermophis* whose range extends into Colombia.

The small *Dermophis* that have few secondary annuli also can be distinguished, now that more material is available. *D. parviceps* is relatively small, lacks dark coloration in its annuli, and is distinguished by its pinkish head in life. The head is pale brown in preservation but contrasts with the body color markedly, unlike the condition in other *Dermophis. D. occidentalis* of the Pacific slopes of Costa Rica has more annuli, but it is a small-sized taxon and is allopatric to *D. parviceps.* Again, more specimens from more localities are necessary to sort out *Dermophis* in Pacific Costa Rica and Panama.

Savage and Wake (2001) currently recognize three groups of species within *Dermophis:* (1) large caecilians with numerous secondary annuli (*mexicanus* and *oaxacae*); (2) moderate-sized forms also with numerous secondaries (*costaricense, glandulosus, gracilior*); and (3) small-sized to moderate-sized forms with few secondaries (*occidentalis, parviceps*). Their analysis provided more characters, and the sample sizes for the taxa are much larger in most cases. In particular, annular counts and other characters are better defined for some of the species of *Caecilia* and *Oscaecilia.*

HISTORICAL BIOGEOGRAPHY

Savage and Wake (1972) hypothesized that the ancestral *Dermophis* stock evolved in the Veracruzian center of differentiation (terminology of Savage

1966, 1982) north of a large marine embayment across central Guatemala, whereas *Gymnopis* is associated with a southern center (Caribbean) in what is now Honduras and Nicaragua, which was isolated south of the marine barrier until after early Pliocene. Subsequently the stock of *Dermophis* crossed over the Isthmus of Tehuantepec on two occasions, one leading to the allopatric speciation of *D. oaxacae,* and the other to that of *D. mexicanus,* which then followed the Pacific lowlands south into the emerging Isthmian Link region of Lower Central America. As the Talamancas of Costa Rica became uplifted, the Pacific population differentiated into *D. parviceps.* The Atlantic population also differentiated somewhat, and late Cenozoic drying reduced various contacts. *Gymnopis* moved northward and southward after the marine barriers were eliminated in the Pleistocene. This straightforward scenario correlated with that of certain other species of amphibians in the region (see Savage and Wake 2001).

New information about caecilians, geology, and biogeography has allowed a more extensive analysis of the historical biogeography of Middle American caecilians. The oldest fossils of caecilians are a limbed form from the early Jurassic of Arizona in the Kayenta formation, and New World fossils include vertebrae from the late Cretaceous of Bolivia, the Paleocene of Brazil and Bolivia, the Miocene of Colombia, and the Quaternary of southern Mexico; Old World fossil vertebrae are from the Cretaceous of the Sudan (fossil information is summarized in Savage and Wake 2001). An early distribution, rather than a strictly Gondwanan one, is implied by the fossil data; *Dermophis* and *Gymnopis* are the only remaining definitive Laurasian representatives. All other caecilians, though, are derived from Gondwanan ancestors and either are restricted to fragments of its landmass or, in the case of the Ichthyophiidae of Southeast Asia, dispersed more widely from the Indian Plate. The current absence of caecilians from Madagascar and Australia likely represents extinction events, and the discovery of fossils from those landmasses is anticipated.

Savage and Wake's (2001) current interpretation of the history of the Central American lineages is based on the above vicariant scenario of separation of the ancestor of the *Dermophis-Gymnopis* clade in tropical North America and the ancestor of the South American caecilians on the insular South American continent by the breakup of the Isthmian Link in the Paleocene and the separation of the continents for the next 30 million years. More recent history is interpreted in terms of new information about the terranes in the region (summarized in Savage and Wake 2001). At the end of Paleocene, the Maya Block was in its present position and would become eastern Mexico and northern Nuclear Central America. The Chortis Block, the future southern Nuclear Central America, was well to the west. Far southwest was a series of volcanic islands that would later coalesce to form the Chorotega and Choco blocks. The Chorotega Block would become today's Costa Rica and western Panama, and the Choco Block eastern Panama and western Colombia. By mid-Eocene, the Chortis was sutured to the southern part of the Maya Block. During late Oligocene and

Miocene, the Chorotega and Choco blocks narrowed the gap between Nuclear Central America and South America. Further uplift of those blocks by subduction of the Cocos Plate caused the reconnection of the continents by the Panamanian Isthmus in middle Pliocene.

The Central American caecilian core areas of distribution strongly correlate with these major tectonic terranes: the Maya Block with *Dermophis* group 1, the Chortis with *Gymnopis,* the Chorotega with *Dermophis* groups 2 and 3, and the Choco with *Caecilia* and *Oscaecilia.* Savage and Wake (2001) suggested that by early Cenozoic, *Dermophis* and *Gymnopis* ancestors were already on the Maya and Chortis blocks, respectively. *Gymnopis* expanded its range across faults after the Chortis Block sutured to the Maya Block in Eocene. Uplift of east–west ranges fragmented the distribution of *Gymnopis* and led to the possible differentiation of *G. syntrema* to the north and *G. multiplicata* to the south of the suture zone. *G. multiplicata* has moved southward onto the emerging Isthmian land bridge and onto the northern part of the Chorotega Block.

Dermophis distribution is apparently the result of a combination of vicariance and dispersal events. The uplift of the central ridge of southern Mexico and Nuclear Central America beginning in Oligocene isolated *mexicanus* to the east and *oaxacae* to the west. We continue to suggest that *mexicanus* dispersed across the Isthmus of Tehuantepec onto the Pacific slope and expanded its range southward and that *oaxacae* gradually extended northward. The southern species of *Dermophis* are principally on the Chorotega Block, with *D. glandulosus* and *D. parviceps* ranging onto the Choco. This dates the origin of our groups 2 and 3 of *Dermophis* to the Miocene. The Nicaragua Depression formed a marine link north of the Chorotega in late Miocene, isolating the ancestors of the southern *Dermophis* from the *mexicanus* lineage. The emergence of the Isthmian Link allowed dispersal of *Dermophis* onto the Choco Block during Pliocene to Recent time. The differentiation of the southern species appears to be the result of local orogenic effects, such as the uplift of the cordilleras.

According to Savage and Wake (2001), *Caecilia* and *Oscaecilia* moved into lower Central America after the closure of the Panamanian seaway in Pliocene. Four species are restricted to the Choco Block (*O. elongata, C. isthmica, C. leucocephala,* and *C. nigricans*), and three occur on the southern part of the Chorotega (*O. osae, O. ochrocephala,* and *C. volcani*), apparently northern dispersals of sister taxa in the two genera. The correlation of caecilian distributions with the major tectonic terranes allows us to construct a more detailed scenario of the historical biogeography of Central American caecilians and generates some potentially testable hypotheses of the timing of divergences of the taxa involved. When Wake (1993) reexamined microcomplement fixation data (Case and Wake 1977), she found that the divergence of *Dermophis* and *Gymnopis* at least by Oligocene, and *Dermophis* from *Caecilia* by the Paleocene, correlated with the biogeographic hypotheses of Savage and Wake (1972), and the times seem to correlate reasonably well with our more detailed analysis also.

Molecular Systematics of Selected Species

New data from the ongoing work on molecular systematics of caecilians provide insights into both caecilian relationships and caecilian biogeography. There are several caveats associated with our work in progress, but we present these as we speculate on the implications of our data and analyses. Several testable hypotheses emerge from our current work, and some correlate with those of the biogeographic scenario. First, we acknowledge that the molecular phylogenetic analysis is based on rather few taxa, so the addition of new taxa might well change substantively the relationships presented in this chapter (as we found with our recent addition of *Microcaecilia* to our data set). Further, our representation of crucial groups is highly incomplete, because of the relative rarity of these animals. Our work is effectively at the generic level, although concepts of species and of genera of caecilians are not well defined and subject to substantial interpretation.

MP analysis of the 16S data set (fig. 2.1) by the use of equal weighting produced a single most parsimonious tree (tree length [TL] = 484 steps; 140 characters were parsimony informative; CI = .576, RI = .528). The basal relationships between clades are unresolved, but there is support for two clades. One clade includes *Caecilia* and *Oscaecilia* as sister taxa (bootstrap value [bs] 98%), with *Typhlonectes* being their sister group (bs 92%); the other is a *Gymnopis-Dermophis* clade (bs 89%) that is sister to a clade containing *Siphonops* and both species of *Microcaecilia* (bs 73%). There is little resolution at the base of the tree, but there is support for *Epicrionops* as the most basal taxon, and surprisingly, *Boulengerula* is the sister taxon to *Ichthyophis* + *Caecilia* + *Typhlonectes*. *Epicrionops* is a member of the northern South American family Rhinatrematidae, which is basal to all other caecilians on morphological grounds (see Nussbaum and Wilkinson 1989), and found to be basal in the Hedges et al. (1993) molecular analysis. (*Ichthyophis* is a member of a separate family, the Ichthyophiidae, which is basal to "caeciliids" on morphological grounds; *Boulengerula* is a caeciliid. See Nussbaum and Wilkinson 1989 and Hedges et al. 1993.)

The 16S tree includes two species of *Caecilia*: an unidentified species, probably South American (included by Hedges et al. 1993; it is their 16S sequence in GenBank that we added to our data), and *Caecilia volcani*, a Panamanian species, as well as *Oscaecilia ochrocephala*, another Panamanian species. The South American *Caecilia* and the Panamanian *Oscaecilia* are sister taxa, with the Panamanian *C. volcani* their sister taxon, suggesting that the relationships of *Caecilia* and *Oscaecilia* are complex. Data for more species of both genera (and their relatives) are needed to understand the relationships. *Typhlonectes natans*, a member of the viviparous, aquatic South American group recognized as a family but long thought by many to be paraphyletic within the Caeciliidae (as illustrated by Hedges et al. 1993), is the sister taxon to the *Caecilia-Oscaecilia* clade in our analysis. That New World clade is deeply divided from a clade that includes

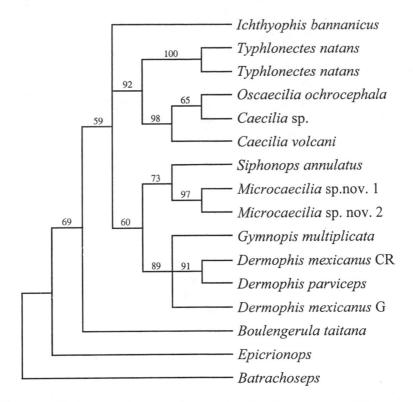

Figure 2.1   Maximum parsimony tree for 14 species of caecilians, based on 16S data. A plethodontid salamander is the outgroup taxon. Bootstrap values are listed. CR, Costa Rica; G, Guatemala.

Siphonops annulatus, widely distributed in South America; *Microcaecilia,* from northern South America; and *Dermophis* and *Gymnopis,* the two Central American genera. Within the *Gymnopis-Dermophis* clade, there is support for *D. mexicanus* from Costa Rica as sister taxon to *D. parviceps* (bs 91%); the relationships of these two taxa to *D. mexicanus* from Guatemala and to *G. multiplicata* are unresolved.

The ML topology (not shown) is very similar to the MP topology. The only difference is that the *Siphonops-Microcaecilia* clade is not the sister to the *Gymnopis-Dermophis* clade. Also, in this analysis *D. mexicanus* from Guatemala is the sister taxon to *G. multiplicata,* whereas *D. mexicanus* from Costa Rica is the sister to *D. parviceps.*

The cyt *b* analysis (fig. 2.2) results in similar topologies to those of 16S, but with less resolution for the basal relationships. The MP analysis resulted in four

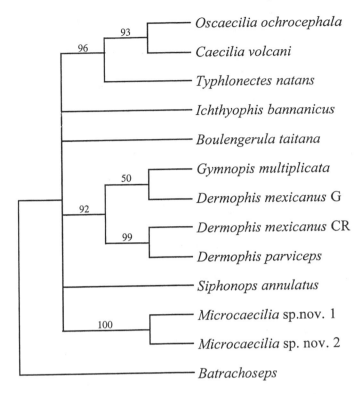

Figure 2.2   Maximum parsimony tree for 12 species of caecilians, based on cyt *b* data. Bootstrap values are listed. CR, Costa Rica; G, Guatemala.

equally parsimonious trees (TL = 722, CI = .497, RI = .370), with support for the *Gymnopis-Dermophis* clade (bs 57%), with the same internal arrangement as that of the 16S MP. There also is support for the *Oscaecilia-Caecilia-Typhlonectes* clade (bs 88%), and these two clades form part of the basal polytomy together with *Ichthyophis, Siphonops, Boulengerula,* and the two species of *Microcaecilia* as sister taxa. We do not have cyt *b* data for *Epicrionops*. All different weighting schemes for transitions and transversions resulted in identical topologies to the unweighted analysis.

The MP analysis of the combined 16S and cyt *b* data sets (fig. 2.3) results in a topology that also is unresolved at the base. In the MP analysis, there is support for the *Oscaecilia-Caecilia-Typhlonectes* clade (bs 96%), and for the *Gymnopis-Dermophis* clade (bs 92%). This topology differs from the 16S topology in that the 16S analysis places *Siphonops* as the sister taxon to the two species of *Microcaecilia* (fig. 2.1), but the combined analysis allies *Siphonops,* the *Gymnopis-*

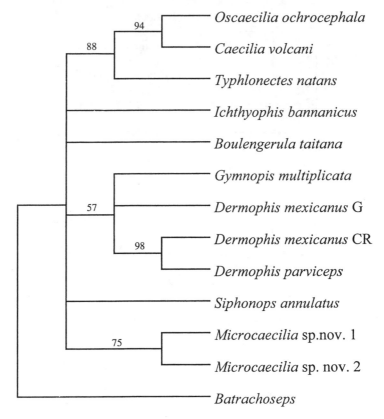

Oscaecilia ochrocephala

Caecilia volcani

Typhlonectes natans

Ichthyophis bannanicus

Boulengerula taitana

Gymnopis multiplicata

Dermophis mexicanus G

Dermophis mexicanus CR

Dermophis parviceps

Siphonops annulatus

Microcaecilia sp.nov. 1

Microcaecilia sp. nov. 2

Batrachoseps

Figure 2.3   Maximum parsimony tree for 12 species of caecilians, based on combined 16S–cyt *b* data. Bootstrap values are listed. CR, Costa Rica; G, Guatemala.

Dermophis clade, and *Microcaecilia* in a multiple polytomy with all the New World clades and with the African *Boulengerula* and the Asian *Ichthyophis*. Therefore most generic relationships are unresolved by the combined analysis.

The ML analysis of the combined data set (fig. 2.4) results in a topology identical to the MP topology, but with more structure at the base. *Siphonops* is sister taxon to the *Gymnopis-Dermophis* clade. The resolution within *Gymnopis-Dermophis* also clusters *D. mexicanus* from Costa Rica as sister to *D. parviceps,* and *D. mexicanus* from Guatemala as sister to *Gymnopis.* In this analysis there is also an *Oscaecilia-Caecilia-Typhlonectes* clade, and curiously, the basal *Ichthyophis* and the east African caeciliid *Boulengerula* cluster as the sister clade to all other caecilian taxa examined.

In all analyses performed, the basal relationships are unresolved, and there is support for at least two clades: the *Gymnopis-Dermophis* clade and the

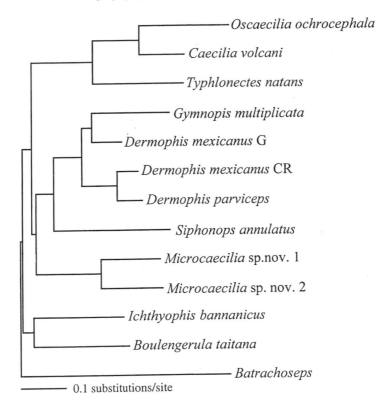

Figure 2.4   Maximum likelihood tree for 12 species of caecilians. A 0.1 substitutions/site scale is indicated to validate branch lengths. CR, Costa Rica; G, Guatemala.

Oscaecilia-Caecilia-Typhlonectes clade. The clades in all the trees are well supported by high bootstrap values. The variation in assessment of relationships among the various trees produced by different analytical methods of different genes points up the problem of noncomparable and incomplete taxon sampling.

Kimura two-parameter distances (tables 2.2 and 2.3) are high among all taxa, including the several New World representatives, suggesting that they are all old, well-defined lineages. Cyt *b* distances are very high; for example, the Guatemalan *Dermophis mexicanus* has a sequence divergence of 22% to the Costa Rican *Dermophis,* and the two species of *Microcaecilia* have a sequence divergence of 23%. Between genera, all the divergences are greater than 21% and as high as 40% (between *Siphonops* and *Typhlonectes*) (table 2.2). Other than that they are large, indicating old lineages; one cannot place much emphasis on these distances. We assume that given the high genetic divergences, the cyt *b* is not informative for resolving the basal relationships of taxa in this study. How-

Table 2.2   Kimura two-parameter distances (16S)

	1	2	3	4	5	6	7	8
1. Ich. bann.	—							
2. Typh. nat. GB	.20142	—						
3. Typh. nat.	.21390	.02977	—					
4. Osc. och.	.22243	.18694	.18682	—				
5. Siph. ann.	.18419	.19472	.17592	.21195	—			
6. Boul. tait.	.18042	.20587	.21164	.19988	.18427	—		
7. Caec. vol.	.19401	.13167	.13306	.10412	.19227	.21178	—	
8. Caec. sp. GB	.25049	.21895	.20440	.11272	.24629	.24346	.09927	—
9. Gym. mult.	.18154	.18652	.17130	.21735	.14916	.17191	.17618	.21941
10. Der. mex. CR	.16169	.18611	.17383	.19987	.11657	.13670	.17254	.21036
11. Der. mex. G	.16912	.17386	.15662	.21744	.12254	.17387	.19212	.24066
12. Mic. sp. nov. 1	.18932	.20898	.19071	.23437	.13305	.17166	.21708	.24865
13. Mic. sp. nov. 2	.22854	.20172	.24024	.14874	.17892	.22451		.26191
14. Der. parv.	.17862	.18728	.17175	.18774	.11598	.14212	.17968	.21147
15. Epic. sp. GB	.21524	.31617	.3216	.30624	.23155	.21594	.31043	.32209

	9	10	11	12	13	14	15
9. Gym. mult.	—						
10. Der. mex. CR	.06812	—					
11. Der. mex. G	.07138	.06334	—				
12. Mic. sp. nov. 1	.13442	.12861	.12944	—			
13. Mic. sp. nov. 2	.16415	.14093	.13900	.09381	—		
14. Der. parv.	.08011	.02056	.06867	.12359	.14305	—	
15. Epic. sp. GB	.23641	.21468	.22442	.24455	.24905	.22801	—

Note: Ich. bann., *Ichthyophis bannanicus;* Typh. nat. GB, *Typhlonectes natans* GenBank; Typh. nat., *Typhlonectes natans;* Osc. och., *Oscaecilia ochrocephala;* Siph. ann., *Siphonops annulatus;* Boul. tait., *Boulengerula taitana;* Caec. vol., *Caecilia volcani;* Caec. sp. GB, *Caecilia* sp. GenBank; Gym. mult., *Gymnopis multiplicata;* Der. mex. CR, *Dermophis mexicanus* Costa Rica; Der. mex. G, *Dermophis mexicanus* Guatemala; Mic. sp. nov. 1, *Microcaecilia* sp. nov. 1; Mic. sp. nov. 2, *Microcaecilia* sp. nov. 2; Der. parv., *Dermophis parviceps;* Epic. sp. GB, *Epicrionops* sp. GenBank.

ever, it does provide information for the clustering of some groups, as indicated earlier in the chapter. The 16S data set does not show evidence of saturation, but the number of informative characters or taxa sampled may be too low to fully resolve the basal relationships.

Our results are similar to those of Hedges et al. (1993) in several ways: *Typhlonectes* and *Caecilia* are sister taxa, and the clade has a high bootstrap value. They found that *Dermophis* and the West African *Schistometopum* are sister taxa, and that that clade is sister to one that includes all of the Seychellean taxa examined, with *Siphonops* the taxon next out (not associated with the *Dermophis-Gymnopis* clade, as we found) but distant from the *Caecilia-Typhlonectes* clade. They did not have *Gymnopis, Oscaecilia, Microcaecilia,* and other taxa available for study, and we did not have *Schistometopum* or include the Seychellean taxa because of our focus on the relationships of the American caecilians. Comparisons of the two analyses point up the difficulties in working with limited sampling at the generic level; the addition or subtraction of any taxon can markedly affect the resulting assessment of relationships. However, we note that analysis of the microcomplement fixation data reported by Hass, Nussbaum, et al. (1993) supports the general relationships found by Hedges et al. (1993) and was in substantial agreement with that of Case and Wake (1977) when the same genera are compared with *Dermophis.* In many ways it is also congruent with our current analysis as well as with Wake's (1992) conclusions about relationships and biogeographic patterns.

Several points of speculation arise from these data.

First, the relationships of *Gymnopis* and *Dermophis* are complex. *Gymnopis multiplicata* is the sister taxon of Guatemalan *D. mexicanus,* our group 1 *Dermophis* representative in our molecular phylogeny; Savage and Wake (2001) have now restricted *D. mexicanus* to northern Central America and have recognized several species in Costa Rica and probably Panama. The two Costa Rican species of *Dermophis* examined in the molecular analysis constitute the sister taxon to Guatemalan *Dermophis* and to *Gymnopis,* the three units forming a polytomy. The close relationship of *Gymnopis* to *Dermophis* had long been known, based on morphological characters (see Taylor 1968). We have yet to identify the Costa Rican voucher specimen by use of the new Savage and Wake characters, but it is apparently a group 2 *Dermophis* representative. In addition, *D. parviceps* was considered by Savage and Wake (2001) to be the sister taxon of all the Costa Rican *Dermophis;* it is a group 3 *Dermophis* representative. Savage and Wake's inference is supported by our molecular data, although we have only the one group 2 species in our data set. Our molecular analysis clearly supports the conclusion of Savage and Wake that the Costa Rican species are distinct from the northern *D. mexicanus* and *oaxacae;* we had no previous idea, though, how distantly related the groups might be, as indicated by the deep divergences shown by the molecular data. The Costa Rican *Dermophis* might well warrant a different generic name to resolve the paraphyly in the clade. The status of *Gymnopis* is enigmatic. More extensive taxon sampling is necessary to clarify relationships of these taxa.

Second, our 16S data illustrate two clades that correlate with the New World components of Taylor's Caeciliinae and Dermophiinae. Further, although the microcomplement fixation data of Case and Wake did not support the recognition of two subfamilies, that of Hass, Nussbaum, et al. (1993) and the sequence data of Hedges et al. (1993) do support recognition of the two units. The

Table 2.3 Kimura two-parameter distances (cyt *b*)

	1	2	3	4	5	6	7	8	9	10	11	12
1. Osc. och	—											
2. Ich. bann.	.31267	—										
3. Boul. tait.	.34295	.24429	—									
4. Gym. mult.	.36474	.35580	.29329	—								
5. Typh. nat.	.34567	.29282	.36406	.35015	—							
6. Der. mex. G	.34247	.26788	.24645	.24235	.35447	—						
7. Der. mex. CR	.38472	.29243	.30889	.30758	.36246	.21892	—					
8. Caec. vol.	.26618	.30742	.35979	.36474	.29670	.32870	.37089	—				
9. Siph. ann.	.35666	.31555	.32351	.33394	.40279	.31682	.31682	.38584	—			
10. Mic. sp. nov. 1	.36758	.26525	.29700	.31043	.34688	.31719	.31997	.33033	.36439	—		
11. Mic. sp. nov. 2	.33157	.25859	.30627	.30925	.30090	.34345	.31340	.33054	.33842	.23047	—	
12. Der. parv.	.39894	.31340	.31379	.27113	.37418	.23686	.14981	.38706	.30587	.33612	.31758	—

Note: Osc. och., *Oscaecilia ochrocephala*; Ich. bann., *Ichthyophis bannanicus*; Boul. tait., *Boulengerula taitana*; Gym. mult., *Gymnopis multiplicata*; Typh. nat., *Typhlonectes natans*; Der. mex. G, *Dermophis mexicanus* Guatemala; Der. mex. CR, *Dermophis mexicanus* Costa Rica; Caec. vol., *Caecilia volcani*; Siph. ann., *Siphonops annulatus*; Mic. sp. nov. 1, *Microcaecilia* sp. nov. 1; Mic. sp. nov. 2, *Microcaecilia* sp. nov. 2; Der. parv., *Dermophis parviceps*.

situation is complicated by our data, though. The addition of *Microcaecilia* to our 16S data set has that genus included in a clade as sister taxon to *Siphonops*, and those two genera are the sister group to the *Gymnopis-Dermophis* clade. However, according to analysis of the combined data set by both MP and ML, *Microcaecilia* is part of a multiple polytomy independent of *Siphonops*, which is the sister taxon to the *Gymnopis-Dermophis* clade (fig. 2.1). Clearly, further work should be undertaken and more taxa included before recognizing subfamilies. However, the deep division of the clades on the basis of molecular evidence supports the recognition of two obvious units, whether they are designated subfamilies or not. We need a much more complete taxon sampling before we can fully define the clades involved. We speculate that thorough analysis of the "Caeciliidae" will result in the recognition of at least three family-level units in the New World and at least two, probably more, in Africa and Southeast Asia. Relationships between South American and African taxa are likely to emerge that will confound resolution. Support, both molecular and morphological, continues to accrue that indicates that the New World Rhinatrematidae is the basal caecilian family, and that it and the Old World Ichthyophiidae constitute the sister group to the higher caecilians. The relationships of the subsets of the Caeciliidae to the Uraeotyphlidae and the Scolecomorphidae are not yet established, although morphological data provide a point of departure; their resolution, however, is not in the purview of this analysis.

Third, the evolution of reproductive modes as suggested in several morphological analyses is supported by the molecular data. Wake (1989, 1993) has indicated that viviparity (maternal retention of the developing embryos and fetuses through metamorphosis and with maternal nutrition via secretions of the oviductal epithelium after the limited yolk supply is exhausted) has arisen at least three times, perhaps more, in caecilians. The basal taxon *Ichthyophis* is an egg layer with a lengthy (approximately one year; see the summary in Himstedt 1996) larval period, and the African *Boulengerula* is apparently an egg layer with direct development (Nussbaum and Hinkel 1994). Among New World clades, the South American *Siphonops* is an egg layer (Goeldi 1899; Gans 1961a), with at least one species that guards and perhaps feeds its altricial young (J. O'Reilly, pers. comm.); it is the sister taxon (singly or with *Microcaecilia*) to the viviparous Central American *Dermophis* and *Gymnopis* (this study). On the basis of fully yolked ovum sizes, Wake has inferred that several *Caecilia* and *Oscaecilia ochrocephala* are oviparous, and that *O. ochrocephala* may have direct development (M. H. Wake, unpub. data). The reproductive modes of most caecilians remain undescribed, including those of *Microcaecilia*. However, so far as is known, the typhlonectids (*Chthonerpeton*, *Potomotyphlus*, and *Typhlonectes*, and perhaps *Atretochoana*) are viviparous (Taylor 1968; Nussbaum and Wilkinson 1989), with maternal nutrition. The relationships of the taxa as indicated by the molecular phylogeny suggest that there are at least two separate origins of viviparity in the New World, in the typhlonectids and in *Gymnopis-Dermophis*. In addition, among Old World taxa, *Geotrypetes* of West Africa is viviparous with maternal care (J. O'Reilly, pers. comm.), and members of the genus *Scolecomorphus* in the African family Scolecomorphidae are also viviparous (Taylor 1968; Nussbaum and Wilkinson 1989), although members of the other genus in the family, *Crotaphatrema*, likely are oviparous (Nussbaum 1985). This suggests minimally a third, and perhaps a fourth, instance of the evolution of viviparity in caecilians, and there may be more, as information on the reproductive biology and life histories of species of caecilians becomes known.

In summary, morphological, molecular, and biogeographic data continue to accumulate for Central American caecilians that shed new light on the biology, relationships, and evolution of the animals. It is essential to continue the search for caecilians throughout the region, because land-use changes continue at an accelerated rate, markedly altering distribution patterns and rendering populations and species extinct. Biologists are just becoming aware of how interesting and important caecilians are biologically, and we run the risk of losing them before we really know anything about them.

Acknowledgments

We appreciate Kris Lappin's early work on the molecular sequences. We thank the Museum of Vertebrate Zoology, University of California at Berkeley, and the Department of Organismic and Evolutionary Biology, Harvard University, for access to their DNA laboratories. We especially thank Maureen Donnelly (Florida International University) for the tissue samples of the two species of *Microcaecilia* and Karen Lips (Southern Illinois University) for the *Caecilia volcani* material. Other samples are from the collection of the Museum of Vertebrate Zoology and that of MHW. MHW's research was supported by NSF IBN 95-27681.

3

Diversity of Costa Rican Salamanders

David B. Wake

Diversification and Species Richness in a Small Area

There are many more species of salamanders in Costa Rica than one would expect from the study of Old World tropical environments, because salamanders have thrived only in the New World tropics. Nearly all are members of a single clade, the supergenus (sg) *Bolitoglossa* (Plethodontidae: Plethodontinae: Bolitoglossini), which is restricted to Middle and South America. Its sister group is the sg *Batrachoseps*, which ranges from Baja California to the Columbia River, mainly west of the Sierra Nevada–Cascades mountain systems; the sister group of these two taxa is the sg *Hydromantes*, which occurs in California and the central Mediterranean region (Jackman et al. 1997). What never ceases to amaze me is that this tropical clade is not only species-rich, containing more than 200 species, nearly half of all currently recognized species of Caudata, but its impressive adaptive radiation is concentrated in Middle America, the smallest part of the Neotropical region. Surprisingly, only about 10% of the members of the clade are South American, and only 2 of the 13 currently recognized genera reach that continent. However, despite the proximity to South America, Costa Rica (and western Panama, which largely shares the same fauna) has a large and diverse salamander fauna. Although it may seem inappropriate to select an apparently arbitrary geographic subdivision for a faunal analysis, in reality Talamancan Central America (montane western Panama with Costa Rica; Wake and Lynch 1976) is an area of high salamander endemism and historical integrity. The older mountain areas and all of the central, eastern, and southern parts of present-day Costa Rica lie on the Chorotega Plate, with antecedents dating to 65 million years before the present (Coates and Obando 1996; Coates 1997; Burnham and Graham 1999). The region has enjoyed a largely separate history, sharply delimited from South America until about 3 million years ago, and having fleeting and intermittent connection to the Chortis section of the complicated Caribbean Plate, from which it is separated largely by the currently land-positive region of the Nicaraguan Depression (fig. 3.1).

Talamancan Central America is the lowest latitude at which salamanders occur in any degree of diversity, yet it is the site of some of the richest local salamander faunas in the world. I believe it has been an important staging ground for salamander evolution.

AN UNLIKELY ADAPTIVE RADIATION

Salamanders are associated with northern continents and have been throughout their long evolutionary history (Savage 1973). All ten salamander families occur in north temperate areas today, and only the Plethodontidae has undergone any significant tropical diversification (Wake 1966, 1970, 1987). Within the Plethodontidae, one subfamily (Desmognathinae), two of three tribes in the other (Hemidactyliini and Plethdontini of the Plethodontinae), and two (*Batrachoseps* and *Hydromantes*) of the three supergenera in the last tribe (Bolitoglossini) are exclusively North Temperate in origin. Thus tropical plethodontids would appear to be an unlikely prospect for adaptive radiation. Compounding the problem is the fact that all bolitoglossines undergo direct development and have abandoned aquatic larvae, thereby restricting themselves to only some of the array of habitats used by other salamanders. For example, in the Appalachian Mountains of eastern North America, salamander diversity is high in large part because species occur in rivers, streams, ponds, springs, and seeps—habitats avoided by tropical bolitoglossines.

Although it is impossible to reconstruct details of the evolutionary history of the tropical bolitoglossines, two factors appear to be associated with their success. The first is direct development, which, while constraining them to only certain ecologies, at the same time removes them from many competitive and high-predation environments. Direct development also makes it possible for these salamanders to avoid breeding migrations and to maintain permanent small home ranges. Not having to move much during their lifetime, tropical bolitoglossines can use such specialized microhabitats as arboreal bromeliads and arboreal moss mats and balls.

A second feature that has been important in the success of tropical salamanders is their extraordinarily specialized feeding system, which involves excellent vision and a long, fast tongue (Wake and Deban 2000). This system combines high visual acuity with biomechanical attributes that enable an animal to feed rapidly, with high accuracy and wide-range directionality, on prey that might be a great distance from the predator's head (approximately 40% of snout-vent length). Perhaps most important, this specialized tongue permits the salamander to remain hidden and thus to feed with stealth.

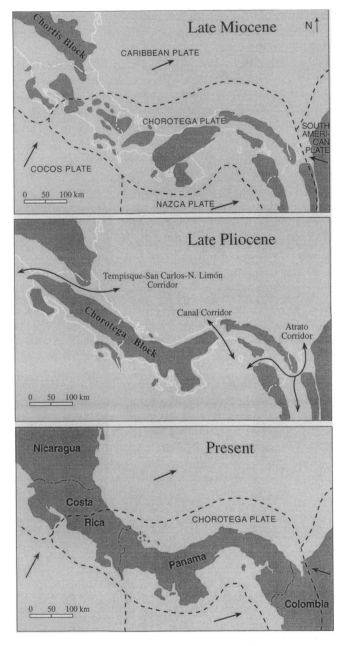

Figure 3.1 Reconstructions of the isthmian region of Lower Central America from Late Miocene to the present. Redrawn with modifications from Coates and Obando 1996.

One might consider each of these features to represent a key innovation. However, direct development evolved at the level of Bolitoglossini, or even lower (Lombard and Wake 1986). The tongue, however, is unique to the sg *Bolitoglossa*. The joint occurrence of direct development and the extraordinary tongue in the ancestral tropical lineage appears in retrospect to have been of special significance. Although both represent high levels of specialization, at the same time they are generalizing in their effects in that they freed the lineage of prior constraints (i.e., life historical) and gave access to environments and habitats, and new ways of using them. Access, followed by ecological innovation, led to new ways of life and to adaptive radiation based on elevational zonation, new patterns of habitat and microhabitat use, and possibly new community interactions (Wake and Lynch 1976).

Other factors were doubtless of significance in the success of the tropical salamanders. As a group, the tropical bolitoglossines have large genomes, which means that they have very large cells (Sessions and Larson 1987; Sessions and Kezer 1991). This in turn means that they have extraordinarily low basal metabolic rates (Feder 1983) so they live life in slow motion, so to speak. Again, this might be seen as an extreme specialization, but the advantage that is general for the group comes about because they require little food. Another factor that seems to have been important to the clade is the presence of tail autotomy, involving a complex of morphological, physiological, and behavioral specializations (Wake and Dresner 1967). All tropical species display tail autotomy, which must have arisen in the immediate ancestral stock because there is no specialized autotomy region in the sequential sister taxa *Batrachoseps* and *Hydromantes,* and the autotomy mechanisms found in the distantly related *Ensatina* and *Hemidactylium* evolved independently (Wake 1991a). The tail is an important organ, used for locomotion (especially in arboreal settings) and for fat storage, and its loss directly affects reproduction (Maiorana 1976), so one must conclude that the universality of autotomy attests to its survival value. The fixed number of vertebrae in all tropical genera but one may be associated with the presence of the specialized region at the tail base. In species with variable numbers of trunk vertebrae, asymmetry is relatively common, and in such instances the tail breakage pattern may be disabled (Frolich 1991). The species of *Oedipina* have extraordinarily long tails that may break at any point; perhaps this feature resulted in relaxation of the former constraint on variation in vertebral numbers, and they alone of the tropical species have more than 14 trunk vertebrae.

WHY SO MANY SPECIES?

Species density in restricted geographic areas is greater for tropical bolitoglossines than that typical of most temperate zone clades of salamanders. If one considers only strictly terrestrial species, there can be substantially more species of salamanders on a rich elevational tropical gradient than on a temperate one, often as many as 14–18 or more local species (Wake and Lynch 1976; Wake 1987; Wake et al. 1992). The main factors related to this species richness are the narrow elevational and geographic ranges of the species. Many of the species are restricted to relatively high cloud forests, and the lower margins of their ranges lie above the levels of the passes in mountain chains. Furthermore, Middle America is a tectonically dynamic area, with much land movement and extensive volcanism having occurred in the last 60–70 million years, especially since mid-Tertiary (Coates and Obando 1996; Coates 1997; fig. 3.1), the time inferred from genetic distances (e.g., García-París and Wake 2000) among taxa to have been the main period of species formation. There have been associated breakups and new formations of cloud forest, and accordingly abundant opportunity for fragmentation of species ranges and local ecological specialization associated with new opportunities (e.g., García-París, Good, et al. 2000). Geographic and alloparapatric species formation has resulted.

GENERAL GEOGRAPHY OF THE ADAPTIVE RADIATION

Salamanders in the Neotropics are concentrated in four areas: the eastern and southern margins of the Mexican Plateau; the mountainous regions of Chiapas, Guatemala, and western and central Honduras (termed Nuclear Central America by Wake and Lynch 1976); the mountainous regions of Costa Rica and western Panama (Talamancan Central America of Wake and Lynch 1976); and less important, the mountains of northwestern South America. Recent molecular phylogenetic studies (Parra-Olea 1999; García-París and Wake 2000; Parra-Olea and Wake 2001) are giving definition to the cladistic structure of the radiation. A major conclusion is that Nuclear Central America contains the largest number of clades and the largest number of endemic clades (e.g., *Dendrotriton, Cryptotriton, Bradytriton, Nyctanolis, Ixalotriton,* and several clades within *Bolitoglossa*), although the southeastern margins of the Mexican Plateau and associated mountains in northern Oaxaca also harbor many clades, including some that are species-rich and endemic. Notable endemics are *Chiropterotriton* and *Thorius,* which may be two of the earliest branches of the tropical bolitoglossines (Parra-Olea 1999). The geographically central and northern regions of tropical America have been more important in the long history of bolitoglossines than have been the more southerly areas of radiation. An attempt (Hendrickson 1986) to reconstruct the historical biogeography of the tropical salamanders was hampered by the lack of a robust phylogenetic hypothesis. At present we have a large database of mtDNA sequences that we expect will give more resolution and aid in future reconstructions of historical biogeography. The most basal lineages in various phylogenetic analyses are always northern, either eastern Mexico or Nuclear Central America (Parra-Olea 1999; García-París and Wake 2000; D. B. Wake, unpub. data), and these areas appear to have been the primary centers for the adaptive radiation of the tropical bolitoglossines.

NO MID-DOMAIN EFFECT

Recently Lees et al. (1999) and Colwell and Lees (2000) identified a mid-domain effect in biogeographic analysis, an artifact of the tendency for ranges of species to pile up in the middle of any preset boundaries. They propose that the mid-domain effect be treated as a kind of null model to test against alternative hypotheses of species richness. This is not the place for a detailed examination of this hypothesis for tropical salamanders, but because the effect is a challenge to any biogeographic hypotheses about species richness, I present here only a few observations. If one takes 20° N and S as the rough outer bounds of distribution of the sg *Bolitoglossa*, only a small fraction (approximately 10%) of its species occur in South America, and no more than 10 species (counting all possible described and undescribed species known to me) occur between 0° and 20° S. Accordingly, the mid-domain hypothesis is rejected for the tropical bolitoglossines as a whole. On a more local scale, for example, along elevational transects such as that presented in preliminary form herein, a more formal analysis of the patterns is required but it may be premature.

Costa Rican Salamanders

A BIT OF HISTORY

Tropical salamanders have been known for at least 180 years, and Costa Rican species have been studied since Keferstein (1868) and Gray (1868) almost simultaneously described the same species (now *Oedipina uniformis*). Later history is reviewed by Taylor (1952a). Dunn made personal visits to Costa Rica and Panama, described some new species, and revised others (summarized in Dunn 1926). Taylor spent a few months in 1947 in continual fieldwork and described many new species, especially in his general review of the salamander fauna (Taylor 1952a). Neither Dunn nor Taylor worked along the Panamanian border region to any extent. Taylor (1952a) recorded 22 species of salamanders from Costa Rica. These were placed in six genera: *Chiropterotriton* (restricted to Mexico by Wake and Elias 1983; Costa Rican species now in *Nototriton*), *Magnadigita* (synonymized with *Bolitoglossa* by Wake and Brame 1963a), *Bolitoglossa*, *Parvimolge* (restricted to Mexico by Wake and Elias 1983; Costa Rican species now in *Nototriton*), *Haptoglossa* (synonymized with *Oedipina* by Brame 1968), and *Oedipina*. Work by Jay Savage and his group on the entire Costa Rican herpetofauna began in 1958 and has continued to the present. Arden Brame revised *Oedipina* (1968), synonymizing eight species recognized by Taylor (1952a) and naming four new ones. Jim Vial's doctoral research (Vial 1968) dealt mainly with the ecology of tropical salamanders, in particular *Bolitoglossa pesrubra* (most of the populations he studied are now placed in *B. pesrubra,* although those from lower elevations belong to an undescribed species). Vial also conducted some systematic studies; he synonymized two of Taylor's species of *Bolitoglossa* (Vial 1966; his conclusion was rejected by García-París, Good, et al. 2000, on the basis of molecular data) and described one new species (Vial 1963). I made my first trip to Costa Rica with the Savage group in 1961 and worked with Brame on some of the materials (Wake and Brame 1963b), but apart from brief trips in 1971 and 1973, I did not return for further studies until the mid-1980s. My associates and I continued fieldwork until 1994, and we hope to continue our investigations in the future.

THE SPECIES OF COSTA RICAN SALAMANDERS

Systematics

At present at least 45 species of salamanders are known to occur in Costa Rica. This total includes seven species of *Bolitoglossa* and one of *Oedipina* that are undescribed but will be soon (table 3.1). Of these undescribed species, three were announced in a preliminary manner by García-París, Good, et al. (2000). The recently described *Bolitoglossa anthracina* occurs within a few kilometers of the Costa Rican border (Brame et al. 2001), and two or three additional Panamanian species (one or two as yet undescribed) occur within 50 km of the border. Thus a realistic estimate of the number of species in the country is 45–48, or more than double Taylor's (1952a) number.

Many cryptic or morphologically very similar species of plethodontid salamanders have been recognized (e.g., Highton 2000). With few exceptions, the species I discuss here are not cryptic but are morphologically, genetically, and ecogeographically distinct. We are just now beginning to recognize cryptic species in tropical America, and the number of species is certain to increase.

If one includes western Panama with Costa Rica, the fauna of the region is surprisingly cohesive and highly endemic, not only at the species level but also at the level of clades of salamanders and other amphibians and reptiles (Savage 1982; Campbell 1999). In large part this is the result of the restriction of many species to upland habitats (the Talamancan herpetofauna of Savage 1982), and the fact that the region has had a history of independence from uplands to the north and west (Nuclear Central America) and to the south and east (the northwestern reaches of the Andes) (Coates and Obando 1996; Coates 1997; fig. 3.1). The Costa Rican species belong to three major clades, *Nototriton,*

Table 3.1   Salamanders known from or anticipated to be found in Costa Rica

Clade name	Described Costa Rican species	Undescribed Costa Rican species	Panamanian species
Nototriton			
picadoi	5		
richardi	2		
Oedipina			
Oedipina	11		
Oedopinola	3	1	2
Bolitoglossa			
alvaradoi	1		
mexicana	2		
subpalmata	11	7	
"Eladinea"	2		2
Total	37	8	4

Oedipina, and *Bolitoglossa,* each well-supported monophyletic groups based on recent published and unpublished molecular and morphological evidence. *Nototriton,* as revised by García-París and Wake (2000), includes seven Costa Rican species in two clades: a *picadoi* group, including five species, and a *richardi* group, including two species (Good and Wake 1993). A sixth member of the *picadoi* group occurs in Nicaragua (Köhler 2002; relationships based on unpublished mtDNA data from G. Parra-Olea and D. B. Wake), but otherwise the clades are endemic to Costa Rica. Although the two clades of *Nototriton* are well supported, their relationship to a third clade, in Guatemala and Honduras, is unresolved. A combined analysis of all available mtDNA data weakly supports a sister group relationship of the Costa Rican clades (García-París and Wake 2000). The presence of two of the three species groups in Costa Rica suggests that this part of Middle America might have been the center of origin and that the northern clade represents a zoogeographic dispersal. This hypothesis is attractive to me because three other clades of miniaturized salamanders (*Bradytriton, Cryptotriton, Dendrotriton*), all related to *Nototriton* but none to its sister taxon, and none of them sister taxa of each other, are endemic to Nuclear Central America. I consider it unlikely that one more miniaturized clade might have independently arisen in that region.

    Oedipina has at least 14 Costa Rican species, plus an undescribed species, by far the most species of this genus that occur in any of the Middle American countries. There are two additional species in the Caribbean lowlands of adjacent Panama. Of these 17 species, 11 are members of the subgenus *Oedipina* and the remainder are in the subgenus *Oedopinola*. The subgenera are clades, distinct on the basis of morphological and allozymic data (Good and Wake 1997) as well as mtDNA (García-París and Wake 2000) evidence (although *Oedopinola* is not as well supported by mtDNA as is *Oedipina*). Both of these clades are represented outside Costa Rica. *Oedipina* ranges from Guatemala to central Panama but is clearly centered on Costa Rica, to which most of the species are restricted. *Oedopinola* is less well known but more widely distributed, ranging from Chiapas, Mexico, to Ecuador. However, the facts that both subgenera occur in the country, and that most species in each subgenus occur in the country, suggest again that this part of Middle America might have witnessed the origin and most of the diversification of this clade. Furthermore, the deepest genetic divergences, between the two subgenera and also within each subgenus, occur in Talamancan Central America (García-París and Wake 2000).

Bolitoglossa, the largest of the tropical genera, has 23 (possibly more) species in Costa Rica. These belong to three main clades, one that includes nearly all of the Costa Rican species and all of the Panamanian and South American species as well, and two other clades with northern affinities, each containing but a single Costa Rican species. Although *Bolitoglossa* is the largest clade of salamanders (81 species currently recognized), the number of species found in Costa Rica is exceptionally high, greater than the number found in any other country. Analysis of mtDNA sequences (M. García-París, G. Parra-Olea, and D. B. Wake, unpub. data) reveals that most of the species are members of a major clade that corresponds well with the informal grouping labeled *Bolitoglossa* alpha (Wake and Lynch 1976). The major difference is that this grouping does not include the *B. mexicana* group, whose exclusion from *Bolitoglossa* alpha was forecast by Papenfuss et al. (1983).

Most Costa Rican and western Panamanian taxa belong to a group that I believe to be monophyletic that contains three clades. One of these includes species allied to *Bolitoglossa subpalmata*, and a second includes a diverse assortment of mainly upland species associated with the Talamancan mountain system. (I refer to these collectively as the *subpalmata* clade in table 3.1.) The third, a southern clade, includes all of the species in South America, several in Panama, and two that reach Costa Rica. This I call the Eladinea clade, a name that is available for this or even a more inclusive clade should *Bolitoglossa* eventually be subdivided (Miranda-Ribeiro 1937; the name *Eladinea estheri*, now considered a synonym of *B. altamazonica*, was assigned to specimens from Utinga at Belém, Pará, Brazil).

The *Bolitoglossa subpalmata* clade has recently been studied by García-París, Good, et al. (2000), who identified five species: *B. subpalmata* itself, which is associated mainly with the Cordillera Central and the Cordillera de Guanacaste and neighboring ranges, and the remaining species, all Talamancan, including the well-known *B. pesrubra* (which until its recent reclassification was known as

B. subpalmata), the most studied and best known of the tropical salamanders. Two more species are associated with the northern part of the Talamancas, *B. gracilis* and an undescribed species. Four other undescribed members occur in the Panama–Costa Rica border region.

The second clade is less cohesive and far more diverse in morphology and ecology. It includes several taxa that occur in sympatry with members of the *subpalmata* group in the northern Talamancas, including three species that are found in sympatry with *Bolitoglossa subpalmata*. In order of decreasing abundance they are *B. cerroensis*, a larger species, *B. sooyorum*, a slightly larger species, and the very much larger but rare *B. nigrescens*. This subclade is morphologically diverse and includes the largest (*B. robusta*) and smallest (*B. diminuta*) species in the entire genus, as well as both fully terrestrial and fully arboreal forms. Several undescribed species are known, including one from the northern portion of the Talamancas and another from the Panama–Costa Rica border region.

Many species from Panama and all of the South American members of the genus belong to the Eladinea clade, which has the most southerly distribution of any in the genus. The northernmost representative of this clade is *Bolitoglossa colonnea,* which ranges as far north as the northeastern lowlands (it is present at La Selva; Donnelly 1994a). Another member of the clade in the Costa Rica–Panama border region is *B. schizodactyla,* which is known from the Bocas del Toro lowlands and ranges east beyond the Panama Canal. The clade might have originated in eastern Panama, where other species occur, and the two representatives in Costa Rica likely arrived via lowland dispersal.

Recent molecular evidence suggests that *Bolitoglossa lignicolor* and *B. striatula* are the southernmost members of the *B. mexicana* clade (García-París, Parra-Olea, et al. 2000). All members of this clade are lowland forms, and the largest number of species occurs in Nuclear Central America. Presumably these species (which are not sister taxa) or their forerunners spread to the south via lowland routes after final connection of the Chortis and Chorotega plates (fig. 3.1).

Bolitoglossa beta of Wake and Lynch (1976), although not as cohesive a unit as they envisioned, includes most of the species in the genus that occur in Nuclear Central America. The Costa Rican *B. alvaradoi,* a poorly known species from middle-elevation cloud forests mainly on the northeastern slopes of the Cordillera Central, is allied (on the basis of sharing a derived tail base morphology and in mtDNA sequences) with more northern members of the assemblage, although it has no close relatives (M. García-París, G. Parra-Olea, and D. B. Wake, unpub. data).

General Ecology

All tropical plethodontids are nocturnal predators, and with one exception (Wake and Campbell 2001), all are strictly terrestrial throughout life. Direct development with no free-living larval stage is the exclusive life-history mode. Ecological differentiation among related species is associated mainly with different elevational limits and differences in locomotion and microhabitat use (Wake and Lynch 1976; Wake 1987; Wake et al. 1992). Throughout Middle America, local elevational transects extending from sea level to over 3,000 m support as many as 18 species, but relatively few species are strictly syntopic (e.g., the maximum number is 7 of the 15 species that occur along the best-studied transect, in western Guatemala; Wake and Lynch 1976). Elevational zonation is evident in the Cordillera Central (Wake 1987) and the Cordillera de Talamancas (Wake 1987; García-París, Good, et al. 2000), the latter region being as rich in species as any place in Middle America. There are 18 species in a relatively local region on the Caribbean slopes of the part of the Cordillera de Talamancas known as Cerro de la Muerte. Recent work has also shown extensive elevational stratification in a broad, less well-studied transect across the continent near the Costa Rica–Panama border (J. Hanken, D. B. Wake, and J. M. Savage, unpub. results; fig. 3.2), which is the southernmost region of relatively high salamander diversity. I estimate that 22 species occur along this transect, although most are poorly known. In northern parts of Middle America, only 2 or 3 species are found in any lowland locality below 500 m in elevation, but in the Costa Rica–Panama border region there are 4 species on the Osa Peninsula and nearby lowlands (*Bolitoglossa colonnea, B. lignicolor, Oedipina pacificensis, O. alleni*), and across the cordillera on the Atlantic side, in the Bocas del Toro region and adjoining Costa Rica, there are 5 species (*B. colonnea, B. biseriata, B. schizodactyla, O. maritima,* and an undescribed *Oedipina*). Thus, at the southernmost limit of high salamander diversity, tropicality is most strongly expressed, and species packing is such that the largest local salamander faunas encountered anywhere in Middle America are found, even though only a small subset of the rich diversity of tropical clades is represented. Conditions at mid-elevations appear ideal for members of the genus *Nototriton,* but they are not known from the region. (One poorly preserved specimen possibly assignable to this genus but not currently identifiable was collected in the Fortuna region.) If this clade were present, there could well be one or two additional cloud forest species.

Salamanders are present to the tops of the highest mountains in the Costa Rica–Panama border zone, and there are three species that occur above 3,000 m. The largest number of species (six) is found in cloud forests between 1,500 and 2,500 m, but the maximal number of strictly syntopic montane species is four (near Las Tablas).

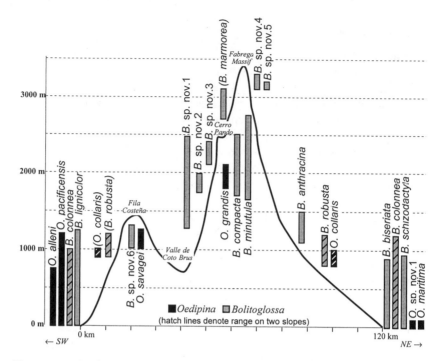

Figure 3.2   Distribution of plethodontid salamanders along an imaginary transect across the continent from the vicinity of Golfito, Costa Rica, on the Pacific Coast (southwest, at left) to the vicinity of Altimira, Bocas del Toro, Panama (northeast, at right). Total known elevational distribution of species in this region is shown. Some species occur on both sides of the mountains, but a number are restricted to upland regions. Undescribed species are indicated by letter designations.

Costa Rican cloud forest salamanders use microhabitats that are often not used elsewhere in Middle America, in particular, leaf litter (*Nototriton richardi, N. tapanti*) and moss mats, often including liverworts and other epiphytes as well (several species of *Nototriton* and *Oedipina, Bolitoglossa gracilis,* and *B. diminuta*). Furthermore, although bromeliads are used (Wake 1987), they are not used as commonly as in cloud forests to the north.

Costa Rican salamander diversity is not widely appreciated because one of the areas of intense research focus, La Selva Biological Station, is relatively depauperate in salamanders, and those present are rarely encountered (Donnelly 1994a). The three species at La Selva, *Oedipina gracilis* (this species, formerly referred to as *O. uniformis,* was resurrected by García-París and Wake 2000), *O. cyclocauda* (previously misidentified as *O. pseudouniformis*), and *Bolitoglossa colonnea*—are all essentially unstudied ecologically (Donnelly 1994a). A fourth species, *B. alvaradoi,* is known from an adjoining site, but it is rarer than the other three. At the higher-elevation Monteverde cloud forest region there is a larger salamander fauna: *Oedipina uniformis, O. poelzi, Nototriton gamezi* (commonly identified as *N. picadoi* or *N. abscondens* but recognized as a new, endemic species by García-París and Wake 2000), *Bolitoglossa subpalmata,* and *B. robusta* (Pounds 2000), but these species have proven difficult to find and study. The species most commonly encountered by biologists in Costa Rica is *B. pesrubra* (formerly known as *B. subpalmata*), which has been studied ecologically (e. g., Vial 1968) and has been seen by many students in the Organization for Tropical Studies (OTS) classes during visits to the Cerro de la Muerte region over the years. What is not so generally appreciated is that this is only one of four species that occur at high elevations on the Cerro, although in recent years *B. cerroensis* has been commonly found at the Querici site used by OTS classes. Salamanders are also uncommon at other biological stations in tropical America, such as Barro Colorado, where only two species are known to occur, both of which are *Oedipina* (Ibáñez et al. 1999). Thus, the common but erroneous impression that tropical salamanders are rare tends to be reinforced.

Patterns of Species Formation

Herpetologists have mainly accepted the standard allopatric (dichopatric) model of species formation, and there is substantial empirical support (e.g., Lynch 1989). The general assumption is that species form as a result of fragmentation of once continuous geographic ranges followed by selection associated with local conditions in the fragments. Stochastic factors also play a role, so that in time two or more differentiates arise from one ancestral form. Vicariance events, as a result of which populations find themselves in changed circumstances because of tectonic movements, climate change, and volcanism, likely play important roles in species formation in salamanders, as they do in frogs (cf. Lynch 1989). Salamanders of the genus *Nototriton* offer a particularly vivid example. Within both the *richardi* and *picadoi* species groups, pairs of species are segregated into northern Talamancan and Central Cordilleran components, with current species in both groups being endemic to one but not both areas, which suggests a common vicariant event (Good and Wake 1993). There has been further species formation in the *picadoi* group, with one species (*abscondens*) associated with the large central volcanos, another to the northwest, in the main mass of the Cordillera de Guanacaste at Monteverde, and two more (*guanacaste* and an undescribed species from central Nicaragua) associated with small volcanos farther to the north and west. A final species (*major*) in the *picadoi* group is found to the southeast, in the upland mass near Moravia de Chirripo. A third species group (*barbouri*) in the genus is known from eastern Guatemala and western Honduras, on the north side of the Nicaraguan Depression. One possible scenario is that a widespread ancestral form was separated on either side of the Nicaraguan Depression, and this led to the establishment of two species groups. A third arose in the south, most likely in association with the Miocene archipelago that was the emergent manifestation of the Chorotega Plate (Coates and Obando 1996). The relationships of the three species groups are unresolved, although most often the *richardi* and *picadoi* groups are considered to be sister taxa (Good and Wake 1993; García-París and Wake 2000). The related but more basal taxa *Cryptotriton* and *Dendrotriton*, as well as the enigmatic but somehow related *Bradytriton*, are restricted to areas north of the Nicaraguan Depression and likely arose in Nuclear Central America. Perhaps the *barbouri* group dispersed northward from Talamancan Central America, and retention of generalized morphology and ecology enabled coexistence with the more specialized resident groups.

There are alternatives in specific cases to dichopatric species formation, however attractive it may be as a general model. The most likely is alloparapatric species formation (Endler 1977), which was recently invoked to explain the local buildup of species in the *Bolitoglossa subpalmata* species group in the northern Cordillera de Talamanca (García-París, Good, et al. 2000). Within this group there has been allopatric species formation, with *B. subpalmata* being separated in the northern mountains of Costa Rica (Cordillera Central, Cordillera de Guanacaste) from the remaining species, in the Cordillera de Talamanca. However, within the latter region there are three species that are parapatrically distributed, each with a distinctive ecological association and a unique elevational distribution. And within the most widespread species, *B. pesrubra*, there is extensive local differentiation in genetic markers and color pattern. These patterns suggest that local adaptation may be the driving force in leading to the differentiation of taxa in association with the dramatic ecological changes that occur with elevational change in the region.

HISTORICAL BIOGEOGRAPHY

On the basis of the large numbers of species now present and the degree of morphological, ecological, and genetic differentiation they display, salamanders must have been in Talamancan Central America for much of the Tertiary. However, many of the lineages present in Nuclear Central America are absent, and most of the clades of salamanders that are found in Costa Rica (e.g., *Nototriton, Oedipina, Oedopinola*, two of the three main clades of *Bolitoglossa*) also have representatives in Nuclear Central America. Thus although there is high species diversity in Talamancan Central America, the fauna is relatively depauperate with respect to major clades and also with respect to major endemic clades (one in *Bolitoglossa*, two species groups in *Nototriton*). This suggests that it was relatively difficult for salamanders to reach the terrain associated with the Chorotega Plate, but once there they thrived. Some dispersal events might have occurred relatively recently. For example, *B. alvaradoi* has its closest relatives (none very close) to the north, and *B. striatula* and *B. lignicolor* are members of the *mexicana* group, widely distributed in Nuclear Central America and Mexico. These species could have dispersed through lowland routes that are not available for most tropical species (i.e., those restricted to montane forests). Although species of *Oedipina* are widely distributed to the north and south of Talamancan Central America, all clades and most species of that genus occur here, and it is likely that the lineage originated here (see also Brame 1968).

PROSPECTS

One would like to think that we are close to having a complete list of species of salamanders of Costa Rica, but I fear that such optimism is premature. Large areas of Costa Rica remain from which salamanders are virtually unknown, even though they are almost certain to be present. I have in mind the Caribbean slopes of the Cordillera de Talamanca in the eastern part of the country, from which only a handful of salamanders have been collected. Yet among these there are at least two specimens I cannot identify to species and that may well represent undescribed taxa. A number of species of *Bolitoglossa* are about to be described, and there is every reason to suspect that more species will be found in other areas that have been as yet relatively little studied. Several species (including undescribed forms) are known from very small samples (e.g., *B. anthracina* from but three specimens), and some are known only from one collecting event. It is difficult to know whether the rarity of certain species (i.e., *B. sooyorum* and *Oedipina paucidentata*, not seen for many years) is the result of recent declines, to loss of habitat, to failure of herpetologists to use specialized searching methods, or simply to the secretiveness of the species. Some species seem to be truly rare, for there has been abundant opportunity for them to have been found by extensive fieldwork in different parts of the country. More than 50 years after its description, *Nototriton richardi* is known from about 15 specimens, and yet individuals continue to appear from time to time. *Bolitoglossa alvaradoi* is known from fewer than 10 specimens over the same time span. In neither case is loss of habitat likely to be the main reason for rarity. However, *B. sooyorum* and *B. nigrescens*, both from Cerro de la Muerte, may well be rare because of habitat modification. In recent years there is the more ominous phenomenon of decline in amphibian populations in general, and especially in Costa Rica and western Panama (Pounds et al. 1996; Lips 1998, 1999; Pounds 2000). Without doubt the once abundant populations of *B. pesrubra* on Cerro de la Muerte have been strongly affected. Whereas these salamanders were once incredibly abundant along the Pan American Highway, they can now scarcely be found at all. For me this is the most dramatic case of declining salamander populations in Costa Rica.

The Talamancan Central America region has been a major staging ground for salamander evolution. The bulk of the salamander fauna can be traced ultimately to more northern zones, but at least one major clade, *Oedipina,* may have originated and undergone its most dramatic radiation here. Importantly, this region supplied the lineages that successfully occupied South America when that continent became physically accessible.

Acknowledgments

I have summarized the results of forty years of investigations, by me and my close associates, in Costa Rica. Jay Savage and Arden Brame encouraged me to study tropical salamanders, and Pedro León, Rodrigo Gámez, the late Douglas Robinson, Luis Diego Gómez, Dan Janzen, and Federico Bolaños have been particularly helpful in the course of my work in Costa Rica. Among many associates who have worked with me on Costa Rican salamanders I particularly thank Arden H. Brame, the late James F. Lynch, James Hanken, David Good, David Cannatella, Kiisa Nishikawa, Gerhard Roth, Andes Collazo, Thomas A. Wake, Marvalee H. Wake, Nancy Staub, Stanley Sessions, Elizabeth Jockusch, Mario García-París, Karen Lips, and Sharyn Marks. Three reviewers made comments that improved the chapter. My work was long sponsored by the National Science Foundation and the Museum of Vertebrate Zoology.

4

On the Enigmatic Distribution of the Honduran Endemic *Leptodactylus silvanimbus* (Amphibia: Anura: Leptodactylidae)

W. Ronald Heyer, Rafael O. de Sá, and Sarah Muller

Most species of the frog genus *Leptodactylus* occur in South America, and all authors who have treated the zoogeography of the genus have concluded that it originated somewhere in South America (e.g., Savage 1982). Savage (1982, 518) summarized the historical herpetofaunal units of the Neotropics as follows: "All evidence points to an ancient contiguity and essential similarity of a generalized tropical herpetofauna that ranged over tropical North, Middle, and most of South America in Cretaceous-Paleocene times. Descendents of this fauna are represented today by the South and Middle American tracks (Elements). To the north of this fauna ranged a subtropical-temperate Laurasian derived unit, today represented by the Old Northern Element (track). By Eocene, northern and southern fragments of the generalized tropical units had become isolated in Middle and South America, respectively. Differentiation in situ until Pliocene produced the distinctive herpetofaunas that became intermixed with the establishment of the Isthmian Link."

The paleogeographical data available to Savage were consistent with the preceding statement. New geological information over the past 20 years has clarified some aspects of the historical relationships between landmasses now comprising Middle and South America, but there is still not a single paleogeographical set of reconstructions accepted by all workers. Iturralde-Vinent and MacPhee (1999) presented cogent arguments that a detailed paleogeographic construction of the Caribbean region (as well as Middle America) will likely never be possible because of the nature of the geological data themselves.

The importance of Savage's biogeographic work in Middle America (e.g., Savage 1982) lies in the conclusion that there has been a relatively old radiation of the herpetofauna in Middle America that developed in situ. The available data do not unambiguously determine how and when the ancestors of this radiation were established in Middle America (see, for example, the introductory sections in Iturralde-Vinent and MacPhee 1999).

The herpetofaunal fossil record for the late Cretaceous through the Pliocene is limited for the New World (Estes and Báez 1985). However, taken at face value (that is, the oldest known fossils represent the first appearances of the taxa in the region), the fossil data suggest that there was limited, ongoing exchange between Middle and South America from the Eocene until the establishment of the Isthmian Link (Estes and Báez 1985, fig. 2). Most of the recent works that include Middle American paleogeographic reconstructions indicate that there is no overwhelming evidence of a land bridge or an island-arc land bridge between North and South America from the Paleocene until the Pliocene establishment of the Panamanian Isthmian Link. However, there is some evidence to suggest that there may have been a brief Panamanian land bridge in the Miocene (e.g., Iturralde-Vinent and MacPhee 1999).

A conservative summary relative to understanding *Leptodactylus* distributions includes the following two elements:

1. Sometime during the Miocene to Eocene, representatives of the South American tropical herpetofauna were on landmasses that now constitute Middle America and began an in situ radiation that resulted in most of the species that now occur in Middle America. An example of this radiation is the many species representing groups or genera of frogs of the family Hylidae endemic to Middle America (see Campbell 1999). How the ancestors of the Middle American radiation arrived on the old Middle American landmasses is not clear.
2. Formation of the Isthmian Link in the Pliocene then allowed two-way dispersals of the Middle American radiation with the South American herpetofauna. Any transport of propagules between the times when the ancestors of the Middle American radiation were in place and the formation of the Panamanian Isthmian Link must have been via overwater dispersal.

The patterns of relationships among the species of *Leptodactylus* that occur in Middle America are rather different from those of some of the Middle American hylid frogs, suggesting different zoogeographic histories and patterns. For example, all species of the hylid genera *Duellmanohyla* and *Ptychohyla* occur only in Middle America, and the relationship of these frogs to other hylid frog groups is unclear. This pattern fits Savage's (1982) model for the Middle American radiation well. However, the Middle American *Leptodactylus* species exhibit a different pattern.

All but one known species of *Leptodactylus* in Middle America have either a widespread distribution in South America, with a modest distribution in Middle America, or extensive geographic distributions in Middle America and occur either along lowland Pacific coastal Colombia and Ecuador or Caribbean coastal Colombia and Venezuela. Within this general characterization, there are four different distribution patterns among Middle American *Leptodactylus* species.

1. *Leptodactylus fuscus,* with a widespread distribution in South America, only extends as far north as the Panama Canal region in Panama.
2. *Leptodactylus insularum* and *L. pentadactylus* each have moderate distributions in Middle America and apparently have close sister-group relationships with species having distributions in South America (Heyer 1998, 2).
3. *Leptodactylus labialis, L. melanonotus,* and *L. poecilochilus* have moderate to extensive distributions in Middle America and have no obvious sister-group relationship either to each other or to any South American species.
4. *Leptodactylus silvanimbus* is known only from one small region of former cloud forest habitat in Honduras.

These patterns could follow a time line of differentiation, with *Leptodactylus fuscus* representing an Isthmian Link distributional extension into Middle America and, at the other extreme, *L. silvanimbus* being the only relictual remnant of the Middle American radiation in Savage's model (1982). Alternatively, *L. silvanimbus* in particular could represent a much more recent cladogenic event, from a common ancestor, with one of the other species of *Leptodactylus* that occur in Middle America.

To understand the zoogeographic history of *Leptodactylus* species in Middle America, we require a much better understanding of their relationships to each other and to all other *Leptodactylus* species. The advent of molecular techniques, in combination with nonmolecular data, gives some hope that it is possible to obtain much better knowledge of phylogenetic relationships among *Leptodactylus* species than is currently available.

In this chapter, we discuss preliminary results by focusing on the relationships of *Leptodactylus silvanimbus* to understand why it has a distribution pattern that differs fundamentally from all other known species of Middle American *Leptodactylus.* Specifically, we are interested in using preliminary data we have gathered to test alternative hypotheses of relationships that have fundamentally different biogeographic correlates. Before framing the hypotheses, we briefly summarize the relationships of *Leptodactylus silvanimbus,* as currently understood. When the species was first described, it could not be clearly allocated to any of the previously defined species groups of *Leptodactylus* (Mc-Crainie et al. 1980). Since then, on the basis of additional data, *L. silvanimbus* has been included as a member of the *Leptodactylus melanonotus* species group (McCrainie et al. 1986; Heyer et al. 1996) or as part of a *L. melanonotus-ocellatus* species group clade (Larson and de Sá 1998). Furthermore, a cladistic analysis of nonmolecular data for a set of taxa that included *L. silvanimbus* demonstrated that the species was a member of the genus *Leptodactylus* but that its relationships to other species in the *L. melanonotus* species group were not conclusive (Heyer 1998). The three hypotheses we wish to test are basically alternative responses to the question Is *L. silvanimbus* a member of the *L. melanonotus* species group?

Hypothesis 1. *Leptodactylus silvanimbus* is a member of the *L. melanonotus* species group that shared a most recent ancestor with *L. melanonotus* itself. If this hypothesis is true, it suggests that the common ancestor of the two species crossed a water barrier from South America to Middle America at some point from the time ancestors of the Middle American radiation were in place to before formation of the Isthmian Link.

Hypothesis 2. *Leptodactylus silvanimbus* is a member of the *L. melanonotus* species group but does not have a sister-group relationship with *L. melanonotus*. This hypothesis suggests that *L. melanonotus* and *silvanimbus*, or their respective ancestors, made independent entries into Middle America from South America, probably before formation of the Isthmian Link but after the ancestors of the Middle American radiation were in place.

Hypothesis 3. *Leptodactylus silvanimbus* does not demonstrate a sister-taxon relationship to any *Leptodactylus* species, including members of the *L. melanonotus* species group. This hypothesis suggests that *L. silvanimbus* is an old species with a relictual distribution, specifically a part of the Middle American radiation in the Savage (1982) model. It would likely not have a close relationship with any other species of *Leptodactylus*.

Materials and Methods

CHOICE OF TAXA

Previous phylogenetic results that included *Leptodactylus silvanimbus* indicated that on the basis of morphological, ecological, and behavioral data (Heyer 1998) and preliminary molecular data for 12S and 16S mitochondrial segments (R. de Sá and W. R. Heyer, unpub. data), the closest relatives of *L. silvanimbus* were members of the *L. melanonotus* and *ocellatus* groups. These same studies demonstrated that members of the genus *Physalaemus* serve as an appropriate outgroup to *Leptodactylus* for phylogenetic analyses.

One of the hypotheses we wish to test is that of a sister-group relationship between *Leptodactylus melanonotus* and *L. silvanimbus*. Beyond these two taxa, we include several other members of the *L. melanonotus* and *L. ocellatus* groups in our inquiry. Because of data availability, we include *L. podicipinus* (*melanonotus* group) and *L. bolivianus, chaquensis,* and *ocellatus* (*ocellatus* group). One member of the *L. fuscus* group (*L. latinasus*) and two members of the *L. pentadactylus* group (*L. pentadactylus* [see next paragraph], *L. rhodomystax*) are included as well. We used *Physalaemus gracilis* as our outgroup taxon because we have complete 12S and 16S sequence data comparable to our *Leptodactylus* data.

As the manuscript was being finalized for submission, WRH realized that he had assumed that a tissue sample of *Leptodactylus pentadactylus* from Peru was the source of the sequence data that RdS had sent to him just prior to RdS leaving on an extensive field trip, rather than a tissue sample from Panama. The nonmolecular data are for the Middle American population. Although the Amazonian and Middle American populations of *L. pentadactylus* represent different species (U. Galatti, pers. comm.), for the purposes of this chapter the results would not change if all the data had come from one or the other of the geographic entities involved.

NONMOLECULAR DATA

Most of the characters used in this analysis are those used previously (Heyer 1998). However, most earlier studies attempted to analyze relationships at the species-group level and higher. We also included color pattern characters and morphometric measurements.

For this study, 51 nonmolecular data characters were screened for phylogenetic analysis (see appendix 4.1 for voucher specimens). Three characters lacked information for three or more of the taxa. Five characters did not vary among the ten taxa. Four characters had a single different state that occurred only in one of the in-group taxa, thus providing no phylogenetic information. The remaining 39 character states are defined in appendix 4.2, following the definitions and rationale used previously (Heyer 1998), except for measurement characters, which are treated in the following paragraph.

The overall approach for defining character states of measurement variables is that described by Thiele (1993). The data for snout-vent length (SVL), head length/SVL, head width/SVL, eye-nostril distance/SVL, tympanum diameter/SVL, thigh length/SVL, shank length/SVL, and foot length/SVL were analyzed separately by sex. None of the measurement characters had equal variances, even when the data were log or log + 1 transformed. Log transformation resulted in most measurement data having lowest *F*-values in the ANOVA analyses and were used in character definition. For the measurement data to be comparable to the other data, we made an a priori decision to recognize no

Table 4.1   Analytic results used to determine the number
of states per character for measurement characters

Log-transformed data	Males	Females
SVL	90	91
Head length/SVL	3	3
Head width/SVL	4	3
Eye-nostril distance/SVL	2	2
Tympanum diameter/SVL	2	3
Thigh length/SVL	3	3
Shank length/SVL	3	4
Foot length/SVL	3	3

Note: Values are ranges of log-transformed variables divided
by the mean of coefficient of variation for all 10 taxa.

more than three character states per measurement character. To determine
whether to recognize two or three states per character, we divided the range of
the variable by the coefficient of variation for the variable (table 4.1). The results
indicate that SVL potentially contains more phylogenetic information than the
other measurement characters. Consequently, three states were recognized for
SVL and two states for all the other characters. The Thiele (1993) procedure uses
a formula that results in numbers that are then rounded to the nearest whole
number, which represents that taxon's state. Most taxa had the same state for
both sexes. In cases in which there was a different state number for the same
species, if the differences between the values resulting from the Thiele (1993)
formula were greater than .05, both states were included in the data set. For
those cases in which the two states differ by less than .05, the value requiring the
least modification to change the whole number assignment was used in order
to recognize a single state for the character involved. The characters and states
involved in this manipulation are as follows: head length/SVL for *Leptodactylus
silvanimbus,* male 0, female 1 → 0; tympanum diameter/SVL for *L. melanonotus* and *podicipinus,* male 1, female 0 → 0; shank/SVL for *L. latinasus, rhodomystax, silvanimbus,* male 1, female 0 → 0; and foot/SVL for *L. podicipinus,*
male 1, female 0 → 0.

The distribution of character states is shown in table 4.2.

MOLECULAR DATA

Tissue samples were obtained from the specimens listed in appendix 4.1. The
DNA extraction procedure followed Hillis et al. (1996). Two segments of the
mitochondrial genome were amplified using polymerase chain reaction (PCR).

Table 4.2 Distribution of character states for nonmolecular data used in phylogenetic analyses

	1	2	3	4	5	6	7	8	9	10	11	12	13
P. gracilis	3	1	3	0	0	1	0	0	0	0	0	0	0
L. rhodomystax	2	0	1	1	0	0	1	2	0,2	1	1	1	0
L. chaquensis	1	0	2	3	2	0	0	0	0	1	0	0	0
L. latinasus	2	0	0	0,3	0	0	0,1	1	0	0	1	0	1
L. bolivianus	0	0	1	2	2	0	0,1	0	0	1	1	0	1
L. pentadactylus	0,3	0	1	1	0	0	0	0	0,1	2	1	1	0
L. podicipinus	0,1	0	2	1	2	0	0,1	0,1	2	0	0	0	1
L. silvanimbus	0	0	2	0	1	0	0,1	0	0	0,1	0	0	0
L. melanonotus	0	0	2	0	2	0	0,1	0	0	0	1	0	0,1
L. ocellatus	0	0	2	3	2	0	0,1	0	0,1	1	0	0	0

	14	15	16	17	18	19	20	21	22	23	24	25	26
P. gracilis	0	0,1	1	1	2	0	0	0	0	1	1	1	0
L. rhodomystax	1	1	0	0	1	1	0	1	1	0	0	0	1
L. chaquensis	1	0,1	0	0	1	1	1	1	1	0	0	0	0
L. latinasus	1	0	0	0	0	0	0	1	1	1	1	0	0
L. bolivianus	0	1	1	1	0	1	1	1	1	0	0	0	0
L. pentadactylus	0	1	0	0	0	1	2	1	1	0	0	0	1
L. podicipinus	0	0,1	0	0	1	1	0	1	1	0	0	0	0
L. silvanimbus	0	1	0	1	1	1	0	1	1	0	0	0	0
L. melanonotus	0	0,1	0	0	1	1	0	1	1	0	0	0	0
L. ocellatus	1	0,1	0,1	0	1	1	1	1	1	0	0	0	0

	27	28	29	30	31	32	33	34	35	36	37	38	39
P. gracilis	1	1	0	0	0	0	0	0	0	4	0	3	1
L. rhodomystax	0	0	1	1	1	0	1	1	3	1	0	1	0
L. chaquensis	0	0	1	0	1	0	0	2	3	0,3	1	1,3	0,1
L. latinasus	0	0	1	0	0	1	2	1	0	2	0	0	0
L. bolivianus	0	0	1	0	1	1	0	2	0	0	1	1	0
L. pentadactylus	0	0	1	1	1	0	1	1	3	2	1	2	0
L. podicipinus	0	0	1	0	0	1	0	1	2	1	0	0	0,1
L. silvanimbus	0	0	1	1	0	0	0	1	4	0	0,1	1	1
L. melanonotus	0	0	1	0	0	0	0	1	1	1	0	0	0
L. ocellatus	0	0	1	0	1	1	1	1	4	0	1	2	1

Note: See appendix 4.2 for definition of characters and states.

A segment of the 12S rRNA of ~350 base pairs (bp) was amplified using primers 12Sa (5′-AAACTGGGATTAGATACCCCACTAT-3′) and 12Sb (5′-GAGGGT-GACGGGCGGTGTGT-3′). A segment of the 16S rRNA of ~500 bp was amplified using primers 16SaR (5′-CGCCTGTTTACCAAAAACAT-3′) and 16Sd (5′-CTCCGGTCTGAACTCAGATCACGTAG-3′). Primer sequences were obtained from Reeder (1995). Double-stranded PCR amplifications were performed in a final volume of 50 µl containing 0.4 µl of each primer, 1.0 µl of each dNTP, 3.0 µl of 25 mM MgCl, and 1.25 units of Taq (*Thermus aquaticus*) DNA polymerase; the reaction was overlaid with 50 µl of mineral oil. PCR conditions were as follows: 94°C for 60 s, 57°C for 60 s, and 72°C for 60 s, with 25 cycles for the 12S amplification and 30 cycles for the 16S amplification. Purification of double-stranded amplified product was performed using Wizard PCR Preps Kit (Promega). Of the purified double-stranded fragment, 0.5 µl were mixed with 1.5 µl of a single IRD-labeled primer, 7.2 µl of sequencing buffer, 1 µl of Sequitherm Excel II (Epicentre Technologies Co.) DNA polymerase, and 6.8 µl of dH$_2$0. Subsequently, 4.0 µl of this mixture was added to each of four tubes containing 2 µl of each nucleotide, respectively. The PCR conditions were as follows (30 cycles): 92°C for 30 s, 55°C for 30 s, and 70°C for 30 s. The single-strand-amplified and infrared-labeled fragments were sequenced in a LI-COR 4200 IR DNA sequencer on 6% acrylamide gels.

Most of the 892 nucleotide positions were easily aligned. Eighty positions had ambiguous alignments and were not used in the phylogenetic analyses (contact authors for aligned sequences).

DATA ANALYSES

The 12S and 16S data are examined for nucleotide substitution saturation by plotting the number of transition changes and number of transversion changes against the total number of nucleotide changes, because we have no reliable information from the fossil record to time date cladogenic events in our data set.

Phylogenetic analyses were conducted using PAUP* test version 4.0a for Macintosh (Swofford 1999).

Four substitution models for the molecular data were explored: Jukes-Cantor (Jukes and Cantor 1969), Kimura two-parameter (Kimura 1980), Hasegawa-Kishino-Yano (Hasegawa et al. 1985), and the general time reversible (REV of Yang 1994a) (see Swofford et al. 1996 for model descriptions). Each substitution model was evaluated assuming (1) no rate heterogeneity among sites; (2) heterogeneity parameter I of Hasegawa et al. (1985), in which a proportion of sites was considered invariable; (3) heterogeneity parameter gamma (Γ) of Yang (1994b), in which all sites follow a discrete gamma distributed rates model; and (4) heterogeneity parameters I + Γ (Gu et al. 1995; Waddell and Penny 1996), in which some sites are considered invariable and variable sites follow a gamma distributed rates model. Rate heterogeneity parameters were optimized under each substitution model: 16 models were tested for both the 12S and 16S, and the 12S and 16S data combined. The likelihood ratio test (Sokal and Rohlf 1969) was used to determine whether models differed significantly in their likelihood scores.

Whether the 12S and 16S data sets should be combined for phylogenetic analysis was examined through two tests: (1) the partition-homogeneity test in PAUP* with heuristic search; and (2) the likelihood ratio test using the best substitution + heterogeneity model for the 12S and 16S data, where χ^2 (approximation) = 2 ($-$Ln likelihood of 12S + 16S data combined $-$ ($-$Ln likelihood of 12S data + $-$Ln likelihood of 16S data)) and the degrees of freedom are equal to the branch lengths ($2T - 3$, where T = number of terminal taxa) + number of substitution model parameters.

The morphological data set was analyzed using maximum parsimony; the molecular data were analyzed using maximum likelihood, with the substitution and heterogeneity models that best fit the data. The morphological data were analyzed with exhaustive search settings. The maximum likelihood analyses were run only with heuristic search settings, because any other search setting took an unacceptable length of time to process (on a Macintosh G4 computer).

All phylogenetic analyses used *Physalaemus gracilis* as the outgroup to provide rooting information.

Results

NUCLEOTIDE SUBSTITUTION SATURATION

The number of transition and transversion changes plotted against the total number of nucleotide changes shows a straight-line relationship for the 16S data (fig. 4.1). The same plots for the 12S data suggest that saturation of transitional changes is present (fig. 4.2). The plot of transversional changes of the 12S data, although not a straight-line relationship, does not give any indication of saturation. Thus, of the two molecular data sets, the 16S data would be expected to resolve older cladogenic events better than the 12S data.

The percentages of nucleotide substitutions for 12S range from 11% to 15% for pair-wise comparisons between *Physalaemus gracilis* and the other species of *Leptodactylus* and from 4% to 12% for pair-wise comparisons among *Leptodactylus* species. The data for 16S are 11%–14% and 3%–11%, respectively.

MOLECULAR MODEL PARAMETERS

Of the 16 substitution and heterogeneity model combinations evaluated, the most complex model gave the best likelihood scores for both the 12S and 16S

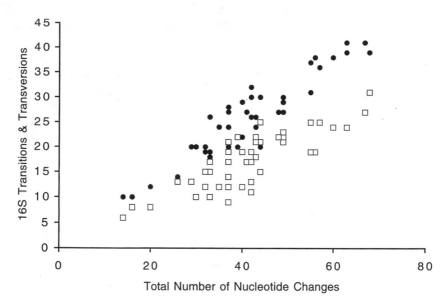

Figure 4.1 Nucleotide changes for 16S data. Filled circles indicate transitions; open squares indicate transversions.

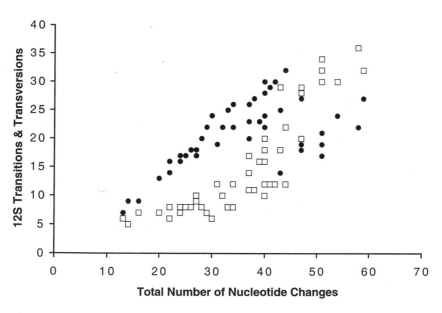

Figure 4.2 Nucleotide changes for 12S data. Filled circles indicate transitions; open squares indicate transversions.

Table 4.3   Likelihood ratio test data for 16 models of molecular evolution

Model	12S data	16S data	12S + 16S data
		−Ln likelihood scores	
JC	1,507.28810	1,895.84292	3,458.99269
JCΓ	1,437.17529	1,809.70338	3,280.90917
JCI	1,439.33569	1,811.45147	3,286.83392
JCI + Γ	1,435.64859	1,807.73041	3,274.52011
K2P	1,481.29020	1,867.32641	3,398.22787
K2PΓ	1,405.98660	1,778.23084	3,212.03986
K2PI	1,408.84496	1,780.13326	3,218.70940
K2PI + Γ	1,403.69033	1,775.53341	3,204.43102
HKY	1,472.70983	1,860.28960	3,387.66330
HKYΓ	1,395.63329	1,768.37781	3,198.73584
HKYI	1,399.26067	1,770.04109	3,205.63591
HKYI + Γ	1,392.81721	1,765.73308	3,190.71719
GTR	1,448.25768	1,843.66716	3,366.63289
GTRΓ	1,374.73904	1,756.60852	3,181.66991
GTRI	1,378.37353	1,758.93003	3,188.24311
GTRI + Γ	1,373.94599	1,753.93739	3,174.97869

Note: JC, Jukes-Cantor model; K2P, Kimura two-parameter model; HKY, Hasegawa-Kishino-Yano model; GTR, general time reversible model; Γ, heterogeneity parameter gamma; I, heterogeneity parameter I; I + Γ, heterogeneity parameters gamma and I.

data (table 4.3). The best model is the general time reversible (GTR) model with heterogeneity parameters I + Γ, and its likelihood score differs statistically from all the other likelihood scores for the other 15 models for both the 12S and the 16S data.

DATA PARTITIONS

The partition-homogeneity test for the 12S and 16S data results in $p = .02$.

The likelihood ratio test using the likelihood scores for the GTR + I + Γ model (table 4.3) has an approximate χ^2 value of 94,119 ($p < .001$ with 27 degrees of freedom).

Both analytic methods yield the same result: the 12S and 16S data are best analyzed separately. Given the inappropriateness of combining the 12S and 16S data, the 12S, 16S, and nonmolecular data sets are analyzed separately.

PHYLOGENETIC RELATIONSHIPS

Nonmolecular Data

Exhaustive maximum parsimony analysis of the nonmolecular data set resulted in 2,027,025 trees evaluated, with a single shortest tree with a length of 90, a single longest tree with a length of 135, and a g_1 statistic of $-.55$. Twenty-seven characters are parsimony informative. The single most parsimonious tree of length 90 has a consistency index excluding uninformative characters of .40, a retention index of .54, and a re-scaled consistency index of .36.

Bootstrap analysis of 100 replicates using default settings in PAUP* results in a moderately resolved tree (fig. 4.3). This tree configuration is used to evaluate the three competing hypotheses.

Molecular Data

Bootstrap analysis of 100 replicates using the GTR + I + Γ model, otherwise with default settings in PAUP* for the 16S data, yielded a completely unresolved tree. Cunningham et al. (1998) indicated that the most complex best-fit maximum likelihood models do not necessarily identify the correct phylogenetic tree. On the basis of analyses of 12S and 16S data in general, two model features seem most important in analyzing these data: (1) separation of transition and transversion parameters into separate classes; and (2) incorporation of the invariable sites parameter. The simplest model that incorporates these features is the Hasegawa-Kishino-Yano (HKY) + I model; the GTR + I model is more complex but contains these two features. For both the 12S and 16S data, bootstrap analyses using 100 replicates were run for the (1) HKY + I, (2) GTR + I, and (3) GTR + I + Γ models.

The 12S data have 51 parsimony-informative characters (estimated from maximum parsimony analysis of the data). All three models resulted in the same partially resolved tree topology (fig. 4.4).

The 16S data have 67 parsimony-informative characters (estimated from maximum parsimony analysis of the data). Both the GTR + I and GTR + I + Γ models yielded completely unresolved trees at 50% level of support. The HKY + I model yielded a partially resolved tree with rather poor support for the few clades that were supported with more than 50% support (fig. 4.5).

Evaluation of Hypotheses

Hypothesis 1. *Leptodactylus silvanimbus* and *L. melanonotus* are sister species. None of the data sets supports this hypothesis. Two of the three data sets do

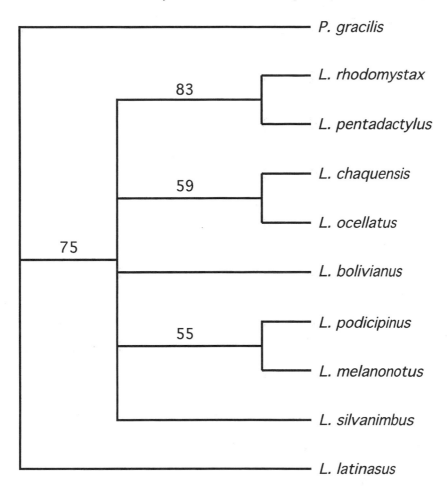

Figure 4.3 Results of parsimony analysis of primarily morphological characters. Numbers indicate bootstrap analysis results.

support a sister-group relationship for *L. melanonotus* and *L. podicipinus* for the taxa analyzed. The results of phylogenetic analysis of the three data sets lead to rejection of this hypothesis.

Hypothesis 2. *Leptodactylus silvanimbus* is a member of the *L. melanonotus* species group. For this hypothesis to be supported by the data, a clade comprising *L. melanonotus, L. podicipinus,* and *L. silvanimbus* in the data set analyzed would have to be recognized. In none of the data partitions is there

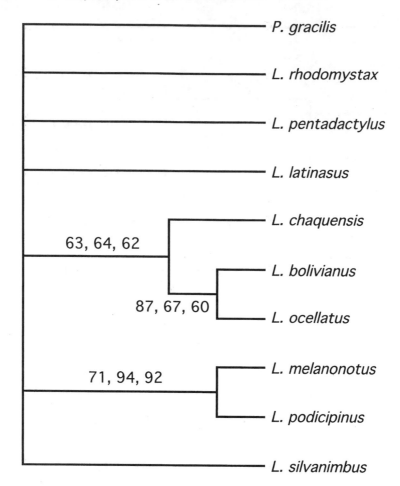

Figure 4.4   Results of maximum likelihood analyses of 12S data. Numbers indicate bootstrap analysis results for HKY + I, GTR + I, and GTR + I + Γ models, respectively.

bootstrap support for this clade at greater than 50% occurrence. In the morphological data set, this clade was found in only 5.5% of the bootstrap replicates. None of the molecular analyses found support for this clade at greater than 5% occurrence. The data are consistent with rejection of this hypothesis.

Hypothesis 3. *Leptodactylus silvanimbus* does not have a close relationship to other species of *Leptodactylus*. The three data partition results are consistent with this hypothesis. Therefore, of the three hypotheses, this one is best supported by our preliminary data.

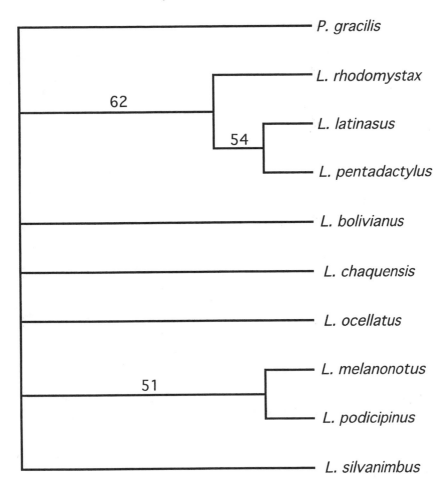

Figure 4.5 Results of maximum likelihood analysis of 16S data for the HKY + I model. Numbers indicate bootstrap analysis results.

We chose to analyze the three data sets separately for what we think are compelling reasons. We recognize that there are arguments for a total evidence, combined data analysis approach. We would have performed combined analyses of our two molecular data sets had they been statistically compatible. Further, some authors argue that the incompatibility tests we used are unreliable for determining the appropriateness of whether to combine data partitions (e.g., Yoder et al. 2001). To evaluate the robustness of our results, we performed a maximum likelihood analysis of the combined three data partitions by using the program in development by D. A. Swofford that applies a likelihood simulation to analyze nonmolecular data. The heuristic analysis resulted in two retained trees. In both of them, *Leptodactylus silvanimbus* demonstrates a basal trichotomy with *Physalaemus gracilis* and all the other *Leptodactylus* species. This result adds additional support to the third hypothesis.

Discussion

The molecular and morphological data are sufficient to demonstrate that if *Leptodactylus silvanimbus* and *L. melanonotus* shared a close sister-group relationship, such a relationship would have been identified. All data partitions strongly reject this hypothesis.

In contrast, additional data could indicate that the second hypothesis is correct rather than the third. Stronger differentiation between these two hypotheses would be facilitated by having strongly supported resolved trees, particularly basally. Our results indicate that although the 12S and 16S ribosomal DNA sequences may help us to understand species relationships, a more slowly evolving molecule is needed to resolve the earlier cladogenic events in the genus. Increasing the number of *Leptodactylus* taxa would increase our confidence in the phylogenetic results by decreasing random error (Swofford et al. 1996, 503). Resolution of relationships is generally better when the ratio of characters to taxa is high. Although there are additional morphological phylogenetically informative characters for *Leptodactylus* that could be added, at this point it is unlikely that we could double the number of such characters, and many of the additional characters would be unknown for many of the known species of *Leptodactylus*. Therefore, the most logical way to add a significant number of new characters to our data set would be through addition of new molecular sequences.

Until such additional data become available, however, the present data best support the conclusion that *Leptodactylus silvanimbus* is the only living member of the genus that is part of Savage's Middle American radiation. *Leptodactylus silvanimbus* currently occurs on the ancient Chortis Block landmass, which has been land positive since the Paleocene (see, for example, Iturralde-Vinent and MacPhee 1999 and references). How long *L. silvanimbus* has occupied the Chortis Block is unknown and likely knowable only if appropriately aged fossil deposits containing the species are discovered.

Appendix 4.1   Voucher Specimens

Tissue Vouchers

Leptodactylus bolivianus. USNM 268966 (USFS 152368). Peru: Madre de Dios; Tambopata Reserve. 12°50′ S, 69°17′ W.

Leptodactylus chaquensis. USNM 319708 (USFS 186524). Argentina: Tucumán; ~40 km southeast of San Miguel de Tucumán at km post 1,253 on International Route 9.

Leptodactylus latinasus. USNM 535969 (RdS 763). Uruguay: Paysandu; Parque Municipal "San Francisco." 34°31′ S, 56°24′ W.

Leptodactylus melanonotus. USNM 535964 (RdS 759). Belize: Cayo; San Jacinto and Spanish Lookout Road, between Webster Highway, Caesar's Hotel.

Leptodactylus ocellatus. USNM 535972 (RdS 755). Uruguay: Rocha; Rocha City, near Bañados de los Indios, Route 14, at 10 km from Route 9, Campo del Sr. Martin, Estancia La Palma Paraje.

Leptodactylus pentadactylus. USNM 268971 (USFS 152237). Peru: Madre de Dios; Tambopata Reserve. 12°50′ S, 69°17′ W.

Leptodactylus podicipinus. USNM 303207 (USFS 053124). Brazil: São Paulo; Fazenda Jatai. 21°33′ S, 47°43′ W.

Leptodactylus rhodomystax. MZUSP 70375 (940333). Brazil: Pará; Serra de Kukoinhokren.

Leptodactylus silvanimbus. USNM 348631 (LDW 10478). Honduras: Ocotepeque; Belen Gualcho. 1,600 m. 14°29′ N, 88°47′ W.

Physalaemus gracilis. RdS 788. Uruguay: Salto; Espinillar.

Morphological Vouchers

The following specimens examined for this study supplement those examined previously (Heyer 1998).

WET METAMORPHOSED SPECIMENS

Physalaemus gracilis. USNM 539179. Pozos Azules, Sierra de Animas, Maldonado, Uruguay (male; jaw, throat, and thigh muscles examined).

DRY SKELETONS

Leptodactylus bolivianus. USNM 298939. Cuzco Amazonico, Río Madre de Dios, Madre de Dios, Peru (female).

L. chaquensis. USNM 227604. Near Embarcación, Salta, Argentina (male).

L. ocellatus. USNM 342484. San Lorenzo, Central, Paraguay (male).

L. *pentadactylus.* USNM 539175. Igarapé Belém, Amazonas, Brazil (male).
USNM 539395. Porto Velho, Rondônia, Brazil (sex unknown, skull only).
L. *podicipinus.* USNM 297780. Estancia Caiman, Mato Grosso do Sul, Brazil (female).
L. *rhodomystax.* USNM 539176. Igarapé Belém, Amazonas, Brazil (sex unknown).

CLEARED AND STAINED SPECIMENS

Leptodactylus latinasus. USNM 227578. Artilleros, Colonia, Uruguay (male).
L. ocellatus. USNM 227598. Near Monte Grande, Buenos Aires, Argentina (juvenile female).
Physalaemus gracilis. USNM 539179. Pozos Azules, Sierra de Animas, Maldonado, Uruguay (male).

Appendix 4.2   Character State Definitions for Nonmolecular Data

See Heyer 1998 for further clarification of state definitions and ordering rationale.

1. Vocal sac: state 0, no vocal sac visible externally but present internally; state 1, indications of lateral vocal folds; state 2, well-developed paired lateral vocal sacs; state 3, well-developed large, single vocal sac. Character states analyzed unordered because PAUP* does not allow multistate taxa for ordered characters.

2. Tympanum visibility: state 0, tympanum well developed, easily seen externally; state 1, tympanum concealed, not visible externally.

3. Male thumb spines: state 0, thumb without modifications; state 1, thumb with one horny spine; state 2, thumb with two horny spines; state 3, thumb with nuptial pads. Character states analyzed unordered because PAUP* does not allow multistate taxa for ordered characters.

4. Dorsolateral folds: state 0, no folds; state 1, 1 short pair; state 2, 1 well-developed pair; state 3, 3–5 well-developed pairs. The state ordering is 0-1-2-3.

5. Toe webbing: state 0, no web or fringe; state 1, weak basal fringes and webbing; state 2, toes with lateral fringes extending length of toes except for tips. The state ordering is 0-1-2.

6. Tarsal decoration: state 0, tarsal fold; state 1, tarsal tubercle and fold.

7. Lip pattern: state 0, upper lip lacking a distinct light stripe; state 1, upper lip with distinct light stripe.

8. Pattern on posterior thigh: state 0, uniform or mottled; state 1, distinct

light stripe on lower portion of posterior thigh; state 2, distinct light spots. Character states analyzed unordered.

9. Belly pattern: state 0, light and/or indistinctly mottled; state 1, distinct bold or anastomotic mottle; state 2, dark with ill- to moderately defined light spots. Character states analyzed unordered.

10–17. Measurement data for SVL (10), head length/SVL (11), head width/SVL (12), eye-nostril distance/SVL (13), tympanum diameter/SVL (14), thigh/SVL (15), shank/SVL (16), and foot/SVL (17). See "Materials and Methods." Character 10 state ordering is 0-1-2.

18. Larval head/body pattern: state 0, uniform and light; state 2, uniform and dark; state 3, mottled. Character states analyzed unordered.

19. Larval tail pattern: state 0, mottled; state 1, rather uniform, pigmented or not.

20. Larval tooth rows: state 0, 2(2)/3 or 2(2)/3(1); state 1, 2/3 or 2/3(1); state 2, 1/2(1). These character states eliminate intraspecific variation in this data set. Character states analyzed unordered.

21. Larval vent position: state 0, dextral; state 1, medial.

22. Depressor mandibulae: state 0, DFSQat; state 1, DFsq or DFsqat.

23. Geniohyoideus medialis: state 0, muscle continuous medially, dividing posteriorly where the posteromedial processes of the hyoid articulate with the body of the hyoid, hyoglossus muscle completely covered ventrally by the geniohyoideus medialis; state 1, muscle divided ventrally, exposing hyoglossus, posterior half of muscle covered ventrally by sternohyoideus.

24. Geniohyoideus lateralis: state 0, no attachment of muscle to hyale; state 1, distinct slip attaches to hyale anterolaterally.

25. Anterior petrohyoideus: state 0, insertion entirely on edge of hyoid apparatus; state 1, insertion entirely on ventral surface of hyoid body.

26. Sternohyoideus origin: state 0, single medial slip originates from meso- and xiphisterna; state 1, medial slip divides in two slips, one originating from anterior portion of mesosternum, another from the posterior meso- and/or xiphisternum.

27. Sternohyoides insertion: state 0, in narrow band near lateral edges of hyoid; state 1, in a narrow band with fibers attached near midline posteriorly.

28. Omohyoideus: state 0, insertion partly on hyoid plate and partly on fascia between the posterolateral and posteromedial processes of the hyoid; state 1, muscle inserts entirely on hyoid plate ventrally.

29. Iliacus externus: state 0, short state of Limeses (1964); state 1, long B state of Limeses (1964).

30. Sartorius: state 0, moderate; state 1, broad.

31. Posterolateral projection of frontoparietal: state 0, no or minimal projection; state 1, distinct projection.

32. Anterior articulation of vomer: state 0, no articulation or overlap with

premaxilla or maxilla; state 1, articulation or overlapping with premaxilla or maxilla.

33. Sphenethmoid and optic foramen relationship: state 0, posterior extent of sphenethmoid widely separated from optic foramen; state 1, posterior extent of sphenethmoid closely approximates optic foramen; state 2, posterior extent of sphenethmoid borders optic foramen. The state ordering is 0-1-2.

34. Pterygoid-parasphenoid overlap: state 0, no overlap in an anterior-posterior plane; state 1, elements overlap but are not in contact; state 2, elements overlap and are in contact. The state ordering is 0-1-2.

35. Advertisement call pulse structure: state 0, single pulse; state 1, each note with 2 consistent, well-defined pulses; state 2, each note with 2–5 strong pulses, one or more of the strong pulses partially pulsed; state 3, each note of more than 6 pulses; state 4, entire note partially pulsed. State 4 unordered; remaining state ordering 0-1-2-3.

36. Call frequency modulation: state 0, none or negligible; state 1, rising frequency modulation, extremely sharp; state 2, rising frequency, moderate; state 3, rising and falling frequencies throughout call; state 4, falling frequencies throughout call. Character states analyzed unordered.

37. Carrier frequencies: state 0, most or all > 1,000 Hz; state 1, most or all < 1,000 Hz.

38. Call duration: state 0, < 0.1 s; state 1, 0.1–0.2 s; state 2, 0.2–0.5 s; state 3, > 0.5 s. Character states analyzed unordered because PAUP* does not allow multistate taxa for ordered characters.

39. Harmonic structure: state 0, none or weak; state 1, distinct.

Acknowledgments

Without James R. McCranie and Larry David Wilson, this report would not have been possible. Their field experience and expertise provided the tissue samples of *Leptodactylus silvanimbus* so that molecular analyses could be performed. The following individuals either provided other tissue samples or facilitated our collecting of same: Andrew Chek, Reginald B. Cocroft III, Ronald I. Crombie, Esteban O. Lavilla, Alejandro Olmos, and Addison Wynn. Dr. P. E. Vanzolini graciously responded to a request for skeletal material of *Leptodactylus pentadactylus* and *L. rhodomystax* by preparing skeletons of those species for this analysis.

WRH thanks the following colleagues at the National Museum of Natural History for their patience in responding to a continuous stream of questions and for the help they provided regarding molecular data and their analyses: Kevin de Queiroz, Jon Norenburg, and James Wilgenbusch.

This research was supported by grants from the F. Jeffress and Kate Miller Memorial Trust (grant J-450 to RdS), the Virginia Academy of Sciences, the Neotropical Lowlands Research Program, Smithsonian Institution (Richard P. Vari, principal investigator), and the National Science Foundation (award 9815787 to RdS and WRH).

5

Chromosomal Variation in the *rhodopis* Group of the Southern Central American Eleutherodactyline Frogs (Leptodactylidae: *Eleutherodactylus*)

Shyh-Hwang Chen

Introduction

In the vast Neotropical rainforests of the Americas, *Eleutherodactylus* is the most diverse frog genus, with more than 500 recognized species (Frost 1985; Duellman 1993). It is the largest genus of vertebrates and is widely distributed in the area between the southern United States and Argentina. All members of the genus are characterized by their T-shaped terminal phalanges. The digit pads are large in the arboreal species and small in terrestrial forms. Another distinctive feature of *Eleutherodactylus* is the reproductive mode. Except for *Eleutherodactylus jasperi* in Puerto Rico, all known eleutherodactyline frogs lay large terrestrial eggs that undergo direct development (without any free-living tadpoles) and hatch as tiny froglets. This reproductive feature seems to be responsible for their extensive radiation into the major Neotropical terrestrial and arboreal habitats (Savage 1975; Miyamoto 1983b). *E. jasperi*, the only known ovoviviparous species in the genus, retains eggs in the oviducts. There the eggs develop into froglets with nutrition from their own yolk (Drewry and Jones 1976; Wake 1978).

During the past 30 years, numerous phylogenetic studies on the genus *Eleutherodactylus* have been conducted. These have been based on external morphology, osteology, myology, allozymes, and karyotypes (e.g., Lynch 1971, 1976, 1986; Heyer 1975, 1984; Miyamoto 1981; Savage 1987; Hedges 1989). None of these authors considered *Eleutherodactylus* a monophyletic genus except Hedges

(1989), who combined all possible sister genera into one genus (*Eleutherodacty-lus*) and subdivided it into five subgenera (*Eleutherodactylus* Duméril and Bibron 1841, *Euhyas* Fitzinger 1843, *Craugastor* Cope 1862, *Syrrhophus* Cope 1878, and *Pelorius* Hedges 1989).

The subgenus *Syrrhophus* is distributed mainly in Mexico and the southern United States, whereas both *Euhyas* and *Pelorius* are found in the West Indies. The subgenus *Eleutherodactylus* is the most widely distributed group, found in South America and northward to Lower Central America and the West Indies. In contrast, the subgenus *Craugastor* (the Middle American clade) is distributed mainly in Mexico and Central America, with a few species in Colombia, on the Pacific coast of Ecuador, and the coast of Venezuela in northern South America (Lynch 1986; Hedges 1989).

STUDY GROUP

A Middle American clade of *Eleutherodactylus* was predicted by Savage (1966, 1973b) to be a distinct monophyletic group. Lynch (1986) studied the variation of jaw musculature between the *m. adductor mandibulae* and the mandibular ramus of the trigeminal nerve within eleutherodactyline frogs. He found that the trigeminal nerve passes either medial ("e") or lateral ("s") to the *m. adductor mandibulae externus superficialis*. All species of *Hylactophryne* and most *Eleutherodactylus* in Mexico, Middle America, and extreme northern South America have the "e" condition of jaw musculature. Lynch (1986) concluded that *Hylactophryne* and most species of *Eleutherodactylus* (including species of the *rhodopis* group) in North and Central America shared a synapomorphy, the "e" condition of jaw musculature, that defined a Middle American clade; he suggested the generic (or subgeneric) name *Craugastor*. Hedges (1989) followed Savage (1987) and treated *Hylactophryne* as a junior synonym of *Eleuthero-dactylus* and placed all species in the Middle American clade in the subgenus *Craugastor*.

The subgenus *Craugastor* in Lower Central America has received much systematic attention. Nearly a half century ago, Taylor (1952b) recognized 7 species of *Microbatrachylus* and 27 species of *Eleutherodactylus* from Costa Rica. Later, the genus *Microbatrachylus* was critically reviewed by Lynch (1965) and synonymized with the genus *Eleutherodactylus*.

Most species of *Craugastor* have been reviewed by species groups. Frogs allied to *Eleutherodactylus bransfordii* were reviewed by Savage and Emerson (1970), who combined all of Taylor's (1952b) taxa in *Microbatrachylus* as junior synonyms of *E. bransfordii*. However, Miyamoto (1983a), using biochemical data, resurrected *E. stejnegerianus*. In addition, a third cryptic species was discovered in southwestern Costa Rica, which occurred sympatrically with *E. stejnegerianus* (at least in the area around Las Cruces).

Eleutherodactylus bransfordii and *E. podiciferus* were placed in the *gollmeri* group by Savage and Emerson (1970) and Miyamoto (1981, 1983a). Lynch (1976) rearranged *Eleutherodactylus* into 17 species groups. He moved *E. bransfordii* to the *rhodopis* group but placed part of the *gollmeri* group (sensu Savage and Emerson 1970; including *E. podiciferus*) into his *fitzingeri* group.

Savage (1987) reviewed the *gollmeri* group and presented a scheme of the Middle American clade (subgenus *Craugastor*) in which the *rhodopis* group was defined as having a diploid chromosome number (2N) of 18 and a fundamental number (NF) of 36; *gollmeri* was included in his *rhodopis* series. With the discovery of *Eleutherodactylus lauraster,* Savage et al. (1996) redefined the *rhodopis* group (series) as having an additional synapomorphy, the DFSQdAT (dfsq*at) jaw muscle configuration, in contrast to those dfsq, DFsqat, or DFSQAT conditions in the other species groups. However, conditions of the *m. depressor mandibulae* of *E. jota, E. saltator,* and *E. sartori* remained unknown.

Lynch (2000) considered *Eleutherodactylus saltator* to be a junior synonym of *E. mexicanus* and resurrected *E. loki* as a valid species in Mexico. Consequently, he proposed revised species groupings for the subgenus *Craugastor.* The fusion of the eighth and sacral vertebrae in all members of Savage's (1987) *gollmeri* group led Lynch (2000) to eliminate the *E. omiltemanus* species group. On the basis of this synapomorphy (fusion of the vertebrae), two species (*E. daryi* and *E. greggi*) of the *omiltemanus* group were added to the *gollmeri* group (sensu Savage 1987). The species *E. omiltemanus* was assigned to the *rhodopis* group (sensu Savage et al. 1996) on the basis of having short legs and other phenetic similarities. However, he transferred *E. occidentalis* from the *augusti* group to the *rhodopis* group without explanation. The monophyly of the *rhodopis* group remained questionable.

CYTOGENETIC RESEARCH

Gross chromosomal data have been suggested to be useful in constructing phylogenies of frogs and other vertebrates (see King 1990 and Sites and Reed 1994 for reviews, but see Kluge 1994 for another view). However, phylogenies constructed strictly from gross karyological morphology have been questioned, because nonhomologous chromosomal rearrangements might have occurred convergently among phylogenetically distant species (Atchley 1972). Using gross chromosomal data is viewed as an art by some (Wiley 1981b) and so must be employed with great care (Bogart 1973). In contrast, detailed chromosome banding patterns make homology determination possible, and consequently, evolutionary relationships inferred from chromosome banding comparisons are considered robust (e.g., King 1990).

Ammoniacal silver stain and C-banding techniques are the most frequently used methods for determining the active nucleolar organizer regions (NORs) and heterochromatin, respectively. Schmid (1982) analyzed the NOR positions of 260 anurans from 23 different genera and suggested that NORs were significant phylogenetic landmarks. Usually, only one pair of NORs is found in each species and is localized in the same region of the same chromosome pair within a closely related species complex or species group (Schmid 1978a; King 1980). The same NOR positions might extend to the generic level without changes (Mahony and Robinson 1986). Exceptions to this rule may indicate that chromosomal rearrangements have occurred in the NOR-carrying chromosome segments during the evolution of the Anura (Schmid 1982).

Five types of secondary constrictions have been found in anuran karyotypes, four of which are related to NORs (King 1980, 1990). It has been suggested that NOR constrictions are species-specific and that their positions are closely correlated with C-positive areas to prevent crossing over in NOR (King 1980). However, their phylogenetic significance remains unclear.

C-band patterns are more variable than are silver-stained NORs (Ag-NORs). No two species have identical C-band patterns (King 1985, 1990). The C-band patterns of eleutherodactyline frogs from Cuba and Puerto Rico consistently show both telomeric and paracentromeric bands and occasional interstitial bands (Bogart 1981), which are not congruent with those of ten Jamaican eleutherodactyline frogs (Bogart and Hedges 1995). Although heterogeneity of heterochromatin has been proposed (Schmid 1978b; King 1990), some C-positive bands shared in Holarctic *Hyla* have proven to be convergent on the basis of restriction enzyme studies (Anderson 1986). The identical C-positive bands of the most closely related species in the species group or recently evolved species may be assumed to be homologous.

Relatively few eleutherodactyline frogs have been karyotyped. Up to the present, there are approximately 13 nominal species in the *rhodopis* group (series) in which 3 species were karyotyped but none were silver-stained or C-banded. The power of C-band and Ag-NOR patterns in phylogenetic analysis remains to be tested. In addition, Costa Rican *E. bransfordii* and *E. diastema* each have been reported as having two different chromosome numbers because of misidentifications (Savage 1987).

Herein I describe the cytogenetics of Lower Central American (Costa Rica and Panama) eleutherodactyline frogs belonging to the *rhodopis* group. The goal is to reveal their chromosome numbers, estimate their phylogenetic relationships, and test the monophyly of the *rhodopis* group.

Materials and Methods

CYTOLOGICAL STUDY

There are five species (including eight karyomorphs) of the *rhodopis* group and four outgroup species of the *diastema* group from Lower Central America available for study (appendix 5.1, table 5.1). I followed the methods of Kezer and Ses-

Table 5.1   Cell plates observed for each type of staining

Taxon	Sex	No. chromosome plates			Locality
		Giemsa	C	Ag	
rhodopis group (subgenus *Craugastor*)					
E. *bransfordii*					
Tapanti karyomorph	2F, 1J	98	23	15	Tapanti, Cartago, C.R.
La Selva karyomorph	1F, 1M	43	8	15	La Selva, Heredia, C.R.
	2F	23	2	0	Sardina de Guapiles, Limón, C.R.
E. *podiciferus*					
Tapanti karyomorph	2F	61	15	19	Tapanti, Orosi, Cartago, C.R.
Palmichal karyomorph	2F	76	10	16	Palmichal, San José, C.R.
E. *stejnegerianus*	1F	44	19	14	Palmichal, San José, C.R.
	1F	8	0	0	San Juan, S. Isidro, San José, C.R.
	1F, 1M	65	27	12	Las Cruces, Punctarenas, C.R.
E. sp. A	3M, 2F, 1J	68	18	16	Fortuna, Chiriqui, Pan.
E. sp. B					
18-karyomorph	3F, 3M	104	20	19	Fortuna, Chiriqui, Pan.
	1F	12	0	0	Nusagandi, Kuna Yala, Pan.
20-karyomorph	2F, 5M	142	44	28	Fortuna, Chiriqui, Pan.
diastema group (subgenus *Eleutherodactylus*)					
E. *diastema*	1M	2	0	0	Nusagandi, Kuna Yala, Pan.
	3M	23	0	0	Pierre, Darien, Pan.
E. *vocator*	3M	15	1	3	Las Cruces, Punctarenas, C.R.
E. sp. D	1M	2	0	0	Guacimo, Limón, C.R.
	1M	5	0	3	Cariblanco, Heredia, C.R.
	1F	23	0	7	Tapanti, Cartago, C.R.
	1F	22	13	2	Sardina de Guapiles, Limón, C.R.
E. sp. E	3F	23	0	1	Nusagandi, Kuna Yala, Pan.

Note: Sex: F, female; M, male; J, juvenile. C, C-banding stained; Ag, silver stained.

sions (1979) and Olert and Klett (1978), with some modifications, to prepare the intestinal chromosomes, and followed Schmid (1978a) to prepare bone marrow chromosomes. The C-banding stain followed the BSG method of Sumner (1972), and the silver staining followed either Goodpasture and Bloom (1975) or Howell and Black (1980).

Table 5.1 summarizes data for the number of cell plates observed for each kind of staining method for each population in the present study. The features of karyotypes, including diploid chromosome number, fundamental number, karyotype groupings, and positions of both secondary constrictions and NORs for each species or karyomorph, are summarized in table 5.2.

The chromosome length is represented by the relative total length (RTL) of the whole genome. The length of the secondary constriction (gap) is not included. The definitions of chromosome types followed Levan et al. (1964), based on the value of arm-ratio value (AR). Chromosome data are summarized for *Eleutherodactylus bransfordii* and *E. stejnegerianus* (see table 5.4 below), for *E. podiciferus* and *E.* sp. A (see table 5.5 below), for both karyomorphs of *E.* sp. B (see table 5.6 below), and for four species of the *diastema* group (see table 5.8 below). All chromosome data are detailed elsewhere (Chen 2001).

Idiograms for all karyomorphs are shown in figure 5.1 (the *rhodopis* group) and figure 5.2 (the *diastema* group).

Table 5.2   Features of karyotypes for each species or karyomorph in the study

| | | | | Positions of | |
| | | | | Secondary | |
Taxon	2N	NF	Chromosome groupings	constrictions	NORs
rhodopis group					
(subgenus *Craugastor*)					
E. *bransfordii*					
Tapanti karyomorph	18	36	6-1-2-0	3p,5p	1pt
La Selva karyomorph	18	36	6-1-2-0	3p,5p	con3p
E. *podiciferus*					
Tapanti karyomorph	18	36	6-3-0-0	3p	con3p
Palmichal karyomorph	18	36	4-5-0-0	4p	con4p
E. *stejnegerianus*	18	36	5-2-2-0	3p,5p	con3p
E. sp. A	18	36	6-2-1-0	4p	con4p
E. sp. B					
18-karyomorph	18	36	6-1-2-0	1p	con1p
20-karyomorph	20	40	7-1-2-0	1p	con1p
diastema group (subgenus					
Eleutherodactylus)					
E. *diastema*	18	36	2-7-0-0	8p	N
E. *vocator*	18	31	5-1-0-3	N	N
E. sp. D	18	36	2-5-2-0	N	N
E. sp. E	18	36	2-5-2-0	9q	N

Note: Chromosome groupings are shown as metacentric-submetacentric-subtelocentric-telocentric. p, short arm; q, long arm; N, not shown; t, terminal; con, NORs positions same as secondary constrictions.

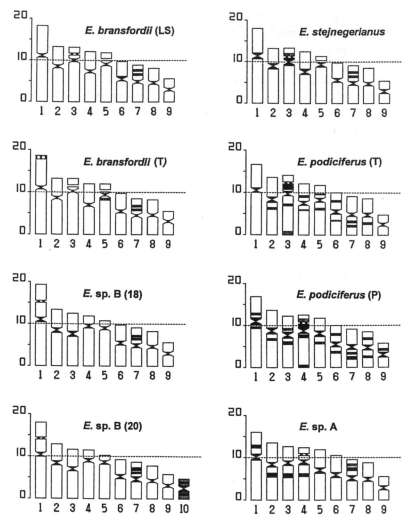

Figure 5.1   Idiograms of karyotypes of Lower Central American eleutherodactyline frogs belonging to the *rhodopis* group. Karyomorphs are indicated in parentheses for La Selva (LS), Tapanti (T), Palmichal (P), 18-karyomorph (18), and 20-karyomorph (20). Chromosome numbers are indicated on the x-axis; RTLs are indicated on the y-axis.

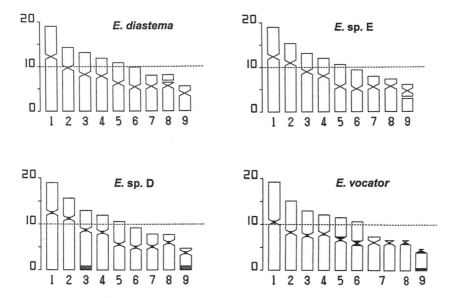

Figure 5.2   Idiograms of karyotypes of Lower Central American eleutherodactyline frogs belonging to the *diastema* group. Chromosome numbers are indicated on the x-axis; RTLs are indicated on the y-axis.

Taxonomic Remarks

STATUS OF *ELEUTHERODACTYLUS BRANSFORDII*

Eleutherodactylus bransfordii is a small, diurnal, leaf-litter dweller that is common and widely distributed in lowland and premontane areas from Nicaragua to Panama (Donnelly 1999). The status of *E. bransfordii* is not clear because there are polymorphic features within or among populations. These features caused several populations to be recognized as different species by Taylor (1952b), but all were placed in the same species, *E. bransfordii,* by Savage and Emerson (1970). Miyamoto (1983a) resurrected *E. stejnegerianus* as a valid species on the basis of his allozyme studies.

In the present study, many *bransfordii*-like frogs from various geographical localities in Costa Rica and Panama were investigated cytogenetically (see fig. 5.7 below). I was able to recognize four different karyotypes from these frogs (fig. 5.1). The karyotype of *Eleutherodactylus stejnegerianus* is different from those of other *bransfordii*-like frogs, a result that corroborates Miyamoto's (1983a) finding that *E. stejnegerianus* is a valid species. Karyotypes of *bransfordii*-like frogs from Fortuna, the Pacific slope highland area of western Panama (1,100–1,300 m), and from the Atlantic slope premontane areas of central Panama, Nusagandi (650 m), have a distinct secondary constriction (same as NOR location) in both short arms of chromosome pair 1. This constriction is unique among all known karyotypes for the *bransfordii*-like frogs. In addition, there are 18- and 20-karyomorphs in the Fortuna population. Therefore, the *bransfordii*-like frogs from Fortuna and Nusagandi (and possibly all localities in between) belong to an undescribed species, *E.* sp. B.

The last two karyomorphs are also distinct from each other. The karyomorph of the Tapanti population is characterized by having a C-banding block associated with the NORs in the terminal regions of both short arms of pair 1. The karyomorph of the La Selva population (all samples from La Selva and Sardina in northeastern Costa Rica) is characterized by having the NOR sites located in the secondary constrictions of both short arms of pair 3. The two karyomorphs are remarkably different and could diagnose two distinct species. Unfortunately, all nominal species of *bransfordii*-like frogs from their type localities have not been investigated cytogenetically. Frogs of the Tapanti karyomorph have not yet been compared morphologically with various named taxa. The specific name that might be applied to either karyomorph remains uncertain. Here I consider both *bransfordii*-like frogs with the La Selva karyomorph from La Selva and Sardina as *E. bransfordii* because these two populations are closer to the type locality (Nicaragua) than is the Tapanti population. However, whether Tapanti's *bransfordii*-like frogs belong to any named synonymized species or represent a totally new form is open to question. The solution to this problem requires further morphological, cytogenetic, and DNA data for resolution of the systematics of the Costa Rican and Panamanian *bransfordii*-like frogs.

STATUS OF *ELEUTHERODACTYLUS PODICIFERUS*

Eleutherodactylus podiciferus is a leaf-litter dweller distributed in the highlands of Costa Rica and western Panama. It generally resembles *E. bransfordii* but differs in having more reddish brown in its dorsal coloration, a pair of distinct hourglass ridges on the dorsum, low and round subarticular tubercles on the hands, and paired vocal slits in adult males that lack nuptial pads. Differences between *E. podiciferus* and *E. bransfordii* are presented by Savage and Emerson (1970). As in *E. bransfordii,* many populations of *E. podiciferus* in Costa Rica and Panama superficially resemble each other. However, the classification of these *podiciferus*-like frogs is still inadequate as a result of character polymorphism. As currently understood, this "species" contains more than one cryptic taxon.

Three populations of *podiciferus*-like frogs (the Tapanti and Palmichal populations in Costa Rica, and the Fortuna population in Panama) were investigated (see fig. 5.8 below). Each karyotype is different. Males of the Fortuna population examined by the author have nuptial pads on the inner-dorsal surface of the thumbs. These were not previously recorded in *E. podiciferus* (Savage and Emerson 1970; Savage and Villa 1986; Savage et al. 1996). The Fortuna form undoubtedly is an undescribed species, *E.* sp. A. In addition, the karyotype of *E.* sp. A is different from both populations (karyomorphs) of *E. podiciferus* in arm ratios, relative lengths, or both. The exceptions are pairs 2 and 8 for *E.* sp. A and *E. podiciferus* (Tapanti karyomorph), and pairs 8 and 9 for *E.* sp. A and *E. podiciferus* (Palmichal karyomorph), which are not significantly different ($p < .01$, Duncan's multiple range test). Both karyotypes and morphological data support *E.* sp. A as a distinct species to be described in the future.

For Tapanti and Palmichal populations, although the relative total length of pairs 3 and 9 and arm ratios of pairs 1 and 2 are significantly different ($p < .01$) in both karyomorphs and there are slight differences in C-banding patterns, only a few individuals were available for study. Without noticing any remarkable structural difference in their karyotypes, I leave their taxonomic status unchanged. Both populations, however, are tentatively regarded as two different karyomorphs of *E. podiciferus* until further morphological comparisons are made.

Chromosome Number

RHODOPIS GROUP

The chromosome numbers of the *rhodopis* group are listed in table 5.3. Except for *Eleutherodactylus* sp. B, all species of the *rhodopis* group are 2N = 18 and NF = 36 (table 5.2). All chromosomes are biarmed. However, Bogart's (1970)

Table 5.3   All known chromosome numbers and fundamental numbers (2N/NF) for the *rhodopis* and *diastema* groups

Group	2N/NF	Species
rhodopis (14 spp.)	18/36	**bransfordii, loki, *podiciferus, *stejnegerianus,*sp. A, *sp. B (18)*
	20/40	**sp. B (20)*
	unknown	*hobartsmithi, jota, lauraster, mexicanus, occidentalis, omiltemanus, pygmaeus, sartori*
diastema (9 spp.)	18/36	**diastema, *sp. D, *sp. E*
	18/31	**vocator*
	unknown	*chalceus, gularis, hylaeformis, scolodiscus, tigrillo*

Note: Contents of species groups were assigned by Lynch (2000) and Savage (1997), respectively, with few modifications made by the author. Chromosome data are based on results of the present study (*) and literature cited elsewhere.

E. *"bransfordii"* with 2N = 20 and NF = 36 is thought to be an unknown *podiciferus*-like frog whose karyotype evolved through centromeric fissions. In contrast, the 20-karyomorph of *E.* sp. B has 2N = 20, NF = 40, and all the chromosomes are biarmed, which is not correlated with any other 20-chromosome species in *Craugastor* (the *biporcatus, rugulosus,* and *gollmeri* groups) but is hypothesized as having evolved by aneuploidy.

DIASTEMA GROUP

The chromosome numbers of the *diastema* group are listed in table 5.3. The chromosome data for four species of the *diastema* group investigated in Lower Central America—*E. diastema, E. vocator, E.* sp. D, and *E.* sp. E—are 2N = 18 (table 5.2), and all but *E. vocator* have NF = 36 with all biarmed chromosomes. The karyotype of *E. vocator* (fig. 5.2) contains two telocentric chromosome pairs (8 and 9) and one heteromorphic pair (7) in which two homologous chromosomes are subtelocentric and telocentric, respectively. Thus, the NF value of *E. vocator* is 31, the lowest NF value known in *Eleutherodactylus.*

C-Banding Patterns

RHODOPIS GROUP

No C-band staining has been previously performed on members of the subgenus *Craugastor.* The C-bands are consistently located in the centromeric regions for all chromosomes in all material studied. The total centromeric C-bands are small (5.79%–11.89% RTL) in all karyomorphs of the *rhodopis* group.

Heavily stained interstitial C-bands are rare in most species of subgenus *Craugastor* (S.-H. Chen, unpub. data). However, there are considerable interstitial C-bands in karyomorphs of *Eleutherodactylus podiciferus* and *E.* sp. A (fig. 5.1). For most of the interstitial C-bands, the locations are so varied that they cannot be treated as homologues to the corresponding bands in the allied species. Contrarily, two interstitial C-bands are consistently located in the long arm of pair 7 in both karyomorphs of *E. podiciferus* and in the short arm of pair 7 in the remaining species. These two bands are unique in all studied species of the *rhodopis* group and cannot be confused with other C-bands in the karyotype. Therefore, the chromosome pair 7 having these two bands is a phylogenetic marker and a synapomorphy of the *rhodopis* group in the subgenus *Craugastor.*

In addition, there are autapomorphic C-bands present in the karyotypes of the *rhodopis* group. A terminal C-banding block in the short arm of pair 1, which is associated with the NOR, appears only in the Tapanti karyomorph of *E. bransfordii* among all species of *Craugastor.*

Pair 10 of the 20-karyomorph of *Eleutherodactylus* sp. B is composed of different sorts of heterochromatin, according to the presence of dense and pale C-bands throughout the whole chromosome. It does not appear in other karyomorphs of the *rhodopis* group or elsewhere in the whole subgenus *Craugastor*. Therefore, pair 10 is an autapomorphic character of the 20-karyomorph of *Eleutherodactylus* sp. B.

DIASTEMA GROUP

Only two species, *Eleutherodactylus vocator* and *E.* sp. D, were successfully C-banded (fig. 5.2). The amount of centromeric C-banding material is extremely small (3.45%–4.29% RTL) in this group. Although many distinctive terminal C-bands were reported for three Cuban *Eleutherodactylus* (subgenus *Eleutherodactylus;* Bogart 1981), no distinctive interstitial or terminal C-bands were found in the karyotypes of the *diastema* group.

Secondary Constrictions

The secondary constrictions are gaps in the chromosome arms, in contrast to the centromere (primary constriction) of the chromosome. Their number and location are varied among species or species groups (table 5.2). The fragile sites appear to be gaps, breaks, acentric fragments, triradials, or a sister chromatid intercrossing, which are common phenomena in eukaryotic animals (Arrieta et al. 1993; Miller and Therman 2001). Only chromosome gaps (i.e., secondary constrictions) are found in the frog karyotypes studied. No cholchicine-induced fragile sites have been reported. However, cold-induced secondary constrictions seem to have been induced frequently during the procedures of preparing the chromosomes (Rudak and Callan 1976; King 1980; Sessions 1982). These are the most likely non-NOR correlated secondary constrictions seen in karyotypes in this study. Thus, a fragile site can remain as a distinctively large gap in a less condensed chromosome, but it may be indistinguishable or even absent in a much condensed chromosome (Bogart 1972; Schmid 1978a). Although many fragile sites (secondary constrictions) are not correlated with NORs, occurrences at a particular site in homologous chromosomes are considered phylogenetically informative when more than one individual or species shows the same site. For example, the secondary constriction in the short arm of pair 5 in three karyomorphs of *bransfordii*-like frogs (i.e., *E. bransfordii* and *E. stejnegerianus*) links them to species in the *rhodopis* group.

The NORs are specific chromosome regions in a mitotic cell that are correlated with ribosomal genes transcribed (18S, 28S, and 5S rRNA) in the nucleolus before metaphase. The location of active NORs in a metaphase karyotype can be easily seen by use of silver staining (Goodpasture and Bloom 1975; Howell and Black 1980), and only active NOR sites in the preceding interphase show the remarkable black spots (silver stains) in the chromosomes (Miller et al. 1976). The silver stainability can be maintained in the NORs throughout the following mitosis, even if the rRNA genes are inactive in particular stages. In situ hybridization provides a tool to allocate the sites of 18S + 28S and 5S rRNA genes in the chromosomes. However, only the 18S + 28S rRNA gene sites can be silver-stained. The fragile sites correlated with the NORs are commonly termed secondary constrictions sensu stricto in usage (Bogart 1972; King 1980, 1990; Miller and Therman 2001). NORs can be seen in a conventional Giemsa-stained karyotype as either secondary constrictions (gaps), faintly stained regions (Miller and Therman 2001; S.-H. Chen, pers. obs.), or without any structural differentiation.

SECONDARY CONSTRICTIONS AND NORS IN THE *RHODOPIS* GROUP

There are as many as two secondary constriction sites located in the short arms of pairs 3 and 5 in *Eleutherodactylus bransfordii* (including both karyomorphs) and *E. stejnegerianus* (fig. 5.1, table 5.2). In contrast, the remaining species have constriction sites found in various locations on the karyotypes. The secondary constrictions of *E. podiciferus* (including both karyomorphs) and *E.* sp. A are located in the short arm of either pair 3 or 4, and in the middle of the short arm of pair 1 in the karyotypes of *E.* sp. B (including both karyomorphs).

Not all secondary constrictions were silver-stained. Within the *rhodopis* group, there is only one pair of NORs located in the secondary constriction of chromosome pair 3 or 4 in *Eleutherodactylus bransfordii* (La Selva karyomorph), *E. stejnegerianus, E. podiciferus,* and *E.* sp. A. The NORs of the Tapanti karyomorph of *E. bransfordii* were shifted to the terminal region of chromosome pair 1, where a large C-banding block is associated. The NORs in *E.* sp. B (18- and 20-karyomorphs) are in the secondary constriction situated in the middle of the short arm of chromosome pair 1. The occurrence of NORs in the Tapanti karyomorph of *E. bransfordii* and *E.* sp. B is autapomorphic for the species group. Not all secondary constrictions in the short arm of pair 5 in *E. bransfordii* and *E. stejnegerianus* are silver-stained. The NORs that occurred in the secondary constrictions of pair 3 or 4 depend on the rank order of the NOR-carrying chromosomes in the karyotype. According to features of the karyotypes, all of the NOR-carrying chromosomes (pairs 3 and 4) in the *rhodopis* group can be regarded as homologous, and these NOR sites are synapomorphic for the *rhodopis* group within *Craugastor*. The NOR sites on pair 1 are only in members of the *bransfordii* subgroup, whereas the NOR sites on pairs 3 and 4 are present in members of both *bransfordii* and *podiciferus* subgroups. The latter are plesiomorphic when compared with the NORs on chromosome pair 1.

Hypothetical changes of the NORs in the *rhodopis* group are proposed in figure 5.6 below and are discussed later in the chapter.

SECONDARY CONSTRICTIONS AND NORS OF THE OUTGROUP SPECIES

Two species, *Eleutherodactylus diastema* and *E.* sp. E, of the *diastema* group show clear secondary constrictions in their karyotypes (fig. 5.2). The secondary constriction is located on the short arm of pair 8 in *E. diastema* and on the long arm of pair 9 in *E.* sp. E. Obviously these secondary constrictions are not homologous when patterns of the karyotypes in both species are compared. No secondary constriction was found in any cell plates of *E. vocator* or *E.* sp. D.

Two West Indian *Eleutherodactylus* with 18 chromosomes have been cytologically studied by Bogart (1981). The features of their karyotypes resemble each other, and every corresponding chromosome pair can be recognized in the two karyotypes. The secondary constrictions are located on the short arms of pairs 4 and 5 in *E. auriculatus* and *E. varians,* respectively. The secondary constriction-carrying chromosome pairs in both species are relatively large ($<$ 10% RTL) and have nearly the same shape in the karyotypes (a larger arm ratio than those of its neighboring pairs). Both chromosome pairs and their secondary constrictions are judged to be homologous. However, the secondary constrictions present in the West Indian 18-chromosome species do not appear in their mainland congeners, the *diastema* group, in Central America. These incongruent features suggest that species of the subgenus *Eleutherodactylus* with 18 chromosomes in the West Indies and in Lower Central America are not closely related and probably evolved independently.

Four species of the *diastema* group were silver-stained to detect the NORs. A pair of silver grains were usually found in the centromeric regions of the chromosome, but no other large silver grains were deposited in the chromosome arms of all examined karyotypes. Because these centromeric silver grains commonly appeared in other silver-stained karyotypes, they cannot be considered NORs. NORs are not found in any species of the *diastema* group, and the secondary constrictions of *E. diastema* and *E.* sp. E are not correlated with the NORs.

TYPES OF NORS

Five types of NORs were found in Australian hylid frogs, genus *Litoria,* according to mitotic behavior and C-banding pattern (King 1980, 1990). In this study, two types of NOR-correlated constrictions were found. Type 2 NORs are found in the Tapanti karyomorph of *E. bransfordii* (pair 1, terminal), and type 1 NORs are present in all other species of the *rhodopis* group (*E. podiciferus, E. stejne-*

gerianus, E. sp. A, and La Selva karyomorph of *E. bransfordii* [all in pair 3 or 4, interstitial] and *E.* sp. B [pair 1, interstitial]). All non-NOR-correlated gaps (e.g., pair 5 of *E. bransfordii*), and both gaps in the karyotypes of *E. diastema* and *E.* sp. E, belong to the type 3 NORs. Because types of NORs in members of each species group are variable, they are not phylogenetically informative.

In contrast, NOR sites are informative. The heritability of NORs by Mendelian inheritance has been reported in various animals (e.g., Henderson and Bruère 1980; Arruga and Monteagudo 1989; Miller and Therman 2001). Species of the same species group usually have NORs in the same location (Schmid 1982). The exception indicates that some kind of evolutionary mechanism(s) had occurred in this particular 18S + 28S rRNA locus (Schmid 1978a). Accordingly, the evolution of the NOR-carrying chromosomes in the *rhodopis* group are discussed later in the chapter.

Karyotypes

Frog species show different karyological features that are not only specific to each species group but also useful in distinguishing two or more sibling species that show only minor external morphological differences. For example, the 22-chromosome form of *Eleutherodactylus berkenbuschii* was resurrected from a 20-chromosome *E. rugulosus* population by Savage and DeWeese (1979). More recently, the resurrection of *E. obesus* and the recognition of a new species, *E. rhyacobatrachus,* from the *E. punctariolus* subpopulation system (Savage 1975) were mainly based on karyological data (Campbell and Savage 2000). Therefore, karyological data are useful in frog taxonomy, especially for the genus *Eleutherodactylus.*

From an evolutionary point of view, the features of the karyotypes are probably fixed in the population. The fitness of a karyotype in a particular situation may work against any unwanted chromosomal rearrangement, thus preventing chromosomal polymorphism to occur in the population or in a lineage (DeWeese 1976). However, using karyological data solely, such as chromosome numbers and NF values, in constructing phylogenies has been challenged because such karyological data may be homoplastic—which is true. But we should keep in mind that the karyological morphologies are heritable along an evolutionary lineage, and the homologous chromosome pairs in this lineage are traceable and show a distinctive pattern. Thus, closely related species must be studied before making any meaningful karyotype comparisons.

When the karyotypes of a large number of closely related species are examined, karyotype patterns usually emerge (Bogart 1973) and some "chromosome markers" are also recognized. Herein, these patterns are termed *model karyotypes.* Of the subgenus *Craugastor,* most species groups have been thoroughly reviewed according to external morphologies, jaw muscles, allozyme comparisons, and even the 2N and NF values. Species groups are thought to be mono-phyletic, which provides a chance to examine their karyotype patterns.

KARYOTYPES OF THE *RHODOPIS* GROUP

Except for the 20-karyomorph of *Eleutherodactylus* sp. B, six species of the *rhodopis* group have been studied cytologically, and the data are 2N = 18 and NF = 36 (see table 5.3). The karyological features of the 18-karyomorph (fig. 5.1) in all species are characterized by pair 1 larger than 15% of the relative total length, pairs 2–5 range between 15% and 10% RTL, and 4 pairs of small meta-centrics. There are no telocentric chromosomes. The karyotype of the 20-karyo-morph of *E.* sp. B resembles that of the 18-karyomorph except for having an ad-ditional heterochromatinized metacentric pair 10.

According to features of the karyotype, members of the *rhodopis* group in Lower Central America can be categorized in two subgroups, the *bransfordii* and *podiciferus* subgroups. The *bransfordii* subgroup includes *Eleutherodactylus stejnegerianus, E. bransfordii* (both karyomorphs), and *E.* sp. B (both karyo-morphs), whereas the *podiciferus* subgroup includes *E. podiciferus* (both karyo-morphs), *E.* sp. A, and probably the northern Central American *E. loki.*

Members of the *bransfordii* subgroup (tables 5.4, 5.6) are characterized by two pairs of large subtelocentric chromosomes, pairs 3 and 5 (on pairs 4 and 5 in both karyomorphs of *Eleutherodactylus* sp. B). Although members of the *podiciferus* subgroup (table 5.5) do not have such telocentric pairs (except for pair 4 in *E.* sp. A), two chromosome pairs having the largest arm ratios in the karyotypes are located on pairs between 3 and 5 (either all submetacentric or having one subtelocentric in these two pairs). These are equivalent to the two subtelocentric pairs in the *bransfordii* subgroup. In addition, chromosome pair 6 is larger than 10% RTL in the Palmichal karyomorph of *E. podiciferus, E.* sp. A, and *E. loki.* This is unknown for other species of *Craugastor* examined. The karyotype of *E. loki* (reported as *E. rhodopis* by DeWeese 1976) from Guatemala has two submetacentric chromosome pairs 4 and 5. These have the largest arm ratios of the karyotype and resemble those of members of the *E. podiciferus* sub-group. *E. loki* probably belongs to the *podiciferus* subgroup, on the basis of karyological data (DeWeese 1976).

The C-banding study shows that the centromeric C-bands of the *rhodopis* group contain only a small amount of the heterochromatin (5.79%–11.89% RTL). There are more interstitial C-bands in the *podiciferus* subgroup than the *bransfordii* subgroup (fig. 5.1). However, the double C-bands of pair 7 that ap-pear in the short arm in most species and in the long arm in *E. podiciferus* are synapomorphic to the *rhodopis* group.

In the *bransfordii* subgroup, two secondary constrictions are located on the short arms of pairs 3 and 5 (*Eleutherodactylus stejnegerianus* and *E. bransfordii*),

Table 5.4 Relative lengths (RTL), arm ratios (AR), and types of chromosomes in *Eleutherodactylus bransfordii* and *E. stejnegerianus*

Chromosome pair	*E. bransfordii* (La Selva) (n = 30)			*E. bransfordii* (Tapanti) (n = 19)			*E. stejnegerianus* (n = 30)		
	RTL	AR	Type	RTL	AR	Type	RTL	AR	Type
1	18.26 ± 1.20	1.56 ± 0.11	m	18.67 ± 1.54	1.46 ± 0.13	m	17.92 ± 1.16	1.74 ± 0.15	sm
2	13.23 ± 0.67	1.82 ± 0.25	sm	13.22 ± 0.57	1.90 ± 0.21	sm	13.17 ± 0.67	2.14 ± 0.23	sm
3	12.70 ± 0.84	3.82 ± 0.65	st	12.68 ± 0.90	4.22 ± 0.60	st	12.84 ± 0.85	3.65 ± 0.62	st
4	12.14 ± 0.82	1.55 ± 0.23	m	12.05 ± 0.97	1.41 ± 0.13	m	12.49 ± 0.79	1.57 ± 0.18	m
5	11.34 ± 0.63	4.12 ± 0.63	st	11.46 ± 0.73	4.56 ± 0.79	st	10.85 ± 0.75	6.34 ± 1.00	st
6	9.76 ± 0.51	1.31 ± 0.10	m	9.72 ± 0.49	1.32 ± 0.13	m	9.74 ± 0.46	1.36 ± 0.18	m
7	8.86 ± 0.58	1.21 ± 0.10	m	8.47 ± 0.69	1.15 ± 0.05	m	9.09 ± 0.59	1.13 ± 0.06	m
8	8.13 ± 0.61	1.33 ± 0.14	m	8.17 ± 0.75	1.30 ± 0.15	m	8.48 ± 0.63	1.30 ± 0.13	m
9	5.58 ± 0.54	1.16 ± 0.08	m	5.55 ± 0.45	1.29 ± 0.10	m	5.42 ± 0.46	1.18 ± 0.09	m

Note: *n* indicates number of cell plates measured. RTLs and ARs are mean ± SD. m, metacentric; sm, submetacentric; st, subtelocentric; t, telocentric.

Table 5.5 Relative lengths (RTL), arm ratios (AR), and types of chromosomes for the Palmichal and Tapanti karyomorphs of *Eleutherodactylus podiciferus* and *E.* sp. A

Chromosome pair	E. podiciferus (Palmichal) (n = 27)			E. podiciferus (Tapanti) (n = 23)			E. sp. A (n = 23)		
	RTL	AR	Type	RTL	AR	Type	RTL	AR	Type
1	16.85 ± 1.05	1.90 ± 0.16	sm	16.59 ± 1.00	1.73 ± 0.12	sm	15.89 ± 0.86	1.74 ± 0.15	sm
2	13.68 ± 0.60	1.79 ± 0.19	sm	13.66 ± 0.62	1.61 ± 0.21	m	13.51 ± 0.67	1.63 ± 0.20	m
3	12.01 ± 0.80[a]	2.73 ± 0.41	sm	13.64 ± 1.02	2.57 ± 0.35	sm	12.63 ± 0.75	2.32 ± 0.41	sm
4	12.32 ± 0.52[a]	1.72 ± 0.41	sm	12.02 ± 0.86	1.64 ± 0.15	m	11.78 ± 0.77[a]	1.46 ± 0.24	m
5	11.76 ± 0.79	2.24 ± 0.25	sm	11.81 ± 0.75	2.37 ± 0.24	sm	11.90 ± 0.88[a]	3.18 ± 0.42	st
6	10.10 ± 0.64	1.22 ± 0.08	m	9.93 ± 0.49	1.22 ± 0.09	m	10.46 ± 0.60	1.27 ± 0.14	m
7	9.07 ± 0.57	1.14 ± 0.06	m	9.02 ± 0.69	1.16 ± 0.07	m	9.53 ± 0.49	1.17 ± 0.09	m
8	8.45 ± 0.50	1.25 ± 0.11	m	8.61 ± 0.60	1.25 ± 0.11	m	8.73 ± 0.44	1.25 ± 0.09	m
9	5.75 ± 0.37	1.09 ± 0.06	m	4.71 ± 0.29	1.16 ± 0.09	m	5.55 ± 0.53	1.09 ± 0.06	m

Note: *n* indicates number of cell plates measured. RTLs and ARs are mean ± SD. m, metacentric; sm, submetacentric; st, subtelocentric; t, telocentric.
[a] Ranking order has been changed for comparison.

Table 5.6   Relative lengths (RTL), arm ratios (AR), and types of chromosomes in the 18- and 20-karyomorphs of *Eleutherodactylus* sp. B

Chromosome pair	18-karyomorph ($n = 24$)			20-karyomorph ($n = 41$)		
	RTL	AR	Type	RTL	AR	Type
1	18.76 ± 0.94	1.42 ± 0.12	m	17.52 ± 1.39	1.46 ± 0.14	m
2	13.41 ± 0.63	1.79 ± 0.25	sm	12.88 ± 1.01	1.85 ± 0.20	sm
3	12.49 ± 0.70	1.57 ± 0.18	m	11.64 ± 0.95	1.52 ± 0.22	m
4	11.90 ± 1.18	4.08 ± 0.68	st	11.52 ± 0.81	3.74 ± 0.61	st
5	10.78 ± 0.98	5.35 ± 0.95	st	10.27 ± 0.73	5.42 ± 0.85	st
6	9.77 ± 0.52	1.29 ± 0.11	m	9.20 ± 0.64	1.30 ± 0.11	m
7	9.02 ± 0.55	1.15 ± 0.08	m	8.69 ± 0.59	1.17 ± 0.12	m
8	8.26 ± 0.62	1.27 ± 0.12	m	7.81 ± 0.72	1.21 ± 0.10	m
9	5.62 ± 0.60	1.18 ± 0.10	m	5.69 ± 0.53	1.19 ± 0.13	m
10	—	—		4.77 ± 0.43	1.17 ± 0.09	m

Note: *n* indicates number of cell plates measured. RTLs and ARs are mean ± SD. m, metacentric; sm, submetacentric; st, subtelocentric; t, telocentric.

and the secondary constrictions on pair 3 are NOR-carrying constrictions. Although there are secondary constrictions on pair 3 in the Tapanti karyomorph of *E. bransfordii,* the NORs are located at the terminus of the short arm of pair 1. The other NOR-carrying secondary constrictions are on the short arms of pair 3 or 4 in members of the *podiciferus* subgroup, and on the short arm of pair 1 in *E.* sp. B.

In summary, the model karyotype of the *rhodopis* group is characterized as follows: (*a*) 2N = 18 and NF = 36; (*b*) pair 1 is larger than 15% RTL; (*c*) pairs 2–5 range between 15% and 10% RTL, with the two largest arm ratio values in pairs 3–5, which are subtelocentric for members of the *bransfordii* subgroup and submetacentric for the *podiciferus* subgroup; (*d*) there are four small (< 10% RTL) metacentric pairs, and pair 6 is larger than 10% RTL in the *podiciferus* subgroup; (*e*) double C-bands are present in the short arm of pair 7, except for *E. podiciferus,* where it is in the long arm; (*f*) the large subtelocentric or submetacentric chromosome pair 3 or 4 has the NOR site, and the small subtelocentric or submetacentric pair 5 does not have a NOR.

DOUBTFUL CHROMOSOME NUMBER OF
ELEUTHERODACTYLUS BRANSFORDII

The karyotypes of *Eleutherodactylus bransfordii* were reported by DeWeese (1975, 1976) as 2N = 18 and NF = 36 and by Bogart (1970) as 2N = 20 with two pairs of telocentric chromosomes (i.e., NF = 36). These data conflict with each other. Although DeWeese did not report the secondary constrictions and NORs, his frogs were from La Selva and Bajo La Hondura, and his chromosome groupings were the same as mine. In particular, his karyotype of *E. bransfordii* shows an elongated chromosome pair 1 that is larger than 15% of the total haploid length, which is in agreement with my findings. Therefore, DeWeese's (1976) karyotype is the same as my karyomorph of the La Selva population. However, Bogart's (1970) karyotype of 2N = 20 with two pairs of telocentric chromosomes was never found in any *bransfordii*-like frogs in the present study. His data were interpreted by DeWeese (1976) as a misidentification of the frogs. What is the most likely frog species examined by Bogart (1970) from Costa Rica? To resolve this puzzle, Bogart's data were compared with all known Costa Rican eleutherodactyline frog species that have been karyotyped.

The characteristics of the karyotype reported by Bogart (table 5.7) showed a relatively short pair 1 (14.9% RTL), a remarkable large subtelocentric chromosome pair among the first five pairs with 14.4% RTL (pair 2, AR = 4.7), a second relatively large telocentric chromosome pair (pair 9, 6.1% RTL), and a last metacentric chromosome pair (pair 10, 5.5% RTL). Frogs with 2N = 20 and NF = 36 in Costa Rica fall into the following species groups: *rugulosus* group, *biporcatus* group, and *gollmeri* group. Except for *E. cuaquero* and *E. catalinae*, the karyotypes of the *rugulosus* group (Campbell and Savage 2000) in Costa

Table 5.7   Comparisons of relative lengths (RTL), arm ratios (AR), and types of chromosomes in *Eleutherodactylus podiciferus* and *E. "bransfordii"* as reported by Bogart (1970)

E. podiciferus				E. "bransfordii"			
Chromosome pair	RTL	AR	Type	Chromosome pair	RTL	AR	Type
1	16.5	1.8	sm	1	14.9	2.0	sm
2	12.6	3.4	st	2	14.4	4.7	st
3	12.6	2.2	sm	(7 + 9)	(14.3)	(1.3)	(m)
				7[a]	8.2	7.01	t
				9[a]	6.1	7.01	t
4	12.6	1.8	sm	3	12.6	1.8	sm
5	11.6	2.9	sm	4	12.2	2.2	sm
6	10.6	1.2	m	5	9.1	1.8	sm
7	9.1	1.0	m	6	8.7	1.2	m
8	8.8	1.2	m	8	8.1	1.2	m
9	5.3	1.1	m	10	5.5	1.2	m

Note: Data of chromosome pairs 7 and 9 of *E. "bransfordii"* are combined and herein considered as homologous to pair 3 of his *E. podiciferus*. m, metacentric; sm, submetacentric; st, subtelocentric; t, telocentric.
   [a] Ranking order has been changed for comparison.

Rica are known. All karyotypes of the 20-chromosome *rugulosus* group species contain a subtelocentric chromosome pair among the first five pairs (on pair 2 of Bogart's frog), but pairs 9 and 10 are metacentric and telocentric, respectively. In addition, there is a distinctly large pair 1 that is larger than 15% RTL and larger than pair 2 by 2.7%–4.07% RTL in all species. Bogart's frogs cannot be allocated to this species group.

A large subtelocentric pair (pair 4) also occurs in four known species of the "*biporcatus*" group (*E. megacephalus, E. rugosus, E. gulosus,* and *E.* sp. F) in Costa Rica and Panama, but its size is in the range of 10.96%–12.19% RTL (S.-H. Chen, unpub. data) and is much shorter than that in Bogart's frogs (14.4% RTL). There is a relatively large pair 1 with 16.82%–17.79% RTL in all species of the "*biporcatus*" group, but not in Bogart's data. Three species of the *gollmeri* group (*E. gollmeri, E. mimus,* and *E. noblei*) are distributed in Costa Rica (Savage 1987), two of which (*E. gollmeri* and *E. mimus*) had been karyotyped (DeWeese 1976; S.-H. Chen, unpub. data). *E. gollmeri,* with 22 chromosomes, may be excluded from consideration.

Although *E. mimus* has a karyotype with 2N = 20, NF = 36, and a relatively short pair 1 (< 15% RTL and not distinctly larger than pair 2), there is no such subtelocentric chromosome pair within the first five pairs in its karyotype (DeWeese 1976; S.-H. Chen, unpub. data). Therefore, if Bogart's data are correct, none of these species is represented by his frogs.

The best alternative is that Bogart's frogs belong to the *rhodopis* group. If we add both telocentric chromosomes together, the relative length of this reconstructed chromosome pair is 14.3% RTL (ranking order 3) and the arm ratio is 1.34, a metacentric chromosome. Then the karyotype has the first five pairs larger than 10% RTL and the next four pairs between 10 and 5% RTL. If the standard derivation and method are considered, Bogart's frogs appear to be *E. podiciferus* because all *bransfordii*-like and *stejnegerianus*-like frogs, including *E.* sp. B, have two pairs of subtelocentric chromosomes in the first five pairs. Although my data for *E. podiciferus* show no such subtelocentric pair, the arm ratios of pair 3 of my data in both karyomorphs of *E. podiciferus* are 2.73 and 2.57, respectively. These are close to 3.0, the arbitrary boundary of subtelocentric chromosomes. Besides, *E. podiciferus* of Bogart (1970) has a subtelocentric pair 2 and many submetacentric chromosome pairs in the first five pairs, which is similar to Bogart's *E. "bransfordii."* The relatively short pair 1 in Bogart's set is probably the result of too much condensation of the chromosomes. Pair 3 of Bogart's *E. podiciferus* has a distinct secondary constriction (fragile site) in the pericentromeric regions and is the most likely place to segregate both chromosome arms. Unfortunately, there is no such centromeric fission present in all populations of *E. podiciferus* investigated in this study. Because there are many distinct geographical *podiciferus*-like karyomorphs in my data, it is possible to get a karyomorph with 20 chromosomes somewhere in Costa Rica. Therefore,

I agree with DeWeese's suggestion that Bogart misidentified his frogs. The probable identity of Bogart's *E. "bransfordii"* with 2N = 20 is an *E. podiciferus* from a population not investigated in the present study.

CHROMOSOME POLYMORPHISM
IN *ELEUTHERODACTYLUS* SP. B

Although both 18- and 20-karyomorphs of the Fortuna population are different in chromosome numbers, they share a unique interstitial secondary constriction in the short arm of pair 1 in which a type 1 NOR is located. The position of the secondary constriction is so unusual that it is also recognized in the Nusagandi population by the only frog (USC 9511) examined, even though it was not silver-stained. In contrast, the unique position of the secondary constriction is unknown in *bransfordii*-like frogs of other populations studied. It is inferred that *bransfordii*-like frogs of both populations (Fortuna and Nusagandi) in Panama belong to the same species, *Eleutherodactylus* sp. B, which has not yet been formally described.

One large metacentric pair 10 in the 20-karyomorph of *Eleutherodactylus* sp. B is totally composed of C-positive materials when tested with the C-banding method and is unique among all C-banded eleutherodactyline frog karyotypes. The function and origin of pair 10 are not clear. It is possible that both complements of pair 10 could be considered as supernumerary chromosomes (B-chromosomes) on the basis of the large amount of heterochromatin, although they paired with each other to form a bivalent chromosome during prophase stage of the first meiosis so they are homologous chromosomes. Besides, in this study there is no specimen having a single supernumerary chromosome in its karyotypes. For example, there is no frog found to be 2N = 19 in karyotype. Thus, the origin of pair 10 of *E.* sp. B is unclear. If we consider the relative total length and arm ratio of pair 10, they most resemble those of pair 9. It is possible that pair 10 of the 20-karyomorph evolved from pair 9 of the ancestral stock of 18-karyomorph by aneuploidy and has undergone a loss of genetic function and heterochromatinization.

KARYOTYPES OF THE OUTGROUP SPECIES
(*DIASTEMA* GROUP)

The *diastema* group (subgenus *Eleutherodactylus*) was employed for outgroup comparisons with the *rhodopis* group. All members of the *diastema* group that have been studied have 18 chromosomes (see table 5.3), the same as in the *rhodopis* group. However, their karyotypes are remarkably different from those of the *rhodopis* group.

The karyotypes of four species of the *diastema* group (table 5.8) are charac-

Table 5.8   Relative lengths (RTL), arm ratios (AR), and types of chromosomes in four Lower Central American *Eleutherodactylus* belonging to the *diastema* group

Chromosome pair	E. diastema ($n = 14$)			E. sp. D ($n = 28$)		
	RTL	AR	Type	RTL	AR	Type
1	18.99 ± 0.84	1.84 ± 0.16	sm	18.99 ± 1.39	1.92 ± 0.26	sm
2	14.21 ± 0.73	2.19 ± 0.20	sm	15.69 ± 1.04	2.59 ± 0.27	sm
3	13.22 ± 0.90	1.69 ± 0.19	sm	13.03 ± 0.66	2.08 ± 0.28	sm
4	11.78 ± 0.21	2.00 ± 0.22	sm	12.01 ± 0.91	2.16 ± 0.23	sm
5	10.93 ± 0.67	1.37 ± 0.22	m	10.59 ± 0.66	1.12 ± 0.09	m
6	9.88 ± 0.50	1.20 ± 0.12	m	9.27 ± 0.61	1.18 ± 0.18	m
7	7.97 ± 0.49	2.19 ± 0.28	sm	7.99 ± 0.52	1.85 ± 0.16	sm
8	7.60 ± 0.54	2.97 ± 0.56	sm	7.82 ± 0.67	3.37 ± 0.38	st
9	5.43 ± 0.56	2.49 ± 0.36	sm	4.64 ± 0.45	3.98 ± 0.60	st

Chromosome pair	E. sp. E ($n = 10$)			E. vocator ($n = 8$)		
	RTL	AR	Type	RTL	AR	Type
1	19.00 ± 0.76	1.88 ± 0.13	sm	19.23 ± 0.71	1.21 ± 0.15	m
2	15.28 ± 0.99	2.56 ± 0.35	sm	15.17 ± 0.85	1.21 ± 0.12	m
3	13.14 ± 0.44	2.21 ± 0.20	sm	13.02 ± 0.70	1.44 ± 0.21	m
4	12.01 ± 0.68	1.97 ± 0.12	sm	12.10 ± 0.84	1.97 ± 0.20	sm
5	10.65 ± 0.60	1.12 ± 0.05	m	11.45 ± 0.66	1.49 ± 0.41	m
6	9.38 ± 0.67	1.19 ± 0.26	m	10.66 ± 0.96	1.31 ± 0.16	m
7	7.94 ± 0.61	2.33 ± 0.25	sm	A7.36 ± 1.18	5.16 ± 2.01	st
				B6.51 ± 0.61	7.01	t
8	7.21 ± 0.79	3.55 ± 0.33	st	6.66 ± 0.88	7.01	t
9	5.39 ± 0.29	3.11 ± 0.31	st	4.76 ± 0.34	7.01	t

Note: n indicates number of cell plates measured. RTLs and ARs are mean $\pm$ SD. m, metacentric; sm, submetacentric; st, subtelocentric; t, telocentric.

terized by having a relatively long chromosome pair 1 that is approximately 19% (the range is 18.99%–19.23%) RTL. Pair 2 is more than 15% RTL in most species examined except for *Eleutherodactylus diastema*, which is 14.21% RTL. The relative total lengths in pairs 1 and 2 are much larger than those in the *rhodopis* group. Pairs 3–5 range between 15% and 10% RTL. There is no subtelocentric chromosome pair located in the first five pairs, among which the largest value of the arm ratios ranks on pair 4 (AR = 1.97) in *E. vocator* or on pair 2 (AR = 2.19–2.59) for the rest of the species. These are unlikely to be homologous to the largest (> 10% RTL) subtelocentric or submetacentric pairs in karyotypes of the *rhodopis* group, according to karyotype patterns. Pairs 6–9 range between 10% and 5% RTL in most species, of which pair 6 of *E. vocator* is larger than 10% RTL and pair 9 in both *E. vocator* and *E.* sp. D is smaller than 5% RTL.

Except for pairs 7–9, all of the chromosome pairs are either submetacentric or metacentric in all members of the *diastema* group. They differ in that the first three pairs are metacentric in *Eleutherodactylus vocator* but submetacentric in the other species. Pairs 8 and 9 vary among species. They are submetacentric in *E. diastema*, subtelocentric in *E.* sp. D and *E.* sp. E, and telocentric in *E. vocator*. Pair 7 is heteromorphic in *E. vocator*, in which two homologous chromosomes are subtelocentric in one homologue and telocentric in the other. These chromosomes are submetacentric in both homologues in all the other species. The NF values are 31 in *E. vocator* and 36 in the other species.

No interstitial C-bands and only minimum centromeric C-band materials were found in two species studied (i.e., *Eleutherodactylus vocator* and *E.* sp. D; fig. 5.2). The C-bands were different from those of the *rhodopis* group.

No synapomorphic secondary constrictions were found for the *diastema* group. The secondary constrictions found in two species are located on the short arm of pair 8 for *Eleutherodactylus diastema* and on the long arm of pair 9 for *E.* sp. E. However, these constrictions are not correlated with the NORs, and none of the members of the *diastema* group shows any NOR in the karyotype. Both constriction-carrying chromosomes might have evolved independently and may not be homologous.

In conclusion, the model karyotype of the *diastema* group is characterized as follows: (*a*) 2N = 18 and NF = 36; (*b*) both pairs 1 and 2 are larger than 15% RTL; (*c*) pairs 3–5 range between 15% and 10% RTL, without a large (> 10% RTL) subtelocentric pair 4; (*d*) all chromosome pairs are metacentric or submetacentric, except for pairs 8 and 9, which are subtelocentric; (*e*) the model contains only a small amount of the centromeric C-bands in two species; and (*f*) it is without any diagnosable Ag-NORs in the karyotype.

There are two major subpatterns of karyotypes in the *diastema* group. These indicate two major phylogenetic lineages within the group, the *diastema* subgroup and the *vocator* subgroup. The karyotype of the *vocator* subgroup is diagnosed as 2N = 18, NF = 31, with a heteromorphic chromosome pair 7 (at least in males), and the last two pairs are telocentric. The karyotype of the *diastema* subgroup is diagnosed as basically the same as that of the model karyotype, except for pairs 8 and 9, which may be either subtelocentric or submetacentric and may have a secondary constriction site present on either of these chromosome pairs. Only *Eleutherodactylus vocator* belongs to the *vocator* subgroup, whereas *E. diastema*, *E.* sp. D, and *E.* sp. E belong to the *diastema* subgroup. None of the other species cited as belonging to the *diastema* group (Savage 1997) has been karyotyped.

REMARKS ON THE TAXONOMIC STATUS
OF SPECIES IN THE *DIASTEMA* GROUP

The karyotype of *Eleutherodactylus diastema* in Panama is similar to that of *E.* sp. D in Costa Rica with slight differences. A distinctive secondary constriction in the short arm of chromosome pair 8 in *E. diastema* is not present in *E.* sp. D, however, and that easily distinguishes them. In addition, the chromosome pairs 8 and 9 are submetacentric in *E. diastema* but are subtelocentric in *E.* sp. D, and pair 9 is greater than 5% of the total haploid length in *E. diastema* but is less than 5% in *E.* sp. D.

Eleutherodactylus sp. D in Costa Rica was considered *E. diastema* by many workers. However, the external morphologies and breeding calls of all Costa Rican *diastema*-like frogs are different from those of Nusagandi and Darien populations in eastern Panama (S.-H. Chen, unpub. data). In addition, the karyotypes of both species are distinct. The status of all *diastema*-like frogs needs review. Because the type locality is in Panama, I temporarily consider *diastema*-like frogs in eastern Panama (Nusagandi and Darien populations) as *E. diastema*. Costa Rican frogs are regarded as an undescribed species.

The chromosome number of *Eleutherodactylus* sp. D from Costa Rica had been reported by León (1970) as N = 10 (i.e., 2N = 20) and from an unknown locality under the name of *E. "diastema"* by Bogart (1970) as 2N = 18 (NF = 36). These reports conflict with each other. Because Bogart's report was based on material without locality and museum number, his data were questioned by Savage (1987) as a misidentification. In contrast, the voucher of León's karyotype (UCR-2396) was deposited in the Museum de Zoología, Universidad de Costa Rica. The voucher (UCR-2396) is a male, 17.2 mm in snout-vent length (SVL), without right tibia and foot, and both vomerine teeth and vocal slits are present. The specimen undoubtedly is *E.* sp. D. Another specimen (UCR-2397) in the same lot measured 17.0 mm SVL and is the same species. Both specimens were collected from Compartamento Tortuga Verde, Tortugero, Limón Province in the Atlantic lowlands of northeastern Costa Rica. My frog (USC 9309) from Sardina used in the present study is from approximately the same area as León's. Its karyotype, 2N = 18, NF = 36, differs from that of León's (1970), but confirms that of Bogart's (1970) as well as of all other populations in the present study. In addition, the chromosome groupings and Bogart's chromosome types are nearly identical to the present data, except for ranking orders of pairs 7 and 8 and a larger relative length in pair 2 (17.1% vs. 15.69%). In contrast to Savage (1987), I suggest that the chromosome data of León's (1970) report of N = 10 is doubtful and is most likely to have involved incorrect labeling during processing.

Eleutherodactylus vocator is an arboreal frog found in southwestern Costa Rica and nearby areas of western Panama. Although its distribution ranges from Costa Rica to Panama, many records are probably based on misidentifications. For example, a series of *vocator*-like frogs collected at Nusagandi (central Panama) were identified as *E. vocator* by local herpetologists, but they proved to be a new species (*E.* sp. E) according to karyotype data. The karyotype of *E. vocator* had not been previously reported. Three males investigated in the present study were collected from the same population (Las Cruces) on different trips. All males are heteromorphic in chromosome pair 7. This is unusual in eleutherodactyline frogs, and the chromosomes seem to be sex chromosomes. An XX-XY sex-determining mechanism is predicted. Unfortunately, no female was successfully karyotyped to test this prediction.

Although *Eleutherodactylus* sp. E superficially resembles *E. vocator* in external morphology, their karyotypes are distinctly different from each other. The major differences in karyotypes between *E.* sp. E and *E. vocator* according to this discussion are as follows: (1) distinct secondary constriction in the short arm of pair 8 in *E.* sp. E but absent in *E. vocator;* (2) chromosome pair 7 homomorphic (both submetacentric at least in females) in *E.* sp. E but heteromorphic (subtelocentric and telocentric at least in males) in *E. vocator;* (3) different chromosome types in pairs 1–3 and 7–9 (table 5.8) between both species; and (4) NF = 36 (female) in *E.* sp. E, but NF = 31 (male) in *E. vocator.*

Obviously, *Eleutherodactylus* sp. E is not the same as Costa Rican *E. vocator* and probably is an undescribed species, on the basis of karyotype data. However, only females of *E.* sp. E were available for study, and the male karyotype is completely unknown. At this time I am not able to predict the sex-determining mechanism in *E.* sp. E.

Phenetic Similarity and Chromosomal Evolution

Although the homologies of chromosome pairs between the *rhodopis* and *diastema* groups are uncertain, the ranking order of each chromosome pair in the karyotype of the two groups has been adjusted so that all homologous chromosome pairs are ranked in the same position within or among species groups (table 5.9). The present alignment seems the most likely based on similarities found in comparing relative total lengths and arm ratios by using a pair-wise Duncan's multiple range test ($p < .01$).

Relative total lengths or arm ratios between two taxa that are not significantly different are valued as 1, following Matsui (1980). The similarity value for a particular chromosome pair ranges from 0 (both relative total lengths and arm ratios are different) to 2 (both relative total lengths and arm ratios are not different), and the similarity value is 1 for either relative total lengths or arm ratios that are not significantly different. Thus, the maximum similarity value for two karyomorphs (or species) without any significant difference in karyotypes is 20 because *Eleutherodactylus* sp. B has 10 pairs of chromosomes, and it has the

Table 5.9   Chromosome types for 12 karyomorphs of the *rhodopis* and *diastema* groups based on the present study

Taxa	Chromosome pair									
	1	2	3	4	5	6	7	8	9	10
E. bransfordii (LS)	m	sm	st	m	st	m	m	m	m	—
E. bransfordii (T)	m	sm	st	m	st	m	m	m	m	—
E. podiciferus (P)	sm	sm	sm/4	sm/3	sm	m	m	m	m	—
E. podiciferus (T)	sm	m	sm	m	sm	m	m	m	m	—
E. stejnegerianus	sm	sm	st	m	st	m	m	m	m	—
E. sp. A	sm	m	st/4	m/5	sm/3	m	m	m	m	—
E. sp. B (18)	m	sm	st/4	m/3	st	m	m	m	m	—
E. sp. B (20)	m	sm	st/4	m/3	st	m	m	m	m	m
E. diastema	sm	sm	sm	sm	m	m	sm	sm	sm	—
E. vocator	m	m	m	sm	m	m	t	t	t	—
E. sp. D	sm	sm	sm	sm	m	m	sm	st	st	—
E. sp. E	sm	sm	sm	sm	m	m	sm	st	st	—

Note: Some chromosome pairs are rearranged in position corresponding to their homologues (a number behind the type indicates original rank order) for further comparisons. Karyomorphs: LS, La Selva; P, Palmichal; T, Tapanti; 18, 18-karyomorph; 20, 20-karyomorph. Types: m, metacentric; sm, submetacentric; st, subtelocentric; t, telocentric.

similarity coefficient of 1.00. The similarity coefficient between two karyomorphs ranges from 0 to 1.

The similarity values and similarity coefficients of karyotypes are listed in tables 5.10 and 5.11. Matrices of the similarity coefficients were analyzed using the unweighted pair group method of averages (UPGMA) to construct phenograms. These analyses were conducted with the NTSYSpc package. A strict consensus tree was constructed if more than one phenogram was produced.

CHROMOSOME EVOLUTION IN THE *RHODOPIS* GROUP

Four UPGMA phenograms (not shown) were constructed for eight members of the *rhodopis* group and four species of the *diastema* group that represented two distantly related subgenera, *Craugastor* and *Eleutherodactylus,* respectively. A strict consensus tree derived from these trees is shown in figure 5.3. All 18-chromosome species are clearly grouped into two distinctive clusters, the *rhodopis* and *diastema* groups. In the *diastema* group, *E. vocator* is outside an unresolved cluster of three taxa. This reflects the unique karyotype of *E. vocator:* a heteromorphic pair 7 in which one homologue is subtelocentric and the other telocentric, the last two pairs (8 and 9) telocentric, the first three pairs metacentric, and a NF = 31.

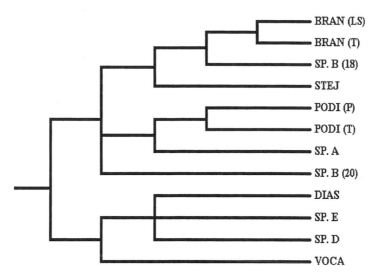

Figure 5.3   Consensus tree for eight karyomorphs of the *rhodopis* group and four species of the *diastema* group derived from four UPGMA trees on the basis of the similarity coefficients shown in table 5.10. Members of both species groups were well separated. However, both karyomorphs of *Eleutherodactylus* sp. B were distantly related. Species: BRAN, *E. bransfordii;* DIAS, *E. diastema;* PODI, *E. podiciferus;* SP. A, B, D, or E, *E.* sp. A, B, D, or E; STEJ, *E. stejnegerianus;* VOCA, *E. vocator.* Karyomorphs: LS, La Selva; P, Palmichal; T, Tapanti; 18, 18-karyomorph; 20, 20-karyomorph.

In contrast, three species of the other subcluster shared characters (e.g., all chromosome pairs homomorphic, NF = 36, and all biarmed chromosomes). However, the differences among these species are remarkable.

The *rhodopis* group is divided into three unresolved clusters. *Eleutherodactylus bransfordii, E. stejnegerianus,* and the 18-karyomorph of *E.* sp. B are grouped in the *bransfordii* subgroup. *Eleutherodactylus podiciferus* and *E.* sp. A form the *podiciferus* subgroup. The 20-karyomorph of *E.* sp. B is unresolved relative to those two groups. The results of the analysis seem counterintuitive because the 18- and 20-karyomorphs of *Eleutherodactylus* sp. B are thought to be from conspecific frogs. However, according to the tree, the karyology of the 18-karyomorph of *E.* sp. B is more similar to that of *E. bransfordii* or *E. stejnegerianus* than to that of the 20-karyomorph of *E.* sp. B.

In this analysis, the chromosome set of the 20-karyomorph of *Eleutherodactylus* sp. B was considered to be the whole genome. Chromosome pair 10 was not considered as supernumerary (B chromosomes). Therefore, the relative total lengths of all chromosome pairs are smaller than those of the other karyotype (18-karyomorph).

Consequently, I recalculated relative total lengths for all chromosome pairs of the 20-karyomorph without including the lengths of pair 10 for the base total.

Table 5.10 Similarity values (upper sector) and similarity coefficients (lower sector) of karyotypes for eight karyomorphs of the *rhodopis* group and four outgroup species of the *diastema* group

		1	2	3	4	5	6	7	8	9	10	11	12
E. bransfordii (LS)	1	—	18	13	9	14	8	15	11	8	7	6	6
E. bransfordii (T)	2	.90	—	14	10	13	9	16	12	9	8	7	8
E. podiciferus (P)	3	.65	.70	—	15	10	13	15	9	8	7	5	6
E. podiciferus (T)	4	.45	.50	.75	—	11	14	11	7	8	8	6	8
E. stejnegerianus	5	.70	.65	.50	.55	—	10	14	8	11	7	6	9
E. sp. A	6	.40	.45	.65	.70	.50	—	9	8	7	6	5	6
E. sp. B (18)	7	.75	.80	.75	.55	.70	.45	—	12	9	7	6	8
E. sp. B (20)	8	.55	.60	.45	.35	.70	.40	.60	—	4	3	5	5
E. diastema	9	.40	.45	.40	.40	.40	.35	.45	.20	—	10	15	16
E. vocator	10	.35	.40	.35	.40	.35	.30	.35	.15	.50	—	9	9
E. sp. D	11	.30	.35	.25	.30	.30	.25	.30	.25	.75	.45	—	16
E. sp. E	12	.50	.40	.30	.40	.45	.30	.40	.25	.80	.45	.80	—

Note: The length of chromosome pair 10 of *E.* sp. B (20) is included in calculating the RTL. Abbreviations for karyomorphs are the same as in table 5.9.

Table 5.11 Similarity values (upper sector) and similarity coefficients (lower sector) of karyotypes for eight karyomorphs of the *rhodopis* group and four outgroup species of the *diastema* group

		1	2	3	4	5	6	7	8	9	10	11	12
E. bransfordii (LS)	1	—	18	13	9	15	8	15	12	8	7	6	10
E. bransfordii (T)	2	.90	—	14	10	13	9	16	12	9	8	7	10
E. podiciferus (P)	3	.65	.70	—	15	10	13	15	12	8	7	5	6
E. podiciferus (T)	4	.45	.50	.75	—	11	14	11	8	8	8	7	8
E. stejnegerianus	5	.75	.65	.50	.55	—	10	14	12	11	8	6	9
E. sp. A	6	.40	.45	.65	.70	.50	—	9	7	7	6	5	6
E. sp. B (18)	7	.75	.80	.75	.55	.70	.45	—	16	9	7	6	8
E. sp. B (20)	8	.60	.60	.60	.40	.60	.35	.80	—	5	5	4	5
E. diastema	9	.40	.45	.40	.40	.55	.35	.45	.25	—	10	15	17
E. vocator	10	.35	.40	.35	.40	.40	.30	.35	.25	.50	—	9	9
E. sp. D	11	.30	.35	.25	.35	.30	.25	.30	.20	.75	.45	—	16
E. sp. E	12	.50	.50	.30	.40	.45	.30	.40	.25	.85	.45	.80	—

Note: The length of chromosome pair 10 of *E.* sp. B (20) is excluded in calculating the RTL. Abbreviations for karyomorphs are the same as in table 5.9.

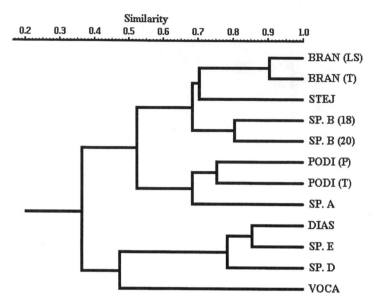

Figure 5.4 A UPGMA tree for eight karyomorphs of the *rhodopis* group and four species of the *diastema* group, based on the similarity coefficients in table 5.11. The pair 10 of the 20-karyomorph of *E.* sp. B was excluded from the entire genome. Both karyomorphs of *E. bransfordii, E.* sp. B, and *E. podiciferus* form a cluster, and two distinctive sublineages were recognized in the *rhodopis* group. The number line across the top of the figure represents overall similarity, where 1.0 indicates identical karyomorphs. Abbreviations are the same as in fig. 5.3.

The new data (table 5.11) were analyzed statistically. The result is a single re-solved UPGMA phenogram (fig. 5.4). Both the *diastema* and *rhodopis* groups are recognized, and the two sublineages in the *diastema* group are confirmed. All studied Lower Central American frogs in the *rhodopis* group fall into two major lineages, the *bransfordii* and *podiciferus* subgroups. The *bransfordii* subgroup includes both the La Selva and Tapanti karyomorphs of *Eleutherodactylus bransfordii* and *E. stejnegerianus,* and both the 18- and 20-karyomorphs of *E.* sp. B. The *podiciferus* subgroup includes both the Palmichal and Tapanti karyomorphs of *E. podiciferus* and *E.* sp. A. The diagnostic differences in external morphologies between *E. bransfordii* and *E. podiciferus* are listed by Savage and Emerson (1970, table 3). Except for the differences in male nuptial pads, which are varied in both subgroups, these features diagnose the *bransfordii* and *podiciferus* subgroups. The outcome of the karyological analysis is congruent with the morphological characters.

Cytologically, using the *diastema* group as the outgroup, we can diagnose the *rhodopis* group (features for the *diastema* group in parentheses) as having the last three chromosome pairs (7–9) metacentric (submetacentric, subtelocentric, or telocentric), double C-bands in one arm of chromosome pair 7 (no distinctive interstitial C-band), and the distinctive secondary constriction in pair 3 or 4 (secondarily lost in *E.* sp. B), which are NORs in most species (no such secondary constrictions).

The karyotypes between subgroups of the *rhodopis* group differ markedly in that pair 5 is submetacentric (a character present in the *diastema* group and other species groups of *Craugastor*) in the *podiciferus* subgroup and subtelocentric in the *bransfordii* subgroup (synapomorphic). The latter also has a feeble non-NOR secondary constriction in the short arm (lost in *E.* sp. B). Except for the double C-bands in pair 7, there are many interstitial C-bands in the proximal regions in the *podiciferus* subgroup; these are rare or absent (also present in the *diastema* group and other species groups of the *Craugastor*) in the *bransfordii* subgroup.

In the *podiciferus* subgroup, both karyomorphs of *Eleutherodactylus podiciferus* share the double C-bands in the long arms of pair 7 (synapomorphy), which are in the short arm (plesiomorphy, also present in the *bransfordii* subgroup) in *E.* sp. A.

In the *bransfordii* subgroup, *Eleutherodactylus stejnegerianus* is the sister of both the La Selva and Tapanti karyomorphs of *E. bransfordii,* and the *E.* sp. B group is the sister of both *E. bransfordii* and *E. stejnegerianus.* The karyotype of the 20-karyomorph of *E.* sp. B has a unique extra heterochromatinized supernumerary chromosome pair 10. Both karyomorphs of *E.* sp. B share characters, the interstitial NOR-correlated secondary constrictions in the short arms of pair 1, and the absence of the secondary constrictions in both pairs 4 and 5. The karyotype of the Tapanti karyomorph of *E. bransfordii* has a unique character, the terminal C-band in the short arm of pair 1 that is NOR-correlated. A unique karyomorph feature of *E. stejnegerianus* is the very short arm of pair 5, with an average value for arm ratio of 6.34. Both karyomorphs of *E. bransfordii* and *E. stejnegerianus* share the same secondary constrictions in pairs 3 and 5. Those in pair 3 are NOR-correlated (except for the Tapanti karyomorph of *E. bransfordii*). This phenogram is convincing because it supports the view that pair 10 in the 20-karyomorph of *E.* sp. B are supernumerary chromosomes.

EVOLUTION OF THE NOR-CARRYING CHROMOSOMES IN THE *RHODOPIS* GROUP

The karyotypes of all sampled species of the *rhodopis* group from Lower Central America and that of *Eleutherodactylus loki* reported as *E. rhodopis* by DeWeese (1976) from Guatemala are unique and cannot be confused with those of any other species group. Although most species of the *rhodopis* group from Mexico and Nuclear Central America remain to be studied, I predict that they all have the model karyotype of the *rhodopis* group. Although patterns of the depressor mandibulae muscles are varied in some Mexican species, which may challenge the monophyly of the *rhodopis* group, all of my specimens except for the new forms from Lower Central America have the muscle formula dfsq*at (Savage et al. 1996). The monophyly of Lower Central American *rhodopis*-like frogs is fully supported by my analysis.

There are three kinds of NOR-carrying chromosomes in the *bransfordii* sub-group (fig. 5.5). The NORs in the secondary constriction of pair 3 (type A) in the La Selva karyomorph of *Eleutherodactylus bransfordii* and *E. stejnegerianus* may be the most primitive. Those in pair 1 (types B and C) are probably derived. The NORs located in pair 1 are restricted to the *bransfordii* subgroup, but those located on pair 3 also appear in the *podiciferus* subgroup.

For both sites of the NORs in pair 1, I consider the interstitial NORs in the short arms of pair 1 (type C), as in *Eleutherodactylus* sp. B, as derived. Those in the distal ends of the short arms of pair 1 (type B), as in the Tapanti karyomorph of *E. bransfordii,* are primitive. Therefore, the evolution of the NOR-carrying chromosomes are hypothesized as the NORs being translocated from chromosome pair 3 (type A) to the distal ends of pair 1 (type B). In turn, the interstitial NORs in pair 1 (type C) were derived by paracentromeric inversions. In this situation, the *E.* sp. B group is the sister of the Tapanti karyomorph of *E. brans-fordii,* and a clade with the La Selva karyomorph of *E. bransfordii* and *E. stejne-gerianus* is the sister to *E.* sp. B plus the Tapanti karyomorph of *E. bransfordii.*

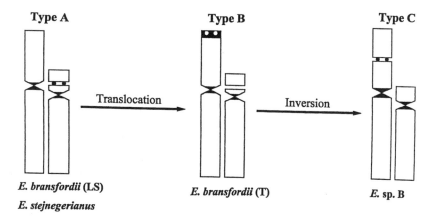

Figure 5.5   Types and hypothetical evolution of the NOR-carrying chromosomes in the *bransfordii* subgroup (*rhodopis* group). Type A has a normal pair 1, and the NORs are located in the secondary constrictions in pair 3. Type B has NORs located at the distal end of pair 1 and a non-NOR secondary constriction in pair 3. Type C has the NORs in the middle of the short arm of pair 1, but secondary constriction was not found in pair 3. Because type A is also present in the *podiciferus* subgroup (a sister taxon of the *bransfordii* subgroup), it is considered the most primitive, and type C is the most derived. LS, La Selva karyomorph; T, Tapanti karyomorph.

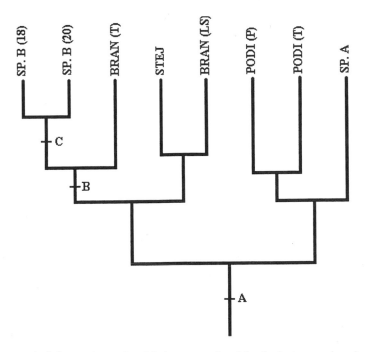

Figure 5.6   A phylogenetic tree for eight karyomorphs of the *rhodopis* group based on the polarity of types of the NORs in fig. 5.5. *E.* sp. B is the sister taxon of the Tapanti karyomorph of *E. bransfordii*, and that clade is the sister group of *E. stejnegerianus* and La Selva karyomorph of *E. bransfordii*. A, B, and C refer to karyomorphs shown in fig. 5.5. Abbreviations are the same as in fig. 5.3.

The topology of the *bransfordii* subgroup in the tree (fig. 5.6) constructed from the NORs data is different from that resulting from the similarity coefficients of the relative total lengths and arm ratios among species (fig. 5.4).

NORs of all species of the *podiciferus* subgroup belong to type A. The double C-bands in the short arms of pair 7 are hypothesized to be primitive, and those in the long arms are hypothesized to be derived.

BIOGEOGRAPHY OF THE LOWER CENTRAL
AMERICAN *RHODOPIS* GROUP

The topology of the latter tree (fig. 5.6) is congruent with geographical distributions in both subgroups. Two populations of the La Selva karyomorph of *Eleutherodactylus bransfordii* were sampled from the northeastern lowlands of Costa Rica (fig. 5.7). The only representative population of the Tapanti karyomorph of *E. bransfordii* was sampled from Tapanti, Cartago Province, Costa

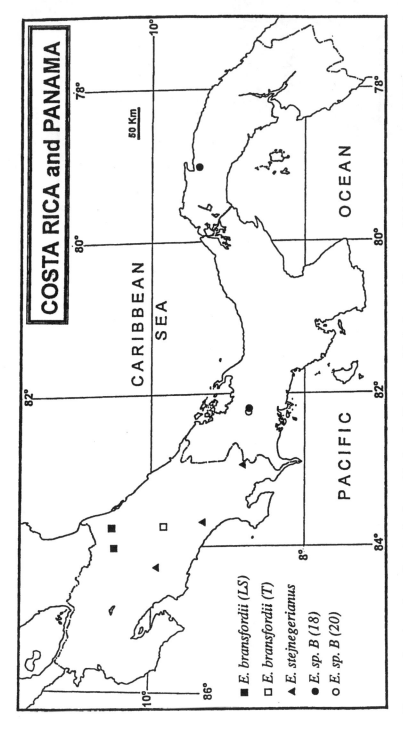

Figure 5.7   Locations of sampling sites for frogs of the *bransfordii* subgroup (*rhodopis* group) karyotyped. LS, La Selva karyomorph; T, Tapanti karyo-morph; 18, 18-karyomorph; 2o, 2o-karyomorph.

Rica, approximately 1,200 m above sea level. Two populations of *E.* sp. B were sampled from Fortuna, western Panama, approximately 1,200 m above sea level, and from Nusagandi, eastern Panama, approximately 600 m above sea level. It is possible that the *bransfordii*-like frogs distributed from western Panama (Fortuna) to eastern Panama (Nusagandi) belong to *E.* sp. B and that all those formerly reported as *E. bransfordii* from central Panama (e.g., Savage and Emerson 1970) are probably conspecific with *E.* sp. B.

Three sampled populations of the *podiciferus* subgroup (fig. 5.8) are the Palmichal karyomorph of *Eleutherodactylus podiciferus* at Palmichal, San José Province, the Tapanti karyomorph of *E. podiciferus* at Tapanti, Cartago Province, approximately 44 km west of Palmichal, and *E.* sp. A at Fortuna, Chiriqui Province, Panama, approximately 220 km southwest of Tapanti. The karyological data indicate that both karyomorphs of *E. podiciferus* are closely related and that the karyological differences could vary interpopulationally. That the karyotype of *E.* sp. A differs from that of *E. podiciferus* is consistent with the presence of nuptial pads in adult males of the former but absent in the latter. Only three species have been recognized in Lower Central America—*E. jota, E. podiciferus,* and *E.* sp. A. All three species are distributed on both slopes of the Cordillera de Talamanca and range from 830 to 2,100 m (Savage and Emerson 1970; Lynch 1980). *Eleutherodactylus podiciferus* in Costa Rica ranges to the Cordilleras Central and Tilaran (Savage and Emerson 1970), including Cerros de Escazu, where the Palmichal karyomorph was found. As in *E. bransfordii,* the taxonomic status of *E. podiciferus*–like frogs are confused by polymorphism of external morphology. Several cryptic species are probably present in the Cordillera de Talamanca (J. M. Savage, pers. comm.). Therefore, I predict that the Cordillera de Talamanca will be recognized as a speciation center of the *podiciferus* subgroup in Lower Central America.

MONOPHYLY OF THE *RHODOPIS* GROUP

The *rhodopis* group is the only species group with 18 chromosomes and one pair of NOR-correlated secondary constrictions in the short arms of one of the first five large chromosome pairs (> 10% RTL). These unique chromosome features make it difficult to determine the chromosome homology for the *rhodopis* group with other species groups of the subgenus *Craugastor.* It suggests that the *rhodopis* group is monophyletic.

Lynch (1993) described the jaw muscle (*m. depressor mandibulae*) for five species of the *rhodopis* group (*Eleutherodactylus bransfordii, E. hobartsmithi, E. mexicanus, E. pygmaeus,* and *E. rhodopis*) as having origins from the dorsal fascia, otic ramus of the squamosal, the *annulus tympanicus,* and with a dorsal flap on the squamosal (dfsq*at). Savage et al. (1996) noted that *E. lauraster, E. podiciferus,* and *E. stejnegerianus* also have the dfsq*at pattern of jaw muscles. Only 18 chromosomes are known from the *rhodopis* group. However, Lynch

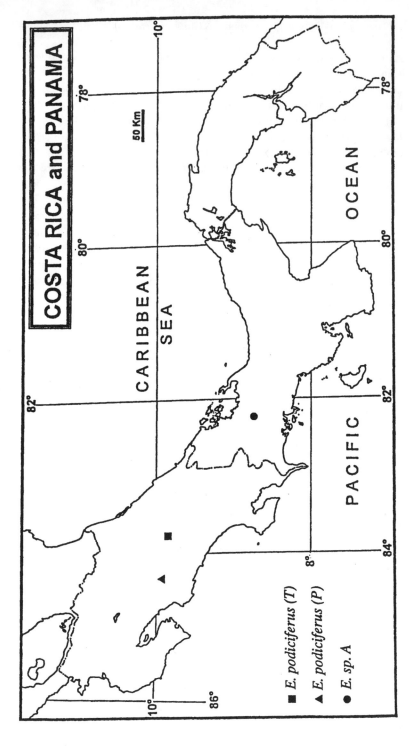

Figure 5.8   Locations of sampling sites for frogs of the *podiciferus* subgroup (*rhodopis* group) karyotyped. P, Palmichal karyomorph; T, Tapanti karyomorph.

(2000), on the basis of external similarities, assigned some confusing species containing different patterns of jaw muscles to the *rhodopis* group, for example, *E. occidentalis, E. omiltemanus,* and *E. sartori.* Among these confusing species, patterns of the jaw muscles are dfsq in *E. occidentalis* (Lynch 1993), and dfsqat in both *E. omiltemanus* (Ford and Savage 1984) and *E. sartori* (Lynch 2000). Both dfsq and dfsqat patterns are common in members of the *fitzingeri* series that have 2N = 20 or 22. According to the chromosome data, I infer that the dfsq and dfsqat patterns are plesiomorphic and that the dfsq*at pattern is apomorphic. This conclusion was reached by Savage (1987, 47). However, Savage (1987) proposed the dfsq*at status to be a synapomorphy of his lineage II. Because it is also found in the *rhodopis* group of his lineage I, the dfsq*at condition for the entire genus *Eleutherodactylus* is homoplastic.

The *rhodopis* group (sensu Lynch 2000) is probably not a monophyletic group. Lynch (2000) argued that "no characters are available to support the hypothesis that the *rhodopis* series (or group) is monophyletic," and his cladogram based on six putative synapomorphies is questionable. There are no chromosome data available for *E. occidentalis, E. omiltemanus,* and *E. sartori,* so to prevent confusion, I exclude all three from the *rhodopis* group (sensu Lynch 2000). Except for *E. omiltemanus,* I do not assign any of them to a species group until detailed chromosome data are available.

Lynch (2000) considered the *gollmeri* group (sensu Savage 1987) as diagnosed by the fusion of the eighth and sacral vertebrae. In addition to seven species of the *gollmeri* group (Savage 1987), two other species, *Eleutherodactylus greggi* and *E. daryi* of the *omiltemanus* group (Ford and Savage 1984), showed this sacral fusion and were transferred to the *gollmeri* group by Lynch (2000). He suggested deletion of the *omiltemanus* group and transferred the only remaining species, *E. omiltemanus,* to the *rhodopis* group because of external similarity. However, the monophyletic status of the *gollmeri* group (sensu Lynch 2000) has been challenged by my chromosomal data (S.-H. Chen, unpub. data). It is premature to split the *omiltemanus* group as Lynch (2000) did. I suggest that the *omiltemanus* group be retained.

Therefore, the *rhodopis* group sensu stricto can be defined as having the following: (1) the normal karyotype of 2N = 18 and NF = 36, except for species having supernumerary chromosomes; (2) double C-bands in chromosome pair 7; (3) NOR-correlated secondary constrictions on the short arms of one of the first five large chromosome pairs (> 10% RTL); (4) the last four chromosome pairs (6–9) metacentric; and (5) dfsq*at + "e" pattern of the jaw muscles.

Contents of the *rhodopis* group sensu stricto are categorized as follows:

rhodopis group sensu stricto (12 spp.):
   bransfordii subgroup (3 spp.):
      E. bransfordii, E. stejnegerianus, and *E.* sp. B

podiciferus subgroup (3 spp.):
  E. jota, E. podiciferus, and *E.* sp. A
not assigned to a subgroup (6 spp.):
  E. hobartsmithi, E. lauraster, E. loki, E. mexicanus, E. pygmaeus,
  E. rhodopis

PHYLOGENETIC RELATIONSHIPS
OF THE *RHODOPIS* GROUP

The similarities between *rhodopis*-like and *gollmeri*-like frogs led herpetologists frequently to confuse or erroneously identify frogs of these groups. Savage (1987) made a thorough revision of the *gollmeri*-like frogs and considered the *rhodopis* group to be closely related to the *gollmeri* group. However, he considered the *rhodopis* group to be a monotypic group of the *rhodopis* series within his lineage 2 (Savage 1987, 27). It was also placed as the sister group to all other members of *Craugastor* lineage 1 (including the *gollmeri* group) plus *Hylactophryne* (Savage 1987, 27). Morphologically, I agree with the close relationship between the *rhodopis* and *gollmeri* groups because the juvenile of *E. gollmeri* (< 20 mm) superficially resembles a medium or large *E. podiciferus.* In addition, unpublished allozyme data (Miyamoto 1981) and my karyological data support a close phylogenetic relationship between the *rhodopis* and the *gollmeri* groups (S.-H. Chen, unpub. data).

SUPERNUMERARY CHROMOSOMES

Supernumerary chromosomes are not common in amphibians. Only 17 species in three families of Caudata and 7 species in four families of Anura exhibit this feature (Green 1991). For the first time, two supernumerary chromosomes in the 20-karyomorph of *Eleutherodactylus* sp. B are reported for the Leptodactylidae.

The supernumerary chromosome is any chromosome in a karyotype that is not required for normal development (Green 1991). This does not mean that a supernumerary does not carry genes. For example, the supernumerary chromosomes may carry sex determination genes in the frog *Leiopelma hochstetteri* (Green 1988, 1990), or carry NOR genes in the salamander *Dicamptodon tenebrosus* and the frog *Scaphiopus hammondii* (Green 1990). However, most supernumerary chromosomes are composed of heterochromatin that is devoid of any genetic activity (John 1981; Green 1990). This is confirmed by the C-banded karyotype of my frogs in which two supernumerary chromosomes are totally heterochromatinized with dark or pale C-banding heterochromatins. Judging from their size, they are probably derived from pair 9 of the normal chromosome sets by aneuploidy. The two additional chromosomes probably underwent Muller's Ratchet (Green 1990, 1991), leading to heterochromatization, gene inactivation, gene multiplication, and gene loss to form the supernumerary chromosomes. Therefore, the supernumerary chromosomes are mostly C-banded and smaller than their progenitor chromosomes. Although the supernumerary chromosomes do not pair with normal chromosomes and are normally random in number (John 1981), two supernumerary chromosomes in my frogs often pair with each other in meiosis I (S.-H. Chen, pers. obs.). The number of supernumerary chromosomes in each frog is either 0 or 2 and is not sex specific. However, no frog was found to have only one supernumerary chromosome, which could be the result of small sample size. Their function could be minimized or completely lost. Function needs to be tested in the future.

Conclusions

Twelve karyomorphs, including five nominal species of the *rhodopis* group and four species of the *diastema* group from Lower Central America, were studied cytologically. The gross chromosomal data of *Eleutherodactylus stejnegerianus*, *E.* sp. A, and *E.* sp. B (the *rhodopis* group) and *E. diastema, E. vocator,* and *E.* sp. E (the *diastema* group) were reported for the first time. All C-banded and silver-stained karyotypes for Lower Central American *Eleutherodactylus* are reported for the first time. The chromosome data and bands of the karyotypes are phylogenetically informative and species-specific.

Two species whose karyotypes had been variously reported (Savage 1987, 43) are corrected: *Eleutherodactylus* sp. D (reported as *E. diastema*) and *E. bransfordii* each are 2N = 18 and NF = 36.

Except for the 20-karyomorph of *Eleutherodactylus* sp. B and *E. vocator*, all have 2N = 18 and NF = 36. Pair 7 of male *E. vocator* (the *diastema* group) is heteromorphic and may be sex chromosomes, but this needs to be confirmed by studying the female karyotype. Pair 10 of the 20-karyomorph of *E.* sp. B has supernumerary chromosomes (B chromosomes) and is the first record of supernumerary chromosomes in the family Leptodactylidae. It is hypothesized that they evolved from pair 9 by aneuploidy and were subject to Muller's Ratchet to form typical heterochromatinized supernumerary chromosomes.

Heterochromatin plays a major role in the repatterning of the chromosomes. All karyotypes have distinctive centromeric C-bands. No interstitial C-bands were found in members of the *diastema* group. Few interstitial C-bands were found in the karyotypes of the *bransfordii* subgroup, which are more distinctive in the *podiciferus* subgroup. The double C-bands in pair 7 are inferred to be synapomorphies of the *rhodopis* group. Three types of NOR-carrying chromosomes are unique to the *rhodopis* group.

A model karyotype for each karyomorph in each species group was established by comparing the gross chromosome data (RTLs and NF values), C-bands, and Ag-NORs, among groups. The phylogeny constructed by comparing the model karyotypes and marker chromosomes among the most closely related species proved to be useful. This study is the first to use chromosome data and banding karyotypes to study the phylogenetic relationships among species in Lower Central American eleutherodactyline frogs.

The present study recognized that the *rhodopis* group sensu stricto is monophyletic with two subgroups, the *bransfordii* and *podiciferus* subgroups. *Eleutherodactylus occidentalis, E. omiltemanus,* and *E. sartori* were excluded from the *rhodopis* group because of different jaw muscle conditions. The *omiltemanus* group (sensu Ford and Savage 1984) was retained and awaits further cytogenetic studies.

Two subgroups, the *vocator* and *diastema* subgroups, are recognized in the *diastema* group. *Eleutherodactylus vocator,* with an XX-XY sex-determining mechanism, is the only species in the *vocator* subgroup, whereas *E. diastema, E.* sp. D, and *E.* sp. E are in the *diastema* subgroup. The phylogenetic relationships among species of the *diastema* group need further study.

Appendix 5.1   Specimens Examined
for the Present Cytological Studies

Abbreviations indicate museum collections of the Costa Rica Expeditions in the Natural History Museum of Los Angeles County (CRE) and personal field series (USC).

rhodopis Group (Subgenus Craugastor)

Eleutherodactylus bransfordii (Cope 1886). *La Selva karyomorph.*—Costa Rica: Heredia Province: La Selva, alt. 30 m, USC 9059 (CRE 10238), USC 9062 (CRE 10241); Limón Province: Sardina de Guapiles, between Río Sardina and Río Chirriposito, alt. 20 m, USC 9277 (CRE 10288), USC 9314. *Tapanti karyomorph.*—Costa Rica: Cartago Province: Tapanti, alt. 1,200 m, USC 8937 (CRE 10133), USC 9082 (CRE 10258), USC 9084 (CRE 10260).
Eleutherodactylus podiciferus (Cope 1876). *Tapanti karyomorph.*—Costa Rica: Cartago Province: Tapanti, alt. 1,200 m, USC 9079–80 (CRE 10255–56). *Palmichal karyomorph.*—Costa Rica: San José Province: Palmichal, along Río Tabarcia, alt. 1,350 m, USC 9047 (CRE 10229), USC 9051 (CRE 10233).
Eleutherodactylus stejnegerianus (Cope 1893).—Costa Rica: San José Province: Palmichal, along Río Tabarcia, alt. 1,350 m, USC 9055 (CRE 10227); San Juan de S. Isidro, alt. 950 m, USC 9023 (CRE 10216); Puntarenas Province:

Las Cruces, 5 km south of San Vito, alt. 1,200 m, USC 9007 (CRE 10173), USC 9208.

Eleutherodactylus sp. A.—Panama: Chiriqui Province: Cunca de la Fortuna, 2 km north of Centro de Investigaciones Tropicales, alt. 1,350 m, USC 9365–67 (CRE 10342–44); trail along the Quebrada Aleman, 1 km south of Centro de Investigaciones Tropicales, alt. 1,200 m, USC 9473 (CRE 10565), USC 9477–78 (CRE 10567–68).

Eleutherodactylus sp. B. *18-karyomorph.*—Panama: Chiriqui Province: Cunca de la Fortuna, Centro de Investigaciones Tropicales, alt. 1,300 m, USC 9373 (CRE 10350), USC 9392 (CRE 10376); trail to the Río Hornitos, near Quebrada Nelson, alt. 1,200 m, USC 9451–52 (CRE 10527–28), USC 9456 (CRE 10531); trail along the Quebrada Aleman, 1 km south of Centro de Investigaciones Tropicales, alt. 1,200 m, USC 9557 (CRE 10574); Kuna Yala: Nusagandi, alt. 350 m, USC 9511 (CRE 10599). *20-karyomorph.*—Panama: Chiriqui Province: Cunca de la Fortuna, 2 km north of Centro de Investigaciones Tropicales, alt. 1,350 m, USC 9370–72 (CRE 10347–49), USC 9374 (CRE 10351); trail to Río Hornitos, near Quebrada Nelson, alt. 1,200 m, USC 9455 (CRE 10532), USC 9559–60.

diastema Group *(Subgenus Eleutherodactylus)*

Eleutherodactylus diastema (Cope 1875).—Panama: Kuna Yala: Nusagandi, alt. 350 m, USC 9534 (CRE 10613); Darien Province: Inrenare Estacion Pirre, alt. ~100 m, USC 9537–38, USC 9545.

Eleutherodactylus vocator Taylor 1955.—Costa Rica: Puntarenas Province: Las Cruces, 5 km south of San Vito, alt. 1,200 m, USC 9016 (CRE 10147), USC 9274, USC 9275 (CRE 10287).

Eleutherodactylus sp. D.—Costa Rica: Limón Province: Río de Guacimo, alt. 130 m, USC 8934; Heredia Province: Cariblanco, alt. 1,000 m, USC 9068 (CRE 10245); Tapanti, alt. 1,200 m, USC 9075 (CRE 10252); Limón Province: Sardina de Guapiles, between Río Sardina and Río Chirriposito, alt. 20 m, USC 9309 (CRE 10301).

Eleutherodactylus sp. E.—Panama: Kuna Yala: Nusagandi, "Ina Igar," near the station, alt. 350 m, USC 9510 (CRE 10589), USC 9518–19 (CRE 10590–91).

Acknowledgments

I express my deepest appreciation to Jay M. Savage for his long-term support, help, discussion, and understanding, and for leading me to his magical New World rainfrogs, *Eleutherodactylus.* I thank Peter V. Luykx and Ken R. Spitze

(University of Miami) and Michael M. Miyamoto (University of Florida) for their valuable comments and positive criticisms.

Special thanks to Federico Bolaños (Universidad de Costa Rica, Costa Rica) and Roberto Ibáñez (Smithsonian Tropical Research Institute, Panama) for organizing and accompanying the field collections in Costa Rica and Panama, respectively, which made my visits to these countries fruitful and enjoyable. The field collections were assisted by Andrew Malk, Rafael Lucas Rodriguez, Roger Bolaños, Miguel A. Rodriguez, Felix A. Garcia, and Deborah A. Clark.

Chromosome preparations were undertaken at the University of Miami (Jay M. Savage), Universidad de Costa Rica (Pedro León), Costa Rica; Smithsonian Tropical Research Institute (Stanley Rand and Roberto Ibáñez), Panama; and in the National Taiwan Normal University, Taiwan, R.O.C. (Jenn-Che Wang and Lu-Hsi Chang). In addition, the Organization for Tropical Studies in Costa Rica and Smithsonian Tropical Research Institute in Panama provided facilities and assistance during my field trips. The bone marrow method was kindly taught by Dr. Michael Schmid. My classmates Karen Lips, Ana R. Young-Downey, Kirsten E. Nicholson, David Bickford, Rob Burgess, and many other graduate students provided different sorts of help, for which I am grateful.

I thank Drs. Michael S. Gaines and Julian C. Lee (University of Miami, Florida) for their support of my graduate study. I also thank all colleagues in the Department of Biology, National Taiwan Normal University, for their encouragement and support. The project was funded in part by the Ministry of Education (Taiwan, R.O.C.) and National Taiwan Normal University (Taiwan, R.O.C.), and the University of Miami. Travel funds were partially supported by grants awarded to Dr. Jay M. Savage.

Funds from the American Society of Ichthyologists and Herpetologists, the Herpetologists' League, and the Society for the Study of Amphibians and Reptiles supported my travel to Baja California in 2000 to participate in the symposium where part of this work was presented.

6

The Physiological Basis of Sexual Dimorphism: Theoretical Implications for Evolutionary Patterns of Secondary Sexual Characteristics in Tropical Frogs

Sharon B. Emerson

The secondary sexual characteristics of anurans are among the most diverse of any vertebrate group. Fangs, glands, prepollical spines, body size, humeral crests, enlarged skulls, vocal sacs, body coloration, and nuptial pads are just a few of the many features in which the sexes may differ (Noble 1931; Duellman and Trueb 1986). Although sexual dimorphisms such as these have been well documented in frogs, there is still much to be learned about the patterns and processes of sexual dimorphism. Few studies on anurans have examined either the proximate physiological mechanisms responsible for the expression of secondary sexual characteristics or the patterns of evolution of sexual dimorphism in most clades. In this chapter I (1) discuss what is known about the physiological basis of sexual dimorphism in frogs (and other vertebrates) and (2) use that information to construct a series of new and testable hypotheses about the evolution of secondary sexual characteristics in some interesting groups of New and Old World tropical frogs. I highlight potential theoretical implications of physiological processes for understanding evolutionary patterns of sexual dimorphism (after Emerson 2000) rather than provide definitive tests of suggested hypotheses.

The discussion focuses on extragenital sexually dimorphic structures in which testosterone and dihydrotestosterone are the active metabolites. There are three reasons for this emphasis. First, androgen-mediated secondary sexual characteristics are the most commonly studied type in frogs. Second, androgens have a greater protein anabolic effect than do estrogens and that hypertrophy affects more tissues than do estrogens (Norris 1997). For example, skeletal muscle is a major site of testosterone action (Bardin and Catterall 1981), and hypertrophied muscles are common sexually dimorphic features of frogs (e.g., Oka et al. 1984). Third, androgen-mediated secondary sexual characteristics are an important component of recent sexual selection models (e.g., Hamilton and Zuk 1982; Folstad and Karter 1992).

Background

The organization–activation theory of sexual differentiation (Phoenix et al. 1959) postulated that sex steroid hormones act at two different times in ontogeny: early in development, often perinatally, to organize secondary sexual characteristics, and later around sexual maturity to activate their expression. The initial formulation of this model suggested that all sexually dimorphic behaviors and morphology required both organization and activation (Young et al. 1964). Subsequently, it has been shown that (1) different types of sexual dimorphism might depend on only organization, only activation, or both organizational and activational effects of hormones for expression (Arnold and Breedlove 1985); and (2) organizational effects are permanent whereas activational effects are temporary (e.g., Arnold and Breedlove 1985; Moore 1991; Hews and Moore 1995).

As commonly presented, the organization–activation theory emphasizes the importance of the ontogenetic stage or relative age of the *organism* for understanding steroid hormone effects (e.g., Arnold and Breedlove 1985). However, more recent endocrine work (including experiments on species from all the major vertebrate classes) has made it clear that at least for some androgen-mediated, morphological, sexual dimorphisms, it may be the relative timing of proliferation and terminal differentiation events of *cells* in tissues rather than the age of the entire organism that is critical (e.g., Marin et al. 1990). The idea is that differences among types of secondary sexual characteristics (organized or activated) may reflect the differing effects of sex steroid hormones on cells, depending on their state of differentiation (Marin et al. 1990; Tobin and Joubert 1991; Fischer et al. 1995; Joubert and Tobin 1995). Variation in cell response is mediated through the numbers and types of steroid receptors that are present in a particular cell type at different times during its ontogeny (Kelley et al. 1989; Young and Crews 1995). Receptors regulate genes that are involved in cell proliferation and differentiation, and it is well established that receptor populations shift ontogenetically (Wilson and McPhaul 1996; Fischer et al. 1998). It is probably because cells of most tissues undergo terminal differentiation relatively early in the ontogeny of an organism that the idea first developed that hormonal effects depended on the age of the organism (Emerson 2000).

Sexually dimorphic morphology can differ between males and females because of differences in cell number (hyperplasia) and/or cell size (hypertrophy) (Bardin and Catterall 1981 and references therein; Tobin and Joubert 1991; Joubert et al. 1994; Catz et al. 1995). At least some androgen-controlled features that result from the organizational actions of hormones appear to vary between the sexes because of differences in cell number (e.g., Sassoon et al. 1986; Marin et al. 1990). These differences in cell number are the result of androgen-mediated proliferation events that take place relatively early in the development of a tissue, preceding terminal differentiation of the cells (Sassoon et al. 1986; Fischer et al. 1995). These hyperplasic effects are permanent. An example of this type of secondary sexual characteristic is the cartilaginous larynx of male *Xenopus* (Kelley 1996).

Conversely at least some androgen-controlled features that result from the activational actions of hormones appear to vary between males and females because of differences in cell size (Greenberg 1942; Hews and Moore 1995; Cooke et al. 1999). These differences in cell size reflect a hormone-mediated, protein anabolic effect on tissue whose cells have already undergone terminal differentiation (Bardin and Catterall 1981). These hypertrophic effects are temporary and occur in adults. Male salamander dentition type (Noble and Pope 1929), frog throat color (Greenberg 1942), tree lizard femoral pores (Hews and Moore 1995), moorhen face shields (Eens et al. 2000), and the rat posterodorsal nucleus of the medial amygdala (Cooke et al. 1999) are examples of this type of sexual dimorphism.

Finally, where it has been examined, those androgen-mediated secondary sexual characteristics that vary between males and females both in cell number and cell size appear to be the result of both organizational and activational actions of hormones. These actions have taken place during two different times of tissue development: during the early, proliferative phase of cell increase and after terminal differentiation of tissue cells (Joubert et al. 1994; Fischer et al. 1995). Such morphological characteristics exhibit permanent differences between males and females that are the result of hyperplasia. Additionally, those differences can be temporarily exaggerated (through hypertrophy) in the terminally differentiated tissue by elevated adult hormone levels. Examples of this type of secondary sexual characteristic are the enlarged forelimb muscles of many breeding male frogs (Muller et al. 1969; Melichna et al. 1972; Oka et al. 1984; Dorlochter et al. 1994; Blackburn et al. 1995), the vocal sacs of male frogs (Greenberg 1942; Hayes and Menendez 1999), and the combs of chickens (Witschi 1961).

In summary, organizational and activational effects of androgens appear to differ, at least in part, because they target different steps in the cell cycle. Organizational effects of androgens reflect a hormone action that takes place during the proliferative stage of tissue development before terminal differentiation of the cells. As a result, organizational effects of androgens produce hyperplasy or an increase in cell number. Such increases in cell number are usually permanent (but for a possible exception, see Smith et al. 1997). Organizational effects of androgens usually occur early in the ontogeny of an organism because that is the most common time of tissue differentiation. In contrast, activational effects of androgens reflect a hormone action that takes place after cell differentiation of a tissue. As a result, activational effects of androgens lead to hypertrophy or an increase in cell size. These increases in cell size are not permanent but fluctuate with shifting androgen levels such as occur between breeding and non-breeding sexually mature male anurans (Delgado et al. 1989).

Linking Physiology and Evolution

The facts that sexually dimorphic morphologies can (1) arise through diverse physiological processes and (2) have different growth characteristics suggest a number of new questions about the evolution of sexual dimorphism. There follow three examples in tropical frogs of interesting hypotheses that emerge from a consideration of these basic biological properties.

EVOLUTIONARY BIAS IN THE GAIN OR LOSS OF SECONDARY SEXUAL CHARACTERISTICS

An increasing number of comparative evolutionary studies map secondary sexual characteristics on phylogenetic hypotheses to examine patterns and predictions of sexual selection theory (e.g., Emerson 1994). An implicit assumption of most of these studies is that the sexual dimorphisms share a common physiological basis. Clearly this does not have to be the case. One possible consequence of the diversity of processes producing sexual dimorphisms is that secondary sexual characteristics produced by different mechanisms may evolve at different rates. Generally, loss of any character is thought to occur more easily and frequently than character gain (Hecht and Edwards 1976). I suggest that, in addition, features produced by hypertrophy and the activational effects of hormones evolve more easily or are lost more easily than are features produced by hyperplasy and the organizational effects of hormones or vice versa. In this chapter I provide preliminary examination of a few sexually dimorphic traits in one clade of tropical frogs to illustrate how one might begin to set up a test of this intriguing hypothesis.

Nuptial pads, vocal slits, and vocal sacs are sexually dimorphic characters that are common in frogs and have evolved or been lost numerous times within clades across families (e.g., Zweifel 1955; Inger 1966; Liem 1970; Lynch 1971; McDiarmid 1971 and references therein) (see fig. 6.1 for an example in one group of *Eleutherodactylus*). Experimental work indicates that nuptial pads and vocal

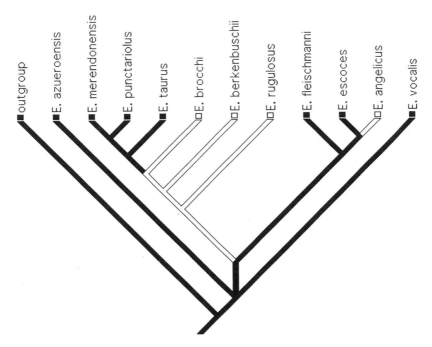

Figure 6.1   Phylogenetic hypothesis of the relationships among members of the *Eleuthero-dactylus rugulosus* species group (after Miyamoto 1983b). The presence (black)/absence (white) of vocal slits has been optimized on the phylogeny. Characters are from Savage 1975.

sacs are mediated (at least in part) by the activational effects of androgens (Emerson et al. 1997; Hayes and Menendez 1999 and references therein). In contrast, other secondary sexual characteristics such as fangs have evolved fewer than ten times across all 4,500 species of frogs (Emerson 1994). These structures are permanent and often begin development before sexual maturity (Emerson and Voris 1992; Katsikaros and Shine 1997), suggesting that they may be mediated by the organizational effects of androgens. In the Southeast Asian fanged frogs belonging to the genus *Limnonectes*, a difference in the evolutionary lability of the sexually dimorphic traits that are the result of activational and (possible) organizational effects of hormones is apparent (fig. 6.2). Those characters mediated by the activational effects of androgens appear to evolve more frequently. Vocal sacs (which are lost at the base of the clade) re-evolve at least three times independently in the clade (fig. 6.2A). Nuptial pads have re-evolved at least once independently as well. In contrast, fangs, which may be the result of the organizational effects of hormones, never re-evolve within the clade even

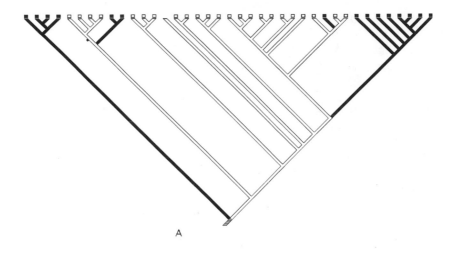

A

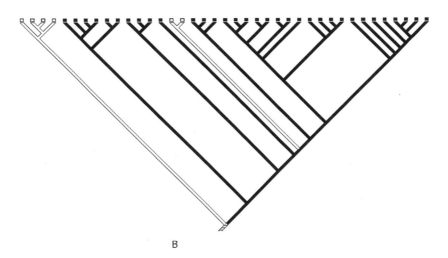

B

Figure 6.2   Phylogenetic hypotheses for the fanged frogs of Southeast Asia. (A) The presence (black)/absence (white) of vocal sacs has been optimized on the phylogeny. (B) The
presence (black)/absence (white) of fangs has been optimized on the phylogeny. Hatching
indicates equivocal character state.

though they are lost in one species group (fig. 6.2B). This example and the common loss and gain of nuptial pads (e.g., Savage 1975) and vocal sacs (e.g., Hayes and Krempels 1986 and references therein) in most frog groups suggest the possibility that these secondary sexual characteristics that result from the activational effects of hormones may be more susceptible to evolutionary gain than those that are under other types of control. To test this hypothesis more broadly, however, we need additional examples of secondary sexual characteristics mapped onto independently derived phylogenies, as well as experimental studies designed to measure how hormones affect anuran secondary sexual characteristics.

INTRASEXUAL COMPETITION AND HYPERPLASIC SEXUAL DIMORPHISMS

Hyperplasic growth provides a greater capacity for structural enlargement than does hypertrophy, where surface-to-volume relationships limit the size individual cells can attain. Some game theory models (Parker 1983) examining male contest competition over females have found that the more intense the male–male competition, the higher the investment in (or the larger the size of) the male armament. This suggests the possibility that male secondary sexual characteristics used in intense intrasexual competition might be (1) likely to have some component of hyperplasic growth and (2) unlikely to be the result of only hypertrophic enlargement. Selection for large-sized armaments could even translate to a bias for sexual dimorphisms that were hyperplasic in origin and thus capable of attaining absolutely larger sizes. In the following paragraphs, I discuss a few examples consistent with the hypothesis that intense male–male competition involves structures that arise, at least in part, through hyperplasy.

Intense male–male competition has been recorded in the gladiator frog, *Hyla rosenbergi*. Male gladiator frogs have prepollical spines that they use to inflict eye and tympanum injuries on other males (Kluge 1981). These structures begin to develop before sexual maturity in males, grow by cell addition (hyperplasy), and are permanent structures. Male tusked frogs, *Adelotus brevis,* also display intense male–male competition, their battles involving vigorous biting (Katsikaros and Shine 1997). Once again, the tusks of the males develop before sexual maturity, grow by cell addition (hyperplasy), and are permanent structures (Katsikaros and Shine 1997).

Three species of rhacophorid frogs have unusual secondary sexual characteristics. The males of *Rhacophorus dulitensis* have flaps of skin that cover the vent area (Inger 1966). *Polypedates otilophus* has a razor-sharp crest located behind the eye (Inger 1966). This sharp, blade-like structure can cut through plastic bags (S. B. Emerson, pers. obs.). In *Boophis albilabris* the males have a ventrally placed vent (Cadle 1995). These three species also have relatively large

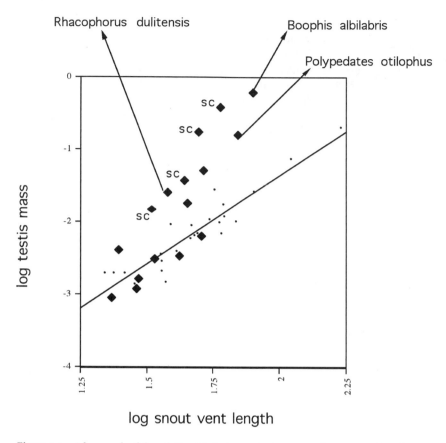

Figure 6.3   A log graph of the relationship between testis mass and snout-vent length for different species of frogs (data from Emerson 1997). Small dots represent values for species in the family Ranidae. The line represents a reduced major axis regression fit to those points. Squares represent values for species belonging to the family Rhacophoridae. Species known to have sperm competition are indicated by SC (next to the square). The squares below the line are species in the rhacophorid genus *Philautus*. These species have an abbreviated, nonfeeding larval stage and a reduced number of eggs (Inger 1966).

testes (fig. 6.3), suggesting that they engage in intense intrasexual competition in the form of sperm competition (Feng and Narins 1991; Kusano et al. 1991; Jennions and Passmore 1993; Emerson 1997). Although the biological role of these unusual secondary sexual characteristics is unknown, I hypothesize that they are features involved in intrasexual competition. I suggest they function to improve the probability of successful insemination in the presence of other males. The prediction for these characteristics (if) involved in intense intrasexual competition is that they would be the result of hyperplasic growth. Although the details of development are not known for any of the characters, the vent flap of *Rhacophorus dulitensis* and the sharpened crest of *Polypedates otilophus* are permanent structures that begin developing before sexual maturity through cell addition rather than cell enlargement.

Many tropical frogs have interesting and unique male secondary sexual characteristics such as those already mentioned. Unfortunately it may be some time before the actual physiological processes mediating the expression of these sexually dimorphic characters are experimentally established. In the meantime a broad interspecific survey of the secondary sexual characteristics in male frogs could test for a correlation of the presence of intense intrasexual competition with the permanence of sexually dimorphic structures.

FEMALE CHOICE AND HORMONALLY MEDIATED SEXUAL DIMORPHISMS

A current controversy in sexual selection centers around the applicability of good genes or indicator models such as that of parasite-mediated selection (Hamilton and Zuk 1982) and the immunocompetence handicap hypothesis (Folstad and Karter 1992). These models are hypotheses about the reliability of signals (reviewed in Andersson 1994). Male secondary sexual characteristics can show considerable phenotypic variation among individuals. Indicator sexual selection models posit that these characteristics may be condition-dependent and reflect phenotypically the quality of a male. According to these models, the female bases her choice of a male on the information provided by the condition-dependent sexual dimorphism. In animals that have repeated reproductive bouts, like many tropical frog species do, condition presumably varies continuously, and the most reliable signals would be those that could reflect those ongoing shifts in adult vigor. For such models to work over the time scale of repeated reproductive bouts, the sexually dimorphic features would have to respond to changing adult hormone levels. That is, the secondary sexual characteristics would have to be the result (at least in part) of activational actions of hormones because only that class of secondary sexual characteristics responds to changes in adult levels of circulating hormones.

Sexual selection by female choice has been demonstrated for some species of anurans (reviewed in Sullivan et al. 1994). In the majority of cases it appears that the females base their decisions on information provided in the advertisement call of the adult male frogs (e.g., Gerhardt and Watson 1995; Welch et al. 1998). Experiments indicate that depending on species, females may prefer higher-intensity, longer-duration, and/or more frequently pulsed calls (reviewed in Runkle et al. 1994; Sullivan et al. 1994; Welch et al. 1998 and references therein). Female choice based on information encoded in the intensity, frequency, and/or duration of male vocalization could be consistent with indicator models of sexual selection (Gerhardt 1991). If this were the case, I predict that those properties of male vocalization would be mediated by the activational effects of androgen hormones and thus capable of reflecting current condition of the male.

Androgen levels vary widely both within and between tropical species of calling male frogs (Emerson and Hess 1996). Moreover, field studies have shown a correlation between calling rates and plasma androgen levels in actively breeding male frogs (Townsend and Moger 1987; Marler and Ryan 1996; Solis and Penna 1997; Emerson and Hess 2001). In parallel, physiological work has demonstrated seasonal changes in the contractile properties of the oblique muscles in temperate frogs that correlate with shifts of androgen levels that occur between breeding and nonbreeding male frogs (Das Munshi and Marsh 1996).

Male advertisement calls are produced by contraction of the external and internal oblique muscles (Pough et al. 1992; Girgenrath and Marsh 1997). These muscles are sexually dimorphic, with males having much larger muscles (Pough et al. 1992). The cross-sectional area of the oblique muscles determines the force of contraction (Peters and Aulner 2000) and thus the intensity of the call. The myosin isozyme of the muscles determines the velocity of muscle contraction (Lannergren 1987; Edman et al. 1988) and the pulse rate of the call (Girgenrath and Marsh 1997). Both muscle size and myosin isozyme type are mediated, at least in part, by the activational effects of androgens (e.g., Gibbs et al. 1989; Catz et al. 1992, 1995; Regnier and Herrera 1993; Blackburn and Bernardo 1998; Emerson et al. 1999). These results are consistent with the prediction for secondary sexual characteristics that are postulated to reflect current condition; that is, they would respond to changes in adult circulating hormone concentrations.

General Implications

Secondary sexual characteristics can arise through cell addition or cell enlargement. Those characteristics that arise through hyperplasy appear to be mediated by the organizational effects of hormones, are permanent structures, and do not respond to adult changes in hormone levels. Those characteristics that arise through hypertrophy are mediated by the activational effects of hormones, fluctuate in structure, and can respond to adult changes in circulating hormone levels.

The differences in the physiological mechanisms responsible for the expression of secondary sexual characteristics may lead to a bias in the evolution of sexual dimorphism. Preliminary data suggest that in frogs, sexual dimorphisms mediated by the activational effects of androgens may be more evolutionarily labile than are secondary sexual characteristics mediated by the organizational effects of hormones.

Secondary sexual characteristics that are involved in intense intrasexual competition are likely to be, at least partially, the result of hyperplasy. Secondary sexual characteristics that are used as indicators of organism quality are more likely to be the result, at least in part, of the activational effects of hormones.

7

Higher-Level Snake Phylogeny as Inferred from 28S Ribosomal DNA and Morphology

Mary E. White, Maria Kelly-Smith, and Brian I. Crother

Early attempts to resolve the evolutionary relationships among families and superfamilies of snakes by use of morphological data mostly disagree with each other, although they find some common ground (e.g., Underwood 1967; Rieppel 1979, 1988; McDowell 1987; Rage 1987). More recent cladistic analyses (Kluge 1991; Cundall et al. 1993; Tchernov et al. 2000) have inferred essentially the same higher-level hypotheses of relationships (fig. 7.1). In general they agreed on a monophyletic Scolecophidia, a paraphyletic Anilioidea, a paraphyletic Booidea, a monophyletic Tropidopheoidea (except for Tchernov et al. 2000), and a monophyletic Colubroidea, at least as these superfamilies were understood by McDowell (1987). As for general relationships, the Scolecophidia were sister to the Alethinophidia, with the latter exhibiting various basal positions of anilioids, followed by booids, tropidopheoids, and colubroids.

Also during the 1990s, attempts at resolving higher-level snake phylogeny with molecular data began to appear. Heise et al. (1995) and Forstner et al. (1995) inferred trees from mtDNA, albeit from different regions (fig. 7.2). Both studies indicated a paraphyletic Scolecophidia, contra the result obtained from morphology. The few exemplar taxa employed in the Forstner et al. (1995) study were resolved consistently with morphology, but the Heise et al. (1995) hypothesis departed greatly from the general conclusions of Kluge (1991), Cundall et al. (1993), and Tchernov et al. (2000). Aside from the paraphyletic Scolecophidia, the Heise et al. (1995) hypothesis suggested a polyphyletic Booidea, with *Python* and *Tropidophis* in a clade with anilioid taxa and *Acrochordus*. In addition to these sequence-based studies, Saint et al. (1998) used the nuclear gene c-*mos* to infer squamate phylogeny, but the focus was on lizard–snake relationships rather than higher-level snake phylogeny. Dowling et al. (1996) attempted to employ allozymes to infer higher-level relationships. The resultant phylogeny was quite peculiar, probably because the divergence times exceeded the limits of resolution for allozyme data (Murphy et al. 1996); we do not consider the Dowling et al. hypothesis in this chapter.

As noted above (and see Zaher and Rieppel 2000), considerable debate remains on the question of snake phylogeny. Our study provides additional empirical data to test higher-level snake phylogeny hypotheses; we employ DNA sequence of the 28S region of the ribosomal DNA (rDNA). Given the ancient age of snakes (macrostomatan fossils of 100 million years ago [mya]; Greene and Cundall 2000; Tchernov et al. 2000), this region was deemed appropriate for the question. The utility and nature of rDNA have been discussed in detail many times (e.g., Mindell and Honeycutt 1990; Hillis and Dixon 1991), so we do not repeat those discussions. In this chapter we report on the inference of snake phylogeny with the 28S DNA sequence alone and in combination with the morphological data from Tchernov et al. (2000). Although we include a number of other lepidosaur taxa in this study, we only report and discuss relationships within snake taxa.

Materials and Methods

We took an exemplar approach to the inference of higher-level snake phylogeny. Included in the study are 14 taxa of snakes and 8 outgroup taxa, including *Sphenodon* (table 7.1). Whole genomic DNA was isolated, and regions of the 28S were amplified using the following primers from Hillis and Dixon (1991): 28dd, 28ee, 28ff, 28mm, 28ll, and 28gg. We also developed a new primer, 28ex (5′-ATCTGAACCCGACTCCCTTTCGA-3′), for amplification in the region near 28ee. The polymerase chain reaction (PCR) primers were purchased from Operon Technologies. The PCR products were cloned and then sequenced using a LiCor automated sequencer. Fluorescent sequencing primers were obtained from LiCor Inc. Contiguous sequences were assembled using the software Sequencer, and sequences were aligned with Clustal X (Thompson et al. 1997) and subsequently modified by eye. All phylogenetic analyses were conducted with PAUP* (Swofford 1999) unless otherwise noted.

When all the operational taxonomic units (OTUs) were included in the analysis, most-parsimonious trees were found using the random addition sequence for stepwise addition 25 times and the heuristic branch-swapping algorithm (TBR, tree bisection and reconnection). We used MULPARS to retain

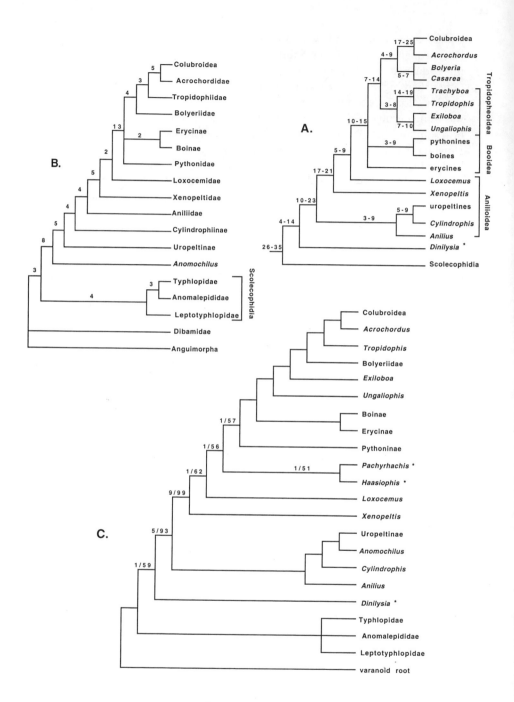

Figure 7.1 (*opposite*) Recent morphology-based hypotheses of higher-level snake phylogeny. (A) Kluge 1991, 36. The values represent "the range of variation in the amount of evidence supporting" each clade. (B) Cundall et al. 1993. Numbers above each branch represent the number of characters that unambiguously change on the branch. (C) Tchernov et al. 2000. Numbers to the left of the slash are Bremer values; to the right are bootstrap proportions out of 1,000 iterations. Asterisks (*) indicate fossil taxa.

Figure 7.2 (*below*) Molecular-based (mtDNA) hypotheses of higher-level snake phylogeny. (A) Heise et al. 1995. The numbers above each branch represent confidence probabilities as calculated in Kumar et al. 1994. (B) Forstner et al. 1995. Numbers represent the range of bootstrap proportions obtained with different outgroups.

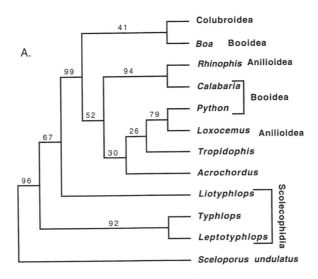

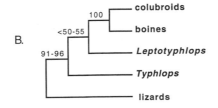

Table 7.1   Taxa included in this study and reference numbers of specimens

Taxa	Material
Ingroup	
Serpentes	
Anilius scytale	H 4362, Louisiana State University (LSU) Museum of Natural Science frozen tissue collection
Epicrates fordii	University of Southern California (USC) 7254
Heterodon platyrhinos	U.S. National Museum (USNM) 1131, gift from Ronald Crombie, USNM
Lampropeltis triangulum	Shed skin gift from Richard Seigel
Leptotyphlops dulcis	Southeastern Louisiana University (SLU) 886, shed skin
Liotyphlops sp.	USNM 186543, gift from Addison Wynn, USNM
Loxocemus bicolor	Carl S. Lieb (CSL) 6201, LSU Museum of Natural Science frozen tissue collection
Python regius	Shed skin gift from David Shepherd
Sanzinia madagascarensis	SLU 894, shed skin gift from Jennifer Pramuk, from individual at the Audubon Zoo, New Orleans
Tropidophis haetianus	Shed skin gift from Jay Wagner, Jr.
Typhlops biminiensis	USC 7871
Ungaliophis panamensis	SLU 893, shed skin gift from Jennifer Pramuk, from individual at the Audubon Zoo, New Orleans
Uropeltis liura	H 2175, LSU Museum of Natural Science frozen tissue collection
Xenopeltis unicolor	H 2582, LSU Museum of Natural Science frozen tissue collection
Outgroup	
Rhynchocephalia	
Sphenodon punctatus	Shed skin gift from Dwight Lawson, University of Texas at Arlington; specimens from Dallas Zoo colony
Sauria	
Anolis carolinensis	Tail from locally collected individual, no voucher retained
Celestus crusculus	USC 7748
Hemidactylus turcicus	Tail from locally collected individual, no voucher retained
Pygopus sp.	DNA, gift from Eric Pianka
Rhineura floridana	pdh 8801, gift from David Hillis, supplied as plasmid
Uromastyx sp.	Shed skin
Varanus bengaliensis	USC 14380, gift from Jay Savage

multiple most-parsimonious trees (mpts), and the maximum number of mpts to be retained was set at 10,000 trees (MAXTREES option). Steepest descent was off, and zero-length branches were collapsed. Accelerated transformation (ACCTRANS option) was used to optimize character state distributions. When subsets of the OTUs were analyzed, most-parsimonious trees were inferred through Branch and Bound. When multiple most-parsimonious trees were found, the trees were combined into strict consensus trees (Sokal and Rohlf 1981) to identify congruent clades and majority-rule consensus trees to identify tendencies of character covariation (Margush and McMorris 1981).

Our phylogeny inference strategy was to employ maximum parsimony (MP) in several sets of analyses. Broadly, three analyses were conducted with (1) all the taxa, DNA only; (2) the subset of taxa represented by morphological data, DNA only; and (3) the subset of taxa represented by morphological data, DNA and morphology combined. The first two analyses were to explore the phylogenetic utility of the 28S data for snakes. The third analysis was for inference of a hypothesis for snake relationships. Equal weights were used in all analyses unless dictated by a skew (not approximately 1:1) in transition/ transversion ratios. Gaps were included alternatively as missing data and as a fifth state. Standard descriptive statistics, consistency index (CI), retention index (RI), and rescaled consistency index (RC), were used to describe the fit of the data to the tree and provide rough measures of homoplasy and synapomorphy. Strength of character support of the internal nodes of the hypotheses was described by the Bremer Support Index (Bremer 1994; with TreeRot.v2, Sorenson 1999), parsimony jackknifing (Jac at 37%; Farris et al. 1996), and the nonparametric bootstrap. The latter two measures were conducted with all the data and with parsimony informative data only to examine the effects of irrelevant characters on bootstrap values as described by Harshman (1994) and Farris et al. (1996).

To check for the level of signal in the data (as inferred through the structure of the data), we measured the g_1 (Hillis 1991; Huelsenbeck 1991) for tree-length frequency distributions. The g_1 statistic is one of a number of phylogenetic signal or support statistics that have been developed (e.g., bootstrap, Felsenstein 1985; permutation tail probability [PTP], Faith and Cranston 1991). These statistics have all been criticized (e.g., Carpenter 1992; Källersjö et al. 1992; Hillis and Bull 1993; Kluge and Wolf 1993; Farris et al. 1994; Slowinski and Crother 1998), and because they all deviate from an ultimate criterion such as minimum length, perhaps their importance should be downplayed (Carpenter 1992; Kluge and Wolf 1993). Although we are fully aware of the potential for the g_1 statistic to mislead interpretation of results (Källersjö et al. 1992), we used the tree-length distribution measure herein. The g_1 was obtained for each data set from a sample of a million random trees. Each g_1 was checked for significance (Hillis and Huelsenbeck 1992). We also calculated the PTP, because although there have been criticisms (as mentioned earlier in the chapter), the measure seems at least to be able to identify the total absence of character covariation (Slowinski and Crother 1998).

After the analysis of the rDNA sequence data, we combined those data with the morphological data set of Tchernov et al. (2000) in a parsimony analysis for a total evidence inference of higher-level snake phylogeny. The taxa in this analysis were restricted to those in common between Tchernov et al. and the rDNA data to reduce missing data problems (Platnick et al. 1991; Wilkinson 1995). The hypotheses were inferred using the exact search method Branch and Bound.

Results

We collected and aligned 1,855 bases for 22 taxa: 14 snakes, 7 lizards, and *Sphenodon*. The individual sequences are available in GenBank, and the aligned sequences plus the PAUP* data files are available at our website (www.selu.edu/Academics/Faculty/bcrother/). Divergence between snakes was low, ranging from 0.28% to 2.2%, with a mean of just less than 1% (0.91%). Including the outgroup taxa, the estimated sequence divergence range increased to approximately 4% with *Sphenodon*. The transition/transversion ratios ranged from 7.0 to 0.25, and the mean was 1.39. And as expected for rDNA, the base frequencies were skewed to GC rich: A = 19.4%, C = 27.8%, G = 36.2%, T = 16.5%.

Signal, as estimated by the g_1 and the PTP, was significant at $p = .01$ for all matrices, except for the reduced DNA matrix, with only taxa represented in the morphological data. The reduced matrix was significant at $p = .05$, as interpolated from Hillis and Huelsenbeck (1992).

DNA ONLY: ALL TAXA

Parsimony analysis of all the taxa with gaps = 5th state yielded one most-parsimonious tree of 585 steps, with the following descriptive statistics: CI = .71, RI = .75, RC = .53 (fig. 7.3A). In general, snakes were monophyletic relative to the included lizard taxa, and this was the most strongly supported clade (53 characters). The (colubroid, tropidophid) booid) anilioid) scolecophidian) relationships were similar to the recent morphological hypotheses. Peculiar clades of (*Python, Loxocemus*) *Leptotyphlops*) and (*Uropeltis, Liotyphlops*) were also recovered. In general, the Jac resampling method resolved more nodes than did the bootstrap, and both methods improved (more resolved nodes, higher values) with the constant and autapomorphic characters removed (fig. 7.3B). Interestingly, in the latter analyses, both methods inferred a basal position of *Liotyphlops* to the rest of the snakes.

Parsimony analysis of all the taxa with gaps = missing yielded 40 mpts at 354 steps, with the following descriptive statistics: CI = .77, RI = .73, RC = .57. In general, the inferred tree (not shown) made little sense; for example, the colubroids were placed near the base of the snakes and the anilioids toward the top of the tree.

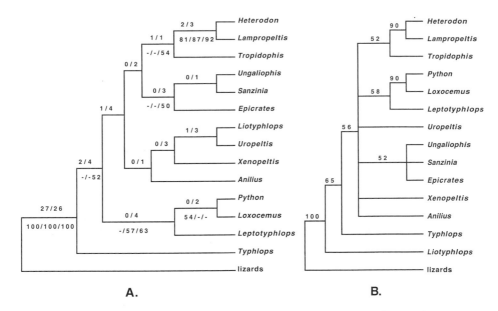

Figure 7.3 (A) Single mpt inferred from DNA sequence only, including all taxa. The numbers above the internal branches indicate total character support and type of character support: unambiguous synapomorphies/ambiguous synapomorphies. For example, the *Heterodon-Lampropeltis* clade is supported by a total of five characters, two with a CI = 1.00 and three with CI < 1.00. The numbers below each branch are bootstrap and Jac resampling proportions for all the data and parsimony-informative characters only: bootstrap all / bootstrap parsimony-informative / Jac all. (B) Jac tree with only parsimony-informative characters.

DNA ONLY: TAXA REPRESENTED BY MORPHOLOGICAL DATA

A pruned analysis (minus *Heterodon, Sanzinia, Sphenodon,* and all the lizards except *Varanus*) of only those taxa included in the morphological study of Tchernov et al. (2000) yielded one mpt at 267 steps, with gaps = 5th state (fig. 7.4). The inferred relationships were consistent with the previous analysis of all the taxa. Treating the gaps = missing yielded another nonsensical hypothesis, with *Tropidophis* and *Lampropeltis* at the base of the tree.

DNA AND MORPHOLOGY COMBINED

With gaps = 5th state, four mpts resulted at 463 steps (CI = .77, RI = .63, RC = .48). The inferred phylogeny is consistent with the morphological hypothesis of Tchernov et al. (2000), except for a paraphyletic Scolecophidia (fig. 7.5A). The

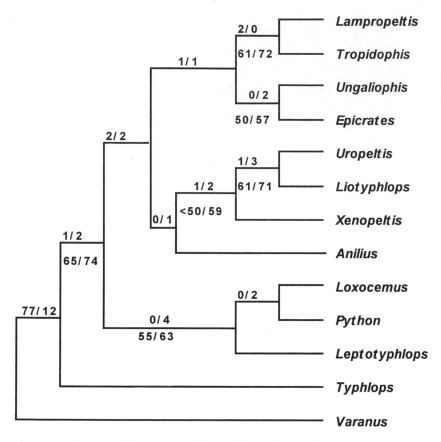

Figure 7.4   Single mpt (CI = .85 [.57], RI = .5, RC = .43) inferred from analysis of the DNA pruned to only those taxa included in the morphological data. The numbers along the internal branches indicate total character support and type of character support: unambiguous synapomorphies / ambiguous synapomorphies (see fig. 7.3). Numbers below the branches indicate bootstrap / Jac resampling proportions from 1,000 iterations. The resampling statistics were calculated with only parsimony-informative characters.

nodes are well characterized as indicated by the Bremer, bootstrap, and Jac indices, except for intra-scolecophidian relationships. However, the separation between the basal scolecophidians and the alethinophidians is well characterized and supported by 25 characters. With gaps = missing, three mpts were recovered at 347 steps (CI = .77, RI = .68, RC = .53). The strict consensus tree (fig. 7.5B) is consistent with the gaps = 5th state tree, with *Python* now unresolved, but resolving *Anilius* and *Uropeltis* as sisters and the Scolecophidia as monophyletic and sister to the rest of the snakes.

Discussion

28S DNA AND SNAKES

Hillis and Dixon (1991) pointed out that the large subunit (LSU) of nuclear rDNA is more variable than the small subunit and especially exhibits high levels of variability in its numerous divergent domain regions. Because of this variability, they noted that the LSU should be useful for phylogeny inference possibly from the Paleozoic, throughout the Mesozoic (~245–65 mya), and into the Cenozoic. Clearly, this is not the case for higher-level snake phylogeny. The mean pairwise divergence was just under 1%, and this is further reflected in the low total number of parsimony informative characters. Out of 1,855 aligned sites, 1,517 were constant, 183 were autapomorphic, and only 155 were informative. Mindell and Honeycutt (1990) reported average divergence between species pairs as 8.1% for 28S, with estimated divergence times at 85 mya. Even within a divergent domain, upon inspection it is easy to see how conserved the snake sequences are relative to lizards (fig. 7.6).

Snake fossils (*Haasiophis* and *Pachyrhachis*) dated from approximately 100 mya are considered to be the sister clade to the crown clade of booids and colubroids (Tchernov et al. 2000), and the earliest positively identified as a snake is *Lapparentophis* (considered an alethinophidian) from the early Cretaceous somewhere from 100 to 140 mya (Rage 1987). The divergence times for the major clades of snakes extends well into the earliest Cretaceous and probably well into the Jurassic (fig. 7.7). It has been argued that *Haasiophis* and *Pachyrhachis* represent the sister lineages to all extant snakes (e.g., Caldwell and Lee 1997; Lee and Caldwell 1998). This scenario, which has been effectively rebutted (Zaher 1998; Zaher and Rieppel 1999a, 1999b, 2000), would make major snake divergences much younger. But even if we assume that Lee and Caldwell's hypothesis is correct, the percentage divergences are still lower than those reported by Mindell and Honeycutt (1990) for similar times.

Why are the divergences so low? Maybe it is simply an artifact of sampling or alignment. However, with the widely divergent taxa we have included, adding taxa would, if anything, reduce divergences between lineages. With alignment, we must assume our hypotheses of homology are robust. More intriguing obvious possibilities are that snakes are much younger than we think or that the rate of change has been slow among snakes. The first hypothesis can be rejected given the age of macrostomatan fossils (Greene and Cundall 2000). The second hypothesis cannot be rejected. To estimate the rate of molecular change, we calculated the percentage divergence/mya by using {100 [1 − (no. identical sites / no. total sites)]} / divergence time (Mindell and Honeycutt 1990). The fossil record of snakes is poor, so rates of molecular change were calculated for only the few comparisons whose estimated divergence times we felt com-

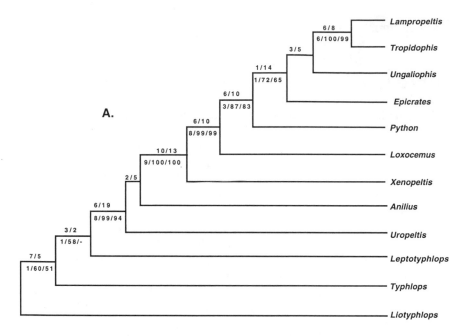

Figure 7.5   (A) (*above*) Results from the analyses of the DNA and morphological data combined with *gaps treated as characters,* showing one of four mpts. (B) (*opposite*) Results from the analyses of the DNA and morphological data combined with *gaps treated as missing,* showing the 50% majority-rule consensus tree of three mpts. Only the nodes of *Ungaliophis* and *Epicrates* were not recovered 100% of the time. In both (A) and (B), the numbers above the internal branches indicate total character support and type of character support: unambiguous synapomorphies / ambiguous synapomorphies (see fig. 7.3). The numbers below the branches are Bremer values / Jac proportions / bootstrap proportions. The Jac and bootstrap proportions are from 1,000 iterations with only parsimony-informative characters.

fortable with. We calculated percentage divergence/mya for the species pairs *Epicrates-Python, Uropeltis-Lampropeltis,* and *Lampropeltis-Heterodon.* On the basis of the fossil record and the timing of the separation of Africa and South America, we use the conservative value of 80 mya for the divergence of *Epicrates-Python.* On the basis of the age of *Pachyrhachis* and its phylogenetic position relative to *Uropeltis* (Tchernov et al. 2000), we used the extremely conservative value of 100 mya for the split of *Uropeltis-Lampropeltis.* The divergence time between *Lampropeltis-Heterodon* is less understood, so we calculated for ages of 20 mya and 40 mya. The three species pairs exhibited similar rates, from 0.012% /mya (*Epicrates-Python*) to 0.0135% /mya (*Lampropeltis-Heterodon* at 40 mya). The striking conclusion is that the rates appear inordi-

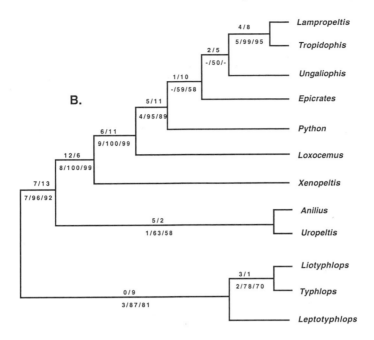

Figure 7.5 (*continued*)

nately slow, even for the conserved 28S region. We compared *Sphenodon* and *Anolis* by using a divergence time of 245 mya (Romer 1956) and found a similar value, 0.0147% /mya. The greatest character separation occurs in the split between snakes and lizards, with 54 characters. Doing the same calculations, using an end of the Jurassic divergence date (144 mya; a conservative estimate but probably too young), we find the highest rate at 0.02% /mya, less than a quarter of the average reported by Mindell and Honeycutt (8.1% /85 mya = 0.095% /mya).

Previous workers (Dessauer et al. 1987; Cadle 1988; Crother and Hillis 1995; Crother 1999) who used molecular data (although not DNA sequence) suggested that the rate of molecular evolution in snakes has been slow. Our 28S sequence data certainly support that conclusion. Martin and Palumbi (1993) reported that percentage sequence divergence/mya in mtDNA was greatly reduced in poikilotherms in comparison with homeotherms, with values all lower than 1%. Unfortunately, the only reptiles included in the comparisons were tortoises, but sea turtles exhibited the lowest values at about 0.5%. None of the previous studies that used DNA sequence for snake phylogeny inference (Forstner et al. 1995; Heise et al. 1995; Saint et al. 1998) discussed or noted any apparent retarded rates of change.

	1310	1320	1330	1340	1351

```
Heterodon       CCTCAGGATAGCTGGCGCTCG-CGGGCAG----AAAGG-------CT----
Lampropeltis    CCTCAGGATAGCTGGCGCTCG-CKGGCAC-----AAGG-------CT----
Tropidophis     CCTCAGGATAGCTGGCGCTCG-CGGGCAC----GA-GG-------CT----
Ungaliophis     CCTCAGGATAGCTGGCGCTCG-CGGGCAC----GA-GG-------CT----
Epicrates       CCTCAGGATAGCTGGCGCTCG-CGGGCAC----GA-GG-------CT----
Sanzinia        CCTCAGGATAGCTGGCGCTCG-CGGGCAC----GA-GG-------CT----
Python          CCTCAGGATAGCTGGCGCTCG-CGGGCAC----GAAGG-------CT----
Loxocemus       CCTCAGGATAGCTGGCGCTCG-CGGGCAC----GAAGG-------CT----
Xenopeltis      CCTCAGGATAGCTGGCGCTCG-CGGGCAC----GA-GG-------CT----
Uropeltis       CCTCAGGATAGCTGGCGCTCG-CGGGCAC-----A-GG-------CT----
Anilius         CCTCAGGATAGCTGGCGCTCG-CGGGCAC----GA-GG-------CT----
Leptotyphlops   CCTCAGGATAGCTGGCGCTCG-CGGGCAC----GAAGG-------CT----
Typhlops        CCTCAGGATAGCTGGCGCTCG-CG-G-AC----G----------CT----
Liotyphlops     CCTCAGGATAGCTGGCGCTCG-C--GC-------AA----CCCCCGT----

Sphenodon       CCTCAGGATAGCTGGCGCTCGTC--------CGAGA-CCCCCCCGT--
Hemidactylus    CCTCAGGATAGCTGGCGCTCG-CGG-------CGAAGG-AAAACCCTCAC
Rhineura        CCTCAGGATAGCTGGCGCTCGTC----ACG--CGAA--------CCCG----
Varanus         CCTCAGGATAGCTGACGCTCGTC----ACCC-CGA---------CCCT---
Celestus        CCTCAGGATAGCTGTCGCTCG-C---GACG-TCGAA-----CCCCC----
Anolis          CCTCAGGATAGCTGGCGCTCGTC-----CG-T----------CCGT---
Pygophis        CCTCAGGATAGCTGGCGCTCGTC---------------------CAC
Uromastyx       CCTCAGGATAGCTGGCGCTCGTC------------------CCCCCTCCT
```

Figure 7.6   Example of a divergent domain region (probably D7a, it corresponds approximately to positions 1850–1900 in *Mus* and *Homo*), where snakes exhibit much less change than do lizards. Even though this is just one possible alignment of the region, it is clear snakes are conserved relative to lizards.

The following conclusions can be reached from this discussion. The rate of change in the nuclear LSU of snakes is slow. Because this rate is slow, the LSU may not be the most appropriate region for elucidating snake phylogeny. It is possible that the large divergent domain (D2) in the first 1,000 bases of the LSU contains greater useful phylogenetic signal (some preliminary data suggest that may be the case). Most of the variation was unique; there were three times as many autapomorphies as synapomorphies (this may be an artifact of sampling: denser sampling may reduce the ratio of autapomorphies to synapomorphies). This resulted in terminal branches being long relative to internal branches, which perhaps caused problems in the parsimony inference with the DNA data.

SNAKE PHYLOGENY

We equate the best inference of phylogeny with the most highly corroborated hypothesis, and we believe that the discovery of the most corroborated hypothesis is related to the inclusion of the total evidence available (e.g., Miyamoto 1985; Hillis 1987; Kluge 1989a, 1997; Barrett et al. 1991). As such, our discussion

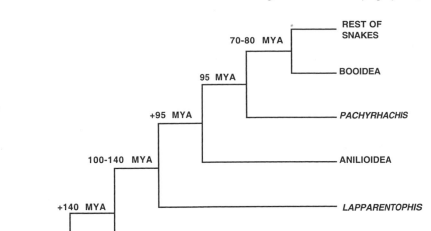

Figure 7.7   Branching diagram depicting a hypothesis for the age of the origin of snakes. The ages are constrained by the relative phylogenetic positions of key groups. Plus (+) signs indicate divergence ages not supported by fossils. For example, representative fossils of anilioid snakes are known from approximately 80 mya, but the position of the fossil taxon *Pachyrhachis* and its known date force the age of the anilioid divergence to at least greater than 95 mya. So with *Lapparentophis* of uncertain age but somewhere between 100 and 140 mya (according to Rage 1987), we extrapolate that the two earlier divergences must be greater (much greater?) than those of the earliest Cretaceous.

of snake phylogeny is focused on the inferred hypotheses from the combined analyses of DNA and morphology (fig. 7.5).

Two consistent hypotheses emerged from the two analyses (1:gaps as characters, 2:gaps as missing data) of the combined data (fig. 7.5). Following the superfamily names of McDowell (1987), the Booidea, Tropidopheoidea, and the Anilioidea were all paraphyletic in both hypotheses. *Anilius* and *Uropeltis* were sister taxa in the gaps = missing analysis but unresolved in the gaps = characters analysis. The status of the Scolecophidia also differed in the two analyses, with the scolecophidians forming a monophyletic group (gaps = missing) or a basal paraphyletic group (gaps = characters). The strict consensus tree of the gaps = characters analysis left *Epicrates* and *Ungaliophis* unresolved with respect to the clade containing *Lampropeltis* and *Tropidophis*. In the gaps = missing result, *Python* was added to that unresolved clade by virtue of the fact that one of the three mpts placed *Python* and *Epicrates* as sister taxa.

These combined hypotheses depart from Tchernov et al. (2000) only in the possibility that the Scolecophidia is paraphyletic. Tchernov et al. (2000) presented a trichotomy for the three scolecophidian families. In both of the combined analyses the clade is resolved, one monophyletic and the other paraphyletic. To choose among the competing hypotheses, it is instructive to look closely at the data that support them (fig. 7.8). The separation of the scolecophidians and the rest of snakes is certain, supported by 25 or 20 characters (see fig. 7.5). The monophyletic hypothesis is supported by nine ambiguous synapomorphies (homoplasies), all of them morphological characters. The paraphyletic hypothesis indicates *Leptotyphlops* as the sister to the rest of the snakes, *Typhlops* next most basal, and *Liotyphlops* the sister to all Alethinophidia. *Leptotyphlops* and *Typhlops* are separated by 5 DNA characters, 3 of which are unambiguous synapomorphies. *Liotyphlops* is separated from the rest of the snakes by 12 characters, 7 of which are unambiguous. Interestingly, of the 17 characters that support a paraphyletic Scolecophidia, 12 of them are DNA and 10 of those are unambiguous synapomorphies. It seems we have finally found the appropriate time frame for the LSU in snake phylogeny inference. Given the type of evidence (ambiguous vs. unambiguous) supporting the two hypotheses, we find the paraphyletic Scolecophidia to be the best corroborated (fig. 7.5A).

Alternatively, one can choose to make decisions about support and corroboration based on resampling approaches such as the bootstrap or jackknife. Interestingly enough, both resampling techniques show higher (significant) frequencies of recovering the monophyletic hypothesis (fig. 7.5). Although we find these resampling proportions illuminating, we consider type of character support (unambiguous vs. ambiguous) more revealing about corroboration than resampling values.

Although Heise et al. (1995) inferred a paraphyletic Scolecophidia from mtDNA, the relationships were quite different in that they found *Liotyphlops* to be the sister clade to the Alethinophidia and a basal clade of *Typhlops-Leptotyphlops* (fig. 7.2A). Forstner et al. (1995), also using mtDNA, also recovered a paraphyletic Scolecophidia that was congruent with our conclusions: *Leptotyphlops* the sister to the rest of snakes and *Typhlops* the basal lineage (they did not include *Liotyphlops;* fig. 7.2B). In a reanalysis of Forstner et al. (1995), Macey and Verma (1997) raised the possibility that the scolecophidians were monophyletic but in the end considered the problem unresolved. Indeed, previous authors have proposed the possibility of a nonmonophyletic Scolecophidia (e.g., Bellairs and Underwood 1951; List 1966; Underwood 1967). Our data suggest that additional debate on this issue is warranted.

Heise et al. (1995) deserves further examination because of its departure from previous ideas of snake phylogeny (fig. 7.2A). Our hypothesis and the hypotheses of Kluge (1991), Cundall et al. (1993), and Tchernov et al. (2000) are largely consistent, and all point to the rejection of most elements of the Heise

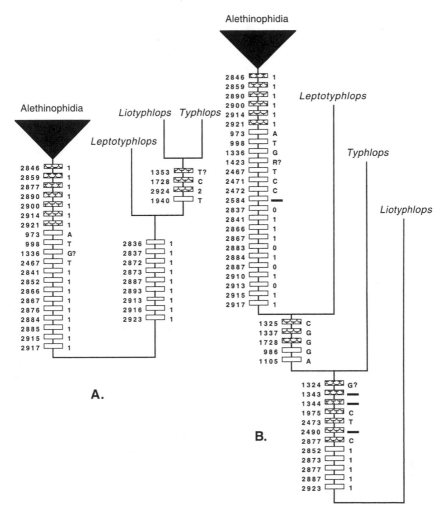

Figure 7.8   A detailed look at the character support for the competing hypotheses of scolecophidian relationships. (A) Monophyletic Scolecophidia inferred when gaps are treated as missing data. (B) Paraphyletic relationships when gaps are treated as characters. The boxes with imbricate scales represent unambiguous synapomorphies (CI = 1.0), and the white boxes represent ambiguous synapomorphies (homoplasies). Here, the morphological character numbered 2826 corresponds to character 1 (Tchernov et al. 2000), so 2846 corresponds to their character 21. The question mark (?) represents a base position where an alternative alignment is possible.

et al. hypothesis. In fact, only our argument for a nonmonophyletic Scolecophidia is similar to the conclusions of Heise et al. (1995). The reasons for the peculiar phylogeny are uncertain, so we do not speculate but simply recognize that the Heise et al. hypothesis is not corroborated by the hypothesis inferred herein or by previous workers.

Our preferred hypothesis (fig. 7.5A) also allows for speculation on the rejection of competing hypotheses of booid and tropidopheoid relationships. Kluge (1991) suggested that pythonines and boines were sister lineages, but Cundall et al. (1993) found evidence that erycines and boines were sister taxa to the exclusion of pythonines. Results from both studies could not resolve the issue of booid monophyly. We did not include an erycine, but we did find that only one of the seven mpts suggested a pythonine–boine relationship and that the majority of characters support the separation of those lineages. As such our analyses indicate a nonmonophyletic Booidea, and our data suggest the separation of pythonines and boines following Cundall et al. (1993).

Kluge (1991) supported McDowell (1987) in the concept of a monophyletic Tropidopheoidea that consisted of *Tropidophis, Trachyboa, Ungaliophis,* and *Exiliboa* (fig. 7.1A). Zaher (1994) argued for the rejection of this monophyletic group from a detailed analysis of four morphological characters. These data pointed to a *Tropidophis-Trachyboa* clade and the inclusion of *Ungaliophis* and *Exiliboa* in the Booidea. Tchernov et al. (2000) found a paraphyletic Tropidopheoidea (fig. 7.1C) but did not support the split suggested by Zaher (1994). Interestingly, the molecular data herein support Zaher's contention and consistently support the position of *Tropidophis* as sister to the advanced snakes (Caenophidia) and *Ungaliophis* in a clade with the booids *Sanzinia* and *Epicrates* or sister to the clade of colubroids *Tropidophis* (figs. 7.3–7.5).

Summary

The nuclear LSU of rDNA is highly conserved among snakes, even at higher-level divergences. Much of the synapomorphic information was present at the deepest nodes, that is, the separations of snakes and lizards and among the scolecophidians at the base of the snake clade. The percentage sequence divergence/mya was surprisingly low, strongly suggesting that the rate of molecular evolution of the LSU has been quite slow in snakes.

The combined analysis of DNA and morphology yielded trees consistent with recent hypotheses, with the exception of the phylogeny by Heise et al. (1995). We argue for a paraphyletic Scolecophidia, a paraphyletic Booidea as proposed by Cundall et al. (1993), and a nonmonophyletic Tropidopheoidea as argued by Zaher (1994). A large-scale study on the scolecophidians is required to test the hypothesis for nonmonophyly. The problem of higher-level snake phylogeny remains a challenge. The details will be trying!

Acknowledgments

Jonathan Campbell, Ronald Crombie, Maureen Donnelly, William Duellman, J. Scott Keogh, Kirsten Nicholson, Dwight Lawson, Eric Pianka, Jennifer Pramuk, Tod Reeder, Juan Renjifo, Jay Savage, Kevin Toal, Addison Wynn, and the Louisiana State University Museum of Natural Science all kindly donated or traded for tissues. Karen Garrett, Kirsten Nicholson, Thomas Ostertag, Travis Taggart, Kevin Toal, Brian Warren, and Steven Werman all contributed in various ways to the project. We thank Hussam Zaher for sharing his knowledge of snake fossils. This project was funded by the National Science Foundation (DEB-9207751).

8

Elapid Relationships

Joseph B. Slowinski and Robin Lawson

Introduction

The snake family Elapidae is a major group of venomous snakes containing nearly 300 species in approximately 60 genera (Golay et al. 1993; herein we use Elapidae in the broad sense to include both terrestrial and marine species, whereas Golay et al. place the marine species in the separate family Hydrophiidae). Despite a number of recent phylogenetic studies, the relationships of elapids both to each other and to other groups of snakes remain poorly understood. In this chapter, we report on efforts to resolve elapid relationships by using sequence data from two evolutionarily independent (sensu Slowinski and Page 1999) DNA regions: mtDNA genes and a nuclear gene. mtDNA sequence data have already been successfully applied to elapid phylogeny (Keogh 1998; Slowinski and Keogh 2000; Slowinski et al. 2001). Here we use an additional 2,697 mtDNA nucleotide positions beyond the 1,116 of the cytochrome *b* (cyt *b*) gene, as well as a part of the nuclear single-copy gene, c-*mos*. These additional mtDNA sequences are derived from the NADH dehydrogenase subunit 1 (ND1), NADH dehydrogenase subunit 2 (ND2), and partial sequence of NADH dehydrogenase subunit 4 (ND4) genes. The c-*mos* gene evolves more slowly than does mtDNA and has been successfully applied to squamate phylogeny (Saint et al. 1998; Harris et al. 1999). We chose the c-*mos* gene because of its conservatism as a way to secure reliable characters that might be used to resolve basal clades within elapids. In comparing the results from the c-*mos* and mtDNA data, we seek to identify areas of congruence. Because of the low probability that a particular monophyletic group will be supported by two independent DNA sequences, when such an outcome occurs it places a high degree of confidence in that grouping (Hendy et al. 1988; Miyamoto and Fitch 1995; Slowinski and Lawson 2002).

OUR CURRENT KNOWLEDGE OF THE PHYLOGENETIC POSITION OF THE ELAPIDS

It is well understood, on the basis of morphological and molecular evidence, that elapids are part of the Colubroidea, along with colubrids, viperids, and atractaspidids (e.g., Rieppel 1988). Rieppel (1988) lists seven morphological synapomorphies supporting the monophyly of colubroids, to which could be added the existence of a serous dental gland (the Duvernoy's or venom gland) if the lack of this condition in colubrids is secondarily derived, as seems likely (Zaher 1999). But little progress has been made in understanding the position of elapids within colubroids. Using immunological data, Dessauer et al. (1987) and Cadle (1988) found tentative evidence for a relationship between the African mole vipers *Atractaspis* and elapids. Later, Heise et al. (1995) found strong support for a connection between *Atractaspis* and elapids by using mtDNA (12S and 16S rRNA genes) sequence data. This was followed by Kraus and Brown's (1998) phylogenetic study of colubroid snakes, which used mtDNA (ND4) sequence data to find a sister-group relationship between elapids and *Atractaspis, Aparallactus,* and the Malagasy genera *Madagascarophis* and *Leioheterodon*. The common thread running through all these studies of a link between elapids and *Atractaspis* is highly intriguing and suggests that the link is real.

OUR CURRENT KNOWLEDGE OF RELATIONSHIPS WITHIN THE ELAPIDS

In the 1970s and 1980s, a series of papers (reviewed in Slowinski et al. 1997) that employed immunological distances and small samples of elapids found that reciprocal tests showed a close relationship between Australo-Melanesian terrestrial elapids and the sea snakes. Three recent molecular studies (Slowinski et al. 1997; Keogh 1998; Slowinski and Keogh 2000) have explored the phylogenetic relationships within elapids by using more comprehensive sampling. All three studies also found support for a monophyletic association of the Australo-Melanesian terrestrial elapids and the sea snakes. Further, these three studies are in agreement that the sea snakes are nonmonophyletic, with the oviparous (*Laticauda*) and the viviparous (*Hydrophis* and related genera) sea snakes having separate origins within the Australo-Melanesian clade. However, the relationships of elapids outside the Australo-Melanesian-marine clade are poorly understood, although Slowinski et al. (2001) discuss strong evidence that the Asian and American coralsnakes are collectively monophyletic.

Methods

TAXONOMIC SAMPLING

For the study of intra-elapid relationships we sampled tissues, usually liver or shed skin, from an Asian and African species from the widespread *Naja* and representative species from every other non-Australo-Melanesian marine elapid genus (except *Micruroides*, which is the sister taxon to *Micrurus* [Slowinski 1995], *Hemibungarus*, sister taxon to *Calliophis* [Slowinski et al. 2001], and *Hemachatus*, which we had trouble sequencing) (table 8.1), as well as a small sample of four Australo-Melanesian marine genera, resulting in a set of 17 elapids. The taxa *Homoroselaps lacteus, Leioheterodon modestus, Malpolon monspessulanus,* and *Pseudaspis cana* were included as outgroups. For analysis of this set of 21 taxa, the complete nucleotide sequence of the mitochondrial protein coding genes cyt *b*, ND1, ND2, and ND4 were used as well as part of the single-copy nuclear protein coding gene c-*mos* (Saint et al. 1998).

To examine in more detail the position of the Elapidae within the Colubroidea, we compared the cyt *b* and c-*mos* gene sequences for a larger set of colubroid taxa than that used for the intra-elapid study. These included representative genera from the colubroid subfamilies Xenodermatinae, Homalopsinae, Boodontinae, Pseudoxyrhophiinae, Colubrinae, Natricinae, Xenodontinae, and the family Atractaspididae (Zaher 1999). In figures 8.1 and 8.2, only the taxa that most closely relate to the systematic position of the Elapidae are identified to species. Members of the subfamilies Colubrinae, Natricinae, Xenodontinae, Homalopsinae, and Pareatinae are each grouped together for representation in figures 8.1 and 8.2. These taxa with their sequence data are taken from works in progress designed to evaluate relationships among the colubroid snakes and snakes as a whole (Lawson and Slowinski 2002; R. Lawson and J. B. Slowinski, unpub. data). We had previously deposited in GenBank cyt *b* sequences for most of the elapids. All remaining sequences used in the section devoted to intra-elapid relationships are now deposited in GenBank (accession numbers AY058923–AY059011).

LABORATORY PROCEDURES

DNA was isolated from tissues by proteinase K digestion in lysis buffer, followed by two rounds of extraction with phenol/CHCl$_3$ and then one with CHCl$_3$ alone (Maniatis et al. 1982). DNA was precipitated by the addition of three volumes of cold 100% ethanol. Preparation of DNA templates for polymerase chain reaction (PCR), PCR product purification, and cycle sequencing followed Burbrink et al. (2000). Primers for amplification and sequencing of DNA segments were as listed in table 8.2. Nucleotide sequence was determined

Table 8.1   Taxa sequenced for this study

Taxon	Origin	Voucher no.
Elapidae		
Australo-Melanesian and		
marine species		
Demansia atra	Australia	Voucher unknown
Drysdalia coronata	Western Australia	SAM R22966
Laticauda colubrina	Indonesia	CAS 220643
Notechis ater	South Australia	SAM R31604
Mambas		
Dendroaspis polylepis	Africa	CAS 220644
Garter snakes		
Elapsoidea nigra	Africa	LSUMZ 56251
Kraits		
Bungarus fasciatus	Ayeyarwady Div. Myanmar	CAS 207988
Cobras		
Aspidelaps scutatus	Africa	LSUMZ 56251
Boulengerina annulata	Zaire	HLMD R-1607
Naja kaouthia	Ayeyarwady Div. Myanmar	CAS 206602
Naja nivea	Africa	No voucher
Paranaja multifasciata	Africa	No voucher
Walterinnesia aegyptia	North Africa	No voucher
King Cobra		
Ophiophagus hannah	Ayeyarwady Div. Myanmar	CAS 206601
Coralsnakes		
Sinomicrurus japonicus	Riu Kiu Islands, Japan	CAS 204980
Calliophis bivirgata	Southeast Asia	LSUMZ 37496
Micrurus fulvius	Florida, USA	CAS 195959
Outgroups		
Homoroselaps lacteus	South Africa	LSUMZ 55386
Leioheterodon modestus	Madagascar	LSUMZ (H 1991)
Malpolon monspessulanus	Spain	MVZ 186256
Pseudaspis cana	Kenya	LSUMZ 37426

Note: CAS, California Academy of Sciences; HLMD, Hessisches Landemuseum Darmstadt; LSUMZ, Louisiana State University Museum of Natural Science; MVZ, Museum of Vertebrate Zoology, University of California; SAM, South Australian Museum.

using either the ABI model 310 or 3100 Genetic Analyzer, and gene segments were edited and assembled using the computer program Sequencher, version 4.0 (Gene Codes Corporation 1999). Alignment of each of these five protein coding genes was easily accomplished by eye, using the computer program Xesee (Cabot and Beckenback 1989).

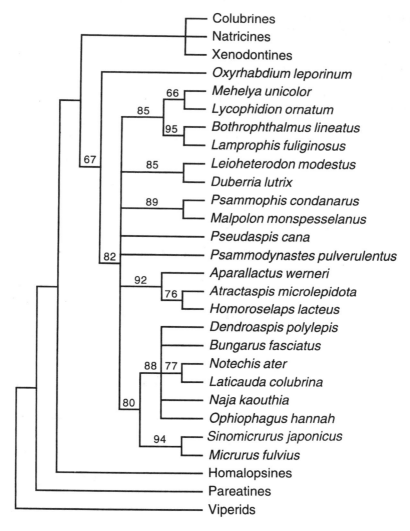

Figure 8.1   A summary of the strict consensus tree of the shortest trees resulting from a parsimony search of the c-*mos* data from a sample of more than 80 colubroid genera. Only the genera related to elapids are depicted. Numbers along the internodes are bootstrap values.

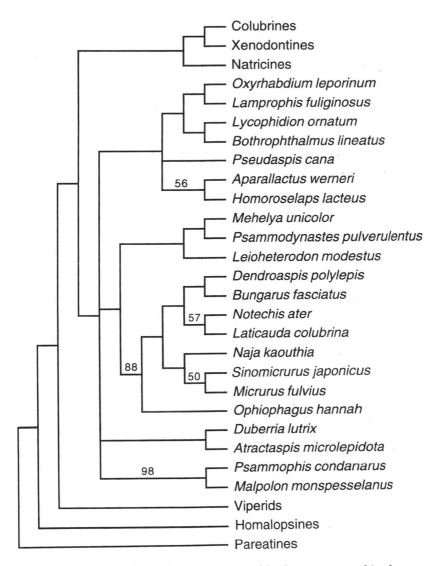

Figure 8.2   A summary of the strict consensus tree of the shortest trees resulting from a parsimony search of the cyt *b* data from a sample of more than 80 colubroid genera. Only the genera related to elapids are depicted. Numbers along the internodes are bootstrap values.

Table 8.2 Primers used for DNA amplification and sequencing

Primer	Primer sequence	Use	Location	Reference
L14910	5' - GAC CTG TGA TMT GAA AAA CCA YCG TTG T -3'	Amp.	tRNA-glu	de Queiroz et al., n.d.
L14919	5' - AAC CAC CGT TGT TAT TCA ACT -3'	Amp./seq.	tRNA-glu	Burbrink et al. 2000
L15584	5' - TCC CAT TYC ACC CAT ACC A -3'	Seq.	cyt *b*	de Queiroz et al., n.d.
H16064	5' - CTT TGG TTT ACA AGA ACA ATG CTT TA -3'	Amp./seq.	tRNA-thr	Burbrink et al. 2000
H15149	5' - CCC TCA GAA TGA TAT TTG TCC TCA -3'	Seq.	cyt *b*	Kocher et al. 1989
H15716	5' - TCT GGT TTA ATG TGT TG -3'	Seq.	cyt *b*	This study
16Sb	5' - ACG TGA TCT GAG TTC AGA CCG G -3'	Amp./seq.	16S rRNA	Palumbi 1996
H3518	5' - CCG TGT CTA CTC TAT CAA GGT AGT CC -3'	Amp./seq.	tRNA-ile	de Queiroz et al., n.d.
L2984	5' - CAA ATY ATY TCA TAY GAM GT -3'	Seq.	ND1	This study
H3097	5' - CAT ATT ATT ATR GCT ARK GGT CA -3'	Seq.	ND1	This study
L4437b	5' - CAG CTA AAA AAG CTA TCG GGC CCA TAC C -3'	Amp./seq.	tRNA-met	Kumazawa et al. 1996
H5877	5' - AAA CTA GKA GCC TTG AAA GCC -3'	Amp./seq.	tRNA-trp	de Queiroz et al., n.d.
L5245	5' - ATY YRA CAT GAC AAA AAA T -3'	Seq.	ND2	This study
H5340	5' - TGR GTT TGG TTT AGG CTG CC -3'	Seq.	ND2	This study
ND4	5' - CAC CTA TGA CTA CCA AAA GCT CAT GTA GAA GC -3'	Amp./seq.	ND4	Arévalo et al. 1994
Leu	5' - CAT TAC TTT TAC TTG GAT TTG CAC CA -3'	Amp./seq.	tRNA-leu	Arévalo et al. 1994
H12405	5' - CAC AGC TTG AYA TTT WTT TAA ATT AC -3'	Seq.	ND4	This study
S77	5' - CAT GGA CTG GGA TCA GTT ATG -3'	Amp./seq.	*c-mos*	R. Lawson and J. B. Slowinski, unpub. data
S78	5' - CCT TGG GTG TGA TTT TCT CAC CT -3'	Amp./seq.	*c-mos*	R. Lawson and J. B. Slowinski, unpub. data

Note: Primers from the current study are named to indicate the position of the 3' nucleotide in the mitochondrial genome of the colubrid snake *Dinodon semicarinatus* (Kumazawa et al. 1998). Primer H15149 has been shortened from its original length (see Kessing et al. 1989). Amp., amplification; seq., sequencing.

PHYLOGENETIC ANALYSIS

The methods of maximum parsimony and maximum likelihood were employed using PAUP* 4.0 (Swofford 2000) to infer phylogenies. For the larger taxon set, the c-*mos* (170 informative characters) and the cyt *b* (629 informative characters) were separately analyzed by maximum parsimony, with all characters weighted equally and uninformative characters excluded. The parsimony analyses were performed by conducting 1,000 sequential tree bisection–reconnection (TBR) heuristic searches with starting trees obtained by random stepwise addition. Because of the large size of these data sets, we did not attempt maximum likelihood analyses. For the smaller taxon set, sequences from the four mitochondrial genes were combined before analyses. This is justified because mitochondrial genes are physically linked and show very little recombination. Thus, the possibility for different branching histories among these genes is unlikely (de Queiroz et al. 1995, 2000). The parsimony searches were conducted as branch and bound searches, with all characters weighted equally and uninformative sites excluded. Before performing the maximum likelihood searches, we used Modeltest (Posada and Crandall 1998) to select the best-fitting model of nucleotide substitution. This resulted in the HKY-85 model, with equal rates across all sites being chosen for the c-*mos* data, and the general–time reversible (GTR) model, with 2.81% sites assumed to be invariant and a gamma rate variation parameter of .56 for the mtDNA data. The maximum likelihood analyses were then conducted as single heuristic searches, with the starting trees found by random stepwise addition. Support for clades was assessed by parsimony and likelihood bootstrapping with 100 replicates. In the case of parsimony bootstrapping, each replicate was analyzed by use of a TBR heuristic search, with the starting tree built by stepwise random addition. In the case of likelihood bootstrapping, each replicate was performed by ASIS stepwise addition only, with no branch swapping.

Results

ELAPID RELATIONSHIPS WITHIN THE COLUBROIDEA

From the c-*mos* sequence data of 80 colubroid taxa, PAUP* found 37,300 trees (the maximum number of trees that can be stored by the program); the shortest trees have 544 steps (retention index [RI] = .726). Figure 8.1 shows a distillation of the strict consensus of these trees, wherein only the genera most closely related to the elapids are shown, the other terminal taxa depicted being subfamilies. Our results show a well-supported link between elapids, atractaspidids (*Atractaspis, Aparallactus*), the enigmatic *Homoroselaps*, and a sample of African, Malagasy, and Asian colubrid genera whose subfamilial allocations have varied from worker to worker. In a recent monograph, Zaher (1999) allocated the Asian *Oxyrhabdium* to the Xenodermatinae, the Asian *Psammodynastes* to the Natricinae, the African *Mehelya, Lycophidion, Bothrophthalmus, Lamprophis, Duberria,* and *Pseudaspis* to the Boodontinae, the Malagasy *Leioheterodon* to the Pseudoxyrhophiinae, and the Africo-Asian *Psammophis* and *Malpolon* to the Psammophiinae.

From the cyt *b* sequence data of the same taxon set, PAUP* found four shortest trees of 10,991 steps (RI = .295). Figure 8.2 shows the consensus tree treated in the same fashion as that in figure 8.1. Again there is a monophyletic group, albeit a poorly supported one, containing elapids and *Atractaspis, Aparallactus, Homoroselaps, Oxyrhabdium, Psammodynastes, Mehelya, Lycophidion, Bothrophthalmus, Lamprophis, Duberria, Pseudaspis, Leioheterodon, Psammophis,* and *Malpolon.* However, beyond the monophyly of the Elapidae, the relationships within this group vary between the two trees (figs. 8.1 and 8.2).

INTRA-ELAPID RELATIONSHIPS

From the c-*mos* sequence data, the parsimony searches found 16 shortest trees of 28 steps (fig. 8.3; RI = .87). The likelihood search found a single tree with a −ln likelihood of 1,510.32855 (fig. 8.4). The parsimony and likelihood c-*mos* trees are similar, the difference being that the likelihood tree has an extra clade not on the parsimony tree, consisting of *Aspidelaps, Walterinnesia,* and *Elapsoidea,* but without significant bootstrap support. Both trees show strong bootstrap support for monophyletic coralsnakes (*Sinomicrurus, Calliophis, Micrurus*) as sister group to all remaining elapids. Further, both trees show strong support for the monophyly of the Australo-Melanesian-marine clade and the monophyly of a subset of the cobras (*Boulengerina, Naja, Paranaja*).

From the combined mtDNA sequence data, the parsimony search found a single shortest tree of 8,729 steps (fig. 8.5; RI = .266). The likelihood search found a single tree with a −ln likelihood of 40,904.37428 (fig. 8.6). The parsimony and likelihood mtDNA trees differ in their placement of the Australo-Melanesian-marine clade and the genera *Calliophis, Dendroaspis, Elapsoidea, Ophiophagus,* and *Bungarus.*

The mtDNA results differ from the c-*mos* results in the placement of the coralsnakes. The c-*mos* data (figs. 8.3 and 8.4) place coralsnakes as sister taxon to all other elapids, but the mtDNA data place them either as sister clade to the non-Australo-Melanesian-marine genera (parsimony; fig. 8.5) or as sister clade to the Australo-Melanesian-marine clade (likelihood; fig. 8.6). The c-*mos* and the mtDNA data do agree on the monophyly of *Sinomicrurus* and *Micrurus*, on the monophyly of a subset of the cobras (*Boulengerina, Naja, Paranaja*), an association between the cobras *Walterinnesia* and *Aspidelaps,* and the monophyly of the Australo-Melanesian-marine taxa.

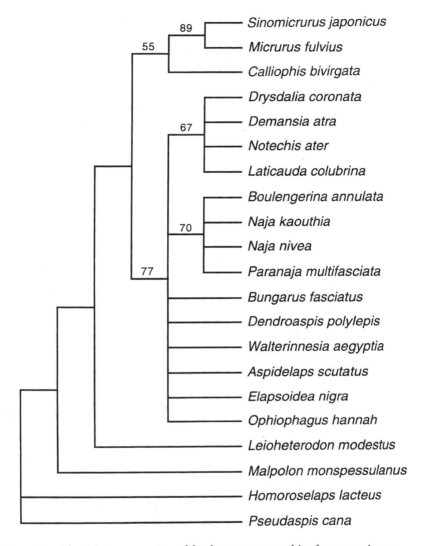

Figure 8.3   The strict consensus tree of the shortest trees resulting from a parsimony search of the elapid c-*mos* data. Numbers along the internodes are bootstrap values.

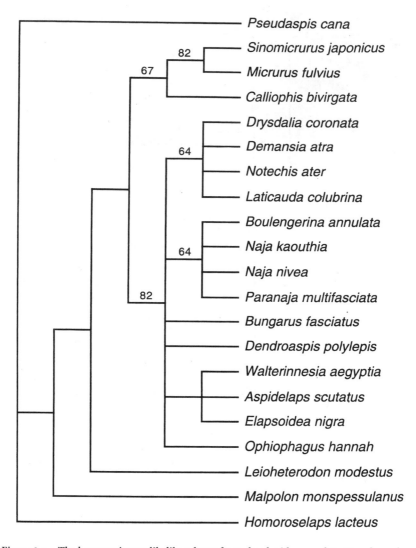

Figure 8.4   The best maximum likelihood tree from the elapid c-*mos* data. Numbers along the internodes are bootstrap values.

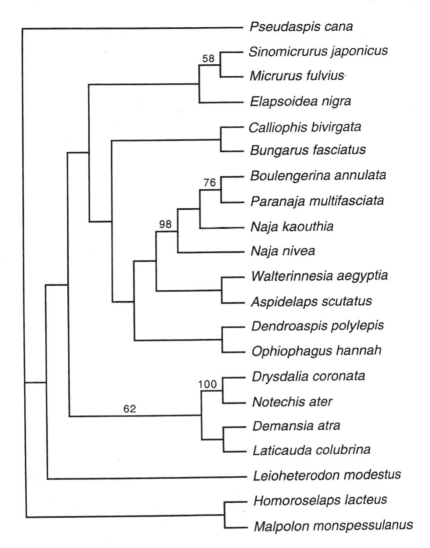

Figure 8.5 The shortest tree resulting from a parsimony search of the elapid mtDNA data. Numbers along the internodes are bootstrap values.

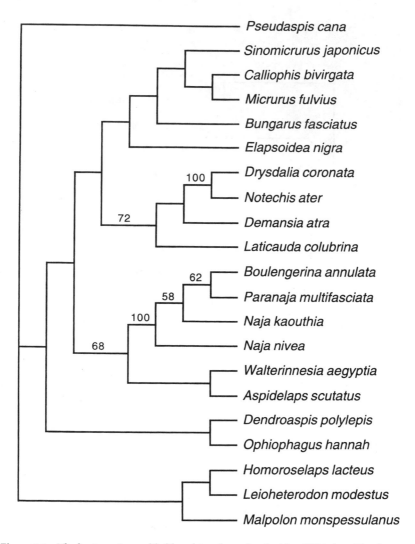

Figure 8.6   The best maximum likelihood tree from the elapid mtDNA data. Numbers along the internodes are bootstrap values.

Discussion

RELATIONSHIPS AMONG THE ELAPIDS WITHIN THE COLUBROIDEA

As we stated at the outset of this chapter, we view areas of agreement between evolutionarily independent genes to provide very strong support for the reliability of monophyletic groups. In this context, it is highly significant that out of more than 80 colubroid genera sampled, the c-*mos* and cyt *b* trees (figs. 8.1 and 8.2) both supported the same relationships between elapids and certain African, Asian, and Malagasy colubrid genera. Collectively, this group is sister taxon to the colubrines, xenodontines, and natricines, with only viperids, homalopsines, and pareatines occupying a more "basal" position.

Note that in figures 8.1 and 8.2, elapids and *Homoroselaps* do not form a monophyletic group. Throughout much of the nineteenth and twentieth centuries, *Homoroselaps* (as *Elaps*) was regarded as part of the Elapidae. However, McDowell (1968) enumerated a number of morphological characters that he argued tied *Homoroselaps* to the aparallactines and *Atractaspis*. Nevertheless, McCarthy (1985) doubted this reallocation, and Underwood and Kochva (1993) argued that *Homoroselaps* was in fact an elapid. The current study shows convincingly that *Homoroselaps* is not an elapid, unless of course the Elapidae is expanded to include additional genera.

Our study has confirmed the link between the mole vipers and elapids stemming from the works of Dessauer et al. (1987), Cadle (1988), and Heise et al. (1995), as well as the paper by Kraus and Brown (1998) that found a link between mole vipers, Malagasy colubrids, and elapids. By using a strategy of sampling a more comprehensive set of colubrid genera, we have shown that this link involves many more taxa than just the mole vipers, Malagasy colubrids, and elapids but that it also includes additional genera placed in the subfamilies Boodontinae, Pseudoxyrhophiinae, Natricinae, and Psammophiinae (sensu Zaher 1999), and the family Atractaspididae (sensu Underwood and Kochva 1993), as well as *Homoroselaps*. Of course there may be additional colubrids not sampled in our study that also share a close relationship with elapids.

INTRA-ELAPID RELATIONSHIPS

Predictably, the mtDNA provide more characters and greater resolution than do the c-*mos* data. However, although the c-*mos* gene provided only 16 parsimony informative characters, the retention index (87%) was high, indicating that the c-*mos* results are fairly reliable. Considering the congruence between the c-*mos* and mtDNA data as well as among these data sets and previous studies, several clades of elapids can be considered strongly supported. First, the

Asian and American coralsnakes form a monophyletic group. Within this clade, certain Asian species (represented by *Sinomicrurus japonicus* in the present study) are clearly more closely related to the American forms than to other Asian coralsnakes. Slowinski et al. (2001) have analyzed the relationships of the Asian and American coralsnakes on the basis of morphology and found that the northern tropical/subtropical Asian species *S. hatori, S. japonicus, S. kellogi, S. macclellandi,* and *S. sauteri* are more closely related to the American coralsnakes than they are to other Asian coralsnakes. Slowinski et al. (2001) hypothesized that the New World coralsnakes are derived from a dispersal of a northern tropical/subtropical Asian coralsnakes over the Bering Land Bridge. The c-*mos* data showed moderate support for the coralsnakes being sister clade to all other elapids. However, this was contradicted by the mtDNA data, so we regard this result as questionable.

Second, the present study and all published molecular analyses of elapid relationships (e.g., Slowinski et al. 1997; Keogh 1998; Slowinski and Keogh 2000 and references therein) support the monophyly of the Australo-Melanesian and marine taxa. This clade is the best-supported clade within elapids and can hardly be doubted.

Third, the present study and Slowinski and Koegh (2000) support the monophyly of the cobras without *Ophiophagus* (*Aspidelaps, Boulengerina, Hemachatus, Naja, Paranaja, Walterinnesia*). Within this group there is strong support from the present study and Slowinski and Keogh (2000) for the monophyly of *Boulengerina, Naja,* and *Paranaja.* Interestingly, in the present study both the c-*mos* and mtDNA data support the nonmonophyly of *Naja,* with the Asian *Naja kaouthia* closer to the African *Boulengerina* and *Paranaja* than to the African *Naja nivea.* This is not necessarily unexpected because, although the monophyly of the Asian cobras has been well documented (e.g., Szyndlar and Rage 1990; Szyndlar and Zerova 1990), we know of no studies supporting the monophyly of all *Naja.*

Unfortunately, the present study as well as previous studies have failed to resolve the relationships of the King Cobra (*Ophiophagus*), the African garter snakes (*Elapsoidea*), the kraits (*Bungarus*), and the mambas (*Dendroaspis*). These are truly enigmatic taxa worthy of further efforts to elucidate their phylogenetic positions.

Below we list what we consider to be the reliably supported clades of elapids and the studies supporting their monophyly:

Australo-Melanesian species/sea snakes
    Slowinski et al. 1997—venom proteins
    Keogh 1998—mtDNA
    Slowinski and Keogh 2000—mtDNA
    This study

Kraits (*Bungarus*)
  Slowinski 1994—morphology
Cobras (*Aspidelaps, Boulengerina, Hemachatus, Naja, Paranaja, Walterinnesia*)
  Slowinski and Keogh 2000—mtDNA
  This study
African garter snakes (*Elapsoidea*)
King Cobra (*Ophiophagus*)
Mambas (*Dendroaspis*)
Asian and American coralsnakes (*Sinomicrurus, Calliophis, Hemibungarus,*
  Micruroides, Micrurus)
  Keogh 1998—mtDNA
  Slowinski et al. 2001—mtDNA and morphology
  This study

Acknowledgments

We thank B. I. Crother, M. A. Donnelly, C. Guyer, M. H. Wake, and M. E. White for organizing the La Paz symposium where this paper was presented. We thank the following people for their role in the procurement of tissues or DNA extracts that were used in this study: R. Anderson, A. Bauer, J. Boundy, F. Burbrink, S. Busack, R. Crombie, H. Dessauer, D. Dittman, P. Frank, J. Gautier, D. Good, G. Haagner, H.-W. Herrman, S. Keogh, R. Murphy, T. Reeder, F. Sheldon, C. Spencer, D. Theakston, M. Toriba, and J. Vindum. We owe Carl Elliger and Eleanor Visser a tremendous debt of gratitude for assisting with the lab work. Funds for the lab work were provided by the California Academy of Sciences. Recent fieldwork in Myanmar by the senior author has provided several of the tissues used in this study. For assistance in the field, thanks are due to H. Robeck, C. Spencer, J. Vindum, and K. Wiseman. Financial support for the fieldwork was provided by the National Science Foundation (DEB-9971861) and the Lindsay Field Research Fund of the California Academy of Sciences. Collecting permits and various other forms of assistance in Myanmar were generously provided by U Uga and U Khin Maung Zaw, past and present directors of the Nature and Wildlife Conservation Division.

9

Wallace and Savage: Heroes, Theories, and Venomous Snake Mimicry

Harry W. Greene and Roy W. McDiarmid

In the dawn of field biology, Alfred Russel Wallace departed for the most distant and dangerous biotic frontiers of the world, carrying with him little formal education but a blessed love of reading and reflective solitude. He sought the insect-ridden Edens of which naturalist explorers dream. His principal lifeline to the English homeland consisted of specimens outbound—birds skinned, insects pinned, plants pressed—and sporadic payments for his treasures inbound. An intense young man, totally focused, awesomely persistent and resourceful, resilient to tropical diseases that might have killed others, and nobly selfless, even to Darwin, who otherwise might have become a bitter rival, Wallace endured, and he triumphed. He succeeded brilliantly because he relished detail while thinking across a wide canvas.

E. O. WILSON (1999)

Among the Bribri, Cabécar, Boruca, Changina, and Chiriquí, when the chicha has been drunk, the night grows late and dark, and the fires die down to burning embers, the wisest old man of the tribe tells his engrossed listeners of a beautiful miraculous golden frog that dwells in the forests of these mystical mountains. According to the legends, this frog is ever so shy and retiring and can only be found after arduous trials and patient search in the dark woods on fog shrouded slopes and frigid peaks. However, the reward for the finder of this marvelous creature is sublime. Anyone who spies the glittering brilliance of the frog is at first astounded by its beauty and overwhelmed by the excitement and joy of discovery. . . . The story continues that those who find the legendary frog find happiness, and as long as they hold the frog happiness will follow them everywhere. . . . Field biologists in particular seem always to be searching for mystical truth and beauty in nature, and frequently at some unperceived level, for the happiness promised by the Indian seers.

J. M. SAVAGE (1970)

The mysteries of mimicry had a special attraction for me. . . . "Natural selection" could not explain the miraculous coincidence of imitative aspect and imitative behavior, nor could one appeal to the theory of "the struggle for life" when a protective device was carried to the point of mimetic subtlety, exuberance, and luxury far in excess of a predator's power of appreciation. I discovered in nature the nonutilitarian delights that I sought in art. Both were a form of magic, both were a game of intricate enchantment and deception.

VLADIMIR NABOKOV (QUOTED IN BOYD AND PILE 2000)

In this chapter, we honor Alfred Russel Wallace and Jay Mathers Savage by discussing venomous snake mimicry, a topic about which they each made fundamental discoveries. As field biogeographers, these men have focused on organisms in nature and thereby contributed to the conceptual advancement of evolutionary biology; each of them also has been outspoken in defense of nature against the onslaught of human activities. We especially admire Wallace for pushing the frontiers of nineteenth-century natural history, for his gracious attitude toward Charles Darwin in the face of their co-discovery of natural selection, and for brilliantly synthesizing across animals as different as insects, carnivores, and snakes (see Quammen 1996; Daws and Fujita 1999; Raby 2001). We praise Savage for flouting traditional boundaries between art and science, for his joyous loyalty to students and friends, and for pursuing the biology of amphibians and reptiles against the backdrop of landscape history, thereby inspiring so many of us to reach higher and farther.

Theories tell us what we know, they suggest where to look and what to measure next, and they encapsulate the state of science. Most of us study nature better when we are guided by theory, and the essence of "the" scientific method is testing alternative hypotheses, ideally via controlled experiments. Nonetheless, organisms are the fundamental packages in which life varies, reproduces, and persists; they are the functioning arenas in which behavioral, physiological, and morphological systems are organized as well as the living building blocks of populations, communities, ecosystems, and biomes (Brooks 2001). Biologists ultimately seek to understand the lives of organisms and patterns of their diversification—not theories and experiments—and thus discoveries of new kinds of organisms and new things about organisms chronically reset the "research cycles" (Kluge 1991) of hypothesis testing that underlie good, progressive science.

In this chapter, we update Pough's (1988a) review of venomous snake mimicry and then show how recent discoveries might modify previous conclusions. Next we describe four macroevolutionary patterns, recognition of which ultimately stems from studies of coralsnake mimicry, and name them for Wallace and Savage. Then we comment on some unanswered questions and promising avenues for future research. By revisiting venomous snake mimicry, we hope to illustrate that in spite of an emphasis in modern science on hypothesis testing and generalization, organisms themselves should always be a central focus of biology (Greene 1986; West-Eberhard 2001). And recalling Nabokov's skeptical comments, quoted in one of the epigraphs at the beginning of this chapter, we take pleasure in linking our two visionary heroes with a phenomenon of such obvious aesthetic as well as scientific appeal. It turns out that in the case of venomous snake mimicry, things may not be quite as they have seemed.

Coralsnake Mimicry Revisited

This beautiful species [*Pliocercus elapoides*] resembles in the distribution of its colors certain Elapses [*sic,* venomous coralsnakes]. . . . It is a beautiful example of analogy of coloring.
E. D. COPE (1860)

In the Vertebrata . . . external form depends almost entirely on the . . . skeleton . . . [and] cannot therefore be rapidly modified by variation. . . . We can hardly see the possibility of a mimicry by which the elk could escape from the wolf, or the buffalo from the tiger. There is, however, in one group . . . such a general similarity of form, that a very slight modification, if accompanied by identity of colour, would produce the necessary amount of resemblance; and . . . there exist . . . species which it would be advantageous to resemble, since they are armed with the most fatal weapons of offence. We accordingly find that reptiles furnish us with a very remarkable and instructive case of true mimicry.
A. R. WALLACE (1870)

THE FIRST CENTURY OF CORALSNAKE MIMICRY

Although Cope (1860, as quoted in an epigraph above) is sometimes credited with first calling attention to coralsnake mimicry, by "analogy" he implied only some shared, unstated function for the external resemblance of his new species of colubrid to venomous coralsnakes (Elapidae). Cope's comments were made shortly before Bates (1862) proposed the concept of mimicry, in which a predator (dupe) avoids a palatable prey (mimic) that it mistakes for an unpalatable species (model) as the result of similar attributes of the latter two organisms. As young men, Wallace and Bates had traveled together in the Neotropics, and the former first clearly described as mimicry the resemblances between non-venomous colubrids and dangerous elapids (Wallace 1867; reprinted and slightly revised in Wallace 1870). Among Wallace's examples of parallel geographic variation in Batesian mimics and models are snake taxa we later analyzed in more detail (Greene and McDiarmid 1981), including the colubrid *Pliocercus* (= *Urotheca*, Savage and Crother 1989), and Wallace (1867) also first offered a compelling explanation for why mimicry is generally rare among vertebrates but common among snakes.

For about a hundred years after Wallace's (1867) report, discussions of the "coralsnake mimicry problem" concerned primarily similarities between venomous elapids and supposedly harmless species (reviewed in Pough 1988a). Early proponents relied on observations of museum specimens and organisms in nature rather than on experiments, and their opponents often resorted to unsubstantiated arguments and assertions. Dunn (1954), Mertens (1956), and Wickler (1968) analyzed model-mimic abundance ratios with the expectation that, because predators will only avoid color patterns if they are usually associated with unpleasant consequences, mimics must be relatively rare. Those data sets were problematic, however, because of difficulties in assessing the availability of snakes to predators, and we now know that interpreting model-mimic ratios is at best complex (Mallet and Joron 1999). Better evidence that a widespread South American colubrid clade (*Erythrolamprus*) does mimic sympatric elapids came from lack of a coralsnake pattern in *Erythrolamprus ocellatus* on Tobago, an island off Trinidad, where there are no venomous models (Emsley 1966; photos in Boos 2001). Critics meanwhile claimed that fatal elapid bites would preclude learning by predators (the "deadly model problem") and that supposedly nocturnal coralsnakes are not accessible to predators with color vision. In fact, the consequences for predators that attack venomous snakes can range from mild discomfort to death (Emsley 1966; Pough 1988a, 1988b). Even early reports documented diurnality in coralsnakes and their presumptive mimics (Strecker 1927; Greene and McDiarmid 1981; Sazima and Abe 1991; Martins and Oliveira 1993, 1998; Gurrola-Hidalgo and Chavez C. 1996; Savage and Slowinski 1996; Smith and Chiszar 1996; Stafford 1999), and color vision is reasonably widespread among vertebrates (e.g., Loop and Crossman 2000).

A frequently expressed alternative to the coralsnake mimicry hypothesis (e.g., Gadow 1911; Brattstrom 1955; Sánchez-Herrera et al. 1981), that similar color patterns were convergently acquired in response to some environmental factor other than predator signaling, is in hindsight surprisingly easy to dismiss. Although independent evolution of similar external appearance among snake species is common, pairs of such taxa generally fall into two groups: those that are ecologically similar but widely allopatric (e.g., the Emerald Tree Boa, *Corallus caninus*, of the Amazon Basin; and Green Tree Python, *Morelia viridis*, on New Guinea), as predicted by the competitive exclusion principle, and those in which one or both members of the pair are venomous and they are sympatric, as expected for mimicry (Greene 1997). A variant of the coralsnake mimicry hypothesis, that brightly colored colubrids escape predation when similarly patterned venomous elapids mistake them as conspecifics, also seems unlikely for these primarily chemosensory organisms. There are numerous records of intrageneric and even intraspecific predation by *Micrurus* (Roze 1996), as well as of predation by those venomous coralsnakes on the presumptive mimics *Cemophora coccinea* (by *M. fulvius,* Heinrich 1996), *Hydrops triangularis* (by *M. lemniscatus,* Roze 1996), *Leptodeira nigrofasciata* (by *M. nigrocinctus,* E. D. Brodie III, pers. comm.), and *Urotheca elapoides* (by *M. bernadi,* Roze 1996; *M. diastema,* H. W. Greene, unpub. obs.). At least one mildly venomous colubrid (*Erythrolamprus aesculapii*) occasionally preys on another (*Oxyrhopus guibei,* Sazima and Abe 1991). In any case, coralsnake mimicry's first century closed with Wickler's (1968) semipopular summary of the problem and a lack of convincing consensus.

PROGRESS AND SYNTHESIS

Twenty years ago we attempted to clarify the coralsnake mimicry problem by distinguishing two questions, of which only the second was dependent on a particular answer to the other: Are color patterns and defensive behaviors of certain venomous snakes aposematic, and are species that resemble those dangerous models in fact mimics (Greene and McDiarmid 1981)? Most early objections to mimicry (e.g., erroneous claims that the models are nocturnal and always deadly) actually had focused on the first question, whereas evidence from model-mimic ratios supposedly dealt with the second (e.g., Dunn 1954; Mertens 1956). Subsequent to Wickler's (1968) review, experimental studies on possible mammalian predators (Gehlbach 1972) and predatory Neotropical birds (Smith 1975, 1977) had addressed aposematic coloration in venomous coralsnakes rather than mimicry per se. In particular, Smith used wooden model snakes to separately control for color and pattern, and thus decisively solved the deadly model problem; she proved that naive individuals of relevant predators do avoid coralsnake color patterns and therefore need not learn the significance of those patterns to avoid harmless mimics. Building on Savage and Vial's (1974) study of Costa Rican coralsnakes, we documented concordant geographic color pattern variation in several putative Batesian mimics (*Atractus, Erythrolamprus, Lampropeltis triangulum,* and *Urotheca elapoides*) with respect to sympatric elapids (Greene and McDiarmid 1981), and thus supported the hypothesis that certain harmless or mildly venomous colubrids are indeed Batesian mimics of deadly elapids. Moreover, *Micruroides euryxanthus* closely resembles a larger venomous coralsnake, *Micrurus distans,* in color pattern only where they are sympatric, implying Müllerian mimicry by those venomous elapids (Greene and McDiarmid 1981).

Smith's (1975, 1977) experiments and our 1981 paper evidently sparked an attitudinal shift among herpetologists in favor of coralsnake mimicry (e.g., cf. Sánchez-Herrera et al. 1981 with Smith and Chiszar 1996; but see Bauer and DeVaney 1987; Beckers et al. 1996), and Pough (1988a, 1988b) linked venomous snakes with general mimicry theory. Two subsequent decades have revealed numerous additional examples of behavioral and color pattern resemblances between colubrids and venomous elapids, reasonably viewed as putative Batesian mimics and models, respectively (e.g., Campbell and Lamar 1989; Marques and Puorto 1991; Sazima and Abe 1991; Martins and Oliveira 1993, 1998; Roze 1996; Wilson et al. 1996; Vogt 1997; Köhler 2001; Zug et al. 2001). Additional cases of presumptive Müllerian mimicry, based on color patterns shifts by one or both venomous species where they are sympatric, include *Micrurus bocourti* and *M. mertensi, M. corallinus* and *M. decoratus, M. dissoleucus* and *M. dumerilii, M. frontalis* and *M. lemniscatus,* and *M. isozonus* and *M. lemniscatus* (Roze 1996; Marques 2002).

Since Pough's (1988a) review, an analysis of specimens from Brazil documented frequency shifts in alternative morphs of *Erythrolamprus aesculapii* across a parapatric range contact between it and two venomous coralsnakes with respectively similar color patterns (Marques and Puorto 1991). In local Costa Rican transects, motmots (the birds studied by Smith 1975) and other predators attacked "normal" plasticine snake models more often than they did those that looked vaguely like coralsnakes, and realistically colored coralsnake models provided even better protection (Brodie 1993; Brodie and Janzen 1995; Brodie and Moore 1995). Savage and his students forged a standardized terminology of coralsnake color patterns (Savage and Slowinski 1990) and more thoroughly documented concordant geographic variation in *Urotheca* and *Scaphiodontophis* (Savage and Slowinski 1996) with respect to sympatric venomous elapids. Beckers et al. (1996) exposed fresh-caught White-nosed Coatis (*Nasua narica*) to sympatric snakes and recorded no avoidance of and one actual attack on *Micrurus,* but their claim that research with model coralsnakes might be irrelevant is unjustified because the latter generally has focused on avian predators, controlled separately for color and pattern, and yielded clearcut results. More recent experimental transects with plasticine models demonstrated that the advantage of a coralsnake pattern in the United States is greater in areas of sympatry with dangerous models than at sites where venomous elapids are absent (Pfennig et al. 2001).

Numerous observations confirm that the palatability spectrum of mimicry theory is continuous or at least multimodal for snakes, and that the concept of Mertensian mimicry (Wickler 1968), in which "mildly venomous" colubrids (e.g., *Erythrolamprus*) are the models and both elapids and nonvenomous colubrids are the mimics, is encompassed by a Batesian-Müllerian mimicry continuum (Emsley 1966; Greene and McDiarmid 1981; Huheey 1988; Pough 1988a, 1988b). In fact, elapid bites are not always fatal to large mammals (e.g., Russell

1967), although they can kill fairly large raptors (Brugger 1989) and even humans (Roze 1996). Moreover, some raptors that do eat rattlesnakes (*Crotalus viridis*) nevertheless prefer nonvenomous colubrids when the latter are common, suggesting that the predators assess the risk of handling dangerous snakes relative to the cost of finding harmless prey (Fitch 1949), rather than regard rattlesnakes as absolutely unpalatable. Finally, some colubrids once thought to be harmless can deliver toxic bites (e.g., *Urotheca elapoides*, Seib 1980), and nonvenomous but powerful constrictors can fatally injure a predator (see Van Heest and Hay 2000), so nonvenomous *Lampropeltis triangulum* and other "harmless" species might even serve as Batesian and/or Müllerian models. Two important implications of these observations are that venomous snake mimicry systems might span punishment levels from temporarily uncomfortable to deadly, and that avoidance mechanisms likely include diverse types of learning as well as genetically based avoidance.

We conclude that as the twentieth century closed, available evidence strongly favored the coralsnake mimicry hypothesis. Although small sample sizes and other logistical problems have limited experimental field studies on that topic, Pough's (1988a, 1988b) emphasis on the special qualities of snake mimicry and Savage and Slowinski's (1992) estimate that 18% of New World non-elapid snakes are coralsnake mimics implied that broader implications might be forthcoming.

Other Venomous Snake Mimicry Systems

Jararacas [*Bothrops*] are not common here . . . but a non-poisonous snake indistinguishable from it in color and pattern seems to be more common. I do not know if the resemblance is based on protective mimicry of the jararaca, or if the coloration was acquired independently by both snakes as a protective resemblance to dead leaves and the like.

FRITZ MÜLLER IN AN 1893 LETTER (QUOTED IN WEST 2003)

[A *P. catenifer sayi*] exhaled in ordinary bullsnake fashion [and] kept up a continuous rattling of its tail among the dried grass and leaves. This combination of sounds greatly resembled the rattling of *Crotalus confluentus* [= *C. viridis*] and our reaction to it was almost as though we were really in the presence of a poisonous serpent.

J. K. STRECKER (1929)

VIPER MIMICS

Gans's (1961b) analysis of geographic color pattern variation in African egg-eating colubrids (*Dasypeltis*) vis-à-vis sympatric vipers (e.g., *Causus*, *Echis*) inspired our study of coralsnake mimicry (Greene and McDiarmid 1981) and remains the only detailed account of mimicry among Old World snakes in general and vipers in particular (for further details, see Young, Meltzer, et al. 1999). As Pough (1988a, 1988b) noted, the color patterns of most vipers and their putative mimics are obviously cryptic, such that alternative hypotheses of concealing rather than aposematic and mimetic coloration are especially plausible. Building on Pasteur's (1982) concept of abstract mimicry, Pough (1988a, 1988b) also pointed out that a generalized resemblance to vipers is relatively common, perhaps because a locally effective cryptic color pattern and behavioral similarity together can provide sufficient mimetic protection—especially given severe punishment for mistakes by predators (e.g., an unusual, sound-producing visual display by the extremely dangerous *Echis* and harmless *Dasypeltis*).

On the basis of specific behavioral and color pattern resemblances, several putative viper mimicry systems are worthy of detailed study, and many Old World catsnakes (*Boiga, Telescopus*) are exemplary candidates (Pough 1988a; see Schleich et al. 1996 for *Telescopus obtusus* as a behavioral mimic of *Echis;* Disi et al. 2001 for photos of *Coluber nummifer* and *Vipera palestinae*). At Tam Dao, Vietnam, freshly collected Chinese Catsnakes (*B. multitemporalis*) reacted to light touch from a human hand by assuming an exaggerated, anterior S-shaped coil and spreading the quadratomandibular joints laterally, then striking repeatedly; they so much resembled in color pattern and behavior a more frequently encountered, sympatric pitviper, the Chinese Habu (*Protobothrops muscrosquamatus*), that in the field we had difficulty distinguishing those two species without close inspection (H. W. Greene and D. L. Hardy Sr., unpub. obs.; voucher specimens in the Museum of Vertebrate Zoology, University of California, Berkeley, MVZ 226520 and MVZ 226628 to MVZ 226638, respectively). Numerous other Old World colubrids are perhaps more abstract viper mimics, and *Psammodynastes pulverulentus* has even long been known as the Mock Viper (Greene 1989).

The colubrid clade *Pituophis,* diagnosed by an epiglottal keel and loud rattle-like hissing (Young et al. 1995), probably originated in the context of rattlesnake mimicry (Sánchez-Herrera et al. 1981). Species of *Pituophis* later diversified in color pattern and behavioral ecology (Rodríguez-Robles and de Jesús-Escobar 2000) and variously resemble rattlesnakes or not in color pattern (Cope 1900; Strecker 1927; Klauber 1956; Benson 1978; Kardong 1980). Sweet (1985) analyzed crypsis and defensive behavior in *Crotalus viridis* and *Pituophis catenifer* across a California habitat gradient and concluded that the latter is at most a behavioral mimic of the former. He also noted that color pattern classes of *Pituophis* in the eastern United States (e.g., Black Pinesnakes, *P. melanoleucus lodingi*) resemble rattlesnakes even less than do their western North American congeners, but perhaps that is because sympatric pinesnakes and Eastern Diamond-backed Rattlesnakes (*C. adamanteus*) often occupy tortoise burrows, so that the former profits primarily from acoustic rather than visual mimicry of the latter.

In response to a predator, Fox Snakes (*Elaphe vulpina*) and many other colubrids vibrate their tails and thereby might resemble rattlesnakes, and perhaps tail vibration by some Old and New World colubrids arose as behavioral mimicry of pitvipers (e.g., Cooper 1859; Hay 1892; Greene 1988, 1997; see sonograms in Kuch 1997a). Other potential North American viper mimics include the Mexican Alpine Blotched Gartersnake (*Thamnophis scalaris*), which has a color pattern reminiscent of sympatric Lance-headed Rattlesnakes (*Crotalus polystictus*), and the Mesoamerican Highlands Gartersnake (*Thamnophis fulvus*), which resembles Godman's Pitviper (*Cerrophidion godmani*) in color pattern and defensive behavior (H. W. Greene, unpub. obs. of *T. fulvus* in Guatemala, voucher specimens at the University of Texas at Arlington; photos of *Thamnophis* in Rossman et al. 1996; photos of *C. godmani* and *C. polystictus* in Campbell and Lamar 1989).

Most species of snail- and slug-eating snakes (*Dipsas, Sibon, Sibynomorphus,* the dipsadinine colubrids) probably are at least abstract mimics of terrestrial vipers, by virtue of similar cryptic color patterns and defensive displays (Peters 1960; Greene and McDiarmid 1981; Sazima 1992; Martins 1996; Martins and Oliveira 1998; Marques et al. 2001, fig. 40; Cadle and Myers 2003), and *Sibon longifrenis* specifically resembles a palm pitviper, *Bothriechis schlegelii* (photos in Greene 1997; Solórzano 2001). Rear-fanged toad-eaters in the colubrid *Waglerophis-Xenodon* clade often exhibit specific color pattern similarities to sympatric venomous pitvipers (*Bothrops, Bothriopsis;* Martins and Oliveira 1998; Marques et al. 2001). In Costa Rica, when threatened by a human, *X. rabdocephalus* flattens its neck and thereby enhances color pattern resemblance to *Bothrops asper* (Pough 1988a); geographic variation in Brazilian *W. merremi* encompasses apparent mimicry of sympatric *Bothrops alternatus, Bothrops itapetiningae,* or *Bothrops jararaca* (Sazima 1992; M. Martins, pers. comm.). *Bothriopsis bilineata* is evidently mimicked by *X. werneri* (Hoogmoed 1985; photos in Starace 1998), and the latter is presumably derived from within the *Waglerophis-Xenodon* clade (Zaher 1999). P. A. Silverstone, a student of J. M. Savage, first described the color in life of *X. werneri:* "very pretty dull blue-green, really a turquoise . . . all dorsal scales have tiny black dots" (Hoogmoed 1985, 85). *Bothriechis schlegelii* and *B. bilineata,* along with *S. longifrenis* and *X. werneri,* respectively, thus represent independent evolutionary shifts to green color patterns and arboreal habitats within clades of drab-colored, terrestrial models and mimics (Martins et al. 2001; Parkinson et al. 2002).

Behavioral head triangulation implies that species of *Dipsas* are at least abstract pitviper mimics, and inter- and intraspecific color pattern variation suggests that in some cases those colubrids mimic particular pitvipers. The widespread *D. indica* looks more like sympatric *Bothrops jararaca* in southeastern Brazil, where the two species are sympatric, than it does elsewhere in South America, and the endemic Atlantic forest *D. albifrons* even more closely resembles *B. jararaca* in color pattern (Sazima 1992). Among other Neotropical colubrids, the shades-of-brown color patterns and defensive behaviors of *Pseustes poecilonotus* (juveniles only), *Tomodon,* and *Tropidodryas* also apparently mimic various *Bothrops* (Sazima 1992; Martins and Oliveira 1998; Marques et al. 2001), whereas the green color and gaping threat displays of colubrid parrot snakes (*Leptophis*) may contribute to abstract mimicry of *Bothriechis* and *Bothriopsis* (Greene 1997).

OLD WORLD ELAPIDS

Cobras, kraits, and other Asian elapids evidently have spawned several mimicry systems among sympatric cylindrophiids, colubrids, and each other. Contrary statements notwithstanding (e.g., Dunn 1954; Roze 1996), Asian coralsnakes (e.g., *Sinomicrurus,* Slowinski et al. 2001) and their putative mimics sometimes are ringed with red, black, and yellow or white. A kukrisnake (*Oligodon cyclurus*) at Tam Dao, Vietnam, possessed the same colorful ringed dorsal pattern as did local *S. macclellandi;* moreover, the kukrisnake thrashed erratically and bit readily when handled, as did *S. macclellandi* at that site (H. W. Greene and D. L. Hardy Sr., unpub. obs.; voucher specimens, respectively, are MVZ 224217, MVZ 224218, and MVZ 226613). Among other Asian colubrid clades, several species of wolfsnakes (*Lycodon*) look very much like sympatric kraits (*Bungarus;* Kuch 1997a; see, e.g., photos in Cox 1991), and color patterns of some reedsnakes (e.g., *Calamaria lumbricoidea, C. schlegelii;* Lim and Lee 1989; Stuebing and Inger 1999) closely resemble those of kraits and/or long-glanded coralsnakes (two species of *Calliophis* formerly referred to as *Maticora;* Slowinski et al. 2001). *Calliophis* (= *Maticora*) *bivirgata* and *B. flaviceps* both have dark blue or black bodies with red-orange or yellow heads and tails, and thus might be Müllerian comimics as well as Batesian models for colubrids (Slowinski 1994; Kuch 1997a; photos in Manthey and Grossmann 1997). The colubrid *Ptyas mucosus* looks like and growls like the King Cobra *Ophiophagus hannah* (Young, Solomon, et al. 1999).

Among other possible Old World snake mimicry systems, some Asian keelbacks (e.g., species of *Pseudoxenodon,* Manthey and Grossmann 1997; Chan-ard et al. 1999) and other natricine colubrids flatten their necks and thus might mimic cobras (*Naja*). At Tam Dao, Vietnam, a Golden Keelback (*Rhabdophis chrysargos*) elevated its head and spread a hood when closely approached, and thereby more closely resembled the overall appearance of sympatric Chinese Cobras (*N. atra;* H. W. Greene and D. L. Hardy Sr., unpub. obs.; voucher specimens respectively MVZ 226578 to MVZ 226582 and MVZ 226617). In some parts of New Guinea the overall shape and color patterns of Viper Boas (*Candoia aspera*) are very similar to those of stout-bodied elapids (species of *Acanthophis,* O'Shea 1996).

MIMICRY OF OTHER VENOMOUS SNAKES

In Mexico the brightly striped *Rhadinaea taeniata aemula* might be a Batesian or Müllerian mimic of sympatric *Coniophanes picevittatus* (Myers 1974), but otherwise the possibility that venomous rear-fanged species serve as models for other colubrids has rarely been considered. One Brazilian example involves *Philodryas viridissimus,* which is capable of seriously envenoming humans, and the nonvenomous *Chironius scurrulus; P. viridissimus* and juvenile *C. scurrulus* are brilliant green (the much larger adults of the latter are brown) and so closely resemble each other in color patterns and defensive threat displays that experienced herpetologists in the field sometimes confuse the two species (Martins and Oliveira 1998; Marques 1999). Another widespread venomous colubrid, *P. olfersii,* is bright green with a middorsal orange stripe only in southeastern Brazil, where it is sympatric with the less offensive, similarly colored, and regionally endemic *Liophis jaegeri* (Di Bernardo 1998; Hartmann 2001). Some venomous atractaspidids might be Batesian and/or Müllerian mimics in that rear-fanged purple-glossed snakes (*Amblyodipsas*) often look remarkably similar to front-fanged stiletto snakes (*Atractaspis*) (photos in Spawls and Branch 1995).

Some Unsolved Puzzles

INVERTEBRATES AS NOXIOUS MODELS?

Most discussions of venomous snakes and mimicry have dealt with them as putative models (e.g., acoustic defensive displays of rattlesnakes and burrowing owls, Rowe et al. 1986), perhaps on the questionable assumption that the other players, usually insects, are not noxious (reviewed in Pough 1988a). Gans (1973), however, suggested that nonvenomous shield-tailed snakes (Uropeltidae) are centipede mimics, on the basis of their similar defensive squirming behaviors and cross-barred ventral patterns, and Vitt (1992) presented compelling circumstantial evidence that some caecilians and elongate, brightly marked lizards are mimics of unpalatable millipedes and/or centipedes. Vitt (1992) also proposed that millipedes are models for coralsnake mimicry, a hypothesis contradicted with field experiments by Brodie and Moore (1995). Other advanced snakes might well be millipede mimics by virtue of a tightly coiled defensive posture coupled with cross-barred ventral (*Contia tenuis,* Leonard and Stebbins 1999) or dorsal coloration (*Xenopholis scalaris,* Zug et al. 2001). Inspired by Vitt (1992), Greene (1997) suggested that because many basal snakes (e.g., *Cylindrophis*) have alternating light and dark barred ventral patterns, myriapod mimicry might have facilitated their early radiation as surface-dwelling rather than burrowing squamates. Because anguid lizards are among the near outgroups of snakes (Lee 1998b), that hypothesis is consistent with extremely similar color patterns among millipedes and South American *Diploglossus* (Vitt 1992). Moreover, juveniles of a North American anguid with somewhat reduced limbs (*Elgaria kingi*) closely resemble sympatric large centipedes (*Scolopendra heros*) in cross-barred color pattern and sinuous locomotor escape behavior (L. J. Vitt, pers. comm.; H. W. Greene, unpub. obs.); the ventral color patterns of some species of *Diploglossus* resemble a coralsnake by having crossbars of red, yellow, and black (Savage 2002).

Several Asian land planarians look like venomous elapids in color pattern, thus potentially adding some surprising complexity to understanding venomous snake mimicry (color illustrations in Graff 1899; Fogden and Fogden 1974; Moffett 1998; photos of coralsnakes in Campbell and Lamar 1989). Among those soft-bodied invertebrates, *Bipalium everetti* of Borneo is approximately 10 cm long and has a banded dorsal body pattern of red-yellow-black-yellow-red like some Asian *Sinomicrurus,* many New World *Micrurus,* and the presumptive Old and New World colubrid mimics of those genera; *B. everetti* even has each end banded only in black and yellow, as is also typical of those elapids. *Bipalium ellioti* has a pattern of black and white bands, thus resembling some New World venomous coralsnakes (e.g., *M. mipartitus*), Old World kraits (e.g., *Bungarus multicinctus*), and putative colubrid mimics of the latter (e.g., some *Lycodon*). Another Asian land planarian, *Dolichoplana harmeri,* has black and light stripes with an orange head and tail, somewhat like an Asian long-glanded coralsnake (*Calliophis* [= *Maticora*] *bivirgata*) and its putative mimics among reedsnakes (*Calamaria,* photos in Lim and Lee 1989; Stuebing and Inger 1999).

An obvious nonadaptive explanation for the similarities between terrestrial Old World flatworms and elapids, especially Neotropical coralsnakes, is coincidence. However, ingestion of *Bipalium kewense* causes instant vomiting in cats and death in chickens, cows, and horses (Winsor 1983), and captive Bornean hornbills of two species rejected *B. everetti* as food (Fogden and Fogden 1974; M. P. Fogden, pers. comm.). Given unpalatability in terrestrial flatworms and evidence that the aposematic color pattern primitive for New World coralsnakes arose in their Asian common ancestor with *Sinomicrurus* (Slowinski et al. 2001), brightly marked planarians might even historically have been Batesian or Müllerian models for some snake mimicry systems. A third possibility is that those flatworms and venomous elapids convergently evolved particular aposematic color patterns because those combinations of hues afford the best signaling characteristics under particular light and predation threats (Hailman 1977).

WHY DOES MIMETIC PRECISION VARY AMONG SNAKES?

A few examples illustrate the wide range of phenotypic matching between putative models and mimics among snakes. Compared with sympatric New World elapids (photos in Campbell and Lamar 1989; Martins and Oliveira 1998), coral-snake mimics can have similar relative band length and identical ring color order (e.g., *Micrurus hippocrepis* and *Urotheca elapoides*), similar relative band length but different ring color order (e.g., *M. fulvius* and *Lampropeltis triangulum*), different relative band length and identical ring color order (e.g., *Micruroides euryxanthus* and *Chionactus occipialis annulatus*), or different relative band length and ring color (e.g., Amazonian *Micrurus* and *Anilius scytale*). Among Brazilian colubrids that mimic *Bothrops* (photos in Marques et al. 2001), resemblance ranges from overall drab, cryptic coloration and defensive head triangulation (e.g., species of *Thamnodynastes*) to strong color pattern similarity (e.g., *Dipsas neivai* and *B. jararaca*). The green dorsums and particular markings of *Sibon longifrenis* and *Xenodon werneri*, putative mimics of *Bothriechis schlegelii* and *Bothriopsis bilineata*, respectively, also imply that specific color pattern components (rather than only abstract viper resemblance) are important in predator deception by those colubrids. Snakes thus exhibit a continuum of resemblance precision, spanning near-perfect or concrete to vaguely abstract mimicry (Pasteur 1982; Pough 1988a, 1988b). Addressing that phenomenon more generally, Edmunds (2000; see also Johnstone 2002) proposed six explanations for what he termed good and poor mimics, each of which might apply to venomous snake mimicry systems:

1. Some venomous models are more noxious than others. Clearly, within a snake fauna and given a range of predator sizes and susceptibilities, the bites of several species of sympatric elapids and viperids can vary greatly in their effects on potential predators.
2. Because of variation in sensory capabilities, what is perceived as poor mimicry by humans might be sufficient against other predators. Birds and mammalian carnivores seem particularly likely to be relevant predators in snake mimicry systems, for example, and some of the former have much more acute vision than many of the latter (Greene 1988; Martins 1996; Tanaka and Mori 2000).
3. Poor mimics might simultaneously signal palatability and danger, and thus confuse predators long enough to permit escape.
4. A poor mimic moving rapidly might be just as effective as a good stationary mimic (see Srygley 1999). The perceived image of color patterns can change when snakes move rapidly (Pough 1988a; Brodie 1992), and this explanation might even explain protective resemblance to relatively slow-moving vipers if the mimics themselves are fast moving (e.g., *Xenodon* and some other colubrids).
5. Poor mimics might be in the process of losing or evolving more precise resemblance. This hypothesis predicts local variation in color patterns, and that closely related populations will exhibit respectively better or poorer mimetic resemblance.

6. A good mimic presumably receives protection only when it is sympatric with a particular model and a relevant predator, whereas a poor mimic might receive some protection involving several models and predators over a large area.

Edmunds (2000) noted that widely foraging predators might encounter a broad spectrum of aposematic models and quickly forget the rarer ones, and that learning abilities of different predators could affect selection for good versus poor mimics. Perhaps the ease with which brightly ringed elapid color patterns versus cryptic viperid patterns are incorporated into avoidance mechanisms by predators also influences whether good or poor resemblance is favored. In any case, with respect to nonmarine, front-fanged taxa (David and Ineich 1999; total number of species in parentheses), there are approximately 46 elapids and 43 viperids in South America (89), 28 elapids and 71 viperids in mainland Asia (99), and 17 *Atractaspis*, 26 elapids, and 51 viperids in Africa (94). Global variation in the species richness and natural history of venomous putative models clearly provides many opportunities for addressing mimetic precision, and Sweet's (1985) analysis of crypsis and putative mimicry of *Crotalus* by *Pituophis* provides an exemplary methodological starting point for field studies of that problem.

STILL MORE QUESTIONS

Why do some colubrids (e.g., *Lystrophis dorbignyi,* Yanosky and Chani 1988) resemble vipers with their dorsal patterns and elapids with their ventral coloration (Martins 1996)? Of what significance are the bizarre "partial coralsnake" color patterns of Bornean *Bungarus flaviceps baluensis* (black and white striped anteriorly, as in the nominate race, but with red, black, and white rings posteriorly; Stuebing and Inger 1999; Kuch and Götzke 2000) and Neotropical *Scaphiodontophis* (coralsnake colors at least anteriorly, sometimes striped or unicolor posteriorly; Savage and Slowinski 1996)? A coralsnake-patterned Brazilian colubrid, *Simophis rhinostoma,* exhibits presumptive viper mimicry (tail vibration) only when threatened at night (Marques 2000), consistent with Savage and Slowinski's (1996) suggestion that species with such puzzling mosaics of defensive traits target different predators at different times and/or in different habitats.

Parallel ontogenetic and color pattern changes between models and mimics do occur (e.g., *Micrurus alleni* and *Lampropeltis triangulum* in southern Costa Rica; Savage and Vial 1974), but what factors account for great size disparity in mimicry systems, which include snake mimics that are far smaller as well as those that are much larger than sympatric models? At La Selva Biological Station, Costa Rica, for example, a geometrid moth larva less than 2 cm long is remarkably similar to sympatric *M. alleni* in color pattern, including the fact that its anterior and posterior are bicolored whereas the body is tricolored (photos in Conniff and Murawski 2001), yet at that site the brightly ringed *L. triangulum* can be 1.5-fold longer than adults of all three sympatric species of *Micrurus* (H. W. Greene, pers. obs.). Perhaps unusually large mimics or exaggerated behavioral displays act as supernormal sign stimuli ("supermimicry"; Howard and Brodie 1973; Brodie 1976) for some predators, and tiny mimics are feasible because the size spectrum of potential prey for motmots and some other small birds (Remsen et al. 1993) encompasses neonate venomous snakes and even shorter insects.

Populations of *Erythrolamprus* vary in color pattern concordant with presence or absence of various sympatric *Micrurus* (Emsley 1966; Greene and McDiarmid 1981; Marques and Puorto 1991); *Micruroides euryxanthus* switches color pattern in sympatry with the larger *Micrurus distans* (Greene and McDiarmid 1981); and the color pattern of *Micrurus decoratus* more closely resembles that of the larger *Micrurus corallinus* than those of close relatives of the former (Marques 2002). Each of those examples implies Batesian mimicry, and granting a palatability spectrum among snakes, detailed studies of size and color-pattern shifts in sympatric South American *Micrurus* (Roze 1996) might clarify whether any venomous coralsnakes are primarily Müllerian co-mimics.

Within butterfly mimicry systems there is substantial overlap of species with similar color patterns in microhabitat and daily activity patterns (DeVries et al. 1999), and the extent to which venomous snake models and mimics overlap in space and time warrants detailed scrutiny. For seasonal activity in southeastern Brazil (Marques et al. 2000), three putative viper mimics (*Sibynomorphus neuwidi, Tomodon dorsatus, Xenodon neuwiedii*) cluster with two species of lanceheads (*Bothrops*) and a putative coralsnake mimic (*Erythrolamprus aesculapii*); conversely, 2 other viper mimics (*Tropidodryas serra, T. striaticeps*) cluster with 7 nonmimetic colubrids, and all 14 of those species cluster to the exclusion of the only elapid (*Micrurus corallinus*). Perhaps raptors and other relatively large predators of those snakes (Martins 1996) are sufficiently aseasonal foragers that they encounter models and mimics regardless of seasonal activity differences in their prey.

Venoms and Macroevolution: The Savage-Wallace Effects

Layla's Paradox, the puzzle as to why competent, powerful, intelligent endothermic predators are thwarted by the defensive displays of essentially harmless ectotherms (Greene 1988, 1997), underlies the evolution of venomous snake mimicry as well as the prevalence of abstract mimicry among snakes, especially vipers. As implied by Wallace (1867) and explicated by Pough (1988a, 1988b), the solution to Layla's Paradox is that sometimes what appears to be a harmless Costa Rican parrot snake (*Leptophis*) turns out to be an Eyelash Pitviper (*Both-*

riechis schlegelii), what looks like a palatable Gophersnake (*Pituophis catenifer*) is actually a Western Rattlesnake (*Crotalus viridis*), and a seemingly odd Asian lepidopteran larvae proves instead to be a juvenile Bamboo Pitviper (*Trimeresurus stejnegeri*). For many predators, much of the time, misidentification of venomous snakes as harmless could have such severe consequences that the risk of an error would be untenable.

Perhaps 25%–35% of nonvenomous snakes worldwide are mimics of elapids and/or viperids (Savage and Slowinski 1992; Greene 1997; additional species in Martins and Oliveira 1998), so a consequence of Layla's Paradox is that as dangerous models, and beyond their own diversification, venomous taxa probably have substantially contributed to overall snake diversity by protecting the lifestyles of harmless mimetic species. Without venomous viper models there likely would be no *Dasypeltis*, a clade of essentially toothless African colubrids whose egg-eating specializations are unmatched by any other snakes (Cundall and Greene 2000). And beyond simple increases in snake species richness, mimetic lineages include adaptive zones that are rare or lacking in the venomous clades themselves. Many mollusk-eating dipsadinines mimic vipers in color patterns and defensive threat displays, yet among the approximately 240 species of ecologically diverse viperids, only *Atheris* (= *Adenorhinos*) *barbouri* parallels that speciose colubrid subclade by preying on soft-bodied invertebrates (Greene 1997; Rasmussen and Howell 1998). Conversely, mimicry is taxonomically far more widespread and common among invertebrates than vertebrates, and the exceptions to that generalization likely reflect intrinsic characteristics of particular clades of organisms (Pough 1988a, 1988b). Accordingly, we summarize four macroevolutionary patterns exhibited by venomous snake mimicry as the Savage-Wallace Effects:

First, mimicry is more likely among closely related organisms that share a common body plan (e.g., among lepidopterans, among fishes; see Seigel and Adamson 1983; Pough 1988b), and thus their specific similarities (e.g., wing color patterns in butterflies) are representative of evolutionary parallelism. Mimicry is less likely among organisms that are grossly different morphologically; evidently even mimicry among major arthropod groups (e.g., salticid spiders that resemble ants; Reiskind 1976) is much less common, for example, than within Insecta. Paradoxically, crypsis often involves similarities among organisms that are taxonomically and morphologically disparate (e.g., katydids whose wings resemble leaves with herbivore damage; E. D. Brodie III, pers. comm.), and perhaps that discrepancy entails differences in the behavioral components of prey discovery (for crypsis) and prey handling (for aposematism and mimicry).

Second, mimicry spanning distantly related organisms, representative of evolutionary convergence, is more likely to involve planarians, myriapods, fishes, snakes, and other groups with relatively simple body forms. Impressive examples are palatable larvae of tropical marine burrfish that look very much like noxious mollusks known as sea hares (Heck and Weinstein 1978), tropical lepidopteran larvae that dramatically resemble pitvipers (Pough 1988a), and legume seed pods that look like caterpillars (Lev-Yadun and Inbar 2002), whereas octopuses are spectacular exceptions to this generalization in that they mimic fish by changing color pattern and even external form (Hanlon et al. 1999).

Third, among vertebrates, snake mimicry is unusually widespread because of (1) and (2), and because venomous species can severely injure or kill predators (Wallace 1867; Pough 1988a, 1988b).

Fourth, the origin of noxious attributes can markedly increase diversity within a clade beyond that encompassed by unpalatable species; dangerous models thereby make otherwise "unprotected niches" possible for harmless relatives, and even for lifestyles not used by the models themselves. Discussions of unbalanced clade diversification typically have focused on key innovations as apomorphies for the groups in question (e.g., Hunter 1998), but the origins of noxious qualities and aposematism within snakes evidently set the scene for repeated evolution of mimetic signals that influenced overall species richness and overall clade diversity far beyond that of noxiousness per se. One way to explore that claim would be to compare sister taxa with and without mimicry against null models for symmetrical cladogenesis (e.g., Slowinski and Guyer 1993); another would be to assess proportions of species in "unprotected niches" for mainland tropical snake assemblages in which elapids and viperids are present, with roughly comparable sites on Madagascar and elsewhere that lack venomous models (cf. Cadle and Greene 1993; Cadle 2003). The latter approach would require far more detailed characterizations of snake assemblages and their predators than are now available (for important exceptions, see Martins 1996; Tanaka and Mori 2000).

Future Prospects

Three general approaches offer much promise for future studies of venomous snake mimicry. First, innovative field and laboratory experiments with models (including models of vipers; Andren and Nilson 1981), detailed analyses of staged encounters with live prey (Oliveira and Santori 1999), and observational studies of free-living predators (Greene 1986) should provide complementary insights into the ethology of feeding and defense as they relate to palatability spectra (for theoretical considerations, see, e.g., Speed and Turner 1999; Rowe and Guilford 2000).

Second, we have scarcely utilized museum materials and other nontraditional approaches to study the evolutionary biology of color pattern variation. A long-standing tradition in vertebrate systematics is to summarize rather than explore variation, and few studies of snakes have assessed local and intraspecific variation in putative models and mimics (e.g., Savage and Crother 1989; Marques and Puorto 1991). Theory predicts, for example, that Batesian models and mimics should vary within populations, whereas Müllerian co-mimics should be monomorphic (e.g., Huheey 1988; Joron and Mallet 1998), and Leenders et al. (1996) suggested that polymorphism for white or red bands within a Costa Rican population of *Micrurus mipartitus* represented attempted escape from Batesian mimics. However, a probable mimic of that coralsnake, *Urotheca euryzona*, is correspondingly dimorphic at La Selva Biological Station (H. W. Greene, unpub. obs., on MVZ 215672, MVZ 215696), approximately 20 km from their site, so that situation invites further study (see Savage and Crother 1989 for details of variation in that species). Museum specimens might also be used to explore the developmental basis for the evolution of mimicry, particularly with respect to locally variable phenotypes (e.g., Meachem and Myers 1961; Martins and Oliveira 1993, 1998; Savage and Slowinski 1996) and theoretical models for color pattern transformations in snakes (Murray and Myerscough 1991; Savage and Slowinski 1996).

Another way in which museum collections can serendipitously contribute to the study of mimicry is illustrated by observations on Scarlet Kingsnakes (*Lampropeltis triangulum elapsoides*) at Archbold Biological Station (ABS; Highlands County, FL). A preserved adult *L. t. elapsoides* (ABS 50) came from the stomach of a road-killed Great Horned Owl (*Bubo virginianus*) and thus confirms predation on this putative mimic of *Micrurus fulvius* (Greene and McDiarmid 1981). Furthermore, among 13 preserved and 1 live *L. t. elapsoides* from ABS, we recorded dorsal black ring fusion counts of 0 for 7 snakes, 1 for 3 snakes, and 2, 3, 4, and 5 for 1 snake each ($x = 1.3$ fusions/snake). The owl's prey item has 10 ring fusions that dorsally obscure some red rings; in that sample of 14 specimens, ABS 50 is thus by far the most unlike *M. fulvius,* just as expected if the typically tricolored pattern of *L. t. elapsoides* facilitates coralsnake mimicry.

Third, aposematic and mimetic attributes that have been fixed for populations, species, and higher taxa can be studied by phylogenetic analysis (Larson and Losos 1996), and with that perspective we might better understand the historical diversification of mimicry assemblages (Turner and Mallet 1996). As Coddington (1988) pointed out, our study of geographic variation in coralsnake mimics (Greene and McDiarmid 1981) would have profited from analyses of their color pattern transformations vis-à-vis cladogenesis in elapid models, as would other snake mimicry systems (for examples with other mimicry systems, see Brower 1996; Zrzavy and Nedved 1999; Symula et al. 2001). There are as yet no detailed phylogenies for all components of any snake mimicry systems (see Wilson and McCranie 1997, regarding Smith and Chiszar 1996), but cladistic analyses are available for some of the model clades (*Bungarus,* Slowinski 1994; New World elapids, Slowinski 1995) and will undoubtedly lead to interesting insights. We know, for example, that the simple tricolored pattern that is primitive for New World elapids arose in their Asian common ancestor with *Sinomi-*

crurus (Slowinski et al. 2001), and therefore that bicolored and more complexly tricolored patterns are derived within a clade characterized by the former (Slowinski 1995). Among colubrids, *Lampropeltis triangulum* (Williams 1988) and *Urotheca* (Savage and Crother 1989) hold special promise for phylogenetic studies of mimetic diversification because of their extensive color pattern variation in and out of sympatry with venomous coralsnakes. The Xenodontini also has excellent potential for historical evolutionary approaches because it includes radiations of coralsnake mimics (*Erythrolamprus*) and viper mimics (*Waglerophis-Xenodon*), both within a larger clade of more generalized cryptic xenodontines (Zaher 1999; Vidal et al. 2000).

Finally, we emphasize that beginning about 135 years ago (Wallace 1867), scientific interest in venomous snake mimicry first encompassed observations from nature (some brightly marked colubrids look like deadly elapids), then indirect efforts at hypothesis testing (model-mimic ratio data); still later came controlled experiments using model snakes and only recently a preliminary theoretical synthesis. Thus far, studies of venomous snake mimicry have made only occasional contributions to our understanding of microevolutionary processes (e.g., see Mallet and Joron 1999; Rowe and Guilford 2000; Jiggins et al. 2001; Pfennig et al. 2001), but snake mimicry is proving unusually interesting in terms of macroevolutionary patterns. Our "discovery" in the natural history literature of toxic, brightly colored flatworms implies that some aspects of the "coralsnake mimic problem" still await explanation, and more generally, recent advances in phylogenetics hold much promise for unraveling the complex history of mimetic evolution in serpentine organisms. We expect that the research cycles initiated by Alfred Russel Wallace in the nineteenth century and invigorated by Jay Mathers Savage in the twentieth century will continue to occupy organismal biologists for some time to come.

Acknowledgments

We thank many friends and collaborators for their input; J. N. Layne and H. M. Swain for facilitating work at Archbold Biological Station; W. G. Eberhard, R. E. Ogren, J. E. Rawlins, and R. Root for advice about invertebrates; J. A. Rodríguez-Robles for information on specimens in the Museum of Vertebrate Zoology; M. P. Fogden for permission to mention his observations of hornbills; and E. D. Brodie Jr., E. D. Brodie III, P. J. DeVries, M. A. Donnelly, G. E. A. Gartner, M. Martins, and K. Schwenk for helpful comments on the manuscript. A grant from the Pacific Rim Foundation to the Museum of Vertebrate Zoology supported HWG's fieldwork in Vietnam, and the Lichen Fund has supported his subsequent research in Florida, Brazil, and elsewhere.

Part II
Ecology, Biogeography, and Faunal Studies

Part 2 begins with studies of single species, builds to analyses of communities and the distributional patterns of particular lineages, and ends with examinations of regional herpetofaunas. The chapters discuss ecology, biogeography, and faunal studies. The tool that unites this section with the previous one is the information encoded in phylogenetic analyses. Those studies whose primary goal is reconstruction of historical relationships appear in the first section; those studies whose primary goal is understanding factors affecting current distributions and/or life history appear in this section.

In chapter 10, an example of current problems in population ecology of tropical amphibians is presented by Karen Lips, who applies multivariate quantitative genetic approaches to explore how selection might act on *Hyla calypsa*, a leaf-breeding frog recently described from Costa Rica. Leaf-breeding frogs are convenient for population studies because quantification of egg production and egg and tadpole survivorship can be performed with relatively great precision. Lips uses a detailed look at a single breeding season to provide insights into a single episode of selection. During this year, morphological and behavioral traits varied significantly among males in the study population, and male reproductive success was skewed toward heavier individuals that had long tenure at the calling sites and called frequently. Of these characteristics, site tenure was the primary factor that predicted male reproductive success, and Lips argues that males with long tenure at the breeding site are more likely to encounter females than are males who visit the breeding site less often. The reproductive patterns described for *Hyla calypsa* are similar to those for four species of tropical frogs with prolonged breeding seasons. Lips provides the first such quantitative analysis of selection for a tropical frog, which should open the door for research into the relative advantages and disadvantages of the leaf-breeding reproductive mode in anurans.

Craig Guyer and Maureen Donnelly, in chapter 11, build on the theme of anuran ecology by exploring patterns of co-occurrence of an assemblage of hylid frogs at a single reproductive site in northeastern Costa Rica. On the basis of studies of temperate-zone anurans, interspecific competition has been suggested to play a role in structuring how calling males of different species associate with each other. Tropical communities have been studied less frequently, but some researchers have suggested that tropical frogs alter the timing of vocalization to prevent temporal overlap and associate in fashions that minimize the pitch of the competing voices of male frogs. Chapter 11 examines these hypotheses by using null models. In it, Guyer and Donnelly describe the species composition of calling choruses of males and emerging groups of froglets and test whether these assemblages can be differentiated from those expected of random associations of species. The patterns observed at this site indicate that calling males create nonrandom assemblages of species but that these associations do not appear to be created in a fashion to maximize the degree of difference in the voices of the frogs. Additionally, juveniles emerge in nonrandom groups that can be used to define prolonged transformers (those that have a few individuals transforming over long periods of time) and explosive transformers (those for which all individuals transform over short periods of time). The disparity between what Guyer and Donnelly describe for a single tropical site and what others have described from among tropical sites and the apparent disconnect between patterns associated with breeding adults and emerging juveniles are identified as important new avenues of research.

In chapter 12, Norman Scott and Luz Aquino continue the theme of examinations of anuran community ecology by describing foraging patterns of frogs of the Gran Chaco of Paraguay. Although most anurans consume small invertebrates, several species in the Chaco consume other frogs, and Scott and Aquino describe this syndrome by using morphological characters, focal-animal observations, and field experiments. Five of the 22 species from the

Chaco eat other frogs, and the 3 species of ceratophrynids and 2 species of leptodactylids that do this all have very large gapes compared with those of species
that do not eat frogs. The ceratophrynids have stiff mandibles and fang-like
teeth that pierce the bodies of anuran prey. Scott and Aquino operationally
define anurophagy and examine why this phenomenon is widespread in the
Chaco region.

As a final approach to examinations of anuran ecology, David Bickford
(chap. 13) describes diversity and conservation of frogs in Papua New Guinea.
This contribution provides a glimpse of how integrated conservation and development projects operate and how such programs can provide useful monitoring data that can be used to address scientific questions. The focus of the
study is a 270,000-ha site in the central cordillera of Papua New Guinea, an area
of incredibly steep topography and exceptionally high rainfall (6–8 m annually). The issues associated with conserving such large and diverse areas represent perhaps the key challenge in tropical conservation. The information in this
chapter serves as a template for the ways that monitoring performed by local
people can make use of their unique insights as well as encourage local control
of conservation efforts. These results will be of interest to those who plan similar projects in tropical countries. Additionally, this contribution includes information on the potential effects of the El Niño Southern Oscillation (ENSO)
on anuran distributions. Frog abundance was observed to increase during the
ENSO event, a finding that was counterintuitive but resulted from movement
of canopy species to lower positions, which improved their chances of being included in monitoring surveys. This chapter links ecological studies of tropical
herpetofaunas to efforts to understand the effects of global climate change, and
those interested in that issue will find important caveats to the proper interpretation of monitoring data as they relate variation in animal population sizes to
environmental changes.

Chapters 14 and 15, which explore historical biogeography of Neotropical
taxa, are linked together because both are derived from evaluations of published phylogenetic hypotheses and lead to differing interpretations of the role
of geological and climatic vicariant events in explaining current distribution
patterns. Chapter 14 explores the historical biogeography of tropical anoles, a
species-rich group of lizards that is well represented in the New World tropics.
Studies of Caribbean species of these lizards have led to important theories of
the role of competition in regulating local species richness. Kirsten Nicholson
focuses on the mainland taxa and uses her recent phylogeny for the group to
explore implications for the origin of the radiation and the likelihood that either vicariance or dispersal has played the primary role in creating the current
distribution of this radiation. Because she fails to find a north-to-south distribution pattern for mainland species groups and because the ancestors of *Norops*
are all from Caribbean islands, she infers that dispersal of an ancestral form

from the Caribbean to the mainland explains the origination of the genus and that the direction of faunal buildup throughout the mainland was as likely to have been from south to north as it was to have been from north to south. All of these features suggest that dispersal of colonists occurred after the major tectonic events of the mainland and that alternative explanations to simple vicariance or random dispersal need to be generated. In chapter 15, Steven Werman reviews the biogeography of the Neotropical pitvipers. Published phylogenies are available for pitvipers, as they are for anoles, and from them likely historical scenarios can be inferred. From these, Werman documents a paraphyletic Middle American assemblage of taxa that includes terrestrial and arboreal clades and a distinct South American assemblage. In contrast to the pattern from anoles, the Middle American pitvipers appear to have differentiated in response to vicariant orogenic and climatic events.

Chapters 16 and 17 present information on how inventory studies are conducted. In chapter 16, Roy McDiarmid and Jay Savage describe a faunal survey of the Osa Peninsula in southwestern Costa Rica. Their approach is traditional in the sense that the primary objective of the inventory is to describe what species are present at a site before it disappears as a result of habitat destruction. Forty-six amphibian and 69 reptilian species are chronicled from this site, and the taxonomic affinities of this fauna are compared with similar surveys at other Neotropical sites. These affinities indicate that the Golfo Dulce region of Costa Rica is an important component of biodiversity of Lower Central America and, therefore, that conservation efforts to preserve representative areas of this region are needed. Sadly, the specific region described by McDiarmid and Savage has been altered in ways that require that such conservation efforts be directed elsewhere.

In chapter 17, Maureen Donnelly, Megan Chen, and Graham Watkins describe a faunal inventory project specifically associated with preserving an area before it is lost. The study area is the Iwokrama Forest, a 360,000-ha forest preserve located in the central part of Guyana. Donnelly and colleagues provide inventory data for the herpetofauna and explore these data to determine ways that such information can be used to determine abundance or rarity of each taxon and, therefore, to estimate the number of taxa present but not recorded in inventory samples. In this chapter, the role of site-to-site variability within large preserves is documented. The analyses and definitions developed in this chapter will be of interest to others working in large and geologically complex areas.

The book ends (chap. 18) with an examination of the herpetofaunal diversity and historical biogeography of the tepuis of South America, a region that has captivated biologists since they were first described by British explorers in the late 1800s. In this chapter, Roy McDiarmid and Maureen Donnelly summarize accumulated inventory lists for these distinctive geological formations that represent habitat islands dispersed over a very broad region. Complex patterns associated with events on both an ecological and evolutionary time scale are required to comprehend how these areas developed the herpetofaunas currently known from them. For this reason, we present it last as a study that combines interpretations and techniques appropriate for both sections of the book. The tepuis are home to 159 species, most of which are highly endemic to the region, if not to specific mountaintops. Because phylogenetic hypotheses largely are lacking for these taxa, other tools must be used to infer historical origins of these faunas. In this chapter McDiarmid and Donnelly use multiple regression to document an unusual association between herpetofaunal diversity and the diversity of primitive plant groups. This feature is used to suggest that the herpetofauna is likely to be ancient and is likely to have been affected by complex changes in the size and shape of the tepuis as they were altered from tectonic activities and erosion.

10

Quantification of Selection and Male Reproductive Success in *Hyla calypsa,* a Neotropical Treefrog

Karen R. Lips

Introduction

Neotropical anurans exhibit the most diverse assortment of reproductive modes of any frog fauna in the world (Duellman and Trueb 1986). Variable environmental conditions acting on demographic and life-history traits of populations likely promoted such diversification (Wells 1977; Crump 1982). Demographic studies that examine environmentally induced variation in reproductive success among populations, or phenotypes within a population, can document selection pressures.

Prolonged breeding seasons often result in male-biased operational sex ratios (Emlen and Oring 1977; Wells 1977), which may cause competition for access to females and result in variation in male reproductive success. Male mating (fertilizations) and reproductive success (offspring) should be directly or indirectly correlated with traits that increase reproductive success by increasing availability to females, by giving an advantage in competition over mates, or by being favored by female choice (Andersson 1994). If traits reliably indicate male quality, selection should favor females that mate with the best males (e.g., good genes, more attractive offspring, greater fertilization success). The "best" males might be determined by body size, site attendance, and territory quality, all of which should indicate either healthier or stronger individuals.

Longitudinal studies of populations in which individuals are followed through their entire lifespan can identify traits that are most important in determining lifetime mating and reproductive success. However, it is not always feasible to follow individuals of an entire population for many years (but see Clutton-Brock 1988). If mating and reproductive success can be quantified for individuals within a breeding season (which constitutes an episode of selection), multivariate analyses can indicate which traits are under sexual selection (Lande and Arnold 1983; Arnold and Wade 1984a, 1984b). This method has been used for studies of phenotypic selection in birds (Price 1984; Weatherhead and Clark 1994), plants (Kalisz 1986; Schemske and Horvitz 1989), insects (Koenig and Albano 1987; Moore 1990; Fairbairn and Preziosi 1996), anurans (Arnold and Wade 1984b; Marquez 1993), reptiles (Hews 1990; Brodie 1992), fish (Warner and Schultz 1992; Fleming and Gross 1994), and mammals (Gibson 1987).

The phenotypic traits measured in my study have been correlated with mating success in a variety of organisms (Andersson 1994). Body size and mass can be important indicators of condition or territory quality (Searcy 1982; Ryan 1985) and may be important in female choice (Davies and Halliday 1977; Gatz 1981a, 1981b; Woodward 1982; Schwagmeyer and Brown 1983; Robinson 1986; Schluter and Smith 1986; Warner 1987; Conner 1988; Zuk 1988; Cuadrado 1999; Fuller 1999). Site attendance is known to increase reproductive success in many anurans (reviewed in Sullivan and Hinshaw 1992; Murphy 1994a) and may reflect the physiological condition of the males. Vocal behaviors that reflect the amount of energy invested in calling (e.g., call duration, frequency, and repetition rate) also might indicate physiological condition and can be subject to female choice (Klump and Gerhardt 1987). Finally, characteristics of the calling site itself can influence female mate choice (Burrowes 2000; Uy and Borgia 2000), and under these conditions males should compete for access to better territories (Ryan 1985). Better territories might be those that have particular kinds of oviposition sites (Howard 1978a), tadpole-rearing sites (Donnelly 1989), or reduced predation (Fincke and Hadrys 2001).

Given the prolonged breeding season characteristic of tropical rainforests and the prevalence of sexual selection in brightly colored tropical birds (Mc-Donald 1989) and marine fish (e.g., Warner 1987), I expected that tropical rainforest frogs would be likely to show uneven reproductive success, perhaps in association with female mate choice. Specifically, I investigated whether any of two morphological (snout-vent length [SVL], mass), two behavioral (site tenure, call rate), or two territory traits (number of leaves, leaf area) were correlated with male reproductive success during one breeding season in a wild population of *Hyla calypsa*, a Neotropical treefrog. Because of the prolonged breeding season, male territoriality, and leaf-breeding habits of this species, I expected to find variation among males in reproductive success to be related to variation in male or territory quality.

STUDY SITE

All surveys were conducted at Finca Jaguar, approximately 18 km north-northeast of La Lucha, Canton Coto Brus, Puntarenas Province, Costa Rica (8°55′ N, 82°44′ W). This private farm is an isolated human residence within the Zona Protectora Las Tablas, a component of the Amistad Biosphere Reserve. Las Tablas supports a high-elevation (1,900 m) cloud forest (Lower Montane Rainforest, in Holdridge 1982) and experiences pronounced wet (May–December) and dry (January–April) seasons, with daily heavy precipitation during the rainy season. Surveys were centered along a 400-m stretch of the headwaters of the Río Cotón, a swiftly flowing permanent stream with a rocky bed, deep pools, and numerous mossy rocks and boulders along the banks. Dense vegetation overhangs the stream. Stream width ranges from 1 to 15 m, and maximum depth reaches 97 cm. Water is cool (approximately 15°C), and its clarity probably indicates nutrient-poor conditions.

STUDY SPECIES

Hyla calypsa (family Hylidae), a small arboreal treefrog, inhabits vegetation over mountain streams in Costa Rica and Panama at elevations above 1,800 m (Lips 1996). *H. calypsa* breed throughout the nine-month rainy season, and males can be found along the stream throughout the year. Males exhibit site fidelity and are found within a 10- to 15-m-long stretch of stream for periods up to 36 months; foreign males rarely enter these areas, even when residents are absent (K. R. Lips, pers. obs.). Additionally, distinctive vocal interactions and scar patterns on the neck and shoulder region characteristic of male–male combat suggested that males vocally and physically defended these areas (K. R. Lips, pers. obs.). Therefore I concluded that these areas are territories. Females visited the stream only to lay eggs on the underside of leaves overhanging the stream, but they laid several clutches during a breeding season. Adults can live at least four years, and individuals of both sexes returned to the same area of the stream multiple times within and among breeding seasons (Lips 1995). Females arrive at the stream asynchronously throughout the breeding season, which results in a male-biased nightly sex ratio and allows for female mate choice among the relatively sedentary males. After my census in 1993, I noted that populations of all stream amphibians at this site began to decline, and by 1996 several species had disappeared (Lips 1998). In 1993, the Finca Jaguar population of *H. calypsa* began to decline and by 1996 had declined to 90% of initial levels; the population was represented by only three males in 1998 (Lips 1998) and may now be extirpated.

Methods

From 1991 through 1996, I monitored the population dynamics and reproduction of this population of treefrogs (Lips 1995). I conducted a mark–recapture study on the adults during nocturnal censuses and quantified clutch characteristics and development during diurnal censuses. Two 400-m transects, separated by 50 m, were delineated along the headwaters of the Río Cotón, with red flagging tied every 10 m. Between July 1991 and August 1996, I ran 118 nocturnal surveys (33 in 1991, 44 in 1992, 17 in 1993, 11 in 1994, and 13 in 1996). I surveyed both transects daily but only sampled the adult population along the upper transect at one- to four-night intervals. Surveys were conducted for the entire breeding season of 1992, but only for the first half of the breeding season in 1994 and 1996, and only for the last half in 1991 and 1993. All animals were identified (sex, age), measured (SVL, mass), and marked with individual toe clips. Recaptures and amplectic pairs were not always weighed or measured. All capture and clutch locations were mapped by estimating the distance to the closest flag.

I defined site tenure as the number of times an individual was present during the 44 survey nights in 1992 and calling rate as the proportion of captures in which males called during the last 10 m of my approach. Both site tenure and calling rate are indices of male activity over the season. The 1992 breeding season extended from 15 April to 15 December, for a total of 244 days. The 44 nights of observations represent 18% of the total breeding season, which is a low proportion of the breeding nights. I may be underestimating variation in site tenure and call rate, which in turn underestimates the importance of these traits in determining reproductive success.

All egg clutches were monitored, and tadpoles were captured upon hatching by suspending a water-filled plastic cup (Hayes 1983) beneath advanced clutches (that is, clutches at or above stage 25 as defined by Gosner 1960). Because males are territorial, site faithful, and fertilize eggs externally during oviposition, I assumed that clutches found in a territory were sired by the resident male, which is defined as a male captured within one month of the oviposition date and within 10–20 m of the clutch. In 1992, I observed the resident male in amplexus during oviposition of 21 (9%) of the 230 clutches produced that year. I assumed that the remaining 91% of clutches were also sired by the resident male. I compared the distribution of oviposition sites to a Poisson distribution (Zar 1984) to determine whether clutches were randomly distributed or not. If clutches were not located at random along the transect, I concluded that females showed a preference for particular males and/or his territory, or that males chose particular call sites and females mated randomly among those males. In a separate study, I enclosed 33 of the 230 clutches with bags of fine mesh to prevent maggot infestation. Although netting could have increased the number of tadpoles that hatched, for this study I assumed that all bagged clutches that produced tadpoles ($n = 24$) would have been successful without the netting, because a χ^2 contingency test indicated that there was no difference in hatching probability between protected and unprotected clutches ($\chi^2 = .0035$, df $= 1$).

Because females almost always oviposit on leaves overhanging the stream, I expected vegetative characteristics to be an important component in females' choice of mates. I further expected that males with high reproductive success would have territories with particular vegetative characteristics. In 1993, I measured the vegetation surrounding 77 clutches and 69 randomly chosen sites to determine whether oviposition sites differed from random sites. For clutches, I vertically centered a 1×1-m grid over the clutch; for randomly chosen sites, I centered the grid over the outermost leaves overhanging the stream. I counted the number of leaves per morphospecies of plant and then measured the length and width of a representative leaf. By multiplying the representative leaf area by the number of leaves and by summing this value for all morphospecies encountered in the grid, I obtained an estimate of the total leaf area (cm^2) per 1×1-m plot. Vegetation coverage between the clutch and random plots was compared by using both the average number of leaves per grid and the average leaf area per grid with one-way analysis of variance (ANOVA). I used the 1993 vegetation data for the quantification of 1992 male territory quality by taking the average of the three nearest vegetation plots, regardless of whether they were a clutch or random site. I calculated two estimates of territory quality: (1) the average number of leaves, and (2) the average leaf area per plot.

To examine reproductive success among males in the population, I plotted the frequency distribution of the three components of reproductive success (clutches, eggs, and tadpoles) for all the males known to be alive during the breeding season. I compared mean values for nonstandardized phenotypic traits of sires to those of nonsires (two-sample t-test) to look for phenotypic differences and then calculated Pearson product-moment correlations between all pairs of phenotypic characters to determine which characters were intercorrelated.

I analyzed individual variance in reproductive success with multivariate regression. First I calculated relative fitness by dividing individual fitness by the mean population fitness for each component (number of clutches, number of eggs/clutch, and number of tadpoles/clutch) and standardized all phenotypic values to a mean of zero. I derived directional selection gradients (β) from regression coefficients of fitness functions (mated or unmated status) to the standardized values of the five male traits. All males are either mated or unmated; thus the residuals of a least-squares regression may not be distributed normally when the dependent variable is bimodal. I regressed the six measurements of male "quality" onto the three multiplicative components of reproductive success and total fitness for the 54 males with complete phenotypic data.

All statistical analyses were performed with *SAS* version 6.08 (SAS Institute 1989).

Results

During the 44 surveys of 1992, I captured 121 males and 89 females; the nightly sex ratio (proportion of males to total captures) was usually male biased (range = 1.0–.45). Male traits varied among the 121 males in the population: male SVL ranged from 28 to 33 mm ($\bar{x} = 30.45$); mass ranged from 1.43 to 2.40 g ($\bar{x} = 1.81$); minimum site tenure ranged from 1 to 19 days ($\bar{x} = 3.90$); calls ranged from 0 to 8 captures ($\bar{x} = 1.28$); number of leaves on the territory ranged from 0 to 89 ($\bar{x} = 40.19$); and leaf area ranged from 0 to 52,816 cm^2 ($\bar{x} = 8,172$).

Regardless of whether data from all 121 males or only the 66 males that fertilized clutches were used, significant correlations existed among most phenotypic characters (table 10.1); the two measures of territory quality were not correlated with any of the other characters but were correlated with each other. Significant correlations indicate that selection acting on any one trait may produce changes in the distribution of other correlated traits, justifying my use of multivariate selection.

The 230 clutches contained 3,453 eggs and produced 1,095 tadpoles. Reproductive success was skewed among the males in the population (fig. 10.1); only 66 males (53%) fertilized a clutch of eggs, and only 55 (45%) sired tadpoles. Variation among the 121 males is expressed by the high coefficient of variation (CV): number of clutches (CV = 81.1), eggs (CV = 77.9), tadpoles (CV = 94.5) sired. Plots centered on clutches had significantly more leaves ($F = 10.13$, $p = .002$, 139 df) and greater leaf area ($F = 7.62$, $p = .007$, 139 df) than did random plots. Clutches were not distributed according to a Poisson distribution

Table 10.1   Pearson correlation coefficients between all pairs of the six phenotypic traits

	Mass	No. captures	No. calls	No. leaves	Leaf area
SVL	.5005***	.3170***	.2613*	−.2113	−.1072
Mass		.2750*	.3409**	−.1604	−.2189
No. captures			.7519***	−.1520	−.2318
No. calls				−.1479	−.1644
No. leaves					.3119**

*$p < .05$; **$p < .01$; ***$p < .001$

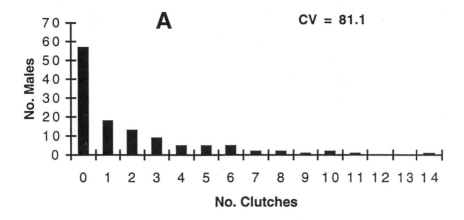

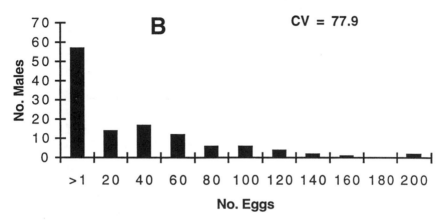

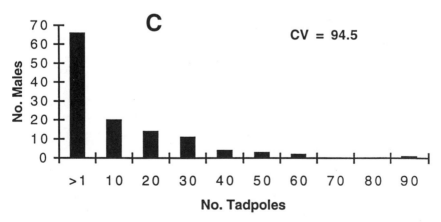

Figure 10.1   Frequency of distribution of the (A) number of clutches, (B) number of eggs, and (C) number of tadpoles for the 121 males present in 1992. Note the increase in the coefficient of variation (CV) from clutches to tadpoles.

Table 10.2 Phenotypic traits for sires and nonsires

Trait	Sires (*n* = 55)	Nonsires (*n* = 66)	*p*
SVL (mm)	30.59 ± 1.06	30.23 ± 1.14	.1510
Mass (g)	1.86 ± 0.21	1.76 ± 0.21	.0369
No. captures	5.29 ± 3.59	2.37 ± 2.11	<.0001
No. calls	1.77 ± 1.67	0.72 ± 0.95	<.0001
No. leaves	39.7 ± 19.8	40.6 ± 18.3	.8380
Leaf area (cm²)	8,683 ± 9,030	7,599 ± 6,660	.4450

Note: The average values of phenotypic traits are ± standard deviation. Sires are males that have sired at least one tadpole. Nonsires are males that sired no tadpoles.

(χ^2 = 336.77, 10 df, $p <$ .001) but were aggregated along particular sections of the stream.

I used *t*-tests to compare male traits at two levels: (1) between males that fertilized clutches and those that did not; and (2) between males that produced tadpoles and those that did not. Males that sired clutches were heavier, had longer site tenures, and were captured more often while calling than were males that did not sire clutches. Similarly, males that produced tadpoles were heavier, had longer site tenures, and were captured more often while calling than were males that did not produce tadpoles (table 10.2). Site tenure and calling behavior were under strong univariate selection in both comparisons, as were mass and calling behavior; SVL and territory quality did not differ between successful and unsuccessful males.

Univariate regressions of male traits indicated that females mated more frequently with those males with longer tenure at the breeding site and that larger males fertilized larger clutches (table 10.3). When indirect selection was removed during multivariate analysis, only site tenure significantly predicted male reproductive success, and it did so for both the number of clutches and the number of tadpoles (table 10.4). The overall models of these two analyses were significant, explaining 19%–20% of the variation in relative fitness (table 10.4). Mass was not used in the regression because of low sample sizes. Low sample sizes caused by missing phenotypic data also restricted analysis of the intermediate fitness components.

Discussion

The skew in male reproductive success within a population is a result of the type of breeding system (Emlen and Oring 1977; Wells 1977). Males of species with explosive breeding, such as *Rana sylvatica*, the Wood Frog, mate indiscrimi-

Table 10.3   Standardized directional selection differentials (*s*)

Trait	No. clutches (*n* = 90)	No. eggs/clutch (*n* = 54)	No. tadpoles/egg (*n* = 54)
SVL (mm)	.2380	*.0641	−.1037
No. captures	.0255	.0063	.1029
No. calls	*.4679	−.2122	−.0614
No. leaves	.2023	.0169	−.0868
Leaf area (cm^2)	.0177	.0069	.0230

Note: These are the slopes of a univariate regression of relative fitness on each individual character.
*$p < .05$

Table 10.4   Results of multiple regression models that explain male mating (number of clutches) and reproductive (number of tadpoles) success on the basis of the five phenotypic traits

	No. clutches (*n* = 90)	No. eggs/clutch (*n* = 54)	No. tadpoles/egg (*n* = 54)	No. tadpoles (*n* = 90)
SVL (mm)	.1965	.0927	−.0311	.1885
No. calls	.0568	−.0244	.0119	.0068
No. captures	*.7198	−.0047	−.0651	*.7616
No. leaves	.1157	.0268	−.1012	.0857
Leaf area (cm^2)	.2981	−.0466	.0677	.3299
F	2.00	.64	.27	1.97
p	.0436	.7691	.9836	.0476
r^2	.2003	.1275	.0587	.1976

Note: Values are the standardized selection gradients (β).
*$p < .05$

nately among the many females that arrive at once to the breeding pond. Males of prolonged breeding species such as *Hyla calypsa* cannot predict exactly when, where, or how many females will arrive on any night. The asynchronous arrival of these females over a long breeding season is expected to produce female mate choice, unequal male reproductive success, and male differences in morphology, vocalization, and/or territory quality (Wells 1977). As expected, males in this population of *H. calypsa* varied in reproductive success, morphology, and territory quality. Clutches were not distributed randomly along the transect, suggesting that females chose among resident males or their territories. Multivariate analysis suggested that females mated with males with longest site tenure.

Site tenure or the duration of attendance at the breeding site has been shown to increase a male's reproductive success (Kluge 1981; Woodward 1982; Trail 1984; Ryan 1985; Gerhardt et al. 1987; Arak 1988; Krujit and de Vos 1988; Högland and Robertson 1990; Dyson et al. 1992; Fincke and Hadrys 2001). Essentially, these males exhibit endurance competition in which persistence determines mating success (Andersson 1994), as occurs in many prolonged-breeding mammals (Gibson 1987; Campagna and LeBouef 1988; Appolonio et al. 1989; Clutton-Brock et al. 1989), birds (Gibson and Bradbury 1985; Lambrachts and Dhondt 1988; Gibson 1989), reptiles (Cuadrado 1999), and frogs (Murphy 1994b). A meta-analysis of 28 lekking species found that attendance and display frequency were the primary correlates to male reproductive success and that territory size and position of territory were not associated with reproductive success (Fiske et al. 1998).

Males that are present at the breeding site more often will obtain more matings because they will encounter more females. Any physiological or behavioral traits that increased male site tenure would be favored, including metabolic efficiency and foraging success (Murphy 1998). If site tenure increases reproductive success, males should be present along the stream as much as possible; but this study and others (Kluge 1981; Ryan 1985; Green 1990; Murphy 1994a, 1994b) have found that males are present for only a fraction of the breeding season. Three hypotheses could explain this absence from the breeding site: mortality, dispersion to other breeding sites, and energy limitations. Male *H. calypsa* are site faithful and may breed on the same territory for up to four years (K. R. Lips, pers. obs.). I suggest that annual or seasonal energy limitations, rather than predation or movement among breeding sites, determine site and probably chorus tenure in this species. This is supported by the similar recapture probabilities for both sexes; recapture probability was 53% (same year), 29% (one year), and 9% (two years) for males and 77%, 19%, and 7%, respectively, for females (Lips 1995).

Both vocal performance and site tenure are energetically costly activities that are limited by individual differences in physiology (Lambrachts and Dhondt 1988) and foraging behavior. A male's energy reserves, physiology, and behavior all interact to determine the length of time spent calling per night and per season. Various aspects of calling behavior (e.g., frequency, complexity, repetition rate) have been shown to be important in attracting females (Bucher et al. 1982; Taigen and Wells 1985; Prestwich 1994). Female *H. calypsa* may have chosen males on the basis of quality or quantity of vocalizations, which I did not measure but which have been shown to affect female mate choice in a variety of frogs (Fellers 1979a, 1979b; Ryan 1985; Given 1988), birds (Catchpole et al. 1984; Searcy and Marler 1984; Eens et al. 1991; Houtman 1992), and insects (Crankshaw 1979; Hedrick 1986; Simmons 1988). Calling is energetically expensive (Pough et al. 1992), and males may be energetically limited in both attendance

at the stream and in calling behavior. I believe this is the case for this population of *H. calypsa* because both male mass and calling frequency were significantly correlated with site attendance.

In cases in which territory quality has an immediate effect on reproductive
success, as would be expected in leaf-breeding frogs, females are expected to
base their choice of mate on territory quality (Ryan 1985). Therefore, males
should compete for access to "good" territories, and male quality should be
correlated with territory quality (Kitchen 1974; Wells 1977; Howard 1978a,
1978b; Searcy 1982; Price 1984; Burrowes 2000). By choosing better territories,
females choose better males. However, a lack of correlation between either of
the measures of territory quality and other male phenotypic traits suggests that
male *H. calypsa* did not compete for territories, that vegetation coverage was
not an appropriate measure of territory quality, or that I did not measure traits
related to territorial defense. The maintenance of intermale spacing, observations of male–male interactions, and scar patterns suggestive of male combat
indicate that these are territorial animals but that the estimate of territory quality may not have been appropriate.

Oviposition sites were nonrandomly distributed throughout the transect
and contained significantly more vegetation than did random sites (Lips 2001),
so I concluded that females preferred oviposition sites with abundant leaves
within a male's territory. I suggest that males active on more nights are more
likely to encounter receptive females, and hence have greater mating success,
than males active on fewer nights. Secondarily, females select the leafiest parts
of an active male's territory in which to oviposit.

It would be interesting to compare selective pressures in other tropical frogs
to determine how ecology, behavior, taxonomy, and habitat influence breeding
systems, but few studies describe variation in offspring among males, and fewer
examine how that variation is associated with male phenotypes. Kluge (1981)
summarized variation in the number of clutches sired among males for 23 populations of 15 species of frogs as gathered from the literature. Surprisingly, I was
not able to locate any additional studies published since that date, so I restrict
my comparison to this compilation. Kluge (1981) found highly skewed reproductive success among males in species of prolonged breeding anurans. Examining his database, I found that patterns of reproductive effort among males of
Hyla calypsa most closely resembled four other tropical stream frogs with prolonged breeding (*Hyalinobatrachium colymbiphyllum, H. fleischmanni, H. valerioi,* and *Hyla rosenbergi*). Males of all four species are territorial, and male–
male combat is common. All of these frogs, except *Hyla rosenbergi*, occur along
streams where females oviposit on overhanging vegetation, and except for
H. calypsa, all show parental care of eggs (McDiarmid 1978; K. R. Lips, pers.
obs.). *Hyla rosenbergi* and other members of the *Hyla boans* group lay eggs in
basins near streams (Kluge 1981). I suggest that the similarity in the distribution

of reproductive success among these species reflects the influence of the prolonged breeding season and the subsequent breeding system (Wells 1977).

I also believe that comparisons of selection at other spatial and temporal scales would be productive. Spatial and temporal variation in selection may exist in this population (e.g., Kalisz 1986; but see Warner and Schultz 1992), but I only had data from one complete breeding season. Given the strong seasonal component of hatching success (Lips 2001), I suspect that selection may vary within the year. Spatial variation in selection may also be important because the distribution of available territories differs along the transect, and selection operating at one end of the transect may differ from that operating elsewhere.

Acknowledgments

I thank Maureen Donnelly and the other organizers of the symposium for their efforts in making this volume happen. I thank the Organization for Tropical Studies and the Ministerio del Ambiente y Energia of Costa Rica for research and collecting permits. This chapter has benefited from comments by and discussions with G. Adams, E. D. Brodie III, P. Cain, M. Donnelly, T. Fleming, C. Horvitz, R. McDiarmid, S. Rand, J. Savage, W. Searcy, K. Spitze, J. Travis, and A. Welch. Funding was provided by National Geographic Society, Organization for Tropical Studies' Pew/Mellon Graduate Fellowship, University of Miami Tropical Biology Fellowship, Explorer's Club, SSAR Grants-in-Herpetology, ASIH Gaige Fund, and the American Museum of Natural History Roosevelt Fund, and by the National Science Foundation (DEB 92-20634 and DEB 94-24595 to Ken Spitze). This is in partial fulfillment of the requirements for the Ph.D. from the University of Miami.

11

Patterns of Co-occurrence of Hylid Frogs at a Temporary Wetland in Costa Rica

Craig Guyer and Maureen A. Donnelly

Choruses created by reproducing anurans are among the most spectacular auditory experiences in the natural world. In north temperate habitats, where patterns of chorusing by members of the family Hylidae have been studied relatively well, frogs typically migrate to temporary wetlands to reproduce (Wilbur 1980). Males of several species may use a single wetland and, therefore, compete to be heard. Inter- and intraspecific competition among males can occur during these chorus events (Duellman and Trueb 1986). In response to intraspecific competition, males of north temperate hylids may use call strength to alter spacing patterns of competing males (Allan 1973), use pitch to assess body size of rival neighbors (Wagner 1989), or alternate calls of nearest neighbors so that individual voices may be heard by females that approach a group of males (Schwartz 1987). In response to interspecific competition, species may differ in the timing of chorusing among seasons and/or within evenings (Wiest 1982). Additionally, species in diverse choruses differ in the pitch and cadence of each call, with the effect of minimizing problems of auditory signals that overlap when they are received by the ear of a female (Gerhardt 1994).

Research on Neotropical hylid frogs has addressed similar issues. Aspects of intraspecific competition involving auditory communication, especially the creation of duets and trios within choruses (Schwartz and Wells 1985) and the use of signal strength to regulate intermale spacing (Wilczynski and Brenowitz 1988), are relatively well studied. Much less is known about patterns of vocalization of Neotropical hylids that might reflect interspecific competition. Individual species are known to alter the timing of their calls to avoid temporal overlap with species of males possessing calls with similar dominant frequencies (Schwartz and Wells 1983a, 1983b). Duellman and Pyles (1983) demonstrated that closely related species of hylids that use the same tropical wetlands have calls that are more dissimilar to each other than they are to the calls of close relatives that do not overlap at breeding sites. These authors concluded that species-rich assemblages of vocalizing hylid frogs have partitioned the airwaves in a fashion that maximizes differences in calls among species, thereby allowing effective communication by the 10–20 species of hylids that may use the same wetland.

Not all species vocalize simultaneously at breeding sites. Instead, on some nights only a single species may be present, whereas on other nights there may be duets, trios, quartets, and more, of species. If the process described by Duellman and Pyles (1983) has been a general feature of evolution in Neotropical hylids, then distinctive patterns of co-occurrence of frogs at reproductive sites would be expected. These patterns might result in nonrandom occurrences of choruses of certain sizes, consistent species composition of choruses, and/or consistent vocal spacing within choruses. Additionally, the pattern of egg deposition associated with the phenology of adults should affect the assemblage of tadpoles that occupy a wetland and therefore the pattern of emergence of those recently metamorphosed tadpoles from the reproductive site (Alford 1999). In this chapter we describe patterns of co-occurrence of hylid frogs at a temporary wetland in Costa Rica and evaluate whether such patterns conform to expectations of competition for the airwaves. Additionally, we explore whether patterns of use of reproductive sites by adult frogs are associated with patterns of emergence of juvenile frogs from the aquatic habitat.

Methods

SAMPLE PROCEDURES

Our study site was a single temporary wetland located at La Selva Biological Research Station in Heredia Province, Costa Rica. We have described this site in detail elsewhere (Donnelly and Guyer 1994) and provide only a brief sketch here.

We refer to our study site as the Research Swamp because this wetland is located on the north side of the Camino Experimental Norte, approximately 250 m from the main laboratory clearing at La Selva. The swamp is approximately 70 m long by 40 m wide and is located in a depression in the old alluvium along the Río Puerto Viejo. This depression fills during heavy rains (after the ground becomes saturated) and occasionally receives floodwaters from creeks that flow into the Río Puerto Viejo drainage during intense storms. The site has an open canopy, likely created by wind-throw of trees with roots weakened by saturated soil. The vegetation of the outer perimeter of the wetland is dominated by an aroid (*Spathiphyllum freidrichsthalii*), and the central portion is dominated by a grass (*Panicum grande*).

We collected data along two 30-m transects. One transect was positioned along the edge of the wetland, to include the region dominated by the aroid, and the other was positioned toward the center of the wetland, to include the region dominated by the grass. We sampled each transect one night a week for 63 consecutive weeks (23 June 1982–31 August 1983). We sampled each transect between 2000 and 2400 h and counted the number of adults and juveniles discovered within 1 m of each side of each transect. Before we started a sample, we recorded all species heard calling.

SPECIES COMPOSITION

Eleven species of hylids were observed in the wetland, of which seven occurred frequently enough to be categorized as residents (*Agalychnis callidryas, Hyla loquax, H. phlebodes, Scinax boulengeri, S. elaeochroa,* and *Smilisca baudinii*). Three of the nonresident species were only observed once. Of these, *Hyla rufitela* and *Smilisca sordida* are known to breed at other sites at La Selva (in shallow swamps and riverine pools, respectively). We do not know where the third species, *Smilisca puma,* reproduces at La Selva, but we did not observe individuals in amplexus or juveniles at our study site; therefore, we presume that the breeding site for this species is in some other habitat. Adults of the fourth nonresident species, *Agalychnis saltator,* were observed on 12 nights. However, this species was recorded in amplexus on only one night, and no newly transformed frogs were found. In samples previous and subsequent to our 1982–83 study, we rarely have seen this species in the Research Swamp and consistently have observed it in a nearby swamp where *Agalychnis callidryas* is uncommon. Therefore, we conclude that the resident species of *Agalychnis* differs between these two swamps and consider *A. saltator* to be a nonresident of the Research Swamp. However, to examine potential effects, we performed all analyses described below with *A. saltator* included and repeated them with this species excluded. Because no differences in conclusions occurred between the two sets of analyses, we present only the latter in this chapter.

PRELIMINARY ANALYSES

We considered each of the 63 sample nights to be independent observations of the hylid assemblages and determined the number of species for which calling males were present and the number of species for which recently transformed juveniles were found in the wetland on each night. From these data, we counted the number of nights with 0, 1, 2, 3, 4, 5, and 6 species present in the calling chorus and with 0, 1, 2, 3, 4, and 5 species emerging as recently transformed juveniles. We compared each of these observed distributions to the binomial distribution in order to test the null hypotheses that calling choruses and groups of emerging juveniles are composed of random assemblages of available species (Pielou 1977).

The most frequent samples of calling species consisted of no frogs or of a single species. Duets, trios, and quartets of species were observed with intermediate frequency, and quintets and sextets were least frequent; at no time did we observe all seven resident species on the same night (fig. 11.1A). Our data document a significant deviation from the binomial distribution ($\chi^2 = 94.6$, $p < .001$). This occurred because nights with no frogs or choruses of solo species were overrepresented and nights with duets or trios were underrepresented. Quartets, quintets, and sextets of species occurred approximately as frequently as expected by the binomial distribution (fig. 11.1A). A significant deviation from the null hypothesis remained when nights without frogs were eliminated from consideration ($\chi^2 = 15.3, p < .005$). In order to characterize the species composition of choruses, we counted the number of nights that each species was present in overrepresented (solo) choruses, underrepresented (duet or trio) choruses, and null-expected (quartet, quintet, or sextet) choruses. Our null hypothesis was that the proportion of observations in each category would be consistent for *Agalychnis callidryas, Hyla ebraccata,* and a pooled category for all other species (pooled to avoid problems with small sample sizes). Our data allowed rejection of this hypothesis ($\chi^2 = 10.6$, $p < .05$; table 11.1). Instead, *A. callidryas* occurred more frequently than all other species as the lone calling species, *H. ebraccata* more frequently than all others as a member of duets or trios, and all other species more frequently than *A. callidryas* or *H. ebraccata* as members of quartets, quintets, or sextets (table 11.1).

Similarly, the number of species of juveniles observed at our study site differed from the pattern expected of a binomial process ($\chi^2 = 29.27$, $p < .001$; fig. 11.1B). The number of nights during which no juvenile was observed was overrepresented, and the deviation from binomial expectation remained when these nights were excluded from the analysis ($\chi^2 = 19.4, p < .001$). Nights with one or two species of juveniles were underrepresented, whereas nights with three, four, or five species of transforming juveniles were represented about as frequently as expected from the binomial distribution. The distribution of observations between nights when one or two species emerged and nights when three or more species emerged differed among species ($\chi^2 = 16.8, p < .001$; table 11.1; data for *Hyla ebraccata, H. loquax, Scinax boulengeri,* and *Smilisca baudinii* pooled). Two species (*Agalychnis callidryas* and *Scinax elaeochroa*) transformed equally frequently on nights with fewer species (1 or 2 species) and nights with more species. All other species transformed most frequently on nights when 3–5 species were transforming (table 11.1). From previous analyses (Donnelly and Guyer 1994), we knew that adults and juveniles segregated from

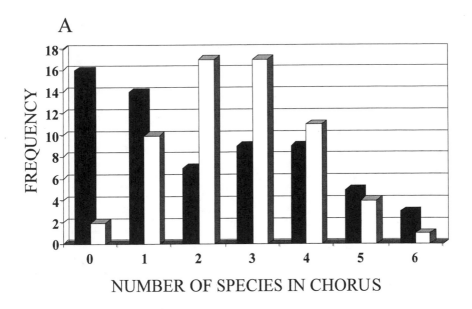

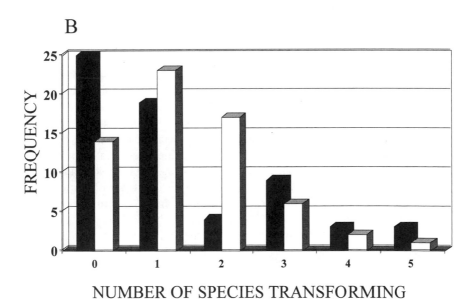

Figure 11.1 Observed (shaded bars) and expected (unshaded bars) numbers of (A) choruses of adult males and (B) assemblages of emerging juveniles within groups of increasing species richness at La Selva, Costa Rica. Expected values are based on binomial
distribution.

Table 11.1   Observed occurrences of adult and juvenile hylid frogs at a temporary wetland at La Selva, Costa Rica

Adults	Species						
	Ac	He	Sb	Hl	Se	Hp	Sm
Overrepresented choruses (1 species)	7	4	2	1	0	0	0
Underrepresented choruses (2–3 species)	8	13	7	6	5	1	1
Null-expected choruses (4–6 species)	14	16	12	14	11	10	2
Juveniles	Se	Ac	He	Sb	Hl	Hp	Sm
Underrepresented groups (1–2 species)	18	4	4	0	0	1	0
Null-expected groups (3–5 species)	15	6	15	2	3	10	3

Note: Categories are for choruses that were underrepresented, overrepresented, or appropriately represented relative to a binomial distribution. Ac, *Agalychnis callidryas;* He, *Hyla ebraccata;* Hl, *Hyla loquax;* Hp, *Hyla phlebodes;* Sb, *Scinax boulengeri;* Se, *Scinax elaeochroa;* Sm, *Smilisca baudinii.*

each other based on rainfall, with adult choruses typically occurring on rainy nights and juveniles typically emerging on nights without rain.

RANDOMIZATION TESTS

Having demonstrated a nonrandom structure of choruses of calling adults and groups of emerging juveniles, we tested whether such structure resulted from significant nesting of species. For example, quintets of calling males or emerging juveniles might be created by adding one new species to an existing quartet. To determine whether choruses or groups of emerging juveniles at La Selva developed in this way, we examined our data for nesting of species within groups (table 11.2). As a measure of nesting of, for example, quartets within quintets, we counted the number of times each observed quartet was found within an observed quintet; these values were summed for all observed quartets. Similar values were calculated for nesting of all possible pairs of choruses of adults or groups of emerging juveniles.

A simple null model of the expected degree of nesting would involve the following steps: (1) determine the composition of all possible group sizes (from solos to sextets); (2) draw at random (with replacement) from this pool of possibilities; (3) measure values of nestedness from each draw; and (4) repeat this procedure many times to create a null distribution from which observed values can be compared.

For a pool of seven species, there are 7 ways to create choruses of soloists or sextets, 21 ways to create duets or quintets, and 35 ways to create trios or quar-

Table 11.2   Composition of choruses at a temporary wetland at La Selva, Costa Rica

Solos	Duets	Trios	Quartets	Quintets	Sextets
			Adults		
Ac (7)	AcHe (2)	AcHeSb	AcHeHlSb (2)	AcHeHlHpSb	AcHeHlHpSbSe (2)
He (4)	AcHl	AcHeSm	AcHeHlSe	AcHeHlSbSe (2)	AcHeHlSbSeSm
Hl	AcSb	AcHeSe	AcHeHpSb	AcHeHlSeSm	
Sb (2)	HeSb	AcHlSb	AcHeHpSe (2)	AcHlHpSeSb	
	HeSe (2)	HeHlSb (3)	HeHlHpSb (2)		
		HeHlSe	HeHlHpSe		
		HeHpSe			
			Juveniles		
Ac	AcHe	AcHeSe (3)	AcHeHpSe	AcHeHpSeSm (2)	
He (2)	AcSe (2)	HeHlSe	HeHlHpSe	HeHlHpSeSm	
Se (16)	HeHp	HeHpSe (4)	HeHpSeSm		
		HeSeSm			

Note: Numbers in parentheses indicate frequency for combinations observed more than once. Ac, *Agalychnis callidryas;* He, *Hyla ebraccata;* Hl, *Hyla loquax;* Hp, *Hyla phlebodes;* Sb, *Scinax boulengeri;* Se, *Scinax elaeochroa;* Sm, *Smilisca baudinii.*

tets. We estimated the expected probability of observing each combination of species by determining the product of an index of abundance of each member species. This was done to account for the fact that certain combinations might be more likely simply because the species comprised by that combination were all more frequently observed at our study site than other combinations. Thus, nesting of species within choruses might result from such differences in abundance. We assumed that the abundance of each species could be estimated from the frequency with which a species occurred in groups of any size (table 11.3) and then adjusted the expected occurrence of each possible group to account for differences in abundance of each species. For example, adults of *Agalychnis callidryas* and *Hyla ebraccata* occurred on 29 and 34 of 63 nights, respectively; adults of *H. phlebodes* and *Smilisca baudinii* occurred on 10 and 3 nights, respectively. Therefore, we estimated that the probability of observing the former pair as a duet $[(29/63) \times (34/63)]$ was greater than the probability of observing the latter pair $[(10/63) \times (3/63)]$. We used this process to determine the null expectation of each possible chorus. We then randomly selected 14 solos, 7 duets, 9 trios, 9 quartets, 5 quintets, and 3 sextets (the observed numbers of these choruses; table 11.2) and counted the number of times each possible pair of choruses displayed nesting of the smaller chorus within the larger chorus. This was repeated 1,000 times to generate a null distribution against which the observed values were compared. We used a Z-score to determine the likelihood that nesting as great or greater than that observed would result from the process used to

Table 11.3   Frequency of occurrence of adults and juveniles
in choruses at a temporary wetland at La Selva, Costa Rica

	Species						
	Ac	He	Hl	Hp	Sb	Se	Sm
Adults	29	34	18	12	21	18	3
Juveniles	10	19	2	11	2	33	3

Note: Ac, *Agalychnis callidryas;* He, *Hyla ebraccata;* Hl, *Hyla loquax;*
Hp, *Hyla phlebodes;* Sb, *Scinax boulengeri;* Se, *Scinax elaeochroa;*
Sm, *Smilisca baudinii.*

create the null pool. Separate analyses were performed for adult choruses and
groups of emerging juveniles.

If interspecific competition for airwaves as described by Duellman and Pyles
(1983) occurred among hylids at La Selva, then we predicted that patterns of vocalizations observed in real choruses would differ from those expected of random assemblages adjusted for differences in abundance. If nesting of species
within choruses resulted from restrictions on vocal characteristics within choruses, then we expected observed choruses would involve a wider total range
of voices (fundamental frequencies) than combinations of species expected
of random assemblages. Additionally, we expected that when members of observed choruses were ranked from the species with the lowest to the highest
fundamental frequency, the mean coefficient of variation of differences between neighboring voices (e.g., the species with the highest voice minus the species with the next highest voice, and so on) would be smaller in observed groups
than random combinations of available species.

We performed randomization tests similar to those described above but using values of fundamental frequency taken for each species by Duellman (1967).
These data were gathered from a wetland located near Puerto Viejo de Sarapiquí, less than 3 km north of the wetland that we studied. Null distributions
were generated from the total vocal range and the variance between neighboring voices within each random chorus, accumulated for 1,000 randomizations.
We used Z-scores to evaluate whether call structure of real choruses differed
from the null distribution.

Finally, we evaluated whether the pattern of emergence of juveniles was associated with the pattern of visitation by adults to the breeding site. Such structure might occur because of consistent patterns of egg deposition by adults followed by consistent larval growth among species. Alternatively, differences in
development times might mask the pattern expected from the reproductive activity of adults. Data on larval periods for the tropical hylids that we studied are
not available. Therefore to estimate development times, we recorded the date

that adult occurrence in the Research Swamp peaked and the date that transforming juveniles peaked. We then used the elapsed time between these two
dates as an estimate of development time. To test the hypothesis that patterns of
co-occurrence of adults were associated with patterns of co-occurrence of juveniles, we used the Jaccard index to measure similarity of occurrences of all possible pairs of species within each age group. The similarity matrices were compared with a Mantel test.

Results

The observed degree of nesting within calling choruses of males was consistently greater than that expected of random choruses adjusted for differences in
abundance among the seven resident species (table 11.4). The observed nestings
of solos within duets and of trios within quartets were sufficiently disparate
from null expectation to reach statistical significance. The other comparisons,
although not statistically significant, gave remarkably similar results and indicated that, on average, only 11% of randomizations resulted in values for nesting as great as or greater than those observed in real choruses (table 11.4).

Table 11.4   Observed (upper half of matrix) and expected (lower half of matrix) degree
of nesting in choruses of adult and juvenile frogs at a temporary wetland at La Selva,
Costa Rica

	Adults					
No. species	1	2	3	4	5	6
1	—	53	75	94	64	42
2	34.1 (.001)	—	19	31	26	21
3	64.3 (.08)	13.0 (.07)	—	25	23	24
4	83.0 (.07)	24.1 (.08)	14.8 (.008)	—	11	21
5	55.8 (.04)	20.9 (.08)	18.8 (.16)	11.3 (.54)	—	11
6	38.5 (.10)	17.2 (.05)	19.8 (.12)	16.6 (.16)	7.2 (.10)	—

	Juveniles				
No. species	1	2	3	4	5
1	—	39	165	55	56
2	36.7 (.38)	—	13	6	9
3	111.1 (.0001)	12.4 (.46)	—	17	20
4	29.6 (.0001)	2.5 (.02)	1.9 (.0001)	—	4
5	50.2 (.10)	8.8 (.46)	15.1 (.14)	0.8 (.0007)	—

Note: Numbers in parentheses indicate proportion of null distribution as extreme as or more
extreme than the observed value.

Table 11.5   Estimated development times for tadpoles at a temporary wetland at La Selva, Costa Rica

Species	Dates of peak no. adults	Dates of peak no. juveniles	Development time (days)
Smilisca baudinii	28 Jul 82	25 Aug 82	30
	10 May 83	2 Aug 83	84
Agalychnis callidryas	9 Nov 82	1 Feb 83	84
	15 Jun 83	13 Aug 83	69
Scinax boulengeri	15 Jun 83	16 Aug 83	62
Scinax elaeochroa	30 Jun 82	31 Aug 82	62
	7 Jun 83	21 Aug 83	56
Hyla ebraccata	16 Dec 82	25 Jan 83	40
Hyla loquax	14 Jul 82	7 Sep 82	55
Hyla phlebodes	30 Jun 82	14 Sep 82	76
	17 May 83	12 Jul 83	56
	28 Jun 83	2 Aug 83	35

Note: Development time is estimated as elapsed time between peak number of adults and peak number of juveniles.

Contrary to the expectation of intraspecific competition, we observed no difference between the range of voices heard in duets, trios, quartets, or sextets and the range expected of random assemblages of identical species richness (fig. 11.2). We did observe that voices within quintets were more disparate than those expected of random assemblages (fig 11.2D). However, the observed distribution of voices within choruses was not unusually uniform, relative to spacing of voices expected of random assemblages of available species (fig. 11.3).

The pattern of nesting of emerging juveniles was not as consistent as the pattern observed in adult choruses (table 11.4). Those species that emerged alone also were significantly nested within trios, quartets, and quintets of emergent species. Similarly, significant levels of nesting were observed for trios within quartets and quartets within quintets. All other comparisons of the observed patterns to expected nesting failed to approach significance. Estimates of development times ranged from 30 days (*Smilisca baudinii*) to 84 days (*Smilisca baudinii* and *Agalychnis callidryas;* table 11.5). We found no significant association between patterns of co-occurrence of adults and juveniles (G = 1.10; $p > .05$).

Discussion

Adults of two species at La Selva, *Agalychnis callidryas* and *Hyla ebraccata,* conformed to expectations of prolonged breeders (Wells 1977). Nights with solo

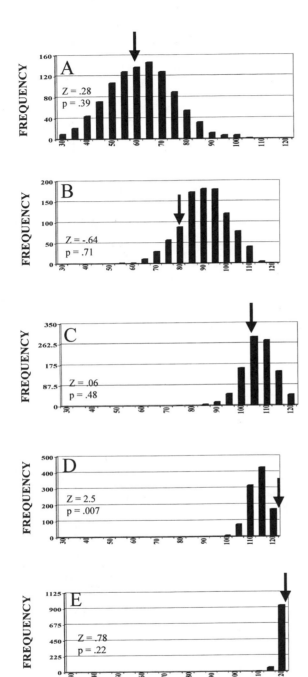

Figure 11.2 Distribution of call range expected by chance for choruses of increasing species richness, of (A) duets, (B) trios, (C) quartets, (D) quintets, and (E) sextets, at La Selva, Costa Rica. Arrows indicate the observed range of calls.

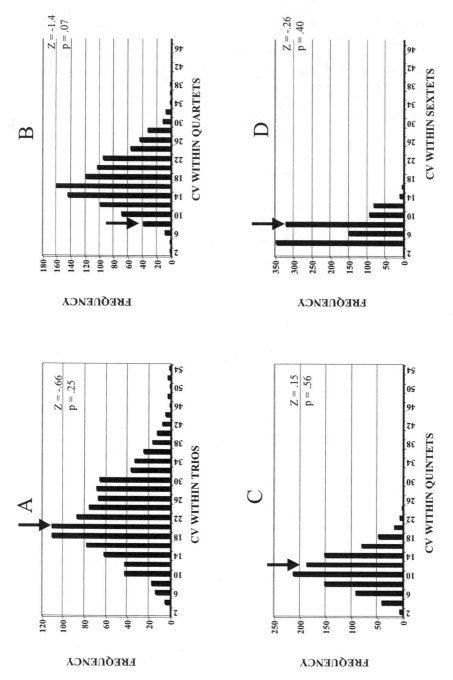

Figure 11.3   Distribution of coefficients of variation expected by chance in choruses of increasing species richness, of (A) trios, (B) quartets, (C) quintets, and (D) sextets, at La Selva, Costa Rica. Arrows indicate the observed coefficients of variation (CV).

choruses were overrepresented relative to the binomial, and these were composed largely of *A. callidryas*, the species most evenly distributed among all chorus sizes. Although duets and trios were underrepresented, *H. ebraccata* was an unusually common member of such choruses and was most frequently associated in these choruses with *A. callidryas, Scinax boulengeri,* and *Hyla loquax.* All four of these species also appeared in choruses of larger size.

Quartets, quintets, and sextets occurred about as frequently as expected by the binomial distribution. These species-rich choruses resulted from addition of a group of five species that entered the wetland only on the relatively few nights when conditions were wettest and, presumably, most favorable. Three species, *Hyla phlebodes, Scinax elaeochroa,* and *Smilisca baudinii,* called and were observed in amplexus only in choruses of large species richness. The other two species, *Hyla loquax* and *Scinax boulengeri,* called in choruses of both small and large richness but appeared to select nights when there were many other species more frequently than they selected nights with fewer species. Thus, the behavior of these five species conformed to expectations of explosive breeders, and the nonrandom appearance of species within choruses resulted in part from addition of explosive breeders, on certain nights, to an assemblage of prolonged breeders.

These findings are consistent with the hypothesis that choruses of increasing species richness are created by sequential addition of new species to preexisting choruses. Such nesting is beyond that expected because some species visit the swamp more frequently than others. This process explains formation of choruses up to those with four species; larger choruses become indistinguishable from random assemblages of available species, and such assemblages appear on the relatively rare nights when nearly all species enter the chorus. Our previous analysis of phenological patterns at La Selva documents that these rare nights are the rainiest ones (Donnelly and Guyer 1994), a pattern that is known for hylids at other Neotropical sites (e.g., Gascon 1991).

Interestingly, no species adopted a strategy of appearing in choruses of low diversity and avoiding choruses of high diversity. Three such species do occur at La Selva (*Agalychnis calcarifer, Hyla rufitela,* and *Smilisca sordida*), but they effect this strategy by selecting reproductive sites where they are the lone hylid present (Marquis et al. 1986; Duellman 1970, 2001). Two of these species (*A. calcarifer* and *H. rufitela*) have calls of extremely low volume (C. Guyer and M. A. Donnelly, pers. obs.). In fact, their voices are so weak that it is difficult to conceive of females of either species being able to hear conspecific males within the choruses of species that we observed at our study site.

Vocalizations of pond-breeding hylids at La Selva do not conform to simple models of competition for the airwaves. This result is difficult to reconcile with the results of Duellman and Pyles (1983) because competitive pressure within a chorus at a particular wetland should be stronger than across geographic regions. Our results suggest that species are added to choruses at La Selva more for their tolerance of environmental conditions associated with reproduction than their ability to avoid overlap of call characteristics with potential competitors. However, the spacing of voices of the seven resident hylids at La Selva seems remarkably uniform. Thus, the evolutionary adjustment of hylid voices across broad geographic scales described by Duellman and Pyles (1983) might result in uniform spacing of voices at local breeding sites. If so, then comparison of spacing of voices within replicate breeding sites should be more uniform than that expected of random assemblages of species drawn from the pool of all species occupying all sites.

Our data indicate structure in the transformation of aquatic tadpoles into terrestrial juveniles. This process appears to involve two types of species. Prolonged transformers (*Scinax elaeochroa* and *Agalychnis callidryas*) appear to have a wide tolerance of environmental conditions during which transformation can take place. These species transform alone or within a pair of species about as frequently as they transform in concert with several other species present in the wetland. Explosive transformers (all other species, but best exemplified by *Hyla ebraccata and H. phlebodes*) appear to have more limited tolerances and emerge only on a few nights and in concert with many other species. Such species-rich assemblages are created by adding the explosive transformers to the set of prolonged transformers. In this respect, the structure of assemblages of transforming juveniles is similar to the structure of choruses of calling adults. Our previous analyses (Donnelly and Guyer 1994) demonstrate a negative association between patterns of adult and juvenile abundance, documenting that species-rich assemblages of transforming frogs occur at La Selva on a relatively few dry nights.

Our data fail to support the simplistic hypothesis that assemblages of juveniles are determined by the patterns of egg deposition exhibited by reproducing adults. In fact, only *Agalychnis callidryas* and *Hyla phlebodes* showed similar phenological patterns between adults and juveniles; both age groups displayed the prolonged condition in *A. callidryas,* whereas both age groups showed the explosive condition in *H. phlebodes.* The most divergent phenology between adults and juveniles was observed in *Scinax elaeochroa,* which was among the most explosive of breeders and the most prolonged of transformers. These observations suggest that development times of tadpoles at La Selva vary within and among species, a feature described for anuran larvae (Alford 1999) but rarely quantified for free-ranging individuals. Our data document no consistent pattern of development times for hylid tadpoles at La Selva. In fact, development times for cohorts of tadpoles of *Smilisca baudinii* may differ by as much as 180%. Similar variability in development has been implied for Neotropical hylid tadpoles because of effects of competition and predation on growth rates (Gascon 1995). Our findings indicate that such variation in juvenile phenologies is sufficient to erase any role that adults might play in shaping patterns of tadpole transformation. Nevertheless, the timing of metamorphosis at La Selva appears to be organized such that an unusually diverse assemblage of juveniles leaves the Research Swamp on a relatively few nights.

Two general classes of explanation for nonrandom aggregations within samples such as ours are available (Pielou 1977). In one class of explanation, species aggregate through independent response to some environmental gradient. The second class of explanation would infer an advantage of participation by species in a group of a given size. Of these two, the former appears to characterize patterns of co-occurrence of adult hylids at La Selva in that each species appears to have its own set point for the degree of moisture required for reproduction. Under this scenario, choruses would increase in size from those with low diversity, established by species with broad moisture tolerances, to those with high diversity by adding species with progressively narrower moisture tolerances. Such choruses would have nonrandom structure, but no specific benefit would be gained by membership within a chorus of a particular size. Two key observations consistent with this interpretation are (1) the close association of observed and expected chorus frequencies as species richness within a chorus increases and (2) the lack of structure to the vocal characteristics of calling males within choruses. These observations suggest that explanations other than competition should be explored for the data described by Duellman and Pyles (1983). However, as indicated earlier in the chapter, past competition might have resulted in unusually consistent spacing of voices within choruses at a particular wetland, a pattern that awaits further analysis. Additionally, the key variable of vocal cadence remains to be examined. The spacing of notes within the call of a species may allow multiple species to communicate on the same night (Gerhardt and Schwartz 1995). If species-rich assemblages at La Selva were organized in a fashion to minimize the overlap of notes among species, then choruses might be structured because of benefits accrued to member species with non-overlapping cadences. Tests of this hypothesis await further examination of call structure among choruses that differ in species richness.

Patterns of co-occurrence of transforming frogs are more difficult to interpret, largely because the tadpole stage is so frequently overlooked in field studies of tropical anurans. Therefore, it is more difficult to infer whether nonrandom aggregations of transforming species are occurring because of independent responses to environmental variables or to advantages earned by group membership. Ctenid spiders are an abundant and obvious source of mortality for transforming frogs at the La Selva Research Swamp, and patterns of transformation might be directed toward simultaneous satiation of arachnid predators, thereby maximizing the number of juvenile frogs that escape predation. Under this scenario, individuals would gain a benefit from group membership. However, it is unclear why such mass exoduses of juveniles should occur on dry nights. If rain dislodges emerging juveniles from their perches, then species-rich assemblages of transforming frogs might occur on dry nights for mechanical reasons. Such an explanation would not invoke a benefit to group membership. Careful examination of developmental rates, environmental conditions, and patterns of aquatic predation is needed to explore how these factors might affect patterns of juvenile transformation in Neotropical hylids.

Acknowledgments

The data described in this study were collected while the authors held predoctoral fellowships from the Jessie Smith Noyes Foundation administered by the Organization for Tropical Studies. David and Deborah Clark, then the station managers of La Selva, extended us many courtesies during our stays at the station, including turning a blind eye to our late-night raids of the station refrigerators after our weekly frog samples. We also thank Harry Greene, Manuel Santana, and Michael and Patricia Fodgen for helping collect some of the information described here and for lengthy discussions about the La Selva herpetofauna. Finally, we thank Marty Crump, Claude Gascon, and James Watling for careful review of this contribution and suggestions for improving it. This chapter is contribution number 77 to the Program in Tropical Biology at Florida International University.

12

It's a Frog-Eat-Frog World in the Paraguayan Chaco: Food Habits, Anatomy, and Behavior of the Frog-Eating Anurans

Norman J. Scott Jr. and A. Luz Aquino

John Graham Kerr's goal for his 1896 expedition to the Gran Chaco was to collect the South American lungfish, *Lepidosiren paradoxa* (Kerr 1950). John Samuel Budgett, a twenty-four-year-old Cambridge University student on his first scientific foray, discovered a frog fauna with characteristics unlike anything that had been described before—notably, the propensity of several species to eat other anurans, and the repellent skin secretions of many members of the fauna (Budgett 1899).

The Gran Chaco straddles the Tropic of Capricorn in the eastern rain shadow of the southern Andes and is located in Bolivia, Argentina, and Paraguay (fig. 12.1). The geological and ecological history of the region is obscured by a 300-m-thick layer of Pleistocene sediments forming the outwash plain of the Andes. Solbrig (1976) claimed that the xeric, subtropical habitats of the Chaco have existed relatively unchanged since the beginning of the Tertiary. According to Solbrig, the core of the Chaco has been an ecologically stable center, producing a relict and endemic biota. Short (1975) wrote that the Chaco is a post-Pleistocene development with biota drawn largely from surrounding xeric formations, such as the Caatinga and Cerrado of Brazil. Bucher (1980) described the present-day ecology of the Chaco.

The Chaco has a hot wet season and a cool dry season. Filadelfia receives

Figure 12.1   Map of South America showing the location of the Gran Chaco (stippled), Paraguay (dark outline), and Filadelfia, Departamento Boquerón (triangle).

about 900 mm of rain annually, most of which falls between October and May. Summer temperatures may exceed 40°C, and winters can have light frosts.

The natural vegetation in the dry Chaco includes a dense thorn scrub of short thorny leguminous trees (mostly *Prosopis* spp.) and tall emergent trees characteristic of the dry Chaco, including quebracho blanco (*Aspidosperma quebracho-blanco;* Apocynaceae), quebracho colorado (*Schinopsis quebracho-colorado;* Anacardiaceae), and palo borracho (*Chorizia insignis;* Bombacaceae). Much of the land in the study area has been converted to agriculture, and we worked in a mosaic of uncut woodland, pastures, plowed fields, and second growth. The area is well drained and not subject to inundation.

Budgett (1899) described the genus *Lepidobatrachus,* with two species, *L. laevis* and *L. asper,* indicating that *Bufo granulosus* (now *B. major*) was their chief food (*L. asper* does not occur at Filadelfia). He found that *Ceratophrys* "lives chiefly off other frogs and toads." Budgett also noted that a captive Paraguayan "grass" snake (probably *Liophis* sp.) lived in a terrarium with, but did not consume, *Melanophryniscus stelzneri* and *Phyllomedusa hypochondrialis,* but the snake readily ate a "*Paludicola signifera*" (probably *Physalaemus albonotatus*).

Four species in the Chaco fauna have aposematic colors; presumably warning coloration in these frogs is combined with repellent skin secretions. The largest and most distinctive is *Leptodactylus laticeps,* with a bright color pattern consisting of discrete black spots with red centers on a yellow or ivory background (plate 12.1A; plates follow page 256). The large microhylid, *Dermatonotus muelleri,* has prominent discrete white spots on a black background on the legs, sides, and ventral surface of the body (plate 12.1B). Both species of *Phyllomedusa* have striking color patterns that may serve an aposematic function; *P. sauvagii* has large, dark-edged white spots on the dark green throat, and *P. hypochondrialis* has a brilliant orange and black-barred pattern on all of the concealed surfaces of the body, legs, and feet (plates 12.1C, 12.1D).

The Inventario Nacional Biológico del Paraguay, a cooperative project initiated in 1980 among the United States Fish and Wildlife Service, the Paraguayan Ministerio de Agricultura y Ganadería, and the United States Peace Corps, enabled the periodic collection of information on the ecology of the Chaco anurans. This chapter describes the results of intensive observations made in the vicinity of the Mennonite town of Filadelfia, Departamento Boquerón, Paraguay (60°5′ W, 22°15′ S; fig. 12.1).

Frog breeding habitats near Filadelfia were created or altered by humans and included stock ponds, drainage ditches, and borrow pits. Occasional forest ponds appeared to be natural, but human activity has resulted in numerous deep ponds with longer hydroperiods than those of forest ponds. Almost all of the ponds dried during the dry season.

Our studies tried to answer several questions: Which species regularly consume frogs? What anatomical characteristics are associated with frog eating?

How do noxious skin secretions affect predator behavior? And, is presumed aposematic coloration associated with effective antipredatory skin secretions?

Materials and Methods

THE FROG FAUNA

The origins and biogeographic history of the Chaco anuran fauna are described by Vellard (1948), Gallardo (1966, 1979), and Cei (1980). Of the 22 anuran species that occur in the Filadelfia area, 8 are largely restricted to the Gran Chaco (table 12.1). Large species (maximum adult snout-vent length [SVL] > 90 mm) that are potential frog predators include *Lepidobatrachus laevis* (plate 12.2A), *Ceratophrys cranwelli* (plate 12.2B), *Leptodactylus laticeps* (plate 12.1A), *Bufo paracnemis* (plate 12.2D), and *Leptodactylus chaquensis* (plate 12.3A). *Chacophrys pierotti* (plate 12.2C; maximum SVL 55 mm) is also a frog predator (Cei 1955) but was not common enough for intensive study, and *Pseudis paradoxa* (plate 12.3D; maximum SVL 66 mm) was a relatively large frog that had the opportunity to eat juvenile frogs.

We performed the studies during the early wet seasons of November 1982 and November and December 1983. Voucher collections are in the Museo Nacional de Historia Natural del Paraguay, Asunción, the United States National Museum, Washington, D.C., and the Museum of Southwestern Biology, University of New Mexico, Albuquerque. Varying numbers of species were examined in each phase of the study, depending on the number of specimens available and the ease with which they could be manipulated in field trials.

Frogs were killed and stomach contents identified under a dissecting microscope. Length, width, and depth of intact prey items were estimated with an ocular micrometer. Hard parts, such as beetle thoraces and elytrae, and snail opercula, apparently accumulated for a long time in the stomach; in these cases the prey items represented by these fragments were not included in the summary unless soft tissues were also present.

Food habits were examined in 113 stomachs of six species: *Ceratophrys cranwelli*, *Lepidobatrachus laevis*, *Leptodactylus chaquensis*, *L. laticeps*, *Bufo paracnemis*, and *Pseudis paradoxa*. These species were selected because they were large enough to eat other frogs and were common enough to provide replicate samples.

We measured SVL and head width with digital calipers on alcohol-preserved specimens of five frog-eating species: *Lepidobatrachus laevis*, *Ceratophrys cranwelli*, *Chacophrys pierotti*, *Leptodactylus chaquensis*, and *L. laticeps*. Mandibular anatomy was examined in live, alcohol-preserved, and skeletal preparations of *Bufo paracnemis*, *Leptodactylus chaquensis*, *L. laticeps*, *L. laevis*, and *C. cranwelli*. Most specimens were from the Filadelfia area, but some were from other parts of the Chaco.

Table 12.1   Frog fauna of the dry Chaco in the vicinity of
Filadelfia, Departamento Boquerón, Paraguay

Taxon	Maximum adult length (mm)
Microhylidae	
Dermatonotus muelleri	67
Elachistocleis bicolor	45
Hylidae	
Hyla raniceps	75
*Scinax acuminatus**	50
S. nasicus	35
Phrynohyas venulosa	70
Phyllomedusa hypochondrialis	42
*P. sauvagii**	70
Bufonidae	
*Bufo major**	67
B. paracnemis	195
Pseudidae	
Pseudis paradoxa	66
Leptodactylidae	
*Leptodactylus bufonius**	69
L. chaquensis	98
L. elenae	50
L. fuscus	52
*L. laticeps**	125
L. mystacinus	60
Physalaemus albonotatus	32
P. biligonigerus	38
Ceratophryidae	
*Ceratophrys cranwelli**	118
*Chacophrys pierotti**	55
*Lepidobatrachus laevis**	139

Note: Frog lengths are maximum adult sizes from this study and
Cei 1980. Asterisk (*) indicates a species endemic to the Chaco.

We conducted 136 field experiments to study prey capture and handling behaviors in three of the larger species (*Bufo paracnemis, Lepidobatrachus laevis,*
and *Leptodactylus chaquensis*) and to assess the palatability of some of the anuran prey items. The predators were chosen because they were large and relatively abundant. Prey species were smaller frogs that were captured as needed,
including *Leptodactylus bufonius* (plate 12.3B) and *Bufo major* (plate 12.3C), as
well as juveniles of larger species.

Tethered frogs were offered at night to free-ranging predatory frogs, and the predator's response was recorded. If the prey was swallowed, we cut the tether, and the predator was observed until we were reasonably certain that the predator was going to retain the prey or until the predator rejected the prey. Forty experimental trials were not analyzed because the predation event was interrupted or the predator did not respond to the tethered prey item.

Each prey item was placed into one of two repellence categories: (1) frogs not secreting noticeably repellent substances, and small juveniles of noxious species; and (2) frogs sometimes secreting large amounts of toxic secretions that made them annoying to handle and that usually killed other frogs in crowded conditions.

Results

FOOD HABITS

Bufo paracnemis and *Pseudis paradoxa* stomachs contained arthropods, a leech, and a snail; the *Leptodactylus* species (*L. chaquensis* and *L. laticeps*) had eaten arthropods, frogs, a snail, and a blind snake; *Ceratophrys cranwelli* stomach contents included frogs and a beetle; and *Lepidobatrachus laevis* stomachs contained frogs, beetles, water bugs, and ampullariid snails (table 12.2). Six species of anurans were identified in the stomach contents (table 12.3). Two species of *Physalaemus* were the most common anuran prey.

There was a marked difference in the median number of food items per stomach between *Bufo paracnemis* (10 items) and the other species (1–4 items; table 12.4). *B. paracnemis* was also the largest frog, but its median prey size (7 mm) was smaller than that for other species (11–35 mm; table 12.4).

FORAGING BEHAVIORS

Foraging behaviors were determined or inferred for *Bufo paracnemis, Pseudis paradoxa,* and all of the frog-eating species except *Chacophrys.* The summaries are based on situations in which we offered the prey to the predator or on observations that we made during other field activities.

Bufo paracnemis could be found anywhere there was exposed ground on wet nights; on dry nights they congregated around pond margins or sat in a pond facing the shore. When prey was presented, *B. paracnemis* usually tried to lap it up with their tongue; frogs did not stick to the tongue well, and they usually slipped off (73% of 15 first snaps). Sometimes a toad would stab at frog prey several times with its tongue without success. If the frog stuck to the tongue, it was ingested with the aid of the forefeet. *B. paracnemis* grabbed a frog with its jaws once. Unacceptable prey were usually rejected as soon as they reached the mouth.

Table 12.2   Number of stomachs and numbers of food items in six species of Chaco frogs, Filadelfia, Departamento Boquerón, Paraguay, 1982–83

Prey type	Predator species					
	BP	CC	LL	LC	LT	PP
Coleoptera						
Adults	144	1	10	45	12	19
Larvae	—	—	1	2	2	—
Hemiptera	1	—	5	14	1	6
Homoptera	—	—	—	1	—	1
Hymenoptera						
Formicidae	56	—	—	5	—	1
Others	—	—	—	2	—	1
Lepidoptera						
Adults	—	—	—	—	1	—
Larvae	33	—	—	2	3	—
Isoptera	1,460	—	—	10	—	—
Odonata						
Adults	—	—	—	—	—	1
Nymphs	—	—	—	—	2	—
Orthoptera	1	—	—	7	4	2
Blattaria	—	—	—	2	1	—
Diplopoda	200	—	—	2	—	—
Opiliones	1	—	—	—	—	—
Araneae	—	—	—	1	—	—
Hirudinea	1	—	—	—	—	—
Gastropoda	1	—	18	—	1	—
Anura	—	7	23	7	3	—
Scolecophidia	—	—	—	—	1	—
Total stomachs	11	11	29	28	19	15
Empty stomachs	0	4	6	3	3	1

Note: BP, *Bufo paracnemis;* CC, *Ceratophrys cranwelli;* LL, *Lepidobatrachus laevis;* LC, *Leptodactylus chaquensis;* LT, *Leptodactylus laticeps;* PP, *Pseudis paradoxa.*

Ceratophrys cranwelli was observed only on rainy nights or on one or two nights after a heavy rain. On wet nights, frogs emerged from shallow burrows and moved to the margin of a frog breeding pond. We did not see them capture prey.

Lepidobatrachus laevis was almost completely aquatic, although juveniles were occasionally seen on land during rainy nights. A frog usually was submerged in water with only the top of its head exposed, and it often spent several hours in the same spot. The frog usually faced a shoreline that was free of

Table 12.3 Numbers of frogs found in stomachs of predatory frogs, Filadelfia, Departamento Boquerón, Paraguay, 1982–83

Prey species	Predatory species			
	Lepidobatrachus laevis	*Ceratophrys cranwelli*	*Leptodactylus chaquensis*	*Leptodactylus laticeps*
Bufo major	3	1	—	2
Pseudis paradoxa	3	—	—	—
Dermatonotus muelleri	10	—	—	—
Leptodactylus spp.	—	1	—	—
Physalaemus albonotatus	4	—	5	—
Physalaemus biligonigerus	2	5	—	—
Unidentified anuran	1	—	2	1
No. stomachs	29	11	28	19

Note: *Dermatonotus muelleri* prey included juvenile frogs, tadpoles, and stages in between.

Table 12.4 Summary of food habits of Chaco frogs, Filadelfia, Departamento Boquerón, Paraguay, 1982–83

	Predator species					
	BP	CC	LL	LC	LT	PP
Stomachs with food	11	7	23	25	16	14
Frog SVL (mm)						
Range	115–195	79–118	60–139	38–98	87–125	44–66
Median	160	97	105	78	105	50
Prey items (%)						
Arthropods	100	12	25	93	84	100
Gastropoda	trace	0	39	0	3	0
Anura	0	88	36	7	10	0
Other	trace	0	0	0	3	0
Prey lengths (mm)						
Minimum	4	12	6	3	10	4
Maximum	110	86	60	38	175	25
Median	7	30	35	12	20	11
Median no. prey	10	1	1	3.5	1	2
Relative food size	0.04	0.31	0.33	0.15	0.19	0.22

Note: Relative food size is the median prey size divided by the median SVL. BP, *Bufo paracnemis;* CC, *Ceratophrys cranwelli;* LL, *Lepidobatrachus laevis;* LC, *Leptodactylus chaquensis;* LT, *Leptodactylus laticeps;* PP, *Pseudis paradoxa.*

aquatic or terrestrial vegetation. Distance from the shore varied between 1 and 25 cm. When startled, *L. laevis* would submerge or would turn and swim toward deep water.

Lepidobatrachus laevis jumped toward and tried to capture proffered prey with a powerful snap of its jaws. It successfully captured the prey in 76% of first feeding attempts ($n = 75$). If the tethered prey item was captured, *L. laevis* either rejected it immediately, tried to swallow it on the surface with aid of the forefeet, or submerged and the prey was either swallowed or rejected. When the first snap was unsuccessful ($n = 18$), *L. laevis* either backed down out of sight (39% of the time) or continued to snap at the prey. On one occasion *L. laevis* captured floating prey from below.

Leptodactylus chaquensis lay in wait in shallow water or on land within a meter of a pond margin. Like *Lepidobatrachus laevis,* they remained immobile for long periods of time. When startled, they jumped into a deeper part of the pond. When *L. chaquensis* was presented with prey, they also snapped, and their success rate (94% of 33 attempts) at first snap was greater than that of *L. laevis.* When *L. chaquensis* captured prey, they occasionally swallowed it immediately or, more commonly, held it in the mouth without moving. Vigorous movements of the forefeet accompanied swallowing. Sometimes the prey item was rejected immediately, but more often even unacceptable prey species were retained in the mouth for several seconds or even minutes. *L. chaquensis* frequently jumped once or twice after prey capture, usually toward deeper water.

Leptodactylus laticeps was usually found sitting on land within a meter from ponds with forested margins. They spent long periods of time in one position, but we did not observe them capture prey. This species was observed on dry as well as wet nights.

Pseudis paradoxa was not observed foraging, but they were seen on the pond surface among floating vegetation. Occasionally we saw them traveling overland on rainy nights.

FEEDING EXPERIMENTS

We conducted 96 feeding trials with three predatory species—*Bufo paracnemis* ($n = 11$), *Leptodactylus chaquensis* ($n = 22$), and *Lepidobatrachus laevis* ($n = 63$; table 12.5). There was a distinct difference between prey species that were accepted and those that were rejected. Prey with a repellence index of 1 were accepted in 39 trials, whereas every prey type with a repellence index of 2 was rejected at least once. Overall, repellence type 2 prey were rejected in 32 of 57 trials. The two *Phyllomedusa* species and *Dermatonotus* were responsible for 78% of the rejections (29 of 37 trials), whereas all other prey items were rejected only 3 times in 60 trials.

Although different numbers and percentages of noxious prey were offered

Table 12.5   Number of prey items accepted and rejected by three predatory frog species in feeding trials, Filadelfia, Departamento Boquerón, Paraguay, 1983

Prey species	Repellence index	Predatory species					
		Bufo paracnemis		*Leptodactylus chaquensis*		*Lepidobatrachus laevis*	
		Accept	Reject	Accept	Reject	Accept	Reject
Dermatonotus muelleri	2	—	—	0	1	1	4
Bufo major	2	1	0	—	—	5	1
B. paracnemis (J)	1	—	—	2	0	3	0
Lepidobatrachus laevis (J)	1	—	—	—	—	4	0
Leptodactylus bufonius	2	1	1	2	0	5	0
L. chaquensis (J)	1	—	—	3	0	8	0
L. fuscus	1	—	—	—	—	4	0
Physalaemus albonotatus	1	—	—	3	0	2	0
P. biligonigerus	1	—	—	3	0	3	0
Phyllomedusa hypochondrialis	2	4	2	0	8	2	9
P. sauvagii	2	—	—	—	—	1	5
Scinax acuminatus	2	0	1	—	—	3	0
S. nasicus	1	1	0	—	—	—	—
Pseudis paradoxa	1	—	—	—	—	3	0

Note: Prey species were adults or juveniles (J).

to the predatory species, *Leptodactylus chaquensis* seemed to be more sensitive to prey secretions than were *Lepidobatrachus laevis* or *Bufo paracnemis* (table 12.5). Repellence category 2 prey items were refused by *L. chaquensis* in 9 of 11 trials (82%); *L. laevis* refused prey in this category in 19 of 36 offerings (53%), and *B. paracnemis* rejected them in 4 of 10 trials.

Predators sometimes exhibited distress, particularly when they swallowed a noxious prey species. For example, two *Bufo paracnemis* that had swallowed *Phyllomedusa hypochondrialis* gaped repeatedly and blinked their eyes; one stuck out a very red, blood-engorged tongue. A small (70 mm) *Lepidobatrachus laevis* grabbed a *P. hypochondrialis* and battled it for 1 min until it was swallowed. The predator writhed, snapped its jaws, and rubbed its mouth with its forelegs. After 13 min, the *Lepidobatrachus* was fairly calm, but it was swollen and floated high in the water. At 15 min, it commenced another bout of writhing and rubbing. Eight minutes later, still swollen and floating, it accepted and swallowed a juvenile *B. paracnemis*.

MORPHOLOGY

Frog Size

Maximum sizes recorded for Chaco frogs are listed in table 12.1. Juveniles of most large species seemed to be scarce in the Filadelfia region. During 1982, we found no *Lepidobatrachus* with a SVL of less than 87 mm, but in 1983, animals with SVL of 50–70 mm were common. In 1982–83, the smallest *Leptodactylus laticeps* we saw was 87 mm, only one juvenile *Ceratophrys* was taken, and the only four *Chacophrys* that we observed were adults. All sizes of *Leptodactylus chaquensis* were abundant. Small juveniles (SVL < 35 mm) of *Bufo paracnemis* were occasionally abundant, but only four or five in the 35–70 mm size range were seen.

Mandible

The mandible of *Bufo paracnemis* has several cartilaginous connections with flexion between the dentary, the mentomeckelian cartilage, and the mentomeckelian bones (fig. 12.2A). The two frog-eating *Leptodactylus*—*L. chaquensis* and *L. laticeps*—have typical anuran mandibles, but theirs are less flexible than those of *Bufo*. The tough cartilage pad at the symphysis is exceptionally well developed (fig. 12.2B). In the three ceratophryid genera, the mandibular symphysis is an immobile joint between the dentaries (fig. 12.2C).

The rounded dorsal rim of *Bufo* mandibles is covered with a thick pad of leathery skin. The dorsal surface of the mandible in ceratophryids is sharp and knife-like and covered by a sheet of thin skin. The leptodactylids have a rounded mandible covered with thick but not leathery skin.

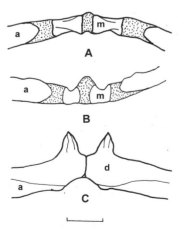

Figure 12.2   Anterior portion of anuran mandibles, lingual view, for (A) *Bufo paracnemis*, (B) *Leptodactylus laticeps*, and (C) *Lepidobatrachus laevis*. Cartilage, or cartilage backed by dentary, is stippled. a, angular bone; d, dentary; m, mentomeckelian bone. Bar = 5 mm.

Maxillary Teeth

The three ceratophryids share the uncommon amphibian trait of having maxillary/premaxillary teeth that are lengthened into sharp, nonpedicellate fangs (Lynch 1971). The two *Leptodactylus* species have relatively normal pedicellate anuran teeth, although they are long and pointed in *L. laticeps*. *Bufo paracnemis* lacks teeth.

Gape

Lepidobatrachus, Ceratophrys, and *Chacophrys* had relatively wider jaws than did *Leptodactylus laticeps,* which in turn had a wider jaw than *L. chaquensis* (fig. 12.3).

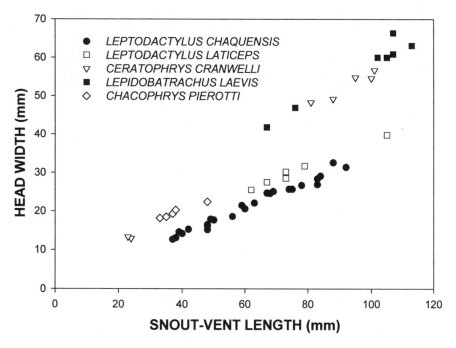

Figure 12.3   Relative head widths of Chaco frogs that eat anurans.

Discussion

ANUROPHAGY

Morphological Characteristics

Our data (table 12.4) and those of others (Budgett 1899; Cei 1955, 1956, 1980; Dure 1999) indicate that 5 of the 22 Chaco species of anurans (*Leptodactylus chaquensis, L. laticeps, Lepidobatrachus laevis, Chacophrys pierotti,* and *Ceratophrys cranwelli*) derive a significant portion of their diet from other frogs.

A ubiquitous characteristic of frog-eating anurans is a large gape (Emerson 1985). In contrast to arthropods, anuran prey can swell up when attacked, effectively doubling their diameter; some frogs extend their forelimbs at right angles to the body to avoid being swallowed (Kerr 1950; Duellman and Trueb 1986). A predator with a large gape can overcome these defenses. In the frog-eating species we studied, gape was consistent with the level of anurophagy (fig. 12.3). The largest gapes were found in the ceratophryids: *Ceratophrys* (a frog-eating specialist), *Chacophrys,* and *Lepidobatrachus* (which consumes frogs and large, globular, ampullariid snails). *Leptodactylus laticeps* has a more varied diet and smaller gape than do the ceratophryids but has a larger gape and probably eats more frogs than does *L. chaquensis.*

In most frogs, there is considerable flexion of the bones of the lower jaw on either side of the cartilaginous mandibular symphysis, and between the symphysis and the mentomeckelian bones and dentary. The dentaries are extremely thin and flexible in this region, and loosely connected to the mentomeckelian complex, allowing the mandibular symphysis to rotate downward (Gans and Gorniak 1982). In addition, the mandibular rami can move independently of each other. In the species that we studied, the mandible of *Bufo* is the most flexible (plate 12.4A), and the two frog-eating *Leptodactylus* have mandibles that are somewhat more rigid. In contrast, the symphyseal region of the mandible of ceratophryids is stiff, formed by a symphysis between the two dentary bones, and the mentomeckelian cartilage pad is replaced by bony, penetrating odontoids (fig. 12.2C, plate 12.4B).

The fang-like teeth of the ceratophryids are similar to those in other anuran-feeding frogs (e.g., *Adelotus, Hemiphractus, Megaelosia,* and *Telmatobius;* Lynch 1971); however, only the ceratophryids have solid nonpedicellate teeth.

Solid maxillary teeth, odontoid processes, a sharp mandibular ridge, and the absence of flexion in the mandible distinguish the ceratophryids from most other anurans. A typical frog's jaws and teeth (or dermal pad in *Bufo*) closely fit the hard and smooth bodies of beetles, and the pedicellate teeth, if present, allow the prey to move inward but not outward. Pedicellate teeth serve as many tiny points of contact on the beetle's surface. The symphyseal pad and the padded mandibular edge function as slightly yielding pressure surfaces to hold the prey and force it upward on the teeth. Mandibular flexion assures maximal contact between the jaw and the prey. Frogs present a different problem. With powerful hind legs, a frog can exert a great deal of force. Tiny folding teeth do not penetrate the soft slippery skin of the frog, and a flexible lower jaw only serves as a weak point in the predator's grasping apparatus.

The teeth of the upper jaw of ceratophryids have been converted into rigid, long, sharp daggers that penetrate soft-bodied prey. These teeth would not hold a large beetle because they provide relatively few pinpoint contact surfaces. Large dagger-like teeth are a liability if they become caught in the joints or penetrate the chitin of the insect's armor.

A similar analysis can be made for the modifications of the lower jaw. The rigid mandibular arch, with its knife-edged dentary and odontoids, firmly drives the body of the prey up onto the long, sharp, rigid maxillary teeth.

Other frog-eating anurans have some of these morphological features, but none has them all. Other frogs that include anurans in their diets (e.g., *Hemiphractus, Megaelosia,* and *Telmatobius,* Lynch 1971; *Pixicephalus adspersus,* Loveridge 1950) have long, pointed teeth and a rigid mandibular symphysis supporting bony odontoids. However, bony odontoids do not always seem to have a dietary significance; in some Asiatic ranids, males have odontoids and females do not, but their diets are similar and do not include frogs (Emerson and Voris 1992; Emerson 1994).

The protrusible tongue in frogs is a striking evolutionary novelty (Regal and Gans 1976). It is as different from the presumed labyrinthodont ancestor as is the synchronous hind leg leap (Gans 1961c), and it may be one of the major reasons for the successful radiation of anurans. The anuran tongue is primarily a device for feeding on small prey items, usually arthropods. The reach and mobility of the tongue is the reason that a 167-mm *Bufo paracnemis* in our sample could have more than 1,400 termites less than 8 mm long in its stomach, each harvested one at a time (or at most two or three) from a nocturnal feeding column. The toad did not have to mobilize its large bulk for each prey capture—only its tongue (Gans and Gorniak 1982).

Bufo paracnemis is probably not an important frog predator (Guix 1993; Dure and Kehr 1996; this study), but it was big enough to feed on them. Examination of their foraging techniques was useful in determining why frogs are not consumed in large numbers. Frogs do not stick well to a tongue surface, and an effective anuran predator must grab prey with its jaws. In our study, the leptodactylids and ceratophryids grabbed anuran prey with their jaws. On only one occasion did we see *B. paracnemis* do this; it usually responded to an anuran prey item as if it were an insect, by lapping with its tongue. Consequently, their

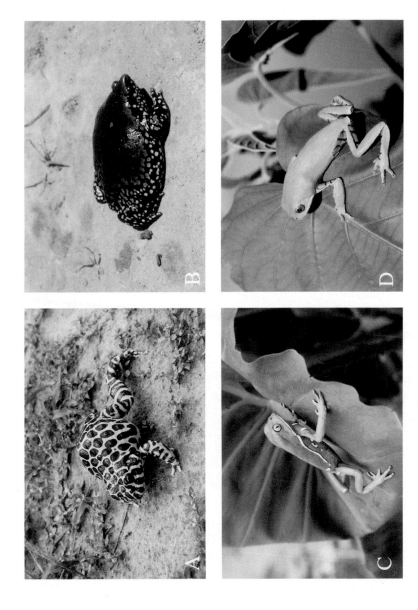

Plate 12.1   Aposematic members of the frog fauna, in Filadelfia, Departamento Boquerón, Paraguay:
(A) *Leptodactylus laticeps*; (B) *Dermatonotus muelleri*; (C) *Phyllomedusa sauvagii*; (D) *P. hypochondrialis*.

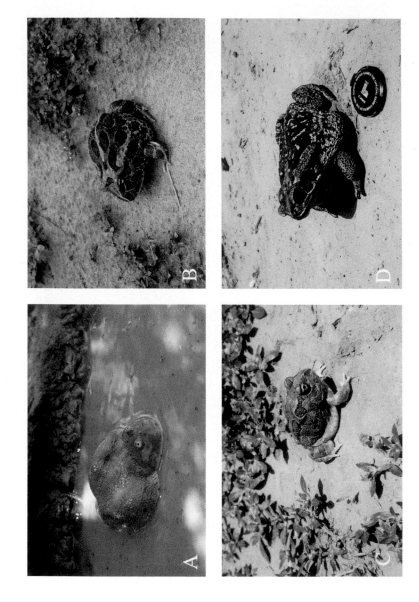

Plate 12.2   Bufonid and ceratophryid species in the frog fauna, in Filadelfia, Departamento Boquerón, Paraguay:
(A) *Lepidobatrachus laevis*; (B) *Ceratophrys cranwelli*; (C) *Chacophrys pierotti*; (D) *Bufo paracnemis*.

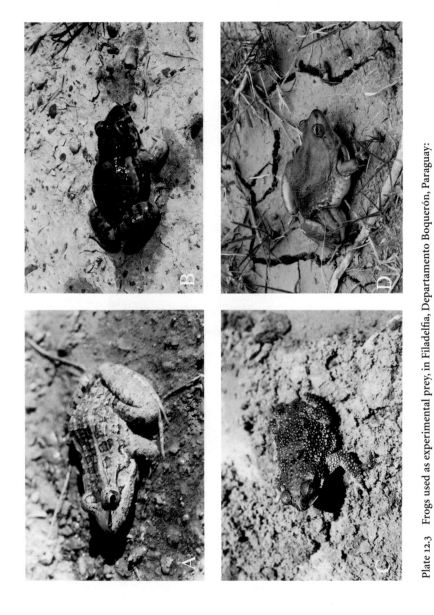

Plate 12.3   Frogs used as experimental prey, in Filadelfia, Departamento Boquerón, Paraguay:
(A) *Leptodactylus chaquensis*; (B) *Leptodactylus bufonius*; (C) *Bufo major*; (D) *Pseudis paradoxa*.

Plate 12.4   Deformation of lower jaws of Chaco anurans:
(A) *Bufo paracnemis*; (B) *Lepidobatrachus laevis*.

success rate at harvesting frogs was very low. The ceratophryids have relatively nonmobile tongues compared with those of *Bufo* and *Leptodactylus,* partly because of the rigid mandible (Regal and Gans 1976). *Leptodactylus chaquensis* and perhaps *L. laticeps* have a protrusible tongue for capturing small arthropod prey, but they also grab frogs with their jaws.

Tolerance of Noxious Secretions

Another component of anurophagy is the ability of the predator to tolerate the noxious compounds secreted by the prey. In our studies, *Lepidobatrachus laevis* and *Bufo paracnemis* were the most tolerant of noxious prey. It is perhaps logical that *L. laevis,* a frog predator, should have evolved resistance to their toxins, but why has *B. paracnemis?* *B. paracnemis* feeds mostly on arthropods. Perhaps the mechanisms that allow anuran predators to tolerate toxins are general enough to function against a wide variety of chemical defenses. Toads are exposed to many potent arthropod defensive secretions; one *Bufo* stomach that we examined had more than 100 pairs of elytrae of an aposematic species of blister beetle (Meloidae) that raised blisters on humans who handled it.

ANURAN DEFENSES

One defense against predators that ingest their prey whole is to grow to a large size. Almost a quarter of the Chaco frog species reach 90 mm, a size that is probably too large to be eaten by another frog, and large size coupled with effective toxicity may make adult *Bufo paracnemis* and *Leptodactylus laticeps* invulnerable to predation.

However, the most common defense by frogs against predators seems to be noxious and toxic skin secretions (Daly et al. 1987). The effects of amphibian secretions on snakes have been widely studied (Sazima 1974; Macartney and Gregory 1981; Barthalmus 1994; Brodie and Brodie 1999), but the interactions between frogs as prey and frogs as predators have not.

Toxins in frog skins consist of a wide array of bioactive compounds (Daly et al. 1987; Erspamer 1994). Our prey choice experiments showed that noxious skin secretions protected adults of certain species, especially the two *Phyllomedusa* and the large microhylid, *Dermatonotus.* The protection was not complete; some individual predators ate both *Phyllomedusa* and *Dermatonotus,* although the majority refused them. Cei and Erspamer (1966) and Anastasi et al. (1966, 1969) found exceptionally high concentrations of at least five powerful, bioactive bradykinin-like polypeptides in the skins of both *P. hypochondrialis* and *P. sauvagii,* and Sazima (1974) showed that skin secretions of a Brazilian species, *P. rohdei,* partially protected it from predation by a snake (*Liophis mil-*

iaris). The several *Dermatonotus* that we found in two *Lepidobatrachus* stomachs were recently transformed juveniles that would not have developed the quantity of noxious secretions that characterized the adults.

Aposematic coloration was the best predictor of unpalatability; the three most unpalatable species, the two *Phyllomedusa* and *Dermatonotus*, were the only aposematic prey we tested.

Several species (*Scinax acuminatus, Bufo major*, small *B. paracnemis, Leptodactylus bufonius, L. chaquensis*) that we judged to be fairly noxious were completely palatable to our experimental predators. Cei et al. (1968) found biogenic amines in the skins of Argentine *B. major*, but the species is regularly eaten by predatory frogs (Budgett 1899; Gallardo 1957; Cardoso and Sazima 1977). Cei (1956, 1980) also stated that *L. laticeps* preys "preferably" on *L. bufonius*. Although *L. bufonius* is relatively unpleasant to handle, Cei et al. (1967) found only small quantities of biogenic amines and polypeptides in its skin.

The biochemistry of many of the skin secretions of amphibians is known (Daly et al. 1987; Erspamer 1994; Daly 1998), and the responses of avian, mammalian, snake, and invertebrate predators to amphibian secretions have been studied (summarized in Daly et al. 1987; Barthalmus 1994). The results of our study of the responses of anuran predators are parallel to the results of studies of snake–amphibian/predator–prey systems (e.g., Spaur and Smith 1971; Garton and Mushinsky 1979; Macartney and Gregory 1981; Shine 1991; Madsen and Shine 1994; Brodie and Brodie 1999) in that we saw a species-specific gradient in the ability of different predators to tolerate amphibian skin toxins.

The presence of several frogs protected by noxious secretions in the Chaco fauna probably has several explanations. The fauna is relatively simple, and the same predator species frequently encounters the same prey. Chaco frogs have coexisted for a long time, and the prey species have had a longer time to evolve more specific toxins to the same predators than they would in unstable environments with diverse and variable suites of predators. The fact that predatory species are able to eat the most strongly protected prey is an indication that this has been a true evolutionary arms race (Brodie and Brodie 1999).

Another reason for the abundance of noxious frogs may be the evolutionary pressures produced by an unpredictable climate. As discussed earlier in the chapter, juveniles are often quite scarce, especially those of the larger species, indicating many years of breeding failure during our study period. A good example of a species that breeds periodically in a hostile environment is the spadefoot (*Scaphiopus couchii*) in the eastern California desert, which may breed on average every five years (Mayhew 1962); it is also quite noxious. The fact that North America frog faunas in a similar climatic regime to the Chaco are composed largely of generally toxic genera (*Scaphiopus, Spea, Bufo*) lends strength to our conclusions (Blair 1979). This hypothesis could be tested by examining frog faunas in similar climates in Africa.

Cannibalism

The three Chaco ceratophryids and probably the *Leptodactylus* species are cannibalistic. Crump (1992) stated that cannibalism is more frequent in ephemeral environments like the Chaco, probably because frogs become more concentrated and only a few will survive until the next rainy season. The frogs that eat conspecifics may gain a survival advantage, whereas the cost in inclusive fitness incurred by eating a relative is low, because the relative probably would not survive anyway, especially if it is smaller.

Acknowledgments

We thank the following people for their encouragement and support: Hon. Hernán Bertoni, minister of the Ministerio de Agricultura y Ganadería, Paraguay; Ing. Pedro Calabrese and Ing. Hilario Moreno of the Paraguayan Servicio Forestal; and J. J. Hurd of the United States Peace Corps, Paraguay. The Mennonites of Filadelfia accepted us and our eccentricities. For field assistance and logistic support, we thank Bill Hahn, Len West, David Norman, and Silvio Espinola. Rayann Robino and Susan Wright expertly handled various stages of the manuscript, and Kent Beaman provided bibliographic assistance. William Magnusson, Robert Reynolds, and one anonymous reviewer provided comments that improved the manuscript.

13

Long-Term Frog Monitoring by Local People in Papua New Guinea and the 1997–98 El Niño Southern Oscillation

David P. Bickford

During the 44 months I spent in Papua New Guinea doing fieldwork and training local people as parabiologists to conduct a long-term frog-monitoring project, I gained many insights that fall into two distinct categories. The first has to do with the social and logistical aspects of training and fieldwork. The initial mandate of my work was to train local people in standard biological monitoring methods. As parabiologists, the local participants would be highly trained individuals who could competently complete tasks, but they would be without formal, time-consuming, expensive education—comparable in some ways to paramedics. In this regard, I was fortunate to work with many of the Pawaia people of the southern parts of Chimbu Province, Papua New Guinea, and they provided the basis for the first category of lessons learned. The second category of lessons learned had to do with counterintuitive changes in the relative abundance of frog species during the 1997–98 El Niño Southern Oscillation event, which caused a catastrophic drought throughout much of Australasia.

I include in this chapter both the social aspects of working with local people as well as the scientific results of the survey in order to bridge the gap that normally exists between conservation biology and "the real world." Conservation biologists often operate in complex local socioeconomic conditions and need to incorporate local perspectives so that they can modify their approaches. This chapter demonstrates one approach that attempted to honor local perspectives during the long-term project. I document the lessons learned and provide recommendations based on those lessons.

The Problem: Declining Amphibian Populations

Amphibian populations around the world have received much attention in the last decade because of their declines (Blaustein and Wake 1990; Wake 1991b; Trenerry et al. 1994; Alford and Richards 1999; Lips 1999), disappearances (Pounds and Crump 1994), or deformities (Johnson et al. 1999). Disappearances are the endpoints of declining populations, whereas deformities may or may not be part of declines. Nonetheless, these phenomena occur simultaneously and in geographically distant locations. Animals in decline or with deformities are often limited in number, difficult to find, or protected by law, making investigations and experiments designed to find causes difficult or impossible at most sites. Two fungal species have been implicated recently as the cause of declines and/or disappearances (Kiesecker and Blaustein 1997; Berger et al. 1998), trematode infestation has been associated with deformities (Johnson et al. 1999), and global warming has been implicated as a factor contributing to declines (Kiesecker et al. 2001). No single cause, however, can explain anuran population dysfunctions across all sites. Whatever the reasons for these phenomena, natural population fluctuations and reactions at the population level to environmental degradation, contamination, weather perturbations, or other periods of stress remain virtually unknown (Pechmann et al. 1991). The lack of data on baseline population dynamics makes it difficult to differentiate human impacts on amphibians from natural disturbances or natural fluctuations in population size.

Habitat destruction and modification are the principal factors in most declines and extinctions, but amphibians are declining and disappearing from seemingly pristine and protected sites. For many reasons, amphibians are good biological indicators of their environment and may be more sensitive to environmental changes than are other organisms (Wake 1991b; Blaustein et al. 1994; see Pechmann and Wilbur 1994 for counterpoints). Understanding amphibian population dynamics is an important and feasible research priority for most ecosystems. Before we understand declines, we must investigate natural populations and describe baseline conditions in order to compare recent declines and disappearances to what may be natural population fluctuations, natural extinctions, or periods of dormancy. The only way to procure these data for comparison is to have long-term monitoring projects under way during periods of natural stresses, weather perturbations, or low-recruitment seasons. Only then can we make a serious attempt at understanding the differences between anthropogenic population declines and the natural variability in populations through time, including extinction events.

Long-Term Monitoring by Local People

I began a long-term frog-monitoring project and training program for local people under the auspices of an Integrated Conservation and Development Project (hereafter ICAD, but known in some literature as ICDP) in Papua New Guinea in September 1995. Integrated conservation and development projects attempt to meet the reality of human needs and desires for development with benefits and revenues based on the conservation of biological diversity. The conservation and development link of the ICAD project is summarized by Pearl (1994). Within the larger ICAD project, the frog-monitoring project is an example of scientific research as a local business. Frog-monitoring provides direct local economic and development benefits through the maintenance of biological diversity because it is only viable if the forest remains intact. The frog-monitoring project's aims were twofold: first, provide data on frog populations from a relatively pristine site; and second, reach the goal of the ICAD project by training local people as parabiologists. Employment as a parabiologist provides cash income for the stakeholders and may serve to promote protection and conservation of biodiversity.

STUDY SITE

The Crater Mountain Wildlife Management Area (CMWMA) is on the southern escarpment of the central cordillera of New Guinea. It includes parts of three provinces (Gulf, Chimbu, and Eastern Highlands provinces) and two local language groups (Pawaia to the south and Gimi to the north). The CMWMA is 270,000 ha and ranges from almost 50 m to more than 3,100 m above sea level (fig. 13.1).

The Crater Mountain Biological Research Station at Wara Sera is located in the center of the CMWMA and is named for its location on the Sera River. The protected study area extends from 850 to 1,350 m above sea level (fig. 13.1). The station lands are owned by one of the local groups in the Pawaia cultural/language group of the Pio-Tura area. Fieldwork by local people has always been a part of the station's purposes and strengths, and in 1995 I initiated a project designed to be fully executed in the field by Pawaia people over a long-term period.

Topography of the area is extreme, with sharp ridges and steep valleys dissected by several rivers and streams. Annual average rainfall is 6.4 m for the Crater Mountain Biological Research Station at Wara Sera (Wright et al. 1997), and there is no dry season (fig. 13.2). A severe El Niño Southern Oscillation (ENSO) event in 1997 and early 1998 dramatically reduced the rainfall received from September to December 1997 (see fig. 13.3 below). The rainfall during this period was not only the lowest ever recorded for those months, but it was also

Figure 13.1 Map of the Crater Mountain Wildlife Management Area (CMWMA) and its location in Papua New Guinea. The CMWMA, which straddles the Eastern Highlands, Gulf, and Chimbu provinces, is indicated by the shaded area on the map of Papua New Guinea.

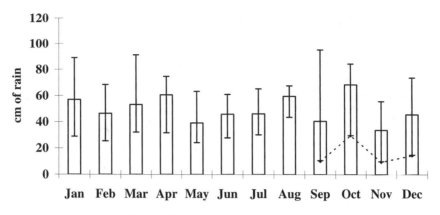

Figure 13.2   Mean monthly rainfall (cm) for the Crater Mountain Biological Research Station at Wara Sera, 1992–98. The drought associated with the El Niño Southern Oscillation event in 1997 is shown by the dashed line. Error bars indicate the minimum and maximum values.

below the 20-cm threshold, what most authors typically define as "dry" for a rainforest environment (Lincoln et al. 1982).

Part One: Parabiologist Training

Establishing a project to be wholly run in the field by the Pawaia was a difficult and lengthy process of trial and error. I learned many lessons during fieldwork and training that have heuristic value for other projects. The cultural, linguistic, and political contexts of working with local people are extremely important and need to be understood and specifically addressed.

Papua New Guinea is unique among developing tropical countries because of traditional land tenure regimes. More than 99% of land in New Guinea is managed privately, with almost no government ownership of land outside of urban areas (Crocombe and Hide 1987). Each group of people (e.g., language group, extended family group) owns its lands in various hierarchies of tenure and usufruct rights, and land is often managed by a group (Brookfield and Brown 1963; Brookfield and Hart 1971). Papua New Guinea has very little rural infrastructure. Many isolated villages in the interior have no road access and little government support for schools and first aid posts. The CMWMA-ICAD project is operated in a remote rural area about 75 km away from the nearest commercial center. The only access to the CMWMA is by small aircraft and/or foot.

THE PAWAIA

Until roughly 25 years ago, the Pawaia were principally a highly mobile population, moving through the forest in small family groups. The Pawaia had intimate and active contact with the forest in which they lived. They had few government services but moved around, lived in, and managed large holdings of land near the southern coast in Gulf Province to the foothills of the dividing range in Chimbu and Eastern Highlands provinces. The Pawaia were illiterate with a rich tradition of storytelling and chanting.

Local conditions for the Pawaia involved in the ICAD currently focus around a central airstrip in the village of Haia (fig 13.1). The first permanent missionary post in the area was in Haia, and a government school was established there in 1991. The Pawaia have been transformed from a largely mobile and rainforest-dwelling people to a sedentary and largely village-based people who do not move through their lands as they once did. Not surprisingly, the Pawaia mirror changes typical of those of indigenous peoples around the world. As the Pawaia have become more connected and influenced by the outside world and different standards of living, they have shown a tendency to replace their forest-based knowledge with new external-based knowledge.

PARABIOLOGIST TRAINING

To ensure data quality and to train and employ as many Pawaia as possible, I used methods that required two to four assistants. During the training period for these assistants, I attempted to provide employment and instructional opportunities regardless of gender or social group. Unfortunately, the Pawaian cultural context made it extremely difficult to include women in the training program. I was not able to hire unmarried women because of cultural taboos, and married women almost always had high demands of motherhood that started immediately after marriage. Although I tried to include married women with children in the training program, it was not feasible for women to simultaneously attend to their children and perform fieldwork. I included two recently married women in the training program, but they soon became pregnant and had to focus on domestic duties. Thus, training was implemented mainly for men, despite my efforts to include women.

I began training the assistants by teaching a series of fundamentals—reading, writing, and basic numeracy skills—and then moved on to techniques more specific to biological fieldwork.

Training Methods—Basic Skills

Because most Pawaia do not have much formal education (the median educational level was less than equivalent to third grade), my priority was to get the basics across—reading, writing, and numeracy—in the context of the monitoring methods I planned to use. Training included small group discussions, workshops, and practice sessions. A few assistants had some formal education, so I trained them to teach their peers. An example of the exercises used by trained assistants to teach their peers was to copy the scientific names of each frog species in the study area and write in the local language the names for each frog species. This often precipitated other learning activities. Other exercises and workshops included introduction of the metric system and standard measuring units. We incorporated independent thinking by estimating the sizes of familiar objects as well as frogs in millimeters and meters to familiarize the Pawaia with the units of the metric system.

By using a variety of training techniques and involving the Pawaia as trainers, we tested the numeracy, literacy, natural history knowledge, and language skills of each assistant. These exercises also provided a medium whereby different generations could teach each other. Often older men had a better grasp of different frog species' names, habitats, and natural history, whereas younger men and boys had better numeracy and written skills. Thus, an exchange was initiated that bridged generations, and sharing of forest-based cultural knowledge was rewarded. The assistants improved their abilities at frog identification, literacy, and numeracy.

To make these exercises less tedious, I incorporated games and exercises found in English-language grade school books and popular magazines. These were helpful and often captured the assistants' attention. Discussions were initiated by the Pawaia assistants on the basis of articles they had read or pictures they had seen in conservation magazines (e.g., *National Geographic, Wildlife Conservation, Nature Conservancy*). I provided the assistants with environmental education articles or pamphlets in Melanesian Pidgin that were produced by the former Christensen Research Institute (based in Madang, Papua New Guinea). These were popular because they were picture-based and provided real-life stories from places the assistants knew within Papua New Guinea. Although not directly related to the frog-monitoring project, these media provided valuable tools to disseminate information about conservation, were helpful in literacy training, and indirectly proved to be valuable for the ICAD project as a whole.

Specific field training required construction of a dichotomous key to identify frogs to species. Because some of the frogs were difficult to identify, we focused on unambiguous characters to distinguish between similar species. Many frogs in the Wara Sera study area did not have common local names (see the section titled "Language and Vernacular") so we invented names. Because I was introducing new names to the Pawaia, I used abbreviated scientific names instead of a "local name–scientific name" mix to alleviate confusion. The first few letters of the genus and species were used as a code name for each frog species. For example, *Sphenophryne schlaginhaufeni* became "Speno slag" and *Hy-*

lophorbus rufescens became "Hylo ruf." The assistants were able to recognize these names and use the codes effectively to identify frogs.

Field Training: Long-Term Monitoring Methods

Rather than document all trials and errors, I focus on the methods that worked and discuss reasons why they worked to facilitate their generality for other projects.

Nonrandom Teams

As the project grew and I was employing 6–10 Pawaia at a time, I realized that there were two general groups of assistants. These were by and large the young group and the old group. The main difference between groups was the amount of formal education in the village school (or more precisely, the amount of time spent in the village). I found that the younger Pawaia who had stayed around the village as youngsters to attend school were generally less knowledgeable about natural history and frog names. The younger group, however, had much better numeracy and literacy skills, and young Pawaia were better at recording data and measuring frogs. The older assistants, on the other hand, were better at identifying frog species and distinguishing between similar frog species. Initially, my random teams sometimes were made up of excellent parabiologists who could read and write extremely well but had problems identifying frogs to species. Likewise, some teams were made up of men who were able to identify all of the frogs found in the area but could not accurately record data. Hence, I paired older and younger Pawaia to balance these different skills. Members of these teams worked together and exchanged knowledge, and proved to be efficient and reliable.

The other nonrandom quality of the assistants that worked with me reflected the ever-present political complications among Pawaia family groups. During each hiring cycle, I included members of different family groups. Although there are ten extended family groups in the village of Haia, not all were included in the beginning years of the station's history. The lack of total group involvement, coupled with the Pawaia history of jealousy and fighting among extended family groups, resulted in a complex situation. To try to give uniform opportunity to all extended family groups, the project sponsored training workshops in which each extended family group elected two members to be fully trained as assistants at the research station. Many of these elected Pawaia were not good assistants or parabiologists, but the project gave the opportunity for training to all extended family groups in an attempt to be fair. After the initial trials of all extended family groups, it became apparent that it was in the project's best interest to keep hiring in a cycle to ensure full inclusion of all extended family groups. If this affected assistant quality, I hired one or two additional highly qualified and experienced assistants to work with the underqualified assistants.

Benefits of Employment
The landholder committee of the ICAD project decides on researcher respon-sibilities and what researchers must provide for their assistants. In the case of our project, assistants hired as employees were expected to receive wages, raingear, flashlights, blankets, candles, food, and medical aid. In this part of Papua New Guinea, jobs are in demand and cash is limited, so there is compe-tition for employment, and the people want the benefits accrued through the CMWMA-ICAD and biological field research. Moreover, the assistants earn respect by being a part of the project because they receive wages and other benefits.

Equal Opportunity to Work
Two-week trial periods (fortnights) were the standard contract for employ-ment. This made wages and turnover of assistants evenly spaced, and equal op-portunity was afforded to each lineage or family group. To provide stability, I tried to avoid a complete turnover of assistants at any single hiring period. I trained the more experienced assistants to be trainers of the less experienced as-sistants. This management strategy was effective for the frog-monitoring proj-ect. Other important repercussions of the employment protocol were the pre-vention of new jealousies among family groups and the amelioration of clashes among lineages of the Pawaia and with their neighbors to the north, the Gimi speakers in the village of Herowana (see Gillison 1993 for historic relations be-tween these groups).

Training of Local Trainers
Training the skilled Pawaia assistants to be trainers was important to the suc-cess of the project. Pawaian trainers shared their experience as parabiologists during the training of less experienced assistants. Employing Pawaian teachers eliminated many of the constraints imposed by translating among three differ-ent languages. In New Guinea, where there are approximately 750 local lan-guages, the lingua franca is Neo-Melanesian Pidgin, and in most urban centers classes are taught in English. The frog-monitoring project involved the many levels of translation that occur between English, Neo-Melanesian Pidgin, and Tɛhōe (the local Pawaia language). One of the most rewarding aspects of the project was seeing the Pawaia train each other in field methods in Tɛhōe and watching older assistants teaching younger ones the Tɛhōe names for the dif-ferent frog species.

Daily Meetings
After about 12 months of fieldwork, I found it difficult to stay on top of all the various things that could go awry with 6–10 field assistants working in two teams. I initiated daily meetings to improve morale and to stay abreast of prob-lems or errors as they occurred. Simply reminding the assistants of principal

points they had learned during training (especially frog identification and measuring protocols) was sufficient and appreciated. These meetings also improved intragroup cohesion. I typically managed two teams of four assistants who would not see each other all morning, so daily meetings ensured a unified working field team for the evening's fieldwork. As the teams became used to these meetings, I allowed opportunities for discussion on other topics. Thus the daily meetings became an avenue for assistants to voice opinions and ask questions, and with experienced teams the meetings were regular social gatherings.

Hands-on Learning

I minimized the amount of formal or classroom learning and maximized hands-on learning by using selected field methods as training tools. This "on-the-job" learning was efficient, and assistants learned quickly while doing the fieldwork. I also used question-and-answer sessions and problem-solving exercises in the field, sometimes employing transit time between sites (e.g., while walking from one leaf-litter plot to the next) or time spent in preparation for fieldwork before the sampling began (e.g., while walking—frog in hand—to the beginning of the visual encounter survey [VES] transect). In the field, assistants appeared to be at ease and were not worried about making mistakes. Informal teaching in the field was more effective than direct questioning typical of classroom situations.

Low-Pressure Tests

Formal teaching and field evaluations were often hindered by performance anxiety. To avoid this, I conducted less formal evaluations for each assistant on a regular basis. I maintained rigor by testing each assistant in the field, either one-on-one or as part of a group. I performed evaluations as part of the field method (e.g., asking an assistant to measure leaf-litter depth at each corner of the plot and then checking his measurements). Sometimes, as part of an evaluation, I would allow the team to work on a problem together (e.g., identification of a particularly difficult or rare frog species). If an assistant was unable to complete a task or made a mistake, instead of calling him out and making an example of him, I simply asked another, more experienced assistant to check his answer. This type of evaluation did not appear to be formal, and the assistants were not anxious about failure when they were corrected by one of their peers. I never attempted classroom written tests because I felt they were not appropriate and would only heighten an already high level of performance anxiety for the Pawaia.

Never "Lose Face"

After inadvertently embarrassing assistants by directly correcting them when they made mistakes, I learned that avoiding embarrassment or humiliation was very important in cross-cultural exchanges with assistants. To provide a working environment in which humiliation was prevented, I used every error as an instructional tool. Instead of telling an assistant that he was wrong and explaining why, I would ask him in a nonthreatening way why he thought his answer was correct. Many times, I never had to specifically mention that his initial answer was incorrect because he would correct himself during a question–answer exchange. For example, if an assistant misidentified a frog, I would ask him why he thought it was that particular species. He would list the attributes or characteristics, which would allow me to ask more specific questions and simultaneously hone his identification skills. If one characteristic was incorrect, I would ask him about it and the range of possible character states in the frog species. By comparing different frog species and characteristics, we could almost always correct a misidentification without causing the assistant to feel humiliated or ashamed.

LIMITATIONS

Many sociocultural limitations affected the field methods on numerous occasions. Because the frog-monitoring project is part of the ICAD project and local people are the primary beneficiaries, cultural acceptance is paramount to success. This means that sacrifices are made and limits are established to both maximize information gathering and minimize cultural disruption or disrespect.

Flexibility is a key factor to any field project that involves working with local people. Project leaders must listen and react to local needs. Becoming intimately familiar with the local perspective is necessary to understand how to be flexible and get the research done. During my work in the CMWMA I was asked for favors that included making personal loans, allowing early termination of a work period to attend a funeral, and buying items in town. Requests like these can be considered but need to be taken care of as they arise so they do not become limiting problems during critical periods of fieldwork. In most cases, there is no tangible effect of cultural beliefs or social customs on methods used in the field. In some cases, however, specific field methods are affected by preexisting beliefs or cultural perspectives that may not be obvious to an outsider.

Legends and Myths: The Masalai

A cultural constraint of fieldwork with the Pawaia became apparent during nocturnal visual encounter surveys (VES) conducted along fixed transects. Many folk tales and legends exist about the pseudomythical *masalai* (or poison-man), who comes out at night and kills people. Although there seems to be a great deal of variability in the strength of belief in these stories, they are perpetuated, and there are instances in which these stories become reality when they elicit violent behavior that is explained without blame or punishment. Because these stories related to murders at night, and we sampled VES transects almost every night, I realized that the unease and lack of morale were not the result of inexperience or arduous conditions. Because of the constant "looking-over-the-shoulder" anxiety related to fear of *masalai,* reliability of data collected at night by local parabiologists was not high initially. However, after the Pawaia told me the stories and I realized the variability of the strength of belief about the myth, we were able to avoid many negative aspects of the *masalai* myth on the monitoring methods. I selected men whose belief in the *masalai* was not strong and worked exclusively with them at night for many months. Because these assistants were never killed by the *masalai,* they would explain to the other assistants that the scientific night work we were doing must be exempt from the actions of the *masalai,* or that the study area was somehow protected from the *masalai.* After grooming a few Pawaia assistants and trainers who effectively addressed the fears of their fellow Pawaia, we were able to overcome this constraint completely.

Legends and Myths: The Singing Worm

One way of tracking frogs—the audio strip transect (AST) method—relies on knowledge of frog calls, so the person recording the data needs to know all the frog calls in the area. As a result of the large number of calls that had to be learned, I found this method impractical, but in any case use of the method was affected by a myth about the singing worm. Frogs of the microhylid genus *Xenobatrachus* are considered to be mythical singing worms by many cultural groups in New Guinea (see Bulmer and Tyler 1968). The perpetuation of this myth stems from the natural history of the frogs. *Xenobatrachus* frogs are fossorial and rarely encountered by humans. They are typically active and calling during or immediately after rain. Their calls are extremely quiet, high pitched, and infrequent "poooop"s emitted from under the soil surface. Worms in New Guinea can be enormous and are also seen during or after rains. Because of the rarity of the frogs, their unusual calls, and their shared fossorial microhabitat with worms, these frogs have become known as the singing worms. I believe, however, that asking local people to disregard myths and stories in their cultural context is not in the best interests of the ICAD project nor the frog-monitoring project and may lead to loss of cultural knowledge over time. Also, in this case the identification method was inappropriate because of the diversity of calling frogs and the difficulty in training, so it was not included.

Language and Vernacular: Frog Identification

Loss of forest-based indigenous knowledge (i.e., loss of the accumulated experience of thousands of generations of people living in the forest) is a tragic reality for most indigenous cultures.

The local Pawaia language, Tɛhōe, is a complex and rich language, with stories and ecological designations that can be intricate and detailed. Partly because young boys attend school in a central village, many are not learning forest-based knowledge that is gained mainly by living and traveling by foot through the forest. Instead, the centralized village is changing the cultural knowledge of the Pawaia. Older men and women are the caretakers of knowledge that is not being passed down.

Although not generally correlated with Latin binomials, the Pawaia Tɛhōe names are useful. The Pawaia sort frogs into two categories: *sion*, frogs you can eat (generally ranids); and *eria*, frogs you cannot eat (they are too small or have toxic skin secretions). In Tɛhōe there are descriptors of color, size, and habitat that modify names in some cases. For example, *paîāmō sion n'hoela hō'ō* is the name for the small edible frog with a short snout (*Nyctimystes cheesmani*). Few frogs have a Pawaian common name that corresponds directly to a Latin binomial. The frogs that do have specific names are the largest, most colorful, most common, have the loudest calls, or are used in local magic potions. Frogs that are eaten, for example, have one-word names such as *sō'ā* (the Tɛhōe name for *Rana grisea*). Because so many frogs did not have names, and because of the generalized and inconsistent Pawaia naming scheme, it was difficult to teach frog identification in local terminology. I used the aforementioned code based on the Latin binomial, which enabled us to name frogs and collect data accurately.

RESULTS AND CONCLUSIONS

Training local people as parabiologists can be a powerful mechanism for including local perspectives in field research and conservation. I recommend that local people be included in all aspects of conservation projects, thereby enabling their participation through education and access to information. Table 13.1 contrasts the field methods I found most useful in Papua New Guinea with typical field protocols.

After nearly four years of training assistants (1995–99), the frog-monitoring project was run for three years by locals in the field at the Crater Mountain Biological Research Station at Wara Sera. It was terminated in 2002, however, until more training could take place because systematic errors were accumulating during data collection. As part of a larger ICAD project, the frog-monitoring program has had many positive results. It provided local cash income for the parabiologists hired to conduct the monitoring. Training local people as parabiologists instilled or augmented an existing stewardship ethic and spent conservation dollars locally where they are needed. Training local people to conduct a long-term frog-monitoring project involves local people in biodiversity conservation. This situation is stable and feasible for long-term conservation

Table 13.1   Contrasting approaches used in field projects

Subject	Traditional approach	Approach used in Papua New Guinea
Training		
Style	Classroom / lecture with fieldwork	Hands-on fieldwork with discussions
Mistakes	Correct mistakes / heuristic display of errors	Use mistakes as learning tools / avoid humiliation
Corrections	Correction from authority figure	Correction from peers
Communication	Translate	Use senior assistants for training in local language
Assessing progress	Formal written and/or oral tests with grades, often in large groups	Informal evaluations with no grades—best in small groups or individually
Working in the field		
Team constituents	Random or select the best workers	Select teams to maximize diversity of skills or knowledge
Choosing workers	Only the best or most productive or reliable	Political representation of all concerned stakeholders
Group meetings	Ranges from often to only when needed	Daily

and may be more effective than forcing local people to uphold lofty conservation ideals that they do not fully understand.

The reality of ensuring success in a long-term monitoring project requires a serious investment of time. In the case of the Pawaia in Papua New Guinea, it took nearly four years of training to produce field teams that can complete monitoring methods by themselves. Even after this lengthy training period, intermittent training needs to be carried out to keep the field teams working effectively. The time investment for other projects may not be as lengthy, but it must be made on the ground with local people. Special attention must be paid to the cultural context and/or the perspectives of local people. Training projects and conservation education programs like the one I have described cannot be administered remotely.

Although it may not be a new paradigm, turning the focus from working *with* local people to working *for* local people is powerful. In a monitoring project that can employ local people as parabiologists, conservation education and training should be priorities. Research projects can flow smoothly, reliable data can be procured, and conservation dollars can be saved by including local people. Inclusion of local people in conservation projects also ensures the project's long-term feasibility.

Part Two: Biological and Scientific Aspects

LONG-TERM FROG MONITORING

I collected baseline data from a pristine site in Papua New Guinea to understand variation in the size of frog populations. New Guinea, the world's largest tropical island, has vast tracks of undisturbed primary forest and is geographically proximate to Australia, where there have been numerous declines and hypothesized extinctions of frogs.

When I designed the monitoring project, I had to balance several factors. I was aware of and concerned with problems of data accuracy. Because of the variation in the quality of data collected by different people, I had to decide which methods would be most rigorous and repeatable. The local people had little formal education, and methods had to be technically simple, effective, and repeatable by different teams at different times.

METHODS

During a trial period, I tested several methods to assess their suitability for training of and long-term use by local people in a tropical, developing country. Some methods (e.g., leaf-litter plots and breeding site surveys) sampled specific microhabitats effectively, whereas others (e.g., VES transects, AST censuses, and general collecting) were more general. A number of methods can be used to measure density (e.g., leaf-litter plots) and others used for indices of density or relative abundances (e.g., VES transects). The methods I used are presented in table 13.2 and are contrasted with other techniques.

Leaf-litter plots were one of the best methods because (a) they sampled a subset of the total frog fauna, which simplified species identification; (b) they incorporated simple but important measures of habitat variables (e.g., temperature, vegetation type, canopy cover); and (c) they involved a team of four assistants, which ensured high morale and an internally consistent way to check data. The AST method (Zimmerman 1994) was the worst method for training because of high frog species diversity (> 40 species) at the study site.

Standard Methods: Part I

Drift Fences and Pitfalls

Drift fencing and pitfall traps are logistically difficult to deploy in a remote site. Moreover, the method is too expensive in terms of materials, transportation, and labor for a grassroots conservation project. Because of financial and logistical constraints, I did not include this method in the monitoring project.

Table 13.2   Standard methods for training local people for long-term frog monitoring

Method	Suitability for training	Data utility for monitoring	Tried (yes or no) and final outcome
Leaf-litter plots	High	Good	Yes—best method
Visual encounter surveys (VES)	High	Good	Yes—problems with species identifications
Audio strip (AST)	Low	Moderate	Yes—impractical in areas with high species richness
Drift fences with pitfall traps	Low	Moderate	No—expensive
Breeding-site surveys	Medium	Moderate	Yes—not suitable in PNG
Tadpole sampling	Medium	Poor	Yes—not suitable
General collecting	Medium	Poor	No—not suitable

Tadpole Sampling and Breeding-Site Surveys

Because less than half of the species at most sites in Papua New Guinea have a free-swimming larval stage, and because most of those that do typically breed in fast-flowing or torrential streams, I did not include tadpole sampling methods in the training program for frog monitoring. Most frogs in Papua New Guinea do not form large aggregations around breeding sites, so surveys at breeding sites are not feasible for a monitoring project.

General Collecting

The method of general collecting is one of the best ways to approach a complete species inventory over a number of years at a single site. General collecting can be one of the only ways to find rare species, but it is not highly suitable for long-term monitoring because it is not quantitative. I did not use general collecting as a frog-monitoring method.

Audio Strip Transects

AST is a form of transect sampling that relies on the ability of an observer to differentiate and identify all frog calls within the study area. It is a powerful method that is useful across a wide range of variables and habitats. Unfortunately, in tropical forests there may be 40–70 frog species in an area, and differentiating and identifying all calls unambiguously is a talent gained only through many hours in the field. Once expertise has been achieved, it is a rapid and useful method to estimate frog relative abundances. I conducted these transects between 1900 and 2100 h, when most frogs were active and calling. To avoid biases (see Zimmerman 1994), I used AST on six preexisting trails that crossed streams, valleys, and ridges but did not follow any waterway or topographic contour for an extended length. For an hour during each AST, I recorded all calling individuals I heard from the trail along with their species name and time I heard them. I also recorded date, time started and finished, approximate length of transect (to 10 m), and weather conditions, as was recorded in the VES transects (see appendix 13.2). I compared the same repeatedly measured transects, as in the VES, across the three time periods. Because of the high species diversity at this site (approximately 40 species), however, AST was not suitable for the training of and long-term monitoring by local people.

Standard Methods: Part II

Leaf-Litter Plots

Leaf-litter plots are a form of quadrat sampling (Scott 1976; Jaeger and Inger 1994) and are one of the best ways to estimate the density of frogs found on the forest floor. I used 5 × 5-m plots because there are many species of terrestrial and fossorial frogs at Wara Sera. Plots larger than 5 × 5 m are difficult to establish because of rugged terrain in the area. We used a random number table (Heyer et al. 1994) to select plot locations. After determining two random two-digit numbers, we paced off the first number along a preselected trail and then took a 90° bearing off that trail, pacing off the second random number perpendicular from the first trail. Unless topography (cliff faces, waterfalls, etc.) prevented it, we would head to the right side off trail for the even numbers and to the left side for odd numbers. Once we located a plot, we delineated the boundaries with a tape measure or pre-measured rope. After the plot was laid out, one person recorded the time and we began to search for frogs (Scott 1976; Jaeger and Inger 1994). Appendix 13.1 has a facsimile of the data sheet we used for leaf-litter plots.

During litter plot sampling, we found frogs and clutches of direct-developing microhylid frogs. Because microhylids dominate the frog fauna of New Guinea and have discrete egg clutches, they are a proxy of frog reproductive activity.

Visual Encounter Surveys

VES is a standard method that is useful across nearly all situations found in the field (Crump and Scott 1994). It has been widely used and remains a mainstay for field herpetologists. The method is standardized by unit effort in people/hours for our use, but it is versatile and can be standardized to time, distance, unit effort, people, or combinations of these. We incorporated nocturnal VES transects along six preselected trails in the study area. We rotated transects so that each transect was sampled approximately once per week. Each group of 3–5 parabiologists carried a flashlight and searched from ground level to approximately 3 m above the forest floor along the trail for one hour (between 1900 and 2100 h). I optimized teams by having one assistant record data and the others search for and capture frogs. Once animals were captured, we recorded the substrate, time, sex, and age and measured snout-vent length (SVL; in mm). After processing the frog, we released the animal where it was found and continued sampling. Appendix 13.2 has a facsimile of the data sheet we used for VES.

Monitoring Methods in Use

Sampling methods that were technically simple, effective, and repeatable and generated direct estimates of density and/or relative abundances were the 5 × 5-m leaf-litter plots and the nocturnal VES transects. Both methods were rated highly among the Pawaia and remained effective throughout training and evaluation periods. Unfortunately, the accuracy of species identification during use of the VES method was not reliable once I left Papua New Guinea. Because of problems associated with species identification, the VES transects are not currently part of the monitoring project. The 5 × 5-m leaf-litter plots are not being used for long-term monitoring at Wara Sera because of a few problems with frog identification. With the institution of additional training and a photo-based identification key, I hope to boost species identification abilities; after the reevaluation of the VES transects and the 5 × 5-m plots, I anticipate reinstating the frog-monitoring project in the future.

Statistics

I analyzed the data from the 5 × 5-m plots, the VES transects, and the AST censuses for three discrete time periods: before, during, and after the ENSO drought. Because the rainfall during the period of September–December 1997 was consistently below the second standard deviation from the mean (except for a brief hiatus in November), I defined this four-month period as the ENSO drought period. I standardized sample sizes for the 5 × 5-m plots, VES, and AST in the three time periods. For example, I compare the 60 VES transects run immediately before the ENSO period, the 60 VES transects run during the ENSO drought, and the 60 VES transects run immediately after the ENSO time period to determine any differences that exist among time periods.

In the 5 × 5-m leaf-litter plot analyses, in which all plots were random and independent of each other, I used a Kruskal-Wallis test to compare frog density among the three time periods (pre-ENSO, ENSO, and post-ENSO; http://www.obg.cuhk.edu.hk/researchsupport/KruskalWallis). I used repeated measures analysis of variance (ANOVA; http://faculty.vassar.edu/lowry/corr3.html) to compare relative abundance for the three time periods because I made repeated observations of the six transects. The null hypotheses for these analyses were that there was no difference in density or relative abundance among time periods. If the differences among time periods were significant, I conducted post-hoc Tukey Honestly Significant Difference tests (http://www .graphpad.com/calculators/tosttest1.ctm) to determine which periods differed from each other.

RESULTS

Because I have data from before, during, and after the ENSO drought of 1997, I can examine how frog populations reacted to a naturally occurring low-rainfall event that lasted nearly four months. I include both methods that were part of the long-term monitoring project (5 × 5-m leaf-litter plots and VES transects) as well as an AST census method that was included and evaluated during the training period but was not part of the long-term monitoring.

Leaf-Litter Plots

I analyzed 972 plots (324 per time period) to make sampling efforts equal in each time period (fig 13.3) and compared the number of clutches per plot across the three time periods. There were significant differences in the number of clutches among the three time periods ($p < .0001$, H = 33.86, df = 80). There were significantly fewer clutches per plot during the ENSO period than during the pre-ENSO period ($p < .05$) and the post-ENSO period ($p < .05$). There was no difference, however, in the number of clutches between the pre-ENSO and the post-ENSO periods.

Nocturnal VES Transects

I compared 180 VES transects (60 per time period) to standardize sampling among the pre-ENSO, the ENSO, and the post-ENSO periods (fig. 13.4) and found significant differences in the number of frogs seen during the three time periods ($p < .0001$, F = 101.93, df = 2, 179). More frogs were observed during the ENSO period than during either the pre-ENSO period ($p < .05$) or the post-ENSO period ($p < .05$). There was no significant difference in the number of frogs seen during the pre-ENSO and the post-ENSO periods.

Audio Strip Transects

I compared 81 one-hour ASTs among the three time periods (27 repeated transects per time period). There were significant differences in the number of frogs heard calling during the three time periods ($p < .0001$, F = 48.63, df = 2, 80). Significantly fewer frogs called during the ENSO period than during the pre-

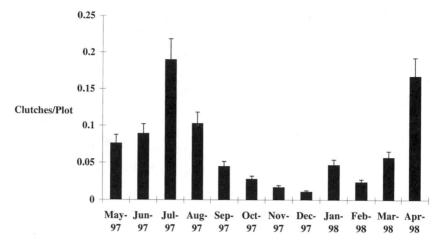

Figure 13.3 Average number of clutches per plot. The El Niño Southern Oscillation event lasted from September to December 1997.

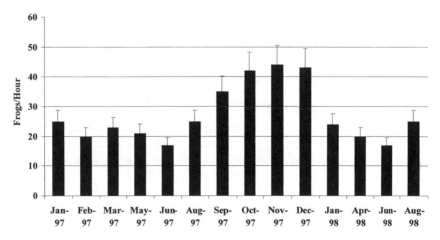

Figure 13.4 Average number of frogs encountered during hour-long nocturnal visual encounter surveys. Error bars = one standard deviation. The El Niño Southern Oscillation event lasted from September to December 1997.

ENSO period ($p < .05$) or the post-ENSO period ($p < .05$). There were also significantly fewer frog calls heard during the post-ENSO period than during the pre-ENSO period ($p < .05$), probably because of the "lag time" of a few weeks after the actual drought had ended when frogs were still responding to the effects of the drought (fig. 13.5).

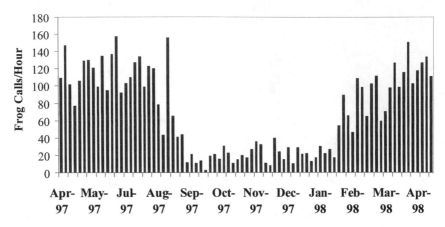

Figure 13.5   The number of frogs heard calling during audio strip transect sampling. The El Niño Southern Oscillation event lasted from September to December 1997.

Conclusions

During training and long-term monitoring, I saw changes in frog calling, number of clutches, and frog density during the severe ENSO drought from September to December 1997. There were many fewer frogs calling, significantly fewer clutches, but higher densities of frogs (during nocturnal VES transects) during the 1997 ENSO. The significantly lower reproductive effort during the ENSO drought, measured both in terms of frog calling as an investment in reproduction and in actual number of clutches per plot, was not surprising. Because frogs are highly susceptible to evaporative water loss, they often change their behavior or activity patterns to avoid physiological stress caused by desiccation. The reduced reproductive effort during the ENSO drought is a typical response for amphibians that experience an extended drought. The higher density of frogs in the nocturnal VES transects, however, was counterintuitive. Although initially confusing, the result makes sense when the species composition and relative abundances are broken down into habitat-use guilds. The significant increase in frogs seen on VES transects during the 1997 ENSO drought is largely explained by a greater number of arboreal frogs encountered (almost 78% of total). I hypothesize that these frogs were simply seeking moisture-rich habitats and fleeing the relatively dry canopy, although I have no direct measures of moisture between the two microhabitats. Anecdotal measurements during (relative humidity = 85%) and after (> 98%) the ENSO drought indicated that it was drier even at ground level.

The other interesting aspect of the counterintuitive result from the VES transects is the implication for interpreting results from a long-term monitoring project. Because we saw significantly more frogs in the VES transects during the drought and encountered fewer frogs after the drought, it superficially appeared that frog populations had crashed or species had disappeared as a result of the 1997–98 ENSO drought. This effect would be particularly notable if the monitoring project had begun during the ENSO, and we could not compare time periods before the ENSO. Although our data from before the 1997–98 ENSO period show that this is not the case (no statistical differences between before and after the 1997–98 ENSO; also see fig. 13.4), it would be easy to make a false hypothesis or jump to the conclusion that there had been a severe population crash or a series of local species extirpations as a result of the low rainfall ENSO without the data collected before the ENSO drought. Cautionary tales such as this are extremely valuable when setting up monitoring projects and analyzing preliminary results.

Appendix 13.1 Sample Data Sheet for 5 × 5-m Leaf-Litter Plots

5 × 5-m Plot
Det (date):
Ples (location):
Taim stat (time begun):
Taim pinis (time finished):
Diwai (trees):
   Bikpela (large, > 50 cm dbh):
   Namel (medium, 50–10 cm dbh):
   Liklik (small, < 10 cm dbh):
Dai diwai (logs on ground):
Hamas lip (leaf-litter depth in mm):

San (amount of sun*):
Ren (amount of rain*):
Shed (canopy cover*):
Wet (amount of soil saturation*):
Ston (amount of bare ground*):
Klaut (amount of cloud*):
Temp (temperature in °C):
Stip: (degree of slope—to nearest 10°):

1)   2)   3)   4)

Rokrok Nem Sians *(frog species code)*	mm *(SVL in mm)*	Ples *(location found)*
.		
.		
.		

Wanem kain ples (general habitat description, e.g., near stream, on ridge with *Pandanus* sp.):
Husat wokim (names of field workers):
Sekim pinis (all categories double-checked):
*Four categories: 0 = no rain, new moon, etc.; 3 = hard rain, full moon, etc.

Appendix 13.2   Sample Data Sheet for Nocturnal VES

VES NAIT

Det (date):   Ren (amount of rain*):
Ples (location):   Mun (amount of moonlight*):
Taim stat (time begun):   Klaud (amount of cloud*):
Taim pinis (time finished):   Temp (temperature in °C):

Rokrok Nem Sians	mm	Ples	Taim
(frog species code)	*(SVL in mm)*	*(location found)*	*(time found)*
.			
.			
.			

Yugo hamas mita long rot (length of transect in m):
Husat wokim (names of field workers):
Sekim pinis (all categories double-checked):
*Four categories: o = no rain, new moon, etc.; 3 = hard rain, full moon, etc.

Acknowledgments

I thank the Pawaia of the Crater Mountain Wildlife Management Area for permission to live and work on their land and for all their help and unwavering support in the field. I thank all of the field assistants with whom I worked during my four years in Papua New Guinea, the Pawaia, the Herowana-speaking and Maimafu Gimi–speaking groups, the Papua New Guineans, and the expatriates. There are literally hundreds of assistants who helped me complete this study. Luke Paro, Lamec Weasoma, Peter Jabare, and Selau Moai were wonderfully helpful and understanding during my first few months in Papua New Guinea. The National Research Institute of Papua New Guinea allowed me to conduct research in that country with institutional support from the University of Papua New Guinea. Financial support came from the University of Miami and the Wildlife Conservation Society. The logistic support of the Research and Conservation Foundation of Papua New Guinea, Missionary Air Fellowship, Seventh Day Adventist Aviation, and Pacific Helicopters was essential. I thank Robert Bino, John Ericho, Mike Hedemark, Paul Hukahu, Paul Igag, Arlene Johnson, Phylis Kaona, Silas Sutherland, and Janine Watson for their help with the frog-monitoring project and for their camaraderie. Arlene Johnson, David Ellis, Paige West, Graham Watkins, Andy Mack, and Debra Wright provided constructive comments on the manuscript. An early version of the manuscript was improved by comments from the Miami Herp Group (Maureen Donnelly, Kirsten Hines, Kirsten Nicholsen, Hardin Waddle, James Watling) and Ted Webb.

The American Society of Ichthyologists and Herpetologists, the Herpetologists' League, and the Society for the Study of Amphibians and Reptiles provided funds that supported my travel to La Paz, Baja California, Mexico, in 2000. These funds allowed me to participate in the symposium at which this work was presented.

14

Historical Biogeographic Relationships within the Tropical Lizard Genus *Norops*

Kirsten E. Nicholson

The phylogenetic and biogeographic relationships of anoline lizards have long been difficult to unravel. Various data types and methods of analysis have been used to elucidate anoline phylogenetics, resulting in, at times, incongruent cladograms (e.g., Etheridge 1960; Gorman, Buth, and Wyles 1980; Gorman et al. 1984; Shochat and Dessauer 1981; Guyer and Savage 1986, 1992; Burnell and Hedges 1990). These alternative phylogenies necessarily demanded alternative biogeographic interpretations. For example, in the controversial field of Caribbean biogeography, anoles were used to forward both dispersalist and vicariance arguments for the origin of Caribbean fauna (Williams 1969, 1983; Guyer and Savage 1986; Hass, Hedges, et al. 1993). Clearly, the phylogenetic and biogeographic issues concerning anoles were far from resolved.

Recent work in molecular systematics has vastly extended our understanding of anole systematics, but some gaps remain. Several studies conducted over the course of several decades have sought to elucidate Caribbean anole relationships (e.g., Gorman and Atkins 1969; Gorman et al. 1969, 1983; Yang et al. 1974; Gorman and Stamm 1975; Gorman and Kim 1976; Gorman, Buth, Soulé, et al. 1980; Gorman, Buth, and Wyles 1980; Wyles and Gorman 1980a, 1980b; Shochat and Dessauer 1981; Burnell and Hedges 1990; Hedges and Burnell 1990; Hass, Hedges, et al. 1993). Only recently have studies been published that are comprehensive in their taxon sampling of Caribbean anoles, and these reveal well-supported and well-resolved cladograms of relationships (Jackman et al. 1997, 1999; Losos et al. 1998). However, these studies examined approximately half of all anoles because mainland species were represented by only a handful of taxa. The mainland anoles (principally the genus *Norops*, excluding the

Cuban and Jamaican representatives) have remained poorly studied overall, and the biogeography of this radiation of anoles has only briefly been explored (Guyer and Savage 1986). Recent work on the molecular systematics of *Norops* (Nicholson 2002) reveals novel relationships compared with those previously proposed. In light of these novel relationships, combined with a general lack of previous biogeographic study, an investigation into the biogeographic relationships of mainland (*Norops*) anoles is warranted. In this chapter I briefly review the phylogenetic relationships of *Norops* and discuss its historical biogeographic relationships. Two aspects of *Norops* biogeography are discussed and treated separately: elucidation of mainland biogeographic patterns and investigation of the origin of Caribbean *Norops* taxa.

Review of Anole Systematics

The earliest phylogenetic work on the anoles was that of Etheridge (1960), who, on the basis of osteological characters, proposed two presumptively monophyletic subdivisions of *Anolis:* an alpha section (found predominantly throughout the Caribbean islands and in South America) and a beta section (found predominantly in Central America but extending into South America, Cuba, and Jamaica and into the Bahamas). Etheridge (1960) proposed subdivisions called series within each section and recognized five series within the beta section (*auratus, fuscoauratus, grahami, petersi,* and *sagrei;* table 14.1). Etheridge (1960) considered alternative relationships among these groups, but he preferred one hypothesis over the others (fig 14.1).

Etheridge's taxonomic arrangement was followed and expanded upon by Williams (e.g., 1974, 1976a, 1976b), who proposed several further subdivisions within the beta series (table 14.1). Williams (1974, 1976b) recognized several species groups within the *N. auratus, N. fuscoauratus,* and *N. petersi* series. He recognized two additional species groups, the *N. onca* and the *N. meridionalis* species groups, which were later elevated to series status by Guyer and Savage (1986) and Savage and Guyer (1989). Williams (1989) published approval of these designations, bringing the total of *Norops* series to seven (table 14.1, far right). On the basis of morphology, karyology, and protein electrophoretic data, Lieb (1981) proposed three subseries within the *N. auratus* series; these were formally named (the *N. auratus, N. laeviventris,* and *N. schiedii* subseries) and recognized by Savage and Guyer (1989). In addition, Lieb (1981) erected nine species groups: three each within the *N. laeviventris* (*N. laeviventris, N. nebuloides,* and *N. subocularis* species groups) and *N. schiedii* (*N. crassulus, N. gadovii,* and *N. schiedii* species groups) subseries, two within the *N. auratus* subseries (*N. cupreus* and *N. sericeus* species groups), and one (*N. nebulosus* species group) that he included in the *N. nebulosus* series. Lieb (1981) attributed the erection of the *N. nebulosus* series to Etheridge (1960), but this was in error.

Table 14.1   Classification schemes for *Norops*

Etheridge and Williams	Lieb	Savage and Guyer
Anolis	*Anolis*	*Anolis*
Alpha section^	Alpha section^	Alpha section^
(six series)^	(six series)^	(six series)^
Beta section^	Beta section^	Beta section^
auratus (*chrysolepis*) series^	*auratus* (*chrysolepis*) series^	*auratus* (*chrysolepis*) series^
auratus species group*	22/23 subseries#	*schiedii* subseries§
humilis species group*	*crassulus* species group#	*crassulus* species group#
lemurinus species group*	*gadovii* species group#	*gadovii* species group#
fuscoauratus series^	*schiedei* species group#	*schiedii* species group#
fuscoauratus species group*	23 subseries#	*laeviventris* subseries§
lionotus species group*	*laeviventris* species group#	*laeviventris* species group#
grahami series^	*nebuloides* species group#	*nebuloides* species group#
petersi series^	*subocularis* species group#	*subocularis* species group#
pentaprion species group*	24 subseries#	*auratus* subseries§
petersi species group*	*auratus* species group*	*humilis* species group*
sagrei series^	*cupreus* species group#	*lemurinus* species group*
meridionalis species group*	*sericeus* species group#	*sericeus* species group#
onca species group*	*humilis* species group*	*cupreus* species group#
	lemurinus species group*	*auratus* species group*
	nebulosus series#	*nebulosus* species group#
	nebulosus species group#	*fuscoauratus* series^
	fuscoauratus series^	*fuscoauratus* species group*
	fuscoauratus species group*	*lionotus* species group*
	lionotus species group*	*grahami* series^

petersi series^
  pentaprion species group*
  petersi species group*
sagrei series^

meridionalis series*
onca series*
petersi series^
  pentaprion species group*
  petersi species group*
sagrei series^

Note: The schemes were proposed by Etheridge, Williams, Lieb, and Savage and Guyer. The alpha section is indicated, but the infra-section classification regarding this section is omitted. Symbols correspond to the author and date of publication of the proposed schemes and indicate the author's contributions. (^), Etheridge 1960; (*), Williams 1974, 1976a, 1976b; (#), Lieb 1981; (§), Savage and Guyer 1989.

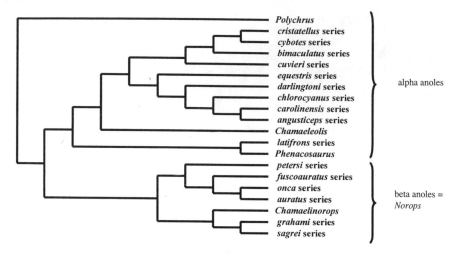

Figure 14.1   Anole phylogeny based on Etheridge (1960). Taxon names represent his proposed series or genera recognized in this work. *Polychrus* is the outgroup.

The *N. nebulosus* species group was later placed in the *N. auratus* subseries by Savage and Guyer (1989) (table 14.1).

A few studies have challenged the monophyletic status of *Norops*, but more recent studies further support its monophyly. Four studies have presented results indicating that *Norops* is not monophyletic (Shochat and Dessauer 1981; Gorman et al. 1984; Burnell and Hedges 1990; Poe 1998). The first two studies used albumin immunologic data and suffered from a lack of sufficient two-way comparisons (Shochat and Dessauer 1981 included a few), nonreciprocity of these comparisons (critiqued by Guyer 1992), taxon sampling problems, and the potential problems associated with phenetic phylogenetic analysis using genetic distances. The results of Burnell and Hedges (1990) are difficult to interpret given that they obtained 2,000 most-parsimonious trees and presented only a bootstrap tree of those results. In Poe's (1998) study, *Norops polylepis* is resolved outside of the remaining *Norops* clade. The explanation for this result is not obvious but perhaps is the result of convergence in the morphological characters. The synthetic work of Guyer and Savage (1986, 1992; fig. 14.2) showed support for *Norops*, and recent studies using mtDNA data (fig. 14.3; Jackman et al. 1997, 1999) strongly support the monophyly of *Norops*. Thus, although there have been challenges, the preponderance of data support the monophyletic status of *Norops*. Because of this, it has been proposed that the beta section of *Anolis* be recognized as the genus *Norops* (Savage and Talbot 1978; Guyer and Savage 1986). This proposition has been criticized (Cannatella

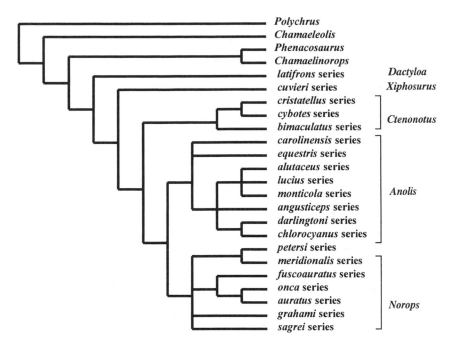

Figure 14.2   Anole phylogeny based on Guyer and Savage (1986, 1992). Taxon names represent genera or series recognized by Etheridge (1960) or Williams (1974, 1976a, 1976b). Names to the right of the brackets represent the genera proposed by Guyer and Savage (1986). *Polychrus* is the outgroup.

and de Queiroz 1989; Williams 1989), and herpetologists are divided on this point. I use the name *Norops* to distinguish this inarguably monophyletic group from the remaining anoles, which are demonstrably paraphyletic. I am using this name more as a matter of convenience rather than in a strict taxonomic sense until all of the monophyletic clades of anoles can be recognized (see Nicholson 2002 for further details on the use of this name). Therefore, I refer to all included beta anoles by using this term (e.g., *Norops auratus*) to distinguish them from the remaining anoles (e.g., *Anolis equestris*).

Few previous studies have examined relationships within *Norops*, and none has explicitly examined the monophyly of the proposed subgroups, but recent work has shed light on their relationships and the status of the current classification. Recent molecular studies (Jackman et al. 1997, 1999) that investigated relationships of Caribbean anoles included 10 species of *Norops* and showed support for two *Norops* series (the *N. sagrei* series from Cuba and the *N. grahami* series from Jamaica) but a lack of support for the *N. auratus* series (fig. 14.3). One study directly tested the monophyly of five of the series of

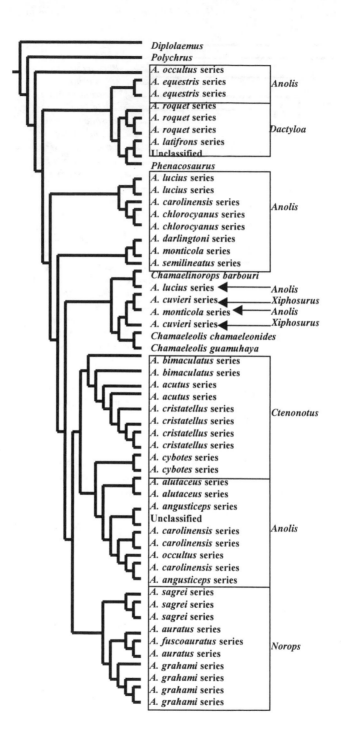

Norops (Nicholson 2002) by using molecular data in the form of nuclear internal transcribed spacer units (ITS-1) and included 52 in-group and 3 out-group species. In-group species were selected to represent replicate taxa within the proposed subgroups of Savage and Guyer (1989; table 14.1) and the geographic range exhibited by the genus. Maximum parsimony analysis of these data (946 aligned base pairs) retrieved nine most-parsimonious trees, the strict consensus of which is shown in figure 14.4. Support was found for the *N. sagrei* (Cuba) and *N. grahami* (Jamaica) series, but there was no support for the mainland *N. auratus, N. fuscoauratus,* or *N. petersi* series. Despite low decay index and bootstrap support for several of the nodes, the monophyly of the *grahami* series was supported, and rejection of the monophyly of the rest of the groups was demonstrated with Templeton tests (table 14.2). In addition, the monophyly of two subseries and four species groups from the mainland (*N. auratus* and *N. laeviventris* subseries, *N. auratus, N. fuscoauratus, N. humilis,* and *N. petersi* species groups) was statistically rejected, whereas two mainland species groups (the *N. crassulus* and *N. laeviventris* species groups) appear to be monophyletic. The lack of support for many of the previously proposed groups is not particularly surprising because they were phenetic groups erected largely on the basis of unique combinations of characters. When these same characters were analyzed cladistically using PAUP* (Swofford 2001), a nearly completely unresolved bush resulted (K. E. Nicholson, pers. obs., results not shown).

Biogeography of Norops

The relationships revealed in the *Norops* phylogeny of Nicholson (2002) lend a unique opportunity to examine the biogeography of a monophyletic group whose geographic range nearly encompasses that of all anoline lizards. Species of *Norops* may be found in any of the habitats within this range, from coastal dry forest to highland cloud forest, and they exhibit extensive morphological and ecological variation.

Figure 14.3   Anole phylogeny based on Jackman et al. (1999). Taxon names have been replaced with the series to which they have been assigned (after Savage and Guyer 1989). *Diplolaemus* and *Polychrus* are the outgroups. Names to the right of the boxes or the arrows reflect the genera proposed by Guyer and Savage (1986). It is interesting that three of their proposed genera (*Ctenonotus, Dactyloa,* and *Norops*) are supported by the Jackman et al. (1999) data. In addition, there is support for a clade including *Chamaelinorops* and *Chamaeleolis,* which could be considered a revised version of Guyer and Savage's (1986) *Xiphosurus.* However, their genus *Anolis* (sensu stricto) is not supported and requires substantial revision.

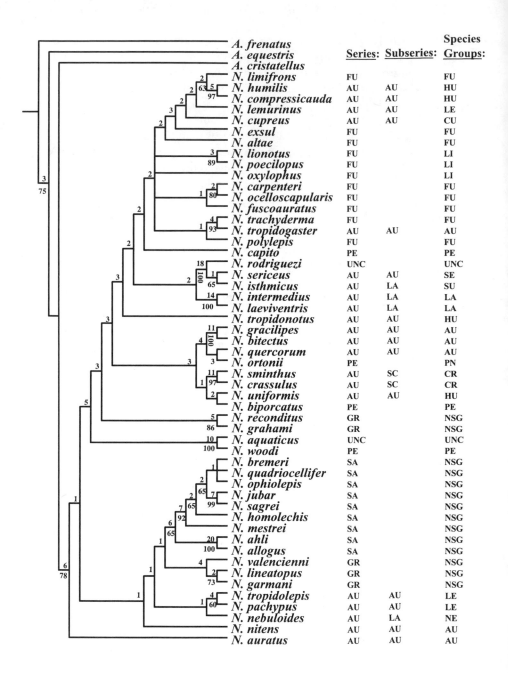

	Series:	Subseries:	Species Groups:
A. frenatus			
A. equestris			
A. cristatellus			
N. limifrons	FU		FU
N. humilis	AU	AU	HU
N. compressicauda	AU	AU	HU
N. lemurinus	AU	AU	LE
N. cupreus	AU	AU	CU
N. exsul	FU		FU
N. altae	FU		FU
N. lionotus	FU		LI
N. poecilopus	FU		LI
N. oxylophus	FU		LI
N. carpenteri	FU		FU
N. ocelloscapularis	FU		FU
N. fuscoauratus	FU		FU
N. trachyderma	FU		FU
N. tropidogaster	AU	AU	AU
N. polylepis	FU		FU
N. capito	PE		PE
N. rodriguezi	UNC		UNC
N. sericeus	AU	AU	SE
N. isthmicus	AU	LA	SU
N. intermedius	AU	LA	LA
N. laeviventris	AU	LA	LA
N. tropidonotus	AU	AU	HU
N. gracilipes	AU	AU	AU
N. bitectus	AU	AU	AU
N. quercorum	AU	AU	AU
N. ortonii	PE		PN
N. sminthus	AU	SC	CR
N. crassulus	AU	SC	CR
N. uniformis	AU	AU	HU
N. biporcatus	PE		PE
N. reconditus	GR		NSG
N. grahami	GR		NSG
N. aquaticus	UNC		UNC
N. woodi	PE		PE
N. bremeri	SA		NSG
N. quadriocellifer	SA		NSG
N. ophiolepis	SA		NSG
N. jubar	SA		NSG
N. sagrei	SA		NSG
N. homolechis	SA		NSG
N. mestrei	SA		NSG
N. ahli	SA		NSG
N. allogus	SA		NSG
N. valencienni	GR		NSG
N. lineatopus	GR		NSG
N. garmani	GR		NSG
N. tropidolepis	AU	AU	LE
N. pachypus	AU	AU	LE
N. nebuloides	AU	LA	NE
N. nitens	AU	AU	AU
N. auratus	AU	AU	AU

Table 14.2 Hypotheses tested using the Wilcoxon Signed-Ranks test

Hypothesis	n	z	p
A. The *auratus* series is monophyletic.	131–141	−2.82 to −3.99	<.0001*
B. The *fuscoauratus* series is monophyletic.	40–55	−3.20 to −3.81	<.0030*
C. The *grahami* series is monophyletic.	43–55	−0.74 to −0.82	>.3600
D. The *petersi* series is monophyletic.	70–80	−2.41 to −2.61	<.0020*
E. The *auratus* subseries is monophyletic.	126–138	−6.11 to −6.36	<.0001*
F. The *laeviventris* subseries is mono-phyletic.	93–99	−3.92 to −4.02	<.0004*
G. The *auratus* species group is mono-phyletic.	67–99	−2.82 to −3.99	<.0339*
H. The *fuscoauratus* species group is mono-phyletic.	38–52	−3.15 to −3.62	<.0036*
I. The *humilis* species group is mono-phyletic.	53–57	−3.01 to −3.10	<.0081*
J. The *lemurinus* species group is mono-phyletic.	144 of 405 (35%) comparisons significant at .05 level or lower; remaining compari-sons were not significant.		
K. The *petersi* species group is mono-phyletic.	72–81	−2.05 to −2.17	<.0200*
L. There are monophyletic island and mainland lineages.	63–74	−1.68 to −1.53	>.0928

Note: The Wilcoxon Signed-Ranks test evaluates whether my phylogenetic reconstruction is significantly shorter than the hypothetical tree(s) listed in the table. Because multiple comparisons were made, ranges for n (the number of characters differing between each paired comparison) and z (the test statistic) are reported. Nonsignificant p values are reported as greater than the lowest p obtained, and significant p values are reported as less than the highest value obtained.

*$p < .05$

Figure 14.4 Strict consensus of nine most-parsimonious trees generated from ITS-1 nuclear DNA sequences. Outgroups are *A. frenatus*, *A. equestris*, and *A. cristatellus*. Numbers above the nodes are decay indices; below the nodes are bootstrap proportions greater than 50%. Previously assigned ranks (series, subseries, and species groups) and subgroups are indicated in columns to the right (see table 14.1). The subgroups are abbreviated as follows: AU, *auratus*; CR, *crassulus*; CU, *cupreus*; FU, *fuscoauratus*; GR, *grahami*; HU, *humilis*; LA, *laeviventris*; LE, *lemurinus*; LI, *lionotus*; NE, *nebuloides*; NSG, no separate group has been proposed below the series level; PE, *petersi*; PN, *pentaprion*; SA, *sagrei*; SC, *schiedii*; SE, *sericeus*; SU, *subocularis*; UNC, unclassified to any subgroup.

GEOLOGY OF THE CARIBBEAN
AND MAINLAND AMERICAS

An investigation of the biogeography of *Norops* is aided by understanding the geology of the region involved. There are no known fossils of *Norops*, so determination of the period during which divergences occurred within this group—and thus a starting point for the geologic context—is difficult. However, the prevailing paradigm (Etheridge 1960; Williams 1969, 1976a, 1983; Guyer and Savage 1986) suggests that a vicariant event resulting in the origin of *Norops* was the separation of North and South America by the eastward movement of an island chain sometimes referred to as the proto–Greater Antilles. The geologic events involving present-day Central America and the Caribbean are poorly understood, and geologists disagree on many details regarding timing of block movements and the exact constituents of the blocks. Thus, I review the geologic history of the region emphasizing the points largely agreed on by geologists (Malfait and Dinkelman 1972; Pindell and Dewey 1982; Sykes et al. 1982; Wadge and Burke 1983; Burke 1988; Ross and Scotese 1988; Donnelly 1989; Dengo and Case 1990; Donnelly et al. 1990; Escalante 1990; Mann et al. 1990; Coates and Obando 1996; Iturralde-Vinent and MacPhee 1999).

The proto–Greater Antillean land block began moving eastward during the late Paleocene to early Eocene (40–80 million years ago [mya]; fig. 14.5A). As this block moved eastward on the Caribbean Plate, it fragmented, the fragments collided, and accretionary blocks formed, eventually resulting in the formation of the Antillean islands (fig. 14.5A). Of particular interest to this discussion is the development of Cuba and Jamaica. Most geologists agree that there was a long and close association between western Cuba and Mexico. It is unclear (i.e., not widely agreed upon by geologists) at what point western Cuba completely separated from the Maya Block; it may have maintained a connection by skirting the edge of the Maya Block, or it may have separated much earlier. Jamaica

Figure 14.5   Diagrams depicting the major geologic events of the Caribbean (Greater Antilles only) and Mesoamerican regions. (A) The geology of the Caribbean, as redrawn from Guyer and Savage 1986. CA, Central America; C Hisp., central Hispaniola; E Cuba, eastern Cuba; J, Jamaica; N & C Hisp., northern and central Hispaniola; N Hisp., northern Hispaniola; PGA, proto–Greater Antilles; PR, Puerto Rico; SH, southern Hispaniola; W & C Cuba, western and central Cuba. (B) The geology of the Mesoamerican region, showing the relative positions of the major block units, as redrawn from Escalante 1990. AF, Atrato Fault; CS, conceptual suture; CSR, conceptual spreading ridge; CT, Colombian Trench; MAT, Middle American Trench; MF, Motagua Fault; NPDB, North Panama Deformed Belt; oCT, ancestral Colombian Trench; MS, Motagua Suture Zone; oMAT, ancestral Middle American Trench; RF, Romeral Fault; RS, Romeral Suture Zone; SA-HE, Santa Elena-Hess Escarpment; SCF, south Caribbean Fault; SES, Santa Elena Suture Zone.

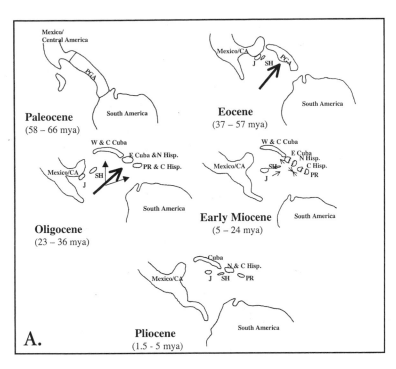

Paleocene
(58 – 66 mya)

Eocene
(37 – 57 mya)

Oligocene
(23 – 36 mya)

Early Miocene
(5 – 24 mya)

Pliocene
(1.5 - 5 mya)

A.

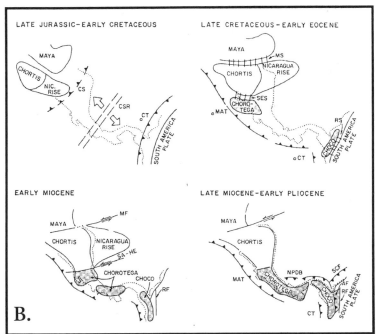

LATE JURASSIC–EARLY CRETACEOUS

LATE CRETACEOUS–EARLY EOCENE

EARLY MIOCENE

LATE MIOCENE–EARLY PLIOCENE

B.

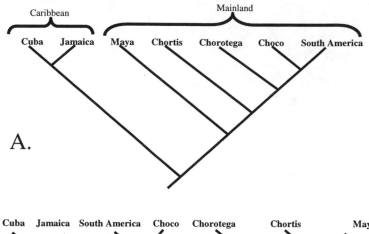

A.

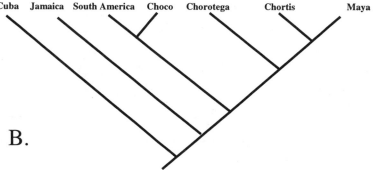

B.

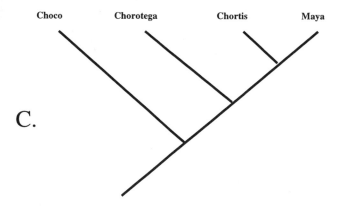

C.

is a volcanic island that arose initially very close to the Maya Block but was carried eastward with the Caribbean Plate. At some point during its eastward trajectory it was submerged, although there is some debate about whether it was completely submerged or some part of it remained land positive. If it was submerged, then by 3–8 mya the island had reemerged.

As the Caribbean islands were developing, other blocks were moving into place, reconnecting North and South America (fig. 14.5B). Three blocks—the Chortis, Chorotega, and Choco—moved from the Pacific (although not from the same direction or area) and reconnected the American plates from north to south, respectively. These blocks were composed of conglomerates of emergent land, although the number of components and their definition are debatable. The closure of the Panamanian Portal completed the reconnection of North and South America and is thought to have occurred approximately 3 mya.

From these geologic events, it is possible to formulate a hypothesis of geologic relationships that may be reflected in the phylogeny of organisms from the region. One possible hypothesis follows logically from the above geologic history and is depicted in figure 14.6A. This cladogram represents a mainland north-to-south pattern such that the Maya Block is located basally in the tree, whereas the southern blocks are nested higher in the tree. Placing the Jamaican and Cuban blocks in this tree is more difficult. The dating of the separation of Cuba from the mainland is unclear, but most geologic reconstructions suggest that Cuba separated from the mainland before the first Pacific block (Chortis) sutured to the Maya Block. Therefore, the Cuba–mainland separation may be represented as a basal separation in the tree. Because Jamaica is thought to have been submerged during its early history, it is assumed that its anoles are the result of over-water dispersal, most likely from Cuba. A sister relationship between Cuba and Jamaica can be hypothesized. Thus, the geological hypothesis depicted in figure 14.6A may be tested against the *Norops* phylogeny.

PREVIOUSLY PROPOSED BIOGEOGRAPHIC
HYPOTHESES FOR *NOROPS*

Previous investigators proposed biogeographic hypotheses that largely agreed with the above geologic hypothesis, with some minor modifications. For the

Figure 14.6   Hypothetical relationship among major regions in Central and South America and the Caribbean and results of BPA. (A) Diagram of a hypothesis of area relationships, based on the generally accepted ideas about the geology of the region (see fig. 14.5). (B) The results of BPA for *Norops*. (C) The combined results for *Norops, Bothriechis,* and the *E. gollmeri* group. The combined analysis only analyzed taxa in common among the studies; therefore, South American and Caribbean taxa were not included in this analysis.

sake of simplicity, I treat these hypotheses as two sets: one is about mainland relationships, the other about Caribbean relationships. Clearly there is some relationship between the mainland and island taxa, but teasing apart this relationship is difficult. Williams (1969, 1976a, 1983) and Guyer and Savage (1986)
hypothesized that mainland *Norops* dispersed north (Maya Block) to south,
eventually invading South America. Williams, who formulated his hypothesis
before the advance of plate tectonic theory for the region, thought that barriers
to southward movement were created by aquatic inundations on the mainland.
The Guyer and Savage (1986) hypothesis was based on emerging geological
models of plate tectonics of Middle America and vicariance biogeographic theory. From these hypotheses, the first question we must ask is this: Is a north-to-
south biogeographic pattern evident in the phylogenetic history of *Norops?*

With respect to the Caribbean, disagreement arises over three issues: the origin of the genus *Norops,* the origin of Caribbean groups within *Norops* (Cuban
and Jamaican), and the putative sister relationship between Cuban and Jamaican *Norops.* Regarding the origin of *Norops,* two hypotheses have been presented suggesting either a mainland or a Caribbean origin for the genus. Several investigators (Etheridge 1960; Williams 1969, 1983; Guyer and Savage 1986,
1992) infer that the genus originated on the mainland after the separation of
North and South America. These authors differed in the precise treatment of
this separation (e.g., pre- or post-tectonic explanations, and vicariance vs. dispersal explanations) but nevertheless agreed that this separation led to a mainland origin of *Norops.* Alternatively, Gorman et al. (1984) and Jackman et al.
(1997) proposed that *Norops* originated from a Caribbean ancestor. They called
attention to the presence of several Caribbean clades that form the sister clades
to *Norops* (Jackman et al. 1997; see also Losos et al. 1998) and inferred a single
dispersal event from the Caribbean to the mainland involving *Norops.* They did
not explain whether this ancestor speciated vicariantly within the Caribbean,
forming the *Norops* lineage, and later dispersed to the mainland, or whether
this ancestor dispersed to the mainland to give rise to the *Norops* lineage there.
These alternative hypotheses for the origin of *Norops* can only be elucidated
with a phylogeny that includes complete or nearly complete taxon sampling for
the outgroups to *Norops.* Therefore, these alternatives are not tested here.

Conflicting hypotheses have been proposed regarding relationships among
Caribbean *Norops.* If Jamaica was submerged during its history, then its anole
fauna must be the result of over-water dispersal. A single dispersal event is required to explain the colonization of Jamaica by *Norops* because the group is
monophyletic (Etheridge 1960; Williams 1976a; Guyer and Savage 1986; Hedges
and Burnell 1990). The question of where this ancestor came from is at issue.
Some investigators have suggested that it dispersed to Jamaica from Cuba
(Etheridge 1960; Guyer and Savage 1986). Other studies have suggested dispersal of the ancestor from the mainland or have lacked the data needed to distin-

guish between the alternatives (Williams 1969, 1983; Burnell and Hedges 1990; Hedges and Burnell 1990 and references therein). These alternatives may be tested by asking a second question: Are Cuban and Jamaican *Norops* sister taxa?

Some investigators have assumed a sister-taxon relationship between Cuban and Jamaican *Norops* but have variously proposed the relationship between Caribbean and mainland *Norops*. As mentioned above, some authors have hypothesized that *Norops* originated on the mainland and later dispersed to the Caribbean (Etheridge 1960; Williams 1969, 1983; Guyer and Savage 1986). Figure 14.7A depicts this hypothesis of relationships. Alternatively, Jackman et al. (1997) hypothesized that *Norops* originated from a Caribbean ancestor; this ancestor may have given rise to the *Norops* lineage in the Caribbean and later dispersed to the mainland, or the ancestor may have dispersed to the mainland,

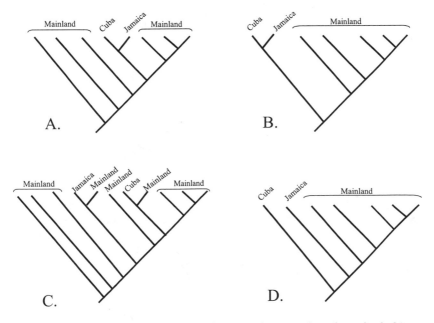

Figure 14.7   Four hypothetical cladograms depicting alternative hypotheses for the biogeographic relationships between Caribbean and mainland *Norops*. (A) A cladogram that supports the sister-taxon relationship of Cuban and Jamaican *Norops* and the origin of Caribbean *Norops* from a mainland ancestor. (B) A cladogram that supports the sister-taxon relationship of Cuban and Jamaican *Norops* and the origin of Caribbean *Norops* from a Caribbean ancestor. (C) A cladogram that supports separate mainland origins for Cuban and Jamaican taxa. (D) A cladogram that hypothesizes a Caribbean origin for Caribbean *Norops* taxa, and specifically a Cuban origin, followed by dispersal to Jamaica, followed by dispersal to the mainland.

given rise to *Norops* there, then dispersed over water back to the Caribbean. The later case should be the same as figure 14.7A. In the former case (origin in the Caribbean), this hypothesis of relationships should appear as in figure 14.7B; this set of relationships would imply dispersal to the mainland from Cuba (in concert with the several Caribbean outgroups to *Norops* shown in Jackman et al. 1999) as well as a later dispersal to Jamaica from Cuba.

An additional alternative should be considered. If a sister-taxon relationship between Cuban and Jamaican *Norops* is not supported, the alterative origins from a Caribbean versus a mainland ancestor will have to be entertained. If they are not sister taxa, and each Caribbean lineage arose independently from separate mainland ancestors, then each Caribbean group should be sister to a mainland taxa/on and the phylogenetic relationships should resemble figure 14.7C. Jackman et al. (1997) retrieved (Cuba (Jamaica, Mainland)) (see fig. 14.7D). This set of relationships (considering also that all the outgroups to *Norops* were Caribbean) might be interpreted as showing a Caribbean origin for *Norops*, followed by dispersal from Cuba to Jamaica, and from there dispersal to the mainland. The *Norops* taxon sampling of Jackman et al. (1997) was low, but it offers another alternative for comparison. These four alternatives can be explored by asking a third question: What are the relationships between Caribbean and mainland *Norops*?

TESTING THE BIOGEOGRAPHIC HYPOTHESES

Mainland Relationships: Is a General North-to-South Distribution Pattern Evident?

The question of a general north-to-south distribution pattern was addressed initially by examining an area cladogram created by replacing taxon names with the geologic blocks on which they currently reside (fig. 14.8). Some taxa inhabit multiple blocks, making it difficult to discern any geologic pattern, and an obvious north-to-south pattern is not observable across the cladogram. This is not surprising, given the species richness within *Norops* (at least 147 species). Alternatively, repeated clades exhibiting a north-to-south pattern might be predicted. Examination of each major clade in figure 14.8 fails to show this predicted pattern clearly.

In an attempt to determine whether a pattern was supported by the data, a matrix was generated from the area cladogram and subjected to Brooks parsimony analysis (BPA; Brooks 1981, 1990), a method in which areas are the independent variables and a phylogeny forms the dependent variable. The BPA method has failings (e.g., Warren and Crother 2001 and references therein) but was used as a first approximation for pattern determination. The power

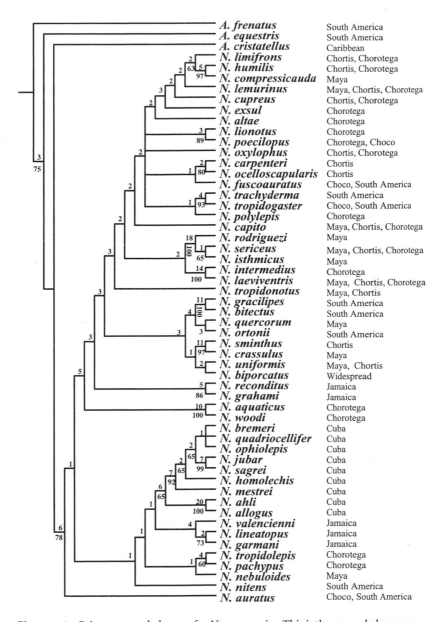

Figure 14.8   Primary-area cladogram for *Norops* species. This is the same phylogeny as that presented in fig. 14.4, with major geologic area names indicated to the right of the taxon names. Area names correspond with those major geologic features discussed in the text. "Widespread" indicates a taxon that occurs on all blocks.

of BPA to elucidate distributional patterns is best discovered when information is used from multiple unrelated clades. However, use of BPA for *Norops* is justifiable given that multiple clades occur within the genus. A parsimony analysis was performed on the area matrix and a single tree resulted (fig. 14.6B). This tree corresponded to the following area relationships: (((((Maya, Chortis) Chorotega) (Choco, South America)) Jamaica) Cuba). For mainland areas, these relationships are opposite to the predicted north-to-south pattern (fig. 14.6A). These results suggest that the selected geological hypothesis is incorrect or that widespread dispersal better explains the mainland distribution of *Norops*.

Few studies have examined the historical biogeography of the Mesoamerican herpetofauna, but two studies shed light on this subject. Two studies involving snakes (*Bothriechis*, Crother et al. 1992) and frogs (*Eleutherodactylus gollmeri* group, Savage 1987) exhibit the same distribution pattern observed in *Norops*. In these cases, they coded for three areas: Maya, Chortis, and Lower Central America (a combination of the Chorotega and Choco blocks). To see if a concordant pattern was observed among our studies, I recoded their matrixes, separating Lower Central America into Chorotega and Choco blocks, and combined the data into a single analysis. With the elimination of the Caribbean and South America taxa, the same single pattern emerges (fig. 14.6C).

Crother et al. (1992) suggested that that the patterns observed between *Bothriechis* and *E. gollmeri* group resulted from the orogenies of the mountain chains (Mexican Sierras, Chiapan and Talamancan highlands), the fragmentation of fault zones, and inundation along the Isthmus of Tehuantepec. These events are much younger and occurred during or after the suturing of the blocks between North and South America. After the blocks moved into place, an ancestral biota could have dispersed southward, forming an initial widespread distribution, which was then fragmented by these younger-aged vicariant events, resulting in the observed distribution patterns. It appears that these same vicariant events may explain the similar patterns observed within *Norops*, and thus the south-to-north distribution pattern. Unlike Crother et al. (1992) and Savage (1987), it is not currently possible to derive a vicariant (or dispersal) explanation for each of the nodes within the current *Norops* phylogeny. A more complete phylogeny for *Norops*, combined with more specific, younger-aged geologic information, is required.

Caribbean Relationships I: Cuban and Jamaican Sister Taxa

Visual inspection of the *Norops* cladogram (fig. 14.8) reveals a common history for species on Cuba and Jamaica, but this relationship is poorly supported. I compared all maximum parsimony trees to all shortest trees not compatible with a sister-taxon relationship for Cuban and Jamaican species and found no significant difference with the observed pattern (results not shown). Therefore, although they appear to be sister taxa in the present cladogram, this topology is not well supported. In comparison, Jackman et al. (1997) included fewer *Norops* taxa but much better node support. Their data did not show a sister-taxon relationship between Cuban and Jamaican *Norops* but instead showed a topology corresponding to that shown in figure 14.7D. It is unknown whether these nodes are statistically supported, but this arrangement is better supported than the nodes (fig. 14.7D). I conclude, therefore, that the sister relationship between Cuban and Jamaican taxa is unresolved and needs further investigation.

Caribbean Relationships II: Caribbean–Mainland Norops Relationship

The question of the Caribbean–mainland *Norops* relationship is difficult to test, but some new clues appear to be evident. From a merely descriptive point of view, visual inspection of the primary area cladogram (fig. 14.8) has already shown us that the Caribbean taxa nest together roughly, and despite being poorly supported, the Cuban and Jamaican taxa are not nesting independently with mainland taxa (e.g., not as in fig. 14.7C). Thus, the hypothesized relationships depicted by figures 14.7C and 14.7D can be tentatively rejected. In particular, the Caribbean taxa do not diverge as the most basal branch. Instead, South American *Norops* represent a basal branch that is sister to the Caribbean clade and all other mainland *Norops*. This arrangement clouds interpretation because it fails to corroborate any of the predications.

In an attempt to gather more satisfying and less ambiguous information, I compared the maximum parsimony trees to all shortest trees compatible with the constraint of two lineages, a Caribbean and a mainland lineage, as depicted in figure 14.7B. No significant difference was observed between these comparisons (table 14.2), and thus the hypothesis of two separate clades of *Norops* cannot be rejected. This finding means that the hypothesis of a Caribbean origin for *Norops* taxa followed by dispersal from the Caribbean to the mainland cannot be rejected.

Conclusion

My biogeographic analysis serves as a first approximation toward understanding *Norops* biogeography. Additional taxa are needed to gain a greater understanding of the biogeography of this group. However, an interpretation is offered given the present findings and results from previous investigators. A once widespread ancestral anole population was broken up when the proto–Greater Antillean land block moved eastward separating North and South America. If anoles were left in North America (= Mexico), then these may have gone extinct. The land block carried with it these anole ancestors and vicariously gave rise to the lineages of anoles as the blocks separated, fused, and accreted. Some vicariant event involving Cuba may have given rise to the *Norops* lineage, eventually developing into the *N. sagrei* group. Some ancestor to this group dispersed to the recently emerged island of Jamaica, eventually resulting in the *N. grahami* group. An ancestor from Jamaica may have dispersed then to the mainland. Once on the mainland, *Norops* dispersed southward as the blocks moved into place, creating a widespread distribution of ancestral taxa. Younger vicariant events, such as the orogenies of the mountain chains and development of the Nicaraguan Depression, resulted in the subdivision of this widespread distribution and the currently observed distribution of species whose biogeographic pattern is reflected in *Bothriechis* and the *E. gollmeri* group.

The implications of this analysis can be summarized as follows:

1. Current data do not support the classic taxonomic arrangement summarized by Savage and Guyer (1989).
2. Relationships among mainland *Norops* do not support a predicted north-to-south biogeographic pattern.
3. Mainland *Norops* biogeographic relationships are concordant with those of other Mesoamerican amphibians and reptiles (e.g., *Bothriechis* and the *E. gollmeri* group), suggesting that younger-aged vicariant events are responsible for their diversity.
4. Caribbean *Norops* taxa (and the genus itself) appear to have originated from a Caribbean ancestor.
5. These data cannot reject a Caribbean origin of Cuban *Norops*, followed by dispersal to Jamaica, followed by dispersal from Jamaica to the mainland.

Acknowledgments

I thank Jay Savage, Brian Crother, Craig Guyer, Jonathan Losos, and Kevin de Queiroz for conversations about biogeography in general and anole biogeography in particular. Although many suggestions and ideas were offered by these people, the ideas contained within this chapter are my own. I owe many thanks to Brian Crother and Mary White for much encouragement and support while I was producing these data in their laboratory. I am very grateful for the loan of tissues from the following people and institutions: Brian Crother, Kevin de Queiroz, Maureen Donnelly, Lee Fitzgerald (via Richard Etheridge and Tom Titus), Julian Lee, Carl Lieb, Randy McCranie, Adrian Nieto Montes de Oca, Louisiana State University, Museum of Vertebrate Zoology (Berkeley, Calif.), and University of Kansas. I thank Jonathan Losos, Jim Schulte, and several anonymous reviewers for editorial comments on drafts of the manuscript. This work formed a portion of my dissertation and was supported by an NSF Dissertation Improvement Grant (9902749), a Gaige Award (American Society of Ichthyologists and Herpetologists), the Graduate School of the University of Miami, and the Department of Biology of the University of Miami. I thank Jonathan Losos and Allan Larson (NSF DEB9982736 to both) for supporting me during the writing of this chapter.

The American Society of Ichthyologists and Herpetologists, the Herpetologists' League, and the Society for the Study of Amphibians and Reptiles provided funds to support my travel to La Paz, Baja California, Mexico, in 2000.

15

Hypotheses on the Historical Biogeography of Bothropoid Pitvipers and Related Genera of the Neotropics

Steven D. Werman

Pitvipers (Viperidae: Crotalinae) occur in temperate and tropical regions of eastern Asia and throughout the Americas. The pitvipers of the New World occur from Canada to southern Argentina and range latitudinally from the Pacific to the Atlantic margins of North, Central and South America (Klauber 1972; Campbell and Lamar 1989; Gloyd and Conant 1990). Pitvipers of the Americas occupy habitats ranging from xeric and mesic lowlands to high-elevation cloud forests (Campbell and Lamar 1989) and are important predators on small vertebrates and occasionally invertebrates (Greene 1992). Cases of accidental human envenomation can result in severe systemic poisoning with significant morbidity and mortality. The impact of pitvipers on human health is of special concern in tropical agricultural areas where appropriate medical attention is not readily accessible (Warrell 1991).

There is a consensus that the New World pitvipers evolved as a single historical radiation. Evidence from studies of DNA sequence variation (Knight et al. 1992; Heise et al. 1995; Kraus et al. 1996; Vidal and Lecointre 1998; Parkinson 1999; Werman et al. 1999; Parkinson et al. 2002) indicates that New World crotalines are monophyletic (contra Burger 1971; Dowling et al. 1996). This chapter concerns those taxa that are predominately Neotropical in distribution, which includes the bothropoid genera (Kraus et al. 1996, 770), *Lachesis,* and the cantils of the genus *Agkistrodon.* The bothropoids are those snakes previously assigned to *Bothrops* (sensu lato) and are recognized as including eight genera (Campbell and Lamar 1989; McDiarmid et al. 1999; Gutberlet and Campbell

2001) (species numbers in parentheses): *Atropoides* (3), *Bothrops* (sensu stricto) (approximately 30), *Bothriechis* (9), *Bothriopsis* (7), *Bothrocophias* (4), *Cerrophidion* (4), *Porthidium* (7), and *Ophryacus* (2). Three species of *Lachesis* (*Lachesis muta* sensu lato), *Agkistrodon bilineatus,* and *A. taylori* are also considered here because they are distributed within or range into the Neotropics. Rattlesnakes (*Crotalus* and *Sistrurus*) that occur south of the Tropic of Cancer or in the Neotropical realm (Müller 1973) were not included.

The importance of the bothropoid pitvipers and related genera for understanding an emerging picture of historical biogeography of the Neotropics is great. Most bothropoids are restricted to either Middle America (Mexico and Central America) or South America, and only five species are found in both regions (*Bothriechis schlegelii, Bothriopsis punctata, Porthidium lansbergi, P. nasutum,* and *Bothrops asper*). *Lachesis stenophrys* is also found in Lower Central America and South America.

In the past 15 years, herpetologists have seen a dramatic increase in their knowledge of phylogenetic relationships among pitvipers, and many new species have been described. Additionally, there is expanded information on distribution patterns. The combination of diversity and extensive geographic distribution, coupled with improved understanding of species relationships, allows for development of biogeographic hypotheses for these snakes. Consequently an array of evidence, including that of geology, phylogeny, distribution, ecology, and comparative biology, can be brought to bear on the question of bothropoid and Neotropical pitviper biogeography. The only important evidence lacking is that of fossil forms.

A large body of previous work (e.g., Savage 1966, 1982; Simpson and Haffer 1978; Duellman 1979a; Whitmore and Prance 1987; Prance 1982; Bush 1994; Haffer 1997; Santiago 2000) has defined and advanced the complexities of Neotropical speciation and geographic patterning in a historical context. It is clear, however, that the biotic history in Middle America is strongly coupled to that of South America, especially with respect to early and recent periods of Neotropical evolution. Neotropical pitvipers represent only a small component of the New World herpetofauna, but an explanation of the history on this group can be a significant piece in the biogeographic puzzle of the Neotropics.

This chapter is part review and part synthesis. Several biogeographic hypotheses concerned with the evolution of certain Middle American and South American pitviper genera have been published (*Bothriechis,* Crother et al. 1992; *Lachesis,* Zamudio and Greene 1997; *Agkistrodon,* Parkinson et al. 2000; *Bothrocophias,* Gutberlet and Campbell 2001). These are incorporated with novel hypotheses for other genera to construct a general overview of pitviper evolution in the Neotropics. This is not an effort to derive a general model of Neotropical evolution on the basis of disparate taxa but rather a thumbnail sketch of the evolutionary events that have taken place within one widely distributed assemblage. Consequently, the hypotheses expressed in this chapter will likely produce more questions than answers.

Phylogenetic Considerations

Cladistic explanations of historical biogeography rely heavily on knowledge of phylogenetic relationships (e.g., Nelson and Platnick 1981; Humphries and Parenti 1986; Morrone and Crisci 1995), requisite for hypotheses of geographic and area relationships. Consequently, an adequate assessment of the phylogeny of bothropoid pitvipers is necessary for the development of biogeographic hypotheses for these snakes. It is clear that species cladograms, and by extension, area cladograms, are useful tools to place the history of a group into proper perspective, relative to physical events of geology and climate and unrelated taxa. The scientific validity of such explanations is argued elsewhere (Morrone and Crisci 1995).

Numerous studies have appeared in the last decade that attempt to ascertain the cladistic and phylogenetic relationships among New World pitviper species (e.g., Werman 1992; Kraus et al. 1996; Salomão et al. 1997; Parkinson 1999; Martins et al. 2001; Parkinson et al. 2002; Gutberlet and Harvey 2002), but fewer than one-half of the recognized bothropoid taxa are represented in these works. Also, there is some disagreement in phylogenetic hypotheses among these studies (Werman 1999; also see figs. 15.1, 15.2). Incongruence of relationships (Werman 1999; Gutberlet and Campbell 2001) is evident and may stem from the use of different data sets derived from phenotypic information of morphology (Werman 1992; Gutberlet 1998), allozymes (Werman 1992), or mtDNA gene sequence data (Kraus et al. 1996; Salomão et al. 1997, 1999; Vidal and Lecointre 1998; Parkinson 1999; Wüster et al. 1999; Taggart et al. 2001; Parkinson et al. 2002) for phylogenetic reconstruction. Not only do the phenotypically based phylogenies differ from those derived from gene sequence data, but the conclusions among the DNA studies that use different mitochondrial genes (i.e., cyt *b*, ND4, 12S and 16S) also exhibit incongruent patterns. In spite of these differences, however, consistent patterns are observed among these studies (see below).

Herein I develop a phylogenetic hypothesis for bothropoid pitviper genera based on a total evidence approach (Kluge 1989a) and compare this hypothesis with others recently developed (Martins et al. 2001; Taggart et al. 2001; Gutberlet and Harvey 2002; Parkinson et al. 2002). Arguments for and against the total evidence approach are outlined by de Queiroz et al. (1995). Congruences between the total evidence hypothesis developed herein and other published results are used to construct biogeographic interpretations of vicariant patterns for pitviper evolution.

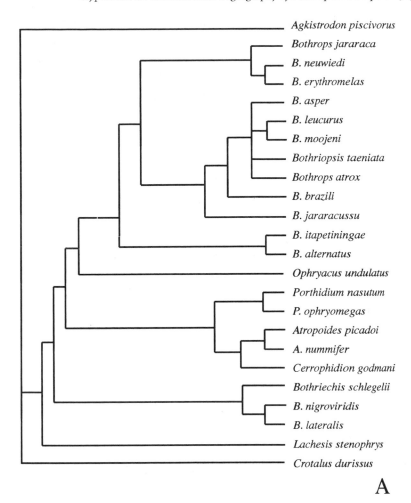

A

Figure 15.1 (A) Maximum parsimony, 50% majority consensus of 15 equally parsimonious trees. Length 232.53; consistency index (CI) = .74; retention index (RI) = .75; rescaled consistency index (RC) = .56; homoplasy index (HI) = .25. All bifurcating nodes have values of 100%. (B) Bootstrap support based on 1,000 replicates.

To gain an overall perspective on the phylogenetic relationships of bothropoid pitvipers, I included 24 species representing most described genera. Information on morphology and allozymes (Werman 1992) was combined with DNA sequence information from GenBank for three mitochondrial gene sequences (12S, 16S, and ND4) to generate cladistic hypotheses. Character infor-

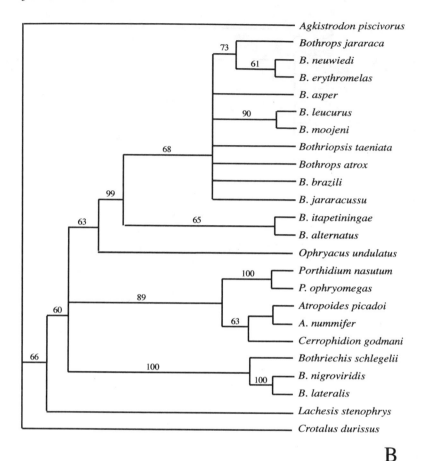

B

Figure 15.1   *(continued)*

mation, aligned DNA sequence data, the input character matrix, and the phylogenetic methods used for hypothesis construction are available online (http://www.press.uchicago.edu/books/donnelly/). The bothropoid taxa were rooted using *Agkistrodon piscivorus* and *Crotalus durissus,* based on Parkinson (1999).

A maximum parsimony, 50% majority rule consensus tree for all taxa is shown in figure 15.1A. Bootstrap support is shown in figure 15.1B. The general topology of this tree depicts a monophyletic *Bothrops/Bothriopsis* and two monophyletic Middle American groups consisting of *Porthidium, Atropoides,* and *Cerrophidion* in one group and three species of *Bothriechis* in another. *Lachesis, Crotalus,* and *Agkistrodon* are basal to the bothropoids. The placement of *Ophryacus* as a sister lineage to *Bothrops* is not well supported (fig. 15.1B) and

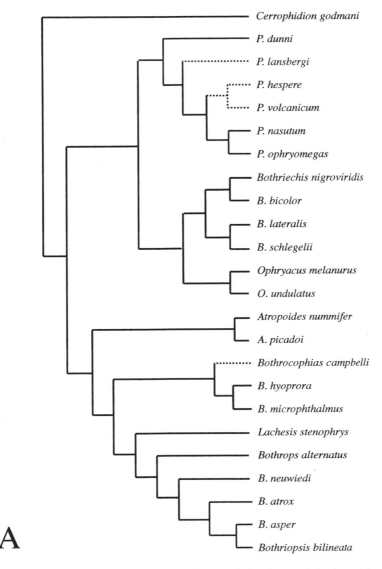

A

Figure 15.2 (A) Representative maximum parsimony solution for certain bothropoid pitvipers (adapted from Gutberlet and Harvey 2002), based on anatomical characters. Dashed lines indicate relationships from alternative solutions. (B) Representative maximum parsimony cladogram for certain bothropoid pitvipers (adapted from Parkinson et al. 2002), based on mtDNA sequence variation. (C) Phylogeny of certain bothropoid pitvipers (adapted from Martins et al. 2001 and references therein), based on mtDNA sequence variation.

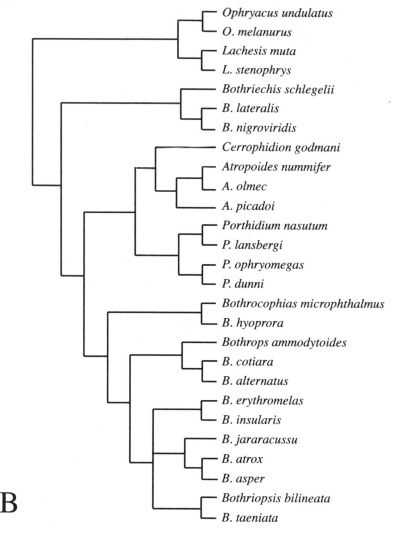

Figure 15.2   *(continued)*

has usually been hypothesized to be a sister group to *Bothriechis* (Crother et al. 1992; Werman 1992; Gutberlet 1998; Gutberlet and Harvey 2002).

Within the South American taxa, three clades are evident (fig. 15.1A). *Bothrops asper, B. brazili, B. atrox, B. leucurus, B. moojeni,* and *B. jararacussu* are part of a terminal clade. *Bothriopsis taeniata,* an arboreal species, is placed within this clade. The position of *B. taeniata,* as nested within or closely allied to *Both-*

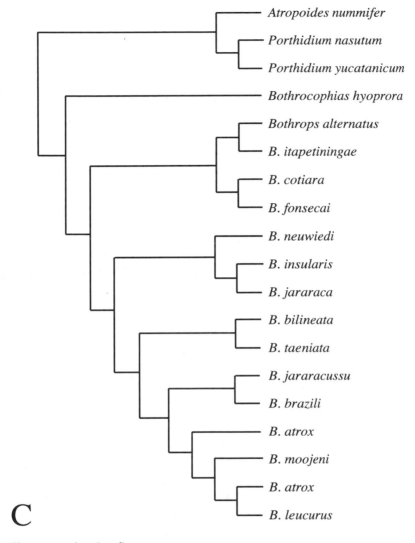

Atropoides nummifer

Porthidium nasutum

Porthidium yucatanicum

Bothrocophias hyoprora

Bothrops alternatus

B. itapetiningae

B. cotiara

B. fonsecai

B. neuwiedi

B. insularis

B. jararaca

B. bilineata

B. taeniata

B. jararacussu

B. brazili

B. atrox

B. moojeni

B. atrox

B. leucurus

C

Figure 15.2 *(continued)*

rops, is consistent with previous studies (Werman 1992; Kraus et al. 1996; Sasa 1996; Salomão et al. 1997; Vidal and Lecointre 1998; Parkinson 1999; Martins et al. 2001; Gutberlet and Harvey 2002; Parkinson et al. 2002). *Bothrops jararacussu* and *B. brazili* fall just outside the "*atrox*" group/complex (sensu Werman 1992; Wüster et al. 1997, 1999), which is consistent with the results of Parkinson et al. (2002; also fig. 15.2B), Salomão et al. (1997, 1999), and Martins et al. (2001;

also fig. 15.2C). *Bothrops jararaca, B. neuwiedi,* and *B. erythromelas* form a clade that is sister to *B. jararacussu* and the *B. atrox* group. Basal in the South American assemblage is *B. itapetiningae* and *B. alternatus.*

The Middle American taxa (*Atropoides, Bothriechis, Cerrophidion,* and *Porthidium;* fig. 15.1A–B) fall primarily into two monophyletic groups of either arboreal or terrestrial forms. The arboreal group includes three species of *Bothriechis* (*B. schlegelii, B. nigroviridis,* and *B. lateralis*) and a terrestrial group that includes a clade comprising *Atropoides picadoi, A. nummifer,* and *Cerrophidion godmani. Porthidium nasutum* and *P. ophryomegas* are a sister group to *Atropoides* and *Cerrophidion.* Basal to the bothropoids is *Lachesis stenophrys.* The hypothesis that *Crotalus* (+ *Sistrurus*) and New World *Agkistrodon* are sister groups to the bothropoid radiation was recently supported by Gutberlet and Harvey (2002) and Parkinson et al. (2002).

Three recent studies have emerged on the phylogenetics of pitvipers that include many New World forms important in the development of the biogeographic hypotheses discussed in this chapter. Gutberlet and Harvey (2002) hypothesized the phylogenetic relationships among Old and New World pitvipers that included members of *Atropoides, Bothriechis, Bothriopsis, Bothrocophias, Bothrops, Cerrophidion, Lachesis, Ophryacus,* and *Porthidium.* One phylogenetic solution is shown in figure 15.2A. They concluded that the bothropoid genera were individually monophyletic but that the relationships among different genera could not be resolved. Gutberlet and Harvey (2002) suggested that *Ophryacus* is the likely sister group to *Bothriechis* and that there was a close relationship among *Lachesis, Bothrocophias, Bothrops, Bothriopsis,* and perhaps *Atropoides* (see fig. 15.2A).

Parkinson et al. (2002), using sequence data from four mitochondrial genes (12S, 16S, ND4, and cyt *b*), investigated the phylogenetic relationships among 56 species of Old and New World pitvipers that included 27 species of bothropoids. One hypothesis for the relationships of the latter taxa is shown in figure 15.2B. As in Gutberlet and Harvey's (2002) study, Parkinson et al. (2002) show rather strong support for the monophyly of the individual Middle America and South American bothropoid genera. *Cerrophidion, Atropoides,* and *Porthidium* consist of a monophyletic assemblage with sister relations to the South America genera: *Bothrops, Bothriopsis,* and *Bothrocophias.* Distant to these two sister groups is *Bothriechis* and a monophyletic *Ophryacus* + *Lachesis* clade that falls basal among the bothropoids. Among the principal South American groups, *Bothrocophias* falls out as a sister group to *Bothrops* + *Bothriopsis.* Among species of *Bothrops,* they identified a lineage composed of *B. ammodytoides, B. cotiara,* and *B. alternatus* as a clade of early divergence within this genus. *Bothrops erythromelas* and *B. insularis* make up a clade closely associated with *B. jararacussu* and two members of the *B. atrox* group (*B. asper* and *B. atrox*). *Bothriopsis* appears closely related to the latter taxa. These arrangements are largely consistent with the hypotheses developed in the present study (fig. 15.1A–B) and those of Martins et al. (2001).

The genus *Bothrops* contains the largest number of species among the bothropoid genera. Martins et al. (2001 and references therein) developed a phylogeny for 12 diverse species from this genus and included related taxa belonging to *Bothriopsis, Bothrocophias, Porthidium,* and *Atropoides.* A phylogeny based on that work is shown in figure 15.2C. This hypothesis depicts *Bothrocophias* as a sister lineage to *Bothrops* + *Bothriopsis.* Within *Bothrops* there is a lineage of early divergence that includes *B. alternatus, B. itapetiningae, B. cotiara,* and *B. fonsecai.* Clades of intermediate divergence relative to a terminal *B. atrox* group include the taxa *B. neuwiedi, B. insularis,* and *B. jararaca* and a clade of *Bothriopsis taeniata* and *B. bilineata.* Finally, *B. jararacussu* and *B. brazili* form a clade of direct sister relation to *B. atrox, B. moojeni,* and *B. leucurus. Atropoides nummifer* and two species of *Porthidium* form a monophyletic clade basal to all the South American representatives. The arrangement of taxa in this analysis (Martins et al. 2001) is strikingly consistent with the relationships identified in the present study (fig. 15.1A), and that of Parkinson et al. (2002), regarding taxa similar to all three studies.

Finally, phylogenetic studies have been conducted on specific genera in which all or most assigned species have been included. These studies include phylogenetic analyses of *Agkistrodon* (Parkinson et al. 2000), *Bothriechis* (Crother et al. 1992; Taggart et al. 2001), *Bothrocophias* (Gutberlet and Campbell 2001), and *Lachesis* (Zamudio and Greene 1997). The results of these investigations are addressed later in this chapter in the biogeographic hypotheses outlined for these genera.

On the basis of the available phylogenetic hypotheses for bothropoid pitvipers, several general patterns have emerged that are likely to remain robust. These patterns include the following: (1) there is evidence for distinct Middle American (albeit paraphyletic) and South American assemblages of Neotropical pitvipers; (2) within the Middle American taxa there exists a monophyletic terrestrial (*Atropoides, Cerrophidion,* and *Porthidium*) group and an arboreal group (*Ophryacus* and *Bothriechis*); (3) the South American bothropoids, with the exception of *Lachesis* in most cases, form a single monophyletic group; (4) *Bothrops* in most cases is paraphyletic, based on the position of *Bothriopsis;* (5) evidence supports the existence of a *B. atrox* group (= complex sensu Wüster et al. 1997); (6) all Middle American and South American genera, as presently recognized (with the possible exception of *Bothrops*), are natural groups of monophyletic taxa; (7) *Agkistrodon* and *Crotalus* are lineages of early divergence in the radiation of New World pitvipers; and (8) *Lachesis* is enigmatic in that it is not consistently allied to any bothropoid genus.

The relationships of many species of bothropoid pitvipers are unknown. However, many of these poorly studied taxa represent species that are rare and have small and restricted distributions. Consequently, given the present under-standing of bothropoid pitviper relationships, hypotheses of vicariant patterns of biogeographic relationships can be developed. Although many hypotheses have been proposed to explain biogeographic diversity (Morrone and Crisci 1995), vicariant hypotheses are considered appropriate here. Studies on specia-tion by Lynch (1989) indicate that with vertebrates, a vicariant model can best explain approximately 70% of cases. In addition, most models explaining the evolution of diversity in rainforests involve an allopatric mechanism (Moritz et al. 2000).

Historical Events Shaping the Neotropics

It is not my intent to present a detailed account of the paleogeographic events or the physiographic and climatic changes that have shaped the present Neo-tropics. However, to properly understand the evolution of the pitvipers that oc-cupy this region, a basic model of historical events is necessary. Consequently, the ideas that follow include hypotheses of the principal events in the evolution of the Neotropics that are relevant to pitviper diversity and speciation through-out the Cenozoic. More detailed accounts of the paleogeography and past cli-mate of Middle America can be found in Savage (1966, 1982) and Campbell (1999), whereas similar historical descriptions for South America can be found in Duellman (1979a, 1999) and Lundberg et al. (1998).

GEOLOGIC EVENTS

Savage (1966, 1982) hypothesized, on the basis of the uniqueness and diversity of the present Middle American herpetofauna, that Nuclear Central America and South America were connected at the beginning of the Paleogene and soon thereafter separated by a geologic event forming the Panamanian portal. This early connection, presumably lost in the late Paleocene to early Eocene, is hy-pothesized to have allowed the in situ differentiation of a Middle American herpetofauna independent from the now isolated herpetofauna of South Amer-ica. The late Cretaceous connection was based on biogeographic information (Savage 1982) because geologic evidence was lacking or controversial. Cur-rently, geologic evidence is still too ambiguous or limited to substantiate a sus-tained subaerial connection or land span between Nuclear Central America and South America during the Paleocene (Iturralde-Vinent and MacPhee 1999). However, it has been suggested that a post-Cretaceous island arc or land bridge may have existed at this time (Malfait and Dinkelman 1972; Rosen 1976; Sal-vador and Green 1980; Iturralde-Vinent and MacPhee 1999).

The development of the connection between Central and South America that closed the Panamanian portal in the early Pliocene (Savage 1982; Kroonenberg et al. 1990; Coates and Obando 1996) was initiated in the Eocene and continued through the Neogene. By the mid-Paleogene, the Chortis Block (present-day Honduras and northern Nicaragua) of Nuclear Central America was already sutured to the Maya Block (present-day Yucatán Peninsula, Guatemala, and Belize) to the north, and the former was moving eastward along the Motagua fault (Escalante 1990). Volcanism east of the Middle American trench in the early Eocene produced subaerial regions of the Chorotega Block (present-day Costa Rica, western Panama) and the Choco Block (present-day eastern Panama and western Colombia). In the middle Miocene, the Chorotega Block was probably represented as a fragmented series of emergent islands, separated from the Choco Block (sutured to western Colombia) by an oceanic portal of abyssal-bathyal depth (Coates and Obando 1996). However, it has also been hypothesized that the entire Chorotega Block or at least the northern portion of it was already sutured to the southern end of the Chortis Block by the late Cretaceous to the late Paleogene (Wadge and Burke 1983; Donnelly 1989; Escalante 1990). The complete closure of the Panamanian portal, with the fusion of the Chorotega and Choco blocks, is estimated to have occurred in the middle Pliocene, 3.5 to 3.1 million years ago (mya) (Coates and Obando 1996), although marine transgressions may have occurred periodically through the early Pleistocene (Beu 2001).

Iturralde-Vinent and MacPhee (1999) hypothesized that a large land span existed in the Eocene–Oligocene transition, connecting northwestern South America to present-day Cuba. This subaerial land span, termed GAARlandia (Greater Antilles–Aves Ridge), is characterized as extending north to a position of close proximity to southeastern North America (present-day Florida) and the Maya Block (present-day Yucatán Peninsula). Continued tectonic and volcanic events in the Caribbean fragmented this land span, which contributed to the Antillean system (Iturralde-Vinent and MacPhee 1999).

Orogenic events in Middle America (Coates and Obando 1996; Marshall and Liebherr 2000) that produced the highland backbone of this region began in the late Miocene and continued through the Neogene and Quaternary, with a major episode occurring during the Pliocene–Pleistocene transition (Savage 1966). According to Savage's model (1966, 1982), the Sierra Madre Occidental and Oriental of southern Mexico and the Sierra Madre del Sur (all west of the Isthmus of Tehuantepec) were elevated first in the Oligocene. This was followed in sequence with the elevation of the Sierra Madre de Chiapas and Guatemala highlands (Miocene) and finally the elevation of the Talamancan range (early Pliocene) in present-day Costa Rica and western Panama. As these upland areas reached their present elevations, they were also fragmented into three major blocks representing the Mexican highlands (west of the Isthmus of Tehuantepec), the highlands of Nuclear Central America (Guatemala, Honduras, and northern Nicaragua), and the highlands of Lower Central America (Costa Rica and Panama). The final orogenic event presumably took place in the "nuclear" highlands (sensu Savage 1982) between the Maya and Chortis blocks at the Polochic-Motagua fault system, separating the highlands of eastern Mexico and northern Guatemala from the adjacent highlands of southern Guatemala and west-central Honduras (see Crother et al. 1992 and references therein).

Concomitant to the events in Middle America, geologic and tectonic forces were also changing the landscape of South America during the Cenozoic. The uplift of the proto-Andes began in the Oligocene and continued with a culmination of intense uplifting in the late Pliocene–early Pleistocene (Duellman 1979c; Kroonenberg et al. 1990). The uplift of the Central and Southern Andes may have reached their present height by the end of the Miocene (Duellman 1979c), whereas the northern Andes were probably uplifted to above 3,000 m relatively recently (approximately 6 mya), coinciding with the closure of the Panamanian portal (Kroonenberg et al. 1990).

Although the majority of South America was above sea level (subaerial) throughout the Cenozoic, there is evidence that northwestern South America (roughly present-day western Colombia and Ecuador) existed as an isolated microcontinent in the Oligocene to early Miocene (Hoorn 1993; Hoorn et al. 1995; Lundberg et al. 1998; Iturralde-Vinent and MacPhee 1999). During this time period, it is hypothesized that marine incursions were present at the Orinoco seaway to the northeast and at the Huancabamba Depression in southern Ecuador. It is unclear as to how far a marine embayment existed along the eastern side of the developing Andean chain or how far it extended into the Amazon basin. The western gate to the Pacific was probably closed in the middle to late Miocene as a result of the continued uplift of the Andes, whereas the Caribbean source of marine transgression was probably abated by the gradual uplift of northwestern Amazonia in the late Miocene to Pliocene (Hoorn 1993; Hoorn et al. 1995; Lundberg et al. 1998). Also, in the middle Miocene, the source input for the Amazon changed from the east (Guiana Shield) to the west (headwaters from the eastern versant of the Andes), resulting in the drainage of the Amazon to the Atlantic (Hoorn 1993; Hoorn et al. 1995; Lundberg et al. 1998). Whether these events, which clearly could have contributed to vicariance and the isolation of the South American herpetofauna, are relevant to pitviper evolution in the region is uncertain. It is evident, however, that the geologic events in the Cenozoic that contributed to the physiography of present-day South America were complex.

CLIMATIC EVENTS

The geologic and orogenic events have had a clear effect on climatic change in Middle America and South America. Historical changes in regional climate have, without question, resulted in the fragmentation of pitviper distributions, have separated populations, or have contributed as effective barriers to organismal dispersal. In any case, climatic change can contribute to vicariant forces affecting species.

The environment of Middle America and South America in the early Cenozoic is characterized as a somewhat continuous mesic tropical climate that remained until the Oligocene (Savage 1966, 1982 and references therein). The cooling and drying trend of this period (Duellman 1999) led to the southward (in North America) and northward (in South America) depression of tropical and subtropical areas along a longitudinal axis. Because the Panamanian portal was present at this time, this led to the probable compression of the tropical herpetofauna into the "peninsula" of Nuclear Central America, isolated from the north by a semi-arid barrier across northern Mexico from the Miocene onward (Savage 1982).

Pivotal to the climatic change in the Neotropics was the continued uplift of the backbone of central mountain ranges in Middle America, the incipient Talamancan range in Lower Central America, and the further elevation of the Andean system in South America. The formation of highland regions led to changes in regional climate and rainfall patterns that are responsible for a striking mosaic and diversity of xeric and mesic habitats that exist today (Wagner 1964; Stuart 1966; Haffer 1987).

In Middle America, the Atlantic versant is dominated by tropical moist and tropical wet forest, except in the Yucatán Peninsula, which is progressively xeric to the north (Vivó 1964; Campbell and Lamar 1989). The Pacific versant of Middle America is savanna and dry forest from the Isthmus of Tehuantepec south to Guanacaste in northwestern Costa Rica. Isolated patches of dry forest occur farther south on the Pacific slopes of western and central Panama, south and east of the Golfo Dulce region of southwestern Costa Rica. The upland regions in Middle America are generally characterized as wet to moist forest, although patches of dry forest occur in the Atlantic foothills of Honduras. To the north, the Sierra Madre system of southeastern Mexico, west of the Isthmus of Tehuantepec, is dominated by temperate pine-oak forest (Wagner 1964; Campbell 1999).

The climate and associated vegetation of South America are strongly influenced by the presence of the Andes and the upland regions of southeastern Brazil and by the amount of rainfall distributed over the continent. The northern and central Andes provide a rain shadow that allows for high precipitation in western Amazonia and the Choco. There is also increased rainfall and wet climate along southeastern Brazil. Xeric areas with vegetation components of the Cerrado, Caatinga, and the Chaco separate the Atlantic forest and the Amazon basin. In addition, there is an abrupt transition of wet to dry climate in Colombia and Venezuela, represented by the Llanos (Haffer 1987; Colinvaux 1996).

Neogene and Quaternary changes in climate as a result of multiple glaciation events likely affected past climate and influenced the resident faunas from a biogeographic perspective. These events precipitated cycles of cooling and drying periods, changes in sea level, atmospheric carbon dioxide levels, and changes in tree-line elevation (Bush 1994). It is clear that recent glaciation cycles may have repeatedly fragmented both mesic and xeric regions, allowing for many organisms to vicariate. In addition, climatic fluctuations and the elevational displacement of montane habitats (lower at glacial maxima) probably affected the distribution and vicariance of Middle American taxa as well.

Biogeography of Middle American Clades

Present evidence suggests that Middle America harbors several genera of monophyletic pitvipers that have had their origins in this region. As mentioned earlier, these genera fall into two groups: one with arboreal representatives (species of *Bothriechis* and *Ophryacus*) and another with terrestrial members (species of *Atropoides, Cerrophidion,* and *Porthidium*). The diversity and geography of these taxa lend support to the hypothesis of Savage (1966, 1982), that the herpetofauna of Middle America developed in situ through much of the Cenozoic. This was the result of isolation from northern elements by climatic barrier(s) that developed in the temperate regions of present-day central Mexico and isolation from South American elements by the Panamanian portal. The Neotropical pitvipers in Middle America are more divergent phenotypically and biochemically (Werman 1992) than their relatives to the south (*Bothrops, Bothriopsis*). *Lachesis,* which also occurs in Middle America, is widely distributed in northern South America, and *Lachesis* spp. are only marginally differentiated phenotypically from one another. Consequently, for Middle American and South American taxa, it is appropriate to examine the distribution and relationships within each genus as an initial step in understanding larger historical patterns for bothropoid pitvipers.

Among Middle American taxa, only the genera *Agkistrodon* (Parkinson et al. 2000), *Bothriechis* (Crother et al. 1992), *Porthidium* (Gutberlet and Harvey 2002), and *Lachesis* (Zamudio and Greene 1997) have been investigated phylogenetically in sufficient detail. This allows for the direct deduction of biogeographic hypotheses. For *Atropoides* and *Cerrophidion,* phylogenetic analyses are either absent or preliminary. Consequently, biogeographic hypotheses for these latter taxa are developed through the application of geographic distribution and area relationships.

BOTHRIECHIS

Members of *Bothriechis* are arboreal species, mostly of moderate- to high-elevation mesic forest, distributed in Middle America from the Isthmus of Tehuantepec (southeastern Oaxaca, Mexico) to Panama (fig. 15.3). The range of one species (*B. schlegelii*), a predominately lowland form, extends farther south

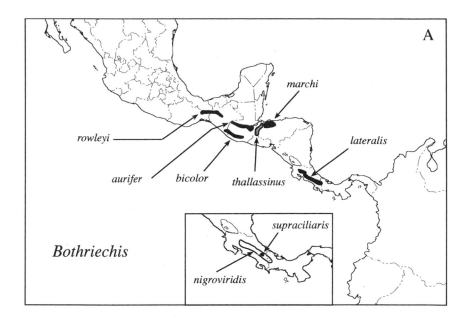

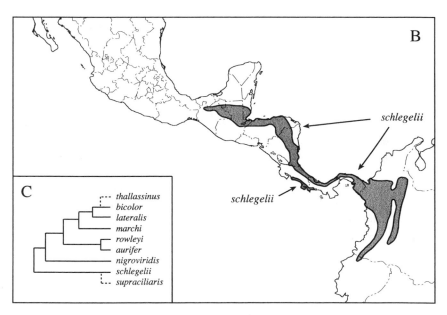

Figure 15.3 (A–B) Approximate distribution of *Bothriechis* in Middle America and northwestern South America (adapted from Campbell and Lamar 1989). (C) Proposed relationships within *Bothriechis* (based on Crother et al. 1992; Taggart et al. 2001). Dashed lines indicate tentative relationships.

into South America and occupies lowland humid forests of western Colombia and northern Ecuador (Campbell and Lamar 1989). Crother et al. (1992), using phenotypic data of morphology and allozymes, derived the phylogenetic relationships for most members of this genus except for *B. supraciliaris* (Solórzano et al. 1998). *Bothriechis supraciliaris* is closely related to *B. schlegelii* and was formerly a subspecies of the latter (Taylor 1954; Werman 1984b). Recently, Campbell and Smith (2000) described *Bothriechis thallassinus* from a region encompassing the border between Guatemala and Honduras. They suggest that *B. thallassinus* is most closely related to *B. bicolor* on the Pacific versant of Chiapas and Guatemala. Consequently, it is included in the hypothesis below and is depicted as a sister lineage to *B. bicolor* (fig. 15.3C).

The phylogeny proposed by Crother et al. (1992) is shown in figure 15.3C. In an equally parsimonious solution, *B. lateralis* is transposed with *B. marchi*. Recently, Taggart et al. (2001) reassessed the phylogenetic relationships of this group with mtDNA sequence data. They concluded that the DNA data were inconsistent with the relationships derived from the phenotypic data of Crother et al. (1992) as a result of lineage sorting or mitochondrial genome transfer between *B. nigroviridis* and *B. lateralis*. Consequently, they concluded that the cladistic relationships based on the phenotypic information (as in Crother et al. 1992) were more robust than those based on the DNA information.

A biogeographic hypothesis adapted from Crother et al. (1992) for *Bothriechis* is summarized in figure 15.4. It is modified here to include *supraciliaris* and *thallassinus*. Crother et al. (1992) proposed that *Bothriechis* evolved in Middle America from a widespread ancestral form that began to diversify in response to geotectonic and orogenic events initiated in the middle to late Paleogene. *Ophryacus*, the putative sister group to *Bothriechis*, diverged west of the Isthmus of Tehuantepec, whereas the latter genus evolved in the east (fig. 15.4A). The initial vicariant event within *Bothriechis* was driven by the development of the highlands and the uplift of the central mountain backbone of Nuclear Central America and Lower Central America (Chiapan highlands, Guatemala-Honduran highlands, and the Talamancan highlands as a block). This allowed for the divergence of a *schlegelii-supraciliaris* ancestor, in the lowlands, from a widely distributed upland ancestor (fig. 15.4B). This presumably requires a gradient hypothesis (Endler 1982) for the differentiation of this latter species. It is possible (but not explicitly stated in Crother et al. 1992) that *B. schlegelii-supraciliaris* existed primarily in the lowland regions of Lower Central America (future Nicaragua and Costa Rica) and later dispersed north into Nuclear Central America and south into the Choco of Colombia and Ecuador. *Bothriechis supraciliaris* likely evolved as an isolated population in the upland regions of southeastern Costa Rica (fig. 15.4F).

The initial fragmentation of the highland forms first occurred between Nuclear Central American highlands and the Talamancas with the formation of

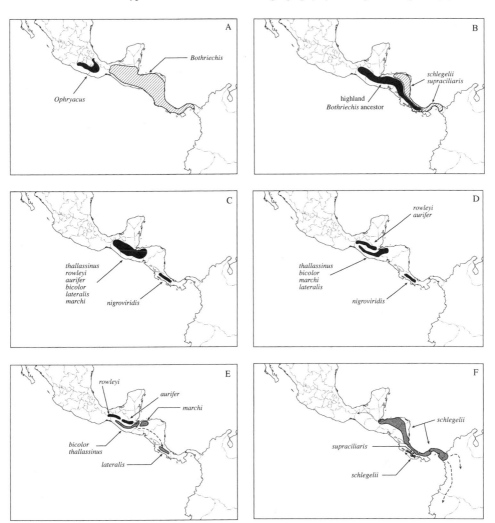

Figure 15.4 (A–F) Proposed biogeographic history for *Bothriechis* and *Ophryacus* (adapted from Crother et al. 1992).

the Nicaraguan Depression (Crother et al. 1992). This led to the divergence of *B. nigroviridis* from the northern highland ancestor of the remaining species (*B. bicolor*, *B. thallassinus*, *B. lateralis*, *B. marchi*, *B. rowleyi*, and *B. aurifer*; fig. 15.4C). This latter clade was divided into a northern highland (Chiapas–west-central Guatemala) ancestor and a southern highland (Chiapas–Guatemala–northern El Salvador and western Honduras) ancestor (fig. 15.4C).

This division was initiated by local range fragmentation and the continued east-ward shift of the Chortis Block relative to the Maya Block at the Polochic-Motagua fault (Crother et al. 1992 and references therein). Continued volcanic activity divided and further fragmented the northern highland population into *B. rowleyi* to the west and *B. aurifer* to the east (fig. 15.4E). *Bothriechis bicolor, B. thallassinus, B. marchi,* and *B. lateralis* apparently underwent recent specia-tion, and their ancestor may have dispersed south into the Talamancas dur-ing climatic fluctuations and the altitudinal depression of highland habitat during the Pleistocene (fig. 15.4E). With population segregation at the Nicara-guan Depression and the fragmentation of the southern highlands in Nuclear Central America by persistent volcanism, *B. marchi, B. bicolor, B. thallassinus,* and *B. lateralis* diversified into distinguishable forms (Crother et al. 1992). It is curious that *B. marchi* is suggested to be the member of earliest divergence in this clade but has an intermediate geographic position between *B. bicolor-thallassinus* and *B. lateralis.*

This hypothesis is dependent on the phylogeny of *Bothriechis* (fig. 15.3C) developed by Crother et al. (1992). The assumption of dispersal, necessary for the origin of *B. lateralis,* is logically derived from the close relationship of this species with *B. bicolor* and *B. thallassinus* as a terminal clade. This dispersal hypothesis of Crother et al. (1992) is further supported by evidence in Savage (1987) that describes the evolution of the *Eleutherodactylus gollmeri* group. In the latter study, the dispersal event of *E. laticeps,* to the north from the Tala-mancas, parallels the proposed dispersal of the *B. lateralis* ancestor, albeit in op-posite directions. If the ancestor to the *B. lateralis-bicolor-thallassinus-marchi* clade dispersed south in the Pleistocene, when climate was favorable and eleva-tion zones were depressed, one might expect that this group would be repre-sented in the Choco of northwestern South America or in the western foothills of the northern Andes as well. At this time, Lower Central America and South America were connected at the Panamanian isthmus. This might lead one to conclude that the evolution of *Bothriechis* preceded the closure of the Pana-manian portal, in contradiction to the hypothesis that within this genus the *B. lateralis-bicolor-marchi* clade underwent speciation rather recently (Crother et al. 1992).

The occurrence of *B. schlegelii* in the Choco of northwestern South America allows for an alternative hypothesis for the divergence within *Bothriechis* that differs slightly with the model just outlined (adapted from Crother et al. 1992). In the previous model, *B. schlegelii* diverges from the upland ancestor of the other taxa in situ in Nuclear or Lower Central America. It is possible that *B. schlegelii* diverged as a unit at the southern extent of "peninsular" Central America (pre-isthmus) after the divergence of *Ophryacus* from *Bothriechis.* The primary assumption of this latter model is that Lower Central America (Chorotega and Choco blocks) was subaerial and sutured to the southern end of the Chortis Block throughout much of the Cenozoic (Donnelly 1989; Crother et al. 1992). As the upland regions in Nuclear Central America and the Talamancas of the Chorotega Block were further elevated, a peninsular effect (Savage 1960) coupled with an upland–lowland vicariance, could have established an ancestor of *B. schlegelii* in the Choco Block at this time (late Miocene to early Pliocene). The uniqueness of the Choco fauna and the close relationship of resident taxa to Middle America have been established (e.g., Cracraft and Prum 1988; Brumfield and Capparella 1996; Santiago 2000; Savage and Wake 2001). After the cladogenesis between *B. schlegelii* and the upland representatives of *Bothriechis,* the Choco Block (subaerial and of low elevation) was then sutured to northwestern South America in the middle to late Pliocene as the Panamanian isthmus was established. The subsequent accretion of the Choco Block to South America would have allowed for *B. schlegelii* to occupy low- and mid-elevation regions of western Colombia and Ecuador. Dispersal into the lowland mesic regions of Middle America, as with *Bothrops asper* (see the section on *Bothrops* later in the chapter), also ensued. The continued uplift of the northern Andes and the xeric climatic barrier to the northeast (Llanos) would have precluded *B. schlegelii* from inhabiting Venezuela. This would explain the presence of *B. schlegelii* in northern South America with a range northward into southeastern Mexico.

OPHRYACUS

The genus *Ophryacus* includes one arboreal and one terrestrial species (*O. undulatus* and *O. melanurus,* respectively), which are restricted to the upland regions of southern Mexico, west of the Isthmus of Tehuantepec (Gutberlet 1998). *O. undulatus* is found in cloud forest and pine-oak forest of the Sierra Madre del Sur and Oriental, whereas *O. melanurus* inhabits pine-oak forest and deciduous dry forest (above 1,600 m) of the Sierra Madre del Sur (Campbell and Lamar 1989).

Species of *Ophryacus* are hypothesized to be the sister group of *Bothriechis,* on the basis of the cladistic analysis of morphological and biochemical phenotypic information (Crother et al. 1992; Werman 1992; Gutberlet 1998; Taggart et al. 2001; Gutberlet and Harvey 2002). Other arrangements have been hypothesized (Parkinson 1999; Parkinson et al. 2002) from the analysis of mtDNA sequence information, and these do not support the close relationship between *Ophryacus* and *Bothriechis.*

Crother et al. (1992) suggested that a vicariant event at the Isthmus of Tehuantepec separated the ancestral population of *Ophryacus* to the west from that of *Bothriechis* to the east (fig. 15.4A). This event was the initial and continued uplift of the Sierra Madre del Sur, Oriental, and Occidental ranges in south-central Mexico starting in the early Paleogene. It is also possible that

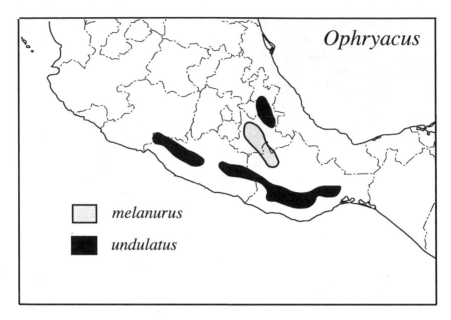

Figure 15.5   Approximate distribution of *Ophryacus* in northern Middle America, demonstrating an upland fragmentation pattern of the Sierra Madre system west of the Isthmus of Tehuantepec (adapted from Campbell and Lamar 1989).

the Isthmus of Tehuantepec was inundated by a marine incursion at this time (Maldonado-Koerdell 1964; Crother et al. 1992), but this is controversial (see Campbell 1999, 119). Thus, the first vicariance among highland biotas in Middle America is hypothesized to have occurred in the region of the Tehuantepec low-lands (fig. 15.4A). This notion is based on the relationships of *Ophryacus* to *Bothriechis* and evidence of other herpetofaunal elements (Savage 1982). This hypothesis is not fully supported and is contradicted by a recent study on the area relationships among montane regions between North America and South America (Marshall and Liebherr 2000). This latter study, based on the distributions and relationships of insects, fishes, reptiles, and plants, suggests that the Talamancan highlands biota was separated first along the highland backbone of Middle America, followed by the Chiapan-Guatemalan highlands and then the Sierra Madre del Sur.

The differentiation of *O. melanurus* from *O. undulatus* probably occurred in the late Pliocene to mid-Pleistocene. The separation of these two species (fig. 15.5) occurred in response to the fragmentation and separation of highland areas in the Sierra Madre del Sur and the Sierra Madre Oriental (or the eastern extent of the Cordillera Transvolcanica), when volcanism was prevalent in this time period (Ferrusquia-Villafranca 1993).

ATROPOIDES

The genus *Atropoides* consists of three species: *A. picadoi, A. nummifer,* and *A. olmec,* and three subspecies within *A. nummifer* (*A. n. mexicanus, A. n. nummifer,* and *A. n. occiduus*). This group ranges widely throughout Middle America and predominately occupies middle- to high-elevation habitats (they are found infrequently in the lowlands). All are found in wet montane forests, although *A. n. occiduus* can be found in dry pine-oak forest habitat on the Pacific versant of southeastern Mexico (Oaxaca and Chiapas), Guatemala, and El Salvador (Campbell and Lamar 1989). The genus does not range into South America and only reaches Panama (fig. 15.6A). The northern limit of the range for *Atropoides* (i.e., *A. n. nummifer*) is southeastern San Luis Potosí, Mexico. In central Guatemala, the distribution of *A. n. occiduus* and northern populations of *A. n. mexicanus* overlap slightly (J. Campbell, pers. comm.). Farther south, in

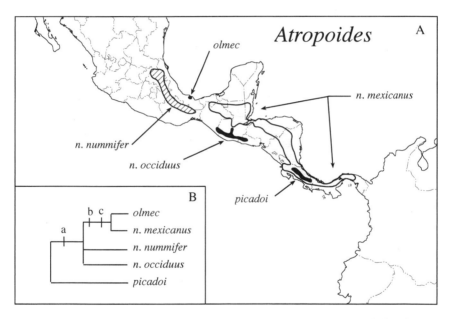

Figure 15.6   (A) Geographic distribution of *Atropoides* (adapted from Campbell and Lamar 1989). (B) Hypothetical relationships within *Atropoides:* (a) the three subspecies of *A. nummifer* and *A. olmec* share a reduced number of ventral scales (< 135) relative to *A. picadoi, Porthidium,* and three species of *Cerrophidion* (> 135); (b) *Atropoides n. mexicanus* and *A. olmec* share derived features that include two (or more) subfoveal rows separating the prelacunal from the supralabials; and (c) nasorostral scales that completely separate the rostral scale from the prenasal on each side (Werman 1984a; Campbell and Lamar 1989; S. D. Werman, unpub. data).

Costa Rica, *A. n. mexicanus* completely circumscribes the geographic distribution of *A. picadoi* and has occasionally been found sympatrically with the latter form (Taylor et al. 1974; Campbell and Lamar 1989).

The phylogenetic relationships within *Atropoides* are largely unknown, although it is evident that *Atropoides* is monophyletic (Werman 1992; Parkinson 1999; Gutberlet and Harvey 2002; Parkinson et al. 2002). A proposed relationship among the members of *Atropoides* is shown in figure 15.6B.

This group is also part of the Middle American herpetofauna that underwent diversification in isolation during the Cenozoic (Savage 1966, 1982), and the important vicariant events were produced by the elevation of the central mountain backbone in Middle America. As with *Bothriechis,* the first interruption of a lowland to moderate-elevation ancestor probably occurred in the region of the Nicaraguan Depression, separating *A. picadoi* from the ancestor of the remaining species (fig. 15.7B). The fragmentation of the northern populations (*A. n. nummifer–n. mexicanus*) at the Isthmus of Tehuantepec and the separation of *A. n. occiduus* on the Pacific versant could have occurred with the uplift and lateral shift of the Nuclear Central American highlands (fig. 15.7C). It is not clear if *A. n. nummifer* and *A. n. occiduus* were separated as a unit or were separated independently from an *A. n. mexicanus–A. olmec* clade. Unlike the vicar-

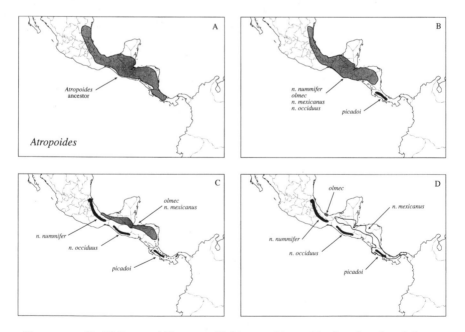

Figure 15.7   (A–D) Proposed biogeographic history of *Atropoides,* based on the relationships in fig. 15.6.

iance between *Ophryacus* and *Bothriechis*, which was affected by the separation of the Mexican highlands (Sierra Madre system) and Nuclear Central American highlands (east of the Isthmus of Tehuantepec), it is possible that *Atropoides* was not because of their propensity for lower elevational regions.

Uplift of the Sierra de los Tuxtlas in Veracruz and habitat changes associated with Neogene glacial episodes isolated the *A. olmec* population from *A. n. mexicanus* as a most recent cladogenic event (fig. 15.7D). The sympatry of *A. n. mexicanus* and *A. picadoi* in Costa Rica is probably best explained by the southward dispersal of *A. n. mexicanus* as a result of the establishment of lower montane corridors through Nicaragua and Costa Rica during the Pleistocene. This latter subspecies has been recorded in the Atlantic lowlands of Costa Rica (Taylor et al. 1974) and has considerable elevational range. Unlike the situation with *Bothriechis lateralis* and *B. nigroviridis*, a dispersal event for *A. n. mexicanus* is supported by records of this species in Nicaragua between Honduran and Costa Rican populations (Campbell and Lamar 1989), which were at one time separated by a marine incursion at the Nicaraguan Depression (Crother et al. 1992 and references therein).

CERROPHIDION

The genus *Cerrophidion* includes four diminutive species (*C. barbouri, C. godmani, C. petlalacalensis,* and *C. tzotzilorum*) distributed throughout the length of Middle America, in moderate- to high-elevation habits (fig. 15.8; Campbell and Lamar 1989; Campbell and Solórzano 1992; López-Luna et al. 1999). One species of *Cerrophidion* was discovered recently (*C. petlalacalensis*). These pitvipers inhabit seasonally dry pine or pine-oak habitat (*C. petlalacalensis* and *C. tzotzilorum*), whereas others have been found in cloud forest (*C. godmani* and *C. barbouri*). *C. godmani* has also been taken in high montane meadows. This group has not been recorded below 2,100 m and, in the case of *C. godmani* and *C. barbouri*, may reach altitudes of approximately 3,300 m (Campbell and Lamar 1989; Campbell and Solórzano 1992).

Cerrophidion is a lineage among terrestrial Middle American pitvipers closely allied with *Atropoides* and *Porthidium* (Werman 1992; Kraus et al. 1996; Parkinson 1999; Parkinson et al. 2002) (figs. 15.1, 15.2B). However, the relationships within *Cerrophidion* have not been fully explored. On the basis of the distribution of these species and the general model of highland Middle American biogeography developed by Savage (1966, 1982), a preliminary hypothesis of their biogeographic relationships can be constructed (fig. 15.9).

If the sequence of fragmentation of major high-elevation regions in Middle America was north to south, as suggested by Savage (1966, 1982), then the first split of an ancestral form would have occurred at the Isthmus of Tehuantepec (fig. 15.9B). The timing of this event is likely to have been coincident with the

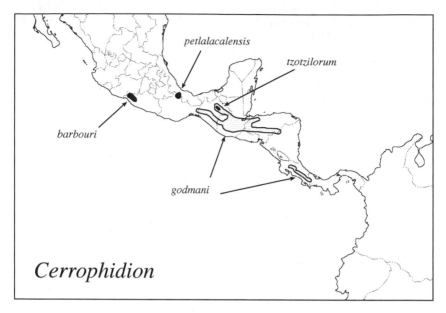

Figure 15.8   Approximate distribution of *Cerrophidion* in Middle America (based on Campbell 1985; Campbell and Lamar 1989; Campbell and Solórzano 1992; López-Luna et al. 1999).

split between *Ophryacus* and *Bothriechis*. The second major vicariant event would have taken place at the Nicaraguan Depression between the Nuclear Central American highlands and the Talamancas of Lower Central America. This event would have occurred in a similar time frame to that of the speciation of *Bothriechis nigroviridis* from the other highland species of Nuclear Central American arboreal pitvipers. The geographic disjunction between Costa Rican and Guatemalan populations of *C. godmani* can be explained as an early vicariant event between the highlands of Nuclear Central American and the Talamancas (cf. Savage 1966, 1982). However, this would require that few divergent phenotypic and genetic changes would have occurred between Nuclear Central American and Costa Rican/Panamanian conspecific populations of *C. godmani* after this vicariance. The alternative explanation would require a recent dispersal event (fig. 15.9C) into the Talamancas of Costa Rica during glacial maxima and elevational habitat depression during these periods. This situation is similar to the scenario of Crother et al. (1992) for explaining the presence of *B. lateralis* in Costa Rica as a recent dispersal event in the evolution of *Bothriechis*. The scenario that *Cerrophidion* was historically confined to the upland regions of southeastern Mexico and Nuclear Central America and that *C. godmani* occurred in the Talamancas of Costa Rica (via dispersal) is consistent with the study by Marshall and Liebherr (2000), in which it is suggested that upland inhabitants in this latter region diverged earlier than the former two (contra Savage 1966, 1982). Because of the congruence of this dispersal hypothesis with a similar explanation for the occurrence of *Bothriechis lateralis* and *Atropoides nummifer* in Lower Central America, this latter hypothesis is favored (fig. 15.9). The alternative would indicate that *C. godmani*, as with *Atropoides picadoi* and *Bothriechis nigroviridis*, was initially isolated in the uplands of Lower Central America when the Talamancan region was first separated from the upland regions of southeastern Mexico and Nuclear Central America. But as *A. picadoi* and *B. nigroviridis* diverged phenotypically and genetically from their congeners to the north, *C. godmani* remained evolutionarily static. The recent establishment of *C. godmani* in Costa Rica (and therefore Panama) is supported by allozyme studies (Werman 1992; Sasa 1997) that show an extreme deficiency in genetic variation among individuals from this region.

The separation and likely contraction of the distributions of *C. barbouri* and *C. petlalacalensis* in southeastern Mexico took place as upland habitats of the Sierra Madre del Sur and the Oaxacan highlands continued to fragment, either through continued volcanic activity or climatic change (fig. 15.9C). The second

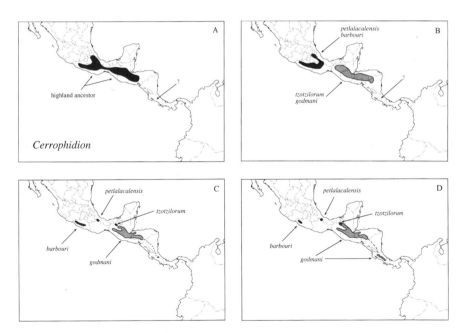

Figure 15.9   (A–D) Proposed biogeographic hypothesis for *Cerrophidion* in Middle America. Question marks (in A and B) indicate that a *godmani* ancestor may or may not have been present in Lower Central America. Dashed arrows (in C and D) indicate potential dispersal events.

segregation in *Cerrophidion* occurred in the northern highlands of Nuclear Central America, resulting in the divergence of *C. tzotzilorum* from *C. godmani* (fig. 15.9C). More recently, populations of *C. godmani* from Honduras dispersed southward and became continuous between Honduras and Costa Rica/Panama, through Nicaragua, during Pleistocene glacial episodes that caused the elevational lowering of montane habitat. Interglacial periods would have been responsible for the hiatus that presently exists in Nicaragua for *C. godmani*.

PORTHIDIUM

The hognosed pitvipers of Middle America, *Porthidium,* consist of a group of small terrestrial species that are restricted primarily to lowland areas of both xeric and mesic habitats. In Middle America most hognosed vipers (*P. dunni, P. hespere, P. lansbergi, P. ophryomegas, P. volcanicum,* and *P. yucatanicum*) are found in tropical deciduous forest of a xeric nature. One species, *P. nasutum,* is found primarily in lowland wet forest situations. The distribution of this group is shown in figure 15.10A.

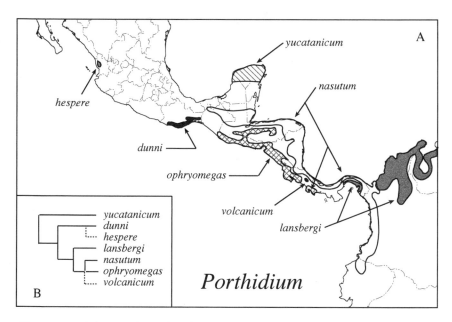

Figure 15.10   (A) Approximate distribution of *Porthidium* in Middle America and northwestern South America (adapted from Campbell and Lamar 1989). (B) Phylogenetic relationships among members of *Porthidium* (based on Gutberlet and Harvey 2002). Dashed lines indicate relationships from alternative solutions.

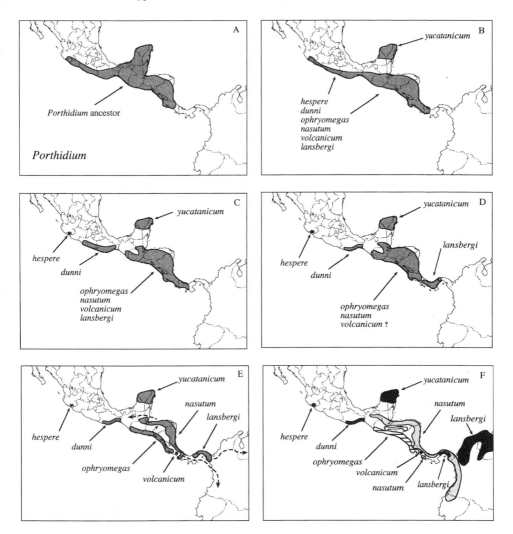

Figure 15.11 (A–F) Biogeographic hypothesis for *Porthidium.* Dashed arrows represent dispersal events.

Gutberlet and Harvey (2002) provided the first hypothesis of the cladistic relationships among members of *Porthidium.* One possible arrangement for these taxa is shown in figure 15.10B. The close association of *P. hespere* and *P. volcanicum* depicted in their hypothesis is difficult to reconcile biogeographically because these two taxa exist in areas separated by considerable distance (fig. 15.10A). Consequently, the model in figure 15.11 depicts a slightly different hypothesis for the divergence of these two species. It is likely that *P. hespere* is sister to *P. dunni,* based on geography, and *P. volcanicum* sister to *P. ophryomegas* or *P. lansbergi.* Solórzano (1994) suggested that *P. volcanicum* shares a close affinity with *P. lansbergi* based on external morphology.

An assumption of the model for the diversification of *Porthidium* (fig. 15.11) is that the ancestor of this group was once widespread in the lowland regions of Middle America. Thus, the first vicariant event, perhaps as a result of emerging climatic changes in eastern Mexico, separated *P. yucatanicum* from the ancestor of the remaining species (fig. 15.11B). Concomitant geologic and climatic changes that occurred in Mexico, west of the Isthmus of Tehuantepec, relative to Nuclear Central America and Lower Central America, led to the fragmentation of northern Pacific coast populations into present-day *P. hespere* and *P. dunni* (fig. 15.11C). Continued separation of xeric regions of the Pacific coast of Mexico resulted in the separation of these two species from the ancestor of the Nuclear Central American and Lower Central American complex of *P. nasutum-ophryomegas-volcanicum-lansbergi.* The fragmentation of xeric habitat in Lower Central America would have led to the differentiation of *P. lansbergi* (fig. 15.11D). The continued uplift and volcanism of the highlands of Guatemala, Honduras, and the Talamancas of Costa Rica would have allowed for the separation of *P. nasutum,* on the mesic Atlantic versant from *P. ophryomegas* on the xeric Pacific versant and *P. volcanicum* from *P. ophryomegas* (fig. 15.11E). In recent times, *P. ophryomegas* could have expanded its range across Nuclear Central America and into the dry forested areas of northern Honduras and the Atlantic versant. Finally, with the complete closure of the Panamanian portal in the Pliocene, *P. nasutum* would have had the opportunity to expand its range into the wet forest regions of the Choco in South America (fig. 15.11E). *P. nasutum* would have also had the opportunity to disperse northward, through mesic lowland forest, into western Guatemala and southeastern Mexico. Panamanian populations of *P. lansbergi* would have similarly been able to disperse and occupy xeric habitat in the extreme northwestern edge of the South American continent (fig. 15.11E–F). The extreme elevation of the northern Andes at this time would have excluded *P. nasutum* from entering the Amazonian basin, whereas *P. lansbergi,* adapted to xeric coastal situations in northern Colombia and Venezuela, would have been unable to enter rainforest habitat of northwestern Amazonia.

AGKISTRODON

Three species of *Agkistrodon* are found in the New World and are considered representatives of the Old Northern Group of Savage (1966, 1982). In North America and Central America, this genus includes *A. piscivorus,* a subtropical, semiaquatic form; *A. contortrix,* a temperate oak forest species; and *A. bilinea-*

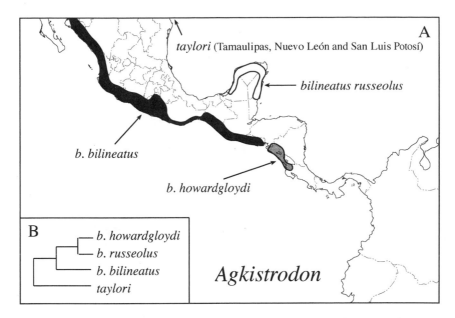

Figure 15.12 (A) Approximate distribution and (B) hypothesized phylogenetic relationships among subspecies of *Agkistrodon bilineatus* and *A. taylori* (adapted from Gloyd and Conant 1990; Van Devender and Conant 1990; Parkinson et al. 2000).

tus, the only member found in tropical Middle America (Gloyd and Conant 1990). The latter species occupies tropical dry forest on both the Atlantic and Pacific coasts of Middle America, and four subspecies have been previously recognized. It has been suggested that *A. bilineatus* has occupied Middle America since the Oligocene and had already diverged from *A. piscivorus* and *A. contortrix* by the late Miocene (Gloyd and Conant 1990; Van Devender and Conant 1990). *Agkistrodon b. bilineatus* is distributed nearly continuously from southern Sonora, Mexico, to southern El Salvador, whereas the subspecies *A. b. howardgloydi* continues southward from El Salvador to northwestern Costa Rica (fig. 15.12A). On the Atlantic coast, *A. b. taylori* (= *A. taylori*) occurs as an isolated population in northeastern Mexico (states of Nuevo León, San Luis Potosí, and Tamaulipas), and *A. b. russeolus* is restricted to the northern Yucatán Peninsula, extreme northern Belize, and the Peten of Guatemala (Campbell and Lamar 1989; Gloyd and Conant 1990).

The relationships among the subspecies (and *A. taylori*) according to Van Devender and Conant (1990) are shown in figure 15.12B. In their account, *A. b. taylori* (= *A. taylori*) diverged from the group in the late Miocene and has since been isolated for a considerable period of time, whereas the remaining sub-

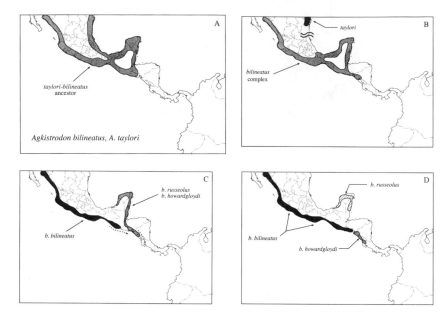

Figure 15.13   (A–D) **Proposed biogeographic hypothesis for the subspecies of *Agkistrodon bilineatus* and *A. taylori*. Parallel lines in (B) indicate the development of a mesic climatic barrier. Dashed arrows in (C) indicate dispersal events.**

species have diverged relatively recently, beginning in the Pleistocene (Van Devender and Conant 1990). Consequently, the biogeographic model for the divergence within *A. b. bilineatus* (fig. 15.13) can be readily constructed from information provided by Gloyd and Conant (1990) and Van Devender and Conant (1990).

This model assumes that during the late Paleogene, *A. b. bilineatus* was widely distributed in Middle America. The population to be separated first was *A. b. taylori* in northeastern Mexico (fig. 15.13B). This event is suggested to have occurred very early (Miocene) in the evolution of this group. Interestingly, the next vicariant event would have had to split and allow the divergence between the west coast *A. b. bilineatus* and the ancestor of *A. b. howardgloydi* plus *A. b. russeolus* (fig. 15.13C). This hypothesis is curious in that the closest relative to *A. b. howardgloydi* in tropical Middle America is also the most geographically distant (*A. b. russeolus*). This requires that the most recent vicariant event separated *A. b. russeolus* (from the Yucatán) from a southern *A. b. howardgloydi* (fig. 15.13D). It may be reasonable to assume that *A. b. bilineatus* developed farther to the north, perhaps west of the Isthmus of Tehuantepec, and has dispersed southward along the Pacific coast to establish parapatry with respect to *A. b. howardgloydi*. At this point, it is clear that the Pacific coast forms and the

Atlantic coast forms do not represent two distinct clades. It is also interesting that given the postulated age of *A. bilineatus* + *taylori* in Middle America, little morphological divergence among the four distinguishable forms has occurred relative to the differentiation within other genera of pitvipers occupying this region.

A recent genetic analysis (mtDNA sequence variation) of the relationships within *A. bilineatus* and the affinity of this species to the other New World *Agkistrodon* (Parkinson et al. 2000) showed strong support for the relationships hypothesized by Van Devender and Conant (1990). One difference, however, is that the DNA data place *A. piscivorus* as the sister lineage to *A. bilineatus*, rather than *A. contortrix*. The sequence divergence estimates that were obtained on the basis of the DNA data also corroborate the predicted divergence times (based on rates in Zamudio and Greene 1997) of *A. bilineatus* as roughly middle Miocene and that of *A. b. taylori* from the other subspecies as roughly late Miocene. Parkinson et al. (2000) also elevated *A. b. taylori* to full species rank on the basis of the relative genetic divergence and geographic isolation from the other subspecies.

The biogeographic hypothesis outlined here (fig. 15.13) is consistent with the results of Parkinson et al. (2000). However, these authors favor the hypothesis that involves a dispersal event into the Yucatán, from the south (i.e., incipient *A. b. howardgloydi*), to establish *A. b. russeolus* in this region. As Parkinson et al. (2000) correctly point out, neither of these two alternatives can be rejected by the phylogenetic hypothesis for this group. The model depicted in figure 15.13 is preferred because it does not include the additional assumption of an independent dispersal event.

Biogeography of South American Taxa

Three genera of bothropoid pitvipers are widely distributed in South America and include *Bothriopsis, Bothrops,* and *Lachesis.* A fourth genus, *Bothrocophias,* is restricted to northwestern South America and the upper Amazon basin (Campbell and Lamar 1989; Gutberlet and Campbell 2001). These taxa appear to have had their origins in this austral continent, although the evidence for *Lachesis* is equivocal. The latter taxon could have originated in Middle America and expanded its range southward across the isthmian link in the Pliocene. Only a single member of *Bothrops* occurs in Middle America (*B. asper*), and this probably resulted from a recent dispersal event. The presence of *Bothriechis* and *Porthidium* in South America is also best explained by a recent dispersal event. These latter two genera are associated with the Middle American herpetofaunal components of Savage (1966). Therefore, it is postulated here that *Bothriopsis, Bothrops,* and *Bothrocophias* evolved in South America when this landmass was isolated from North America and Middle America throughout much of the Cenozoic.

BOTHROCOPHIAS

Bothrocophias was recently erected by Gutberlet and Campbell (2001) to include the toad-headed pitvipers of western South America. This genus comprises *B. campbelli, B. hyoprora, B. microphthalmus,* and a new species, *B. myersi.* One species, *B. hyoprora,* was formerly assigned to *Porthidium* on the basis of similarity of external morphology (e.g., upturned snout, entire subcaudals, etc.). Both *B. hyoprora* and *B. microphthalmus* have been documented to constitute the sister lineage to *Bothrops* (Martins et al. 2001; Gutberlet and Harvey 2002; Parkinson et al. 2002). Thus, the toad-headed pitvipers may represent a lineage of early divergence in the evolution of South American pitvipers. Although *Bothrocophias* has also been allied with *Lachesis* and *Atropoides* (Gutberlet and Harvey 2002), these relationships are tenuous.

The relationships among the four species of *Bothrocophias* and their distributions, as depicted by Gutberlet and Campbell (2001), are shown in figure 15.14A. Both *B. myersi* and *B. campbelli* occur on the western versant of the Andes, whereas *B. microphthalmus* and *B. hyoprora* occur on the Amazonian versant. Both *B. myersi* and *B. hyoprora* are low-elevation species, whereas *B. microphthalmus* and *B. campbelli* are montane forms (Gutberlet and Campbell 2001). Consequently, these authors suggest that the elevation of the northern and middle Andes resulted in the separation of a *B. hyoprora-microphthalmus* ancestor from western versant congeners in the late Pliocene. Gutberlet and Campbell (2001) further suggest that elevational separation of these two ancestral populations resulted in the separation of *B. hyoprora* from *B. microphthalmus* and *B. campbelli* from *B. myersi.* If this scenario were correct, then one would expect that the latter two species would represent a single clade. However, the relationships, based on the parsimony analysis of anatomical characters (Gutberlet and Campbell 2001) among *B. campbelli, B. myersi,* and eastern Andean species *B. microphthalmus* and *B. hyoprora,* are unresolved (fig. 15.14B). Thus, it is possible that either *B. myersi* or *B. campbelli* was the first divergent species in this clade. Additional effort is needed to resolve the polytomy and to ascertain the correct sequence of divergence.

BOTHRIOPSIS

The arboreal pitvipers of *Bothriopsis* (fig. 15.15) are a small group of seven species (*B. bilineata, B. medusa, B. oligolepis, B. peruviana, B. pulcher, B. punctata,* and *B. taeniata*) restricted almost entirely to South America (*B. punctata* reaches extreme eastern Panama [Campbell and Lamar 1989]). They are presently treated as a monophyletic genus on the basis of morphological characteristics (Burger 1971; Campbell and Lamar 1989, 1992; Kuch 1997b). Few molecular data are available to support this hypothesis, although Salomão et al. (1999), Parkinson (1999), Martins et al. (2001 and references therein), and

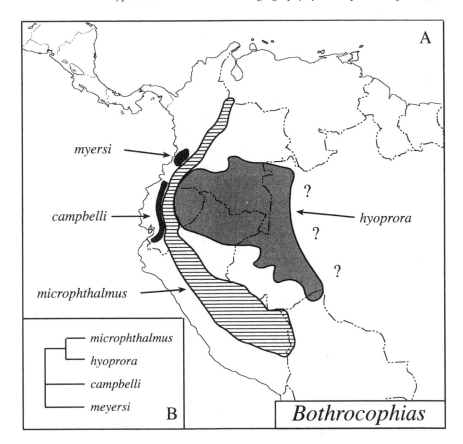

Figure 15.14 (A) Approximate distribution of *Bothrocophias* in South America and (B) proposed phylogenetic relationships (based on Gutberlet and Campbell 2001; U. Kuch, pers. comm.).

Parkinson et al. (2002) have shown, on the basis of mtDNA sequence variation, that *B. bilineata* and *B. taeniata* are sister taxa nested within *Bothrops*. In general, the phylogenetic relationships among members of *Bothriopsis* are unknown, although recent work on this genus will likely result in moving *Bothriopsis pulcher* to *Bothrops* (U. Kuch, pers. comm.). The lack of a phylogeny for the majority of *Bothriopsis* precludes the delineation of vicariant biogeographic hypotheses for this group. Therefore, the comments that follow are heuristic.

Three species of *Bothriopsis* inhabit low-elevation wet forest—*B. punctata, B. bilineata,* and *B. taeniata*—whereas the remaining taxa are found in high-elevation cloud forest up to 3,000 m (Campbell and Lamar 1989). The distribution of the members of *Bothriopsis* is shown in figure 15.15.

It is suggested here that three geographic units of *Bothriopsis* exist: one lowland Choco (*B. punctata*), one lowland Amazonian (*B. bilineata* and *B. tae-*

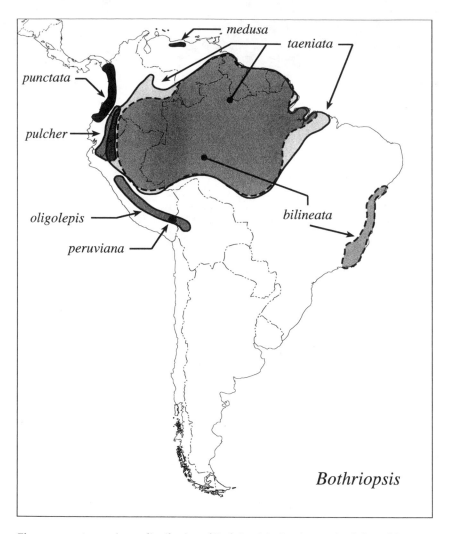

Figure 15.15   Approximate distribution of *Bothriopsis* in South America (adapted from Campbell and Lamar 1989; U. Kuch, pers. comm.).

niata), and a geographically intermediate montane cloud forest group (*B. medusa, B. oligolepis, B. peruviana,* and *B. pulcher*). Among the lowland forms, *B. bilineata* and *B. taeniata* appear to be close relatives (Parkinson 1999; Martins et al. 2001; Parkinson et al. 2002), relative to *Bothrops,* and have a *cis-*Andean distribution through Amazonia. *B. bilineata* also occurs in the Atlantic forests of eastern Brazil.

Speciation within *Bothriopsis* probably began in the late Miocene concomitant with the elevation of the northern and central Andes, after the retreat of the marine embayment of northwestern South America. The ancestor to the modern species of this group is postulated to have been widely distributed in mesic forest regions of South America (fig. 15.16A). The continued uplift of the Andean chain, especially the central and northern Andes, allowed for elevational vicariance separating members of the three groups (fig. 15.16B). The northern Andean orogeny would clearly allow for the separation and differentiation of *B. punctata* in the Choco from *B. bilineata* and *B. taeniata* in the Amazon basin and eastern Brazil. The closest relatives of these three lowland forms could very well be geographically proximate Andean *Bothriopsis* or *Bothrops*, as hypothesized for lowland and highland *Liolaemus* (Schulte et al. 2000). Alternatively, the fragmentation of high-elevation cloud forest habitat into an island chain, which has been suggested for speciation among Andean birds (Hackett 1993) and sigmodontine mice (Patton and Smith 1992), could have precipitated the divergence of *B. pulcher, B. peruviana,* and *B. oligolepis* as a monophyletic group within *Bothriopsis* (fig. 15.16C). The sequence of cladogenesis in the latter taxa is yet to be determined. However, according to Schulte et al. (2000) it is also possible that Andean *Bothriopsis* is not monophyletic and that it evolved by multiple upland invasions and vicariance from geographically local lowland ancestors.

After the establishment of the Panamanian isthmus in the mid-Pliocene, *B. punctata* could have dispersed northward into lower Panama. The development of xeric habitat in northern Colombia and western Venezuela likely precluded *B. punctata* from the Amazon basin. *Bothriopsis medusa* could have affinities to *B. punctata,* the other Andean species, or to *B. bilineata* and *B. taeniata*. The divergence between *cis*-Andean *B. bilineata* and *B. taeniata* most likely occurred during climatic cycling and the repeated contraction and expansion of wet forest habitat in the Pleistocene. Different Amazonian animals have responded differently to the array of local isolation barriers that formed in the Neogene (Haffer 1997; Moritz et al. 2000). A biogeographic analysis of several vertebrate groups (Cracraft and Prum 1988; Santiago 2000) provides evidence, for anurans, lizards, and birds, that upper Amazonian and lower Amazonian clades are closely related and that the Atlantic forest of Brazil forms a geographic clade with regions of the lower Amazon (Santiago 2000). Therefore it is possible that a Pleistocene habitat contraction led to the separation between *B. taeniata* in the upper Amazon and an eastern *B. bilineata* ancestor (fig. 15.16C). At this time *B. bilineata* would have been distributed across southeastern Brazil and as far eastward as the Atlantic. Subsequent to the speciation of these two species and the expansion of mesic forest during a recent interglacial period, *B. taeniata* likely dispersed eastward in Amazonia, and *B. bilineata* expanded its range north and west to become sympatric in recent times. The expansion of xeric habitat between Amazonian *B. b. bilineata* and the Atlantic forest *B. b. smaragdina* (fig. 15.16D) is probably a recent event because the observed

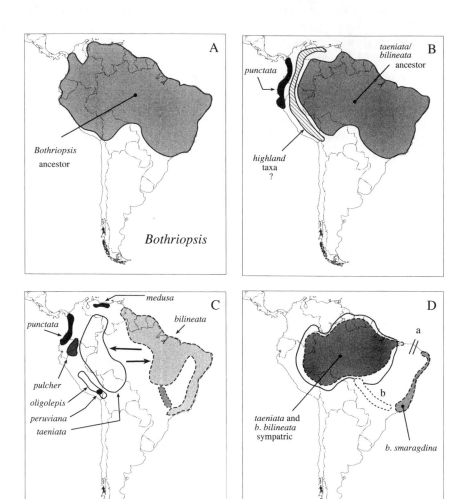

Figure 15.16 (A–D) Proposed biogeographic hypothesis for the evolution of *Bothriopsis* in South America. (A) *Bothriopsis* ancestor. (B) *B. taeniata* and *B. bilineata* ancestor. (C) Range (large arrows) of expansion events producing sympatry of *B. bilineata* and *B. taeniata* in Amazonia. (D) Expansion of xeric habitat between Amazonian *B. b. bilineata* and the Atlantic forest *B. b. smaragdina*. The double bars (a) depict the development of a habitat disjunction between *B. b. bilineata* and *B. b. smaragdina*. The lightly shaded area (b) indicates the loss of a possible forest corridor.

morphological divergence between these two populations has allowed only for the recognition of subspecies (also see *Lachesis* below).

BOTHROPS

Bothrops is the largest genus of Neotropical pitvipers, with approximately 30 species that range from eastern Mexico (*B. asper*) to southern Argentina (*B. ammodytoides*) (fig. 15.17A). *Bothrops* is South American in distribution, except for *B. asper* (fig. 15.18), which presumably invaded Middle America after the closure of the Panamanian portal. Within South America, seven faunal zones for *Bothrops* are proposed here that have biogeographic unity and significance (fig. 15.17B). These areas include (1) the Choco of western Colombia; (2) the west coast lowlands below the Huancabamba Depression, from Colombia to southern Peru; (3) the northwestern xeric lowlands; (4) the moderate- to high-elevation Andean arc, extending from the eastern extent of the Cordilleras de Mérida and de la Costa in Venezuela to the central Andes in southern Peru; (5) the Amazonian rainforest; (6) the relatively extensive xeric belt of the continental east and south, from eastern Brazil to southeastern Argentina; and (7) the mesic Atlantic forest. It is interesting that the xeric belt of the southeastern portion of the continent is also the region that holds the highest diversity for this genus. By comparison, only two species of *Bothrops* inhabit the lowland Amazonian rainforest (*B. atrox* and *B. brazili*). Thus, *Bothrops* is a clade dominated by species adapted to arid and semi-arid tropical regions, unlike *Lachesis,* which seems to have evolved a propensity for mesic forest situations.

Although *Bothrops* has the largest number of species among the Neotropical pitviper genera, it is also the least understood regarding phylogenetic relationships. The most extensive investigations into this genus using cladistic methods have been those of Salomão et al. (1997, 1999), Werman (1992), and Martins et al. (2001 and references therein), and for the *B. atrox* group (= complex), Wüster et al. (1997, 1999). Fewer than one-half of the taxa assigned to *Bothrops* (including four forms restricted to islands: *B. alcatraz, B. lanceolatus, B. caribbaeus,* and *B. insularis*) have been investigated. Thus, a complete and comprehensive hypothesis of the biogeographic history of the entire group is not currently possible. In light of this situation, the ideas presented below are for the purpose of identifying general patterns of the biogeographic history of *Bothrops,* based on relatively diverse taxa.

There is a consensus that the *B. atrox* group minimally includes *B. asper, B. atrox, B. isabelae* (= *B. asper;* see Campbell and Lamar 1989, 195), *B. leucurus* (= *B. prado;* see Puorto et al. 2001), *B. marajoensis,* and *B. moojeni* (Werman 1992; Wüster et al. 1996, 1997, 1999; Salomão et al. 1999). The distribution of these species is shown in figure 15.18. Closely related to the *B. atrox* group are *B. brazili, B. jararacussu, B. pirajai* (Werman 1992; Salomão et al. 1999; Wüster

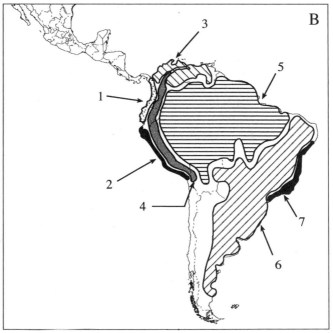

et al. 1999; Martins et al. 2001), *B. sanctaecrucis,* and the newly described *B. muriciensis* (Ferrarezzi and Freire 2001). The latter three species are not considered here because their cladistic relationships are unknown, although all five may be closely related (Campbell and Lamar 1989; Ferrarezzi and Freire 2001). With the exception of *B. moojeni,* the *B. atrox* group conforms to three geographic groupings: northwestern South America, Amazonia, and Atlantic forest. *B. moojeni* occupies a somewhat intermediate region of semi-arid tropical savanna, between the Atlantic forests of eastern Brazil and the Amazon basin (Campbell and Lamar 1989).

Wüster et al. (1999) investigated the relationships within the *B. atrox* group by using mtDNA sequence variation and wide geographic sampling. A summary of their findings is shown in figure 15.19. Wüster et al. (1999) report that *B. atrox* from Ecuador has genetic affinity to conspecifics of the central Amazon (Itacoatiara) and the eastern Amazon (*B. atrox* and *B. marajoensis*). Of interest is that *B. atrox* north of Manaus has a close relationship with *B. atrox* of Guyana and Suriname. This latter group is more closely related to taxa on the Atlantic coast of southeastern Brazil and the southern Amazon (*B. leucurus, B. atrox*) than it is to central Amazonian populations (fig. 15.19). *B. moojeni* appears to have a dual affinity. The southern populations are in a unique clade (*B. leucurus* from western Bahia is also in this clade, but other information [Campbell and Lamar 1989] shows that this species is restricted to coastal areas), whereas the northern populations of *B. moojeni* show affinity to east Amazonian *B. atrox* and *B. marajoensis.* A multivariate analysis of morphological characters (Wüster et al. 1996) that included *B. atrox* and *B. moojeni* showed that these are probably valid species, although a hybrid zone occurs where these two species meet. As stated by Wüster et al. (1999), the mitochondrial clades of the *B. atrox* group do not conform to conventional species designations. It is also interesting that the genetic affinities within these clades roughly conform to several forest refuge centers postulated for glacial periods during the Quaternary (Haffer 1985, 1987; Prance 1982, 1987; Santiago 2000). The *B. atrox* complex is an assemblage of primarily forest species, except for *B. marajoensis* and *B. moojeni.* It is possible that forest contractions and expansions have played

Figure 15.17 (A) Approximate distribution of *Bothrops* in Middle America and South America, excluding island forms (adapted from Campbell and Lamar 1989). (B) Principal biotic units of *Bothrops* in South America. (1) Choco of western Colombia; (2) west coast lowlands below the Huancabamba Depression, from Colombia to southern Peru; (3) northwestern xeric lowlands; (4) moderate- to high-elevation Andean arc, extending from the eastern extent of the Cordilleras de Mérida and de la Costa in Venezuela to the central Andes in southern Peru; (5) Amazonian rainforest; (6) relatively extensive xeric belt of the continental east and south, from eastern Brazil to southeastern Argentina; and (7) mesic Atlantic forest.

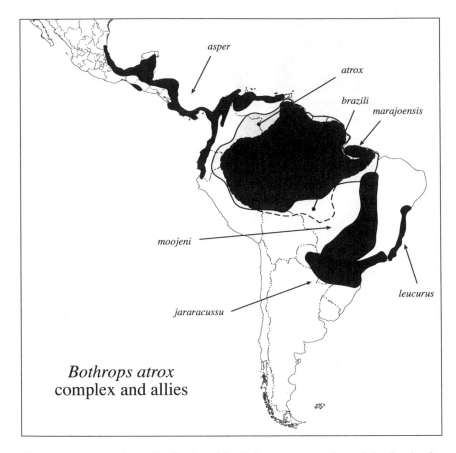

Figure 15.18 Approximate distribution of the *Bothrops atrox* complex and closely related forms in Middle America and South America (adapted from Campbell and Lamar 1989).

a role in the differentiation within this group. However, because the evolution of the mitochondrial genome may not track phenotypic evolution in these taxa (Wüster et al. 1999), it is possible that changes in the mtDNA offer only a distorted reflection of Quaternary evolution shaping the *B. atrox* group. Lineage sorting and hybridization events, problematic with regard to the mitochondrial genome (de Queiroz et al. 1995), may have confounded the evolutionary picture of species relationships in other taxa as well (e.g., Taggart et al. 2001).

The apparent incongruence between the mtDNA relationships and the phenotypic relationships within the *B. atrox* group may not completely preclude the development of a biogeographic scenario for the evolution of the *B. atrox* group and the related taxa *B. brazili* and *B. jararacussu*. The first events of di-

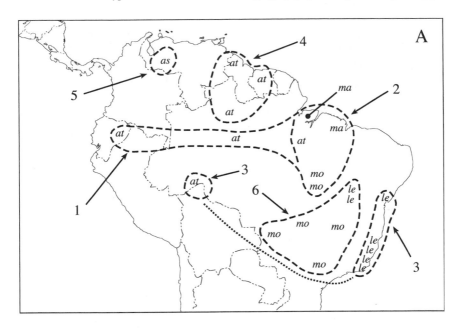

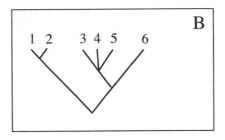

Figure 15.19 (A) Summary of the mtDNA clades (numbered arbitrarily 1–6) for the *Bothrops atrox* complex identified by Wüster et al. (1999). as, *B. asper* (= *B. isabelae*); at, *B. atrox*; le, *B. leucurus*; ma, *B. marajoensis*; mo, *B. moojeni*. (B) Proposed relationships among the mtDNA clades, as numbered in (A) (adapted from Wüster et al. 1999).

vergence (fig. 15.20) likely took place in the late Miocene, before the establishment of the isthmian link, with the differentiation of *B. jararacussu* and *B. brazili* from the *B. atrox* group (fig. 15.20B). The Neogene development of continuing aridity in the southeastern part of the continent could have separated *B. jararacussu* (tropical rainforest, semi-deciduous, and evergreen forest) from the ancestor of *B. brazili* and incipient *B. atrox* group. Present-day *B. atrox* and *B. brazili* are sympatric, and therefore some vicariant event occurred within the Amazon basin to allow for their divergence. This could have been similar to the events separating *Bothriopsis taeniata* from *bilineata* in the upper and lower

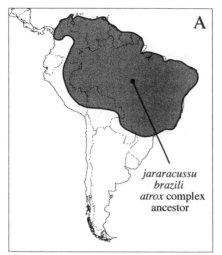
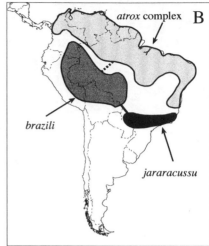
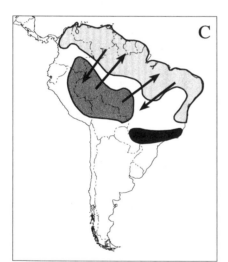
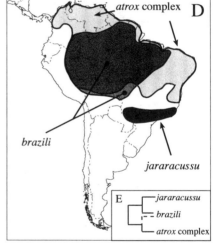

Figure 15.20 (A–E) Proposed biogeographic hypothesis for the divergence between the *atrox* complex and other closely related taxa (*B. brazili* and *B. jararacussu*). Dashed and solid lines in (B) reflect potential affinities in (E). Arrows in (C) indicate possible dispersal necessary for current sympatry in Amazonia between the *B. atrox* complex ancestor and *B. brazili*.

Amazon basin, respectively. As mentioned earlier in the chapter, this is a general pattern for other vertebrates, including anurans and lizards (Santiago 2000). It is likely that *B. brazili* was isolated to the south of the lower Amazon basin and a *B. atrox* group ancestor to the north and west in the upper Amazon basin. Subsequently these two forms expanded their ranges and became sympatric throughout much of the Amazon basin (fig. 15.20C–D).

Differentiation within the *B. atrox* group was probably driven by Pleistocene climatic fluctuations and oscillating fragmentations of mesic habitat. This group of species has probably differentiated recently among South American pitvipers. Because of the cycling of climate and changing distribution patterns during glacial and interglacial periods, the *B. atrox* group may exhibit a palimpsest history (sensu Heyer and Maxson 1982) in which only the most recent patterns are discernable. It is likely that the evolution of the *B. atrox* group (fig. 15.21) was initiated in the Pliocene, after the closure of the Panamanian portal. Thus, the ancestor to *B. asper* was expanding northward into Lower Central America at this time. The persistent drying trend in southeastern South America and the development of glacial cycles likely fragmented the *B. atrox* group ancestor into scattered populations that were perhaps syntopic with some of the proposed centers of endemism (Haffer 1985, 1987; Prance 1987; Santiago 2000; fig. 15.21B). It seems clear that an ancestral *B. moojeni* population and an Atlantic forest population (*B. leucurus*) would have undergone opposing oscillations. The return of interglacial mesic conditions likely allowed for the mixture of populations with some genetic distinctness but overall great morphological similarity. This would have allowed for the accretion of the western and eastern Amazonian clades, the union of the coastal populations, around a mesic corridor to form a circumferential distribution, and an expanding ancestral population of *B. moojeni* and *B. marajoensis* in the open regions of southeastern Brazil (fig. 15.21C). This latter population could have expanded during dry periods to separate the Amazonian and northwestern continental populations from the Atlantic forest forms. With the terminal elevation of the northern Andes and the presumed adaptation by *B. asper* to a variety of habitat types, this species differentiated north of a climatic barrier (Llanos of Venezuela and Colombia) and continued to invade Middle America through mesic corridors to its present range. The expansion of Amazonian rainforest to the east provided for the disjunction between present-day *B. moojeni* and *B. marajoensis,* and the maturation of the Caatinga likely eliminated the forest corridor on the northern Brazilian coast and confined *B. leucurus* to the east coast of Brazil (fig. 15.21D).

Several species of *Bothrops* are distributed along an extensive and generally xeric band running south to northeast along the southern Atlantic side of the continent. This region is interrupted by a small area of humid subtropical climate and *Araucaria* forest, located in extreme southern Brazil and southeastern

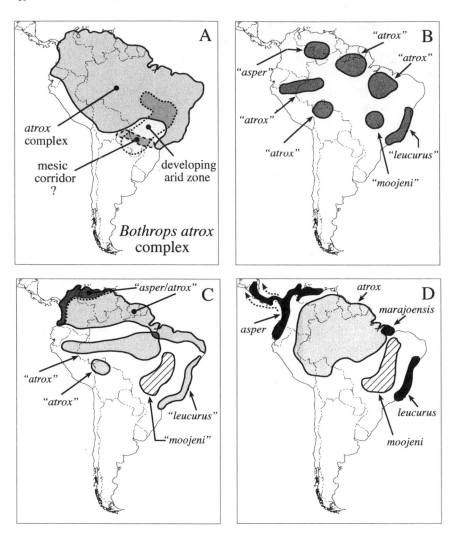

Figure 15.21   (A–D) Model for the evolution of the *B. atrox* complex. Smaller dashes in (A) indicate a region of developing aridity; longer dashes indicate a possible mesic corridor. Dashed arrows in (D) represent dispersal of *B. asper* into Middle America.

Paraguay (Haffer 1987). Because this xeric region extends from the equator to southern Argentina, the number of vegetation zones occupying this band is great. The distribution of most taxa in this region is shown in figure 15.22. In this chapter I refer to these taxa as the "xerophilous" *Bothrops*. Two species, *B. barnetti* and *B. pictus*, are restricted to the xeric coast of Peru and are in-

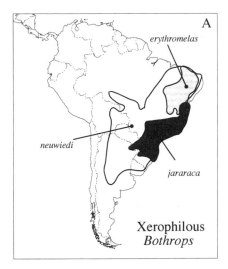 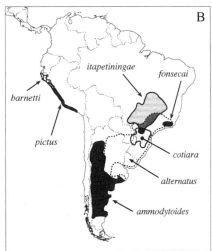

Figure 15.22 (A–B) Approximate distributions of "xerophilous" *Bothrops* of southeastern South America (adapted from Campbell and Lamar 1989).

cluded in this designation. To date there is no clear phylogeny for all the species inhabiting this region, but the affinities of some are known (see figs. 15.1, 15.2B–C). Werman (1992) proposed the *B. neuwiedi* group to minimally include *B. alternatus, B. itapetiningae, B. erythromelas,* and *B. neuwiedi.* Recent phylogenetic analyses have determined that these former taxa and others may constitute at least two distinct lineages. On the basis of the relationships of the present study (fig. 15.1A) and those of Parkinson et al. (2002; also fig. 15.2B), Salomão et al. (1997, 1999), and Martins et al. (2001; also fig. 15.2C), two groups are evident, one composed of *B. jararaca, B. neuwiedi,* and *B. erythromelas* and another that includes *B. cotiara, B. alternatus, B. fonsecai, B. ammodytoides,* and *B. itapetiningae.* With the exception of *B. jararaca,* all members lack a lacuolabial or it is variable (i.e., *B. ammodytoides*). Gutberlet and Harvey (2002) found a close relationship between *B. neuwiedi* and *B. alternatus* (see fig. 15.2A) but did not include the other taxa listed above in their study. This phylogenetic information is used for the development of a general hypothesis of the biogeographic history of the aforementioned taxa (see figs. 15.23–15.25).

The emergence of this relatively xeric expanse was initiated in the late Paleogene with the rain shadow effects of the developing Andes and the general trend of cooling and drying in the Cenozoic. This drying trend was pronounced in the southern (temperate) region of the continent and allowed for the diversification of southern species (e.g., *B. ammodytoides, B. alternatus,* etc.) from a more northern and eastern group (fig. 15.23B). With increasing aridity in devel-

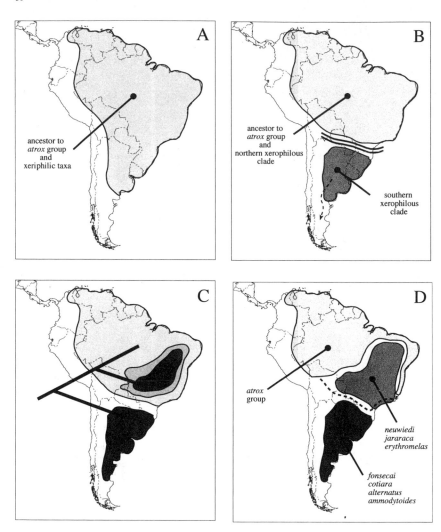

Figure 15.23   (A–D) Proposed biogeographic hypothesis for the evolution of the south-eastern xerophilous taxa from the *B. atrox* group. Double lines in (B) represent a climatic barrier. Phylogram in (C) is based on figs. 15.1A and 15.2C. Dotted line in (D) represents loss of a mesic corridor.

oping xeric regions of the Caatinga and Cerrado, a second group (*B. neuwiedi, B. erythromelas, B. jararaca*) was segregated from the *B. atrox* group and *B. jararacussu* (fig. 15.23C–D). The separation of these three lineages would have been enhanced by subsequent south and eastern expansions of Amazonian forest corridors during mesic episodes connecting forests of the Mato Grosso with the Atlantic forest (Duellman 1982) and the northern Atlantic forest with the lower Amazon (fig. 15.23C). These corridors would have allowed for the diversification of a somewhat northeastern *B. jararaca, B. neuwiedi,* and *B. erythromelas* lineage and a southern *B. ammodytoides, B. alternatus, B. itapetiningae, B. fonsecai,* and *B. cotiara* lineage. This pattern of fragmentation and the likelihood of corridors may also be responsible for the diversification of coralsnakes in this region (i.e., *Micrurus ibiboboca* from *M. frontalis* and the disjunct range of *M. lemniscatus;* cf. Campbell and Lamar 1989 and Slowinski 1995). The retraction of forest corridors probably allowed for the northward dispersal of *B. alternatus* and the southern expansion of *B. neuwiedi.*

The separation between *B. jararaca* and *B. neuwiedi-erythromelas* (fig. 15.24) probably resulted from habitat fragmentation. *Bothrops jararaca,* confined to mesic regions of southeastern and coastal Brazil, would have segregated from *B. neuwiedi* and *B. erythromelas,* which are inhabitants of more xeric habitats (fig. 15.24B). The mechanism leading to the differentiation between *B. erythromelas* and *B. neuwiedi* is not clear, but this could have been the result of habitat differentiation (i.e., continued maturation of the Caatinga) or the segregation of an ancestral distribution by a mesic forest corridor connecting the Amazon basin to the Atlantic forest (fig. 15.24C–D).

The vicariance between *B. ammodytoides* and the other members of this southern group (*B. alternatus, B. itapetiningae, B. fonsecai,* and *B. cotiara*) is also related to ecological vicariance. *B. ammodytoides* inhabits landscapes of harsh aridity and temperature and is the southernmost pitviper in the Western Hemisphere (Campbell and Lamar 1989). The differentiation of *B. cotiara-fonsecai* and *B. alternatus-itapetiningae* probably occurred as a consequence of habitat separation and local adaptation (fig. 15.25). *B. alternatus* seems to have undergone considerable range expansion in recent time, because it is now sympatric with the other species in this group.

Representatives of the genus *Bothrops* also occur in moderate- to high-elevation habitat (fig. 15.26), montane dry forest (*B. lojanus*), cloud forest (*B. venezuelensis*), and moderate-elevation to lower-montane wet forest (*B. andianus, B. colombianus, B. sanctaecrucis*), as tandemly distributed Andean species (Campbell and Lamar 1989, 1992). The relationships and affinities to other members of the genus are unknown, and it would be premature to suggest that this group was monophyletic without supportive evidence. However, it is possible that some of these species share a common ancestry and have differentiated from a contiguous Andean population (or populations) fragmented by the uplift of these mountains through the Neogene. This situation has apparently occurred to other upland forms in Middle America (e.g., *Cerrophidion, Bothriechis*) and is well known for other Andean taxa (Duellman 1979c). Consequently, the possibility that these species of *Bothrops* evolved as a group with the uplift of the Andean chain cannot be discounted. One alternative is that the Andean species share an ancestry with lowland forms of proximate geographic distribution and underwent speciation through a gradient mechanism.

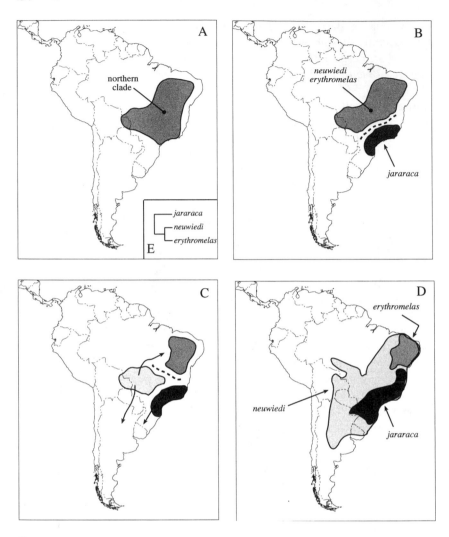

Figure 15.24   (A–E) General hypothesis for the differentiation of the northern clade of xerophilous *Bothrops*. Relationships in (E) are based on fig. 15.1A.

Two xerophilous species, *Bothrops barnetti* and *B. pictus* (= *B. roedingeri* [Campbell and Lamar 1992]), inhabit the lowland regions of the Pacific coast of South America (fig. 15.22B). These taxa are isolated along the xeric western coast of Peru and occupy arid desert scrub (*B. barnetti*) to arid or semi-arid rocky foothills (*B. pictus*) (Campbell and Lamar 1989). It is also possible that the presence of these species on the western coast predated the orogenesis of the

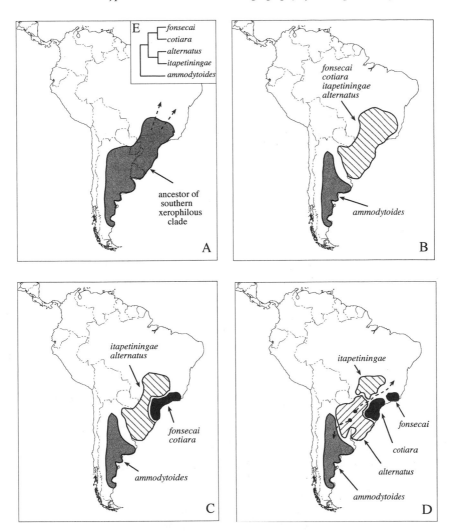

Figure 15.25   (A–E) Hypothesis for the differentiation of the southern xerophilous *Bothrops* clade. The phylogram (E) is based on figs. 15.1A and 15.2B–C. Dashed arrows indicate dispersal or range expansions.

Andean chain and share affinities with *cis*-Andean species of southern South America (e.g., *B. ammodytoides* and *B. alternatus*).

It is evident that *Bothrops* has species of recent divergence (*atrox* group) and members that could have differentiated early in the history of this genus (*B. barnetti* and *B. pictus*). *Bothriopsis* is likely an internal clade within *Bothrops*, thus

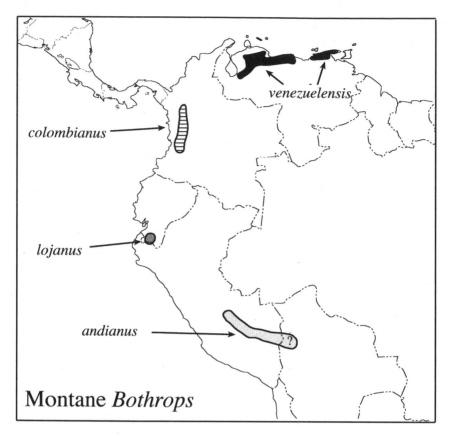

Figure 15.26   Approximate distribution of some montane (Andean) species of *Bothrops* (adapted from Campbell and Lamar 1989).

rendering the latter genus paraphyletic. Of intermediate divergence are taxa found in the xeric belt of the southeastern region of the continent that comprise at least two paraphyletic groups. Present evidence supports *Bothrocophias* as a clade of early divergence and the sister group to the former two genera (fig. 15.27).

LACHESIS

The bushmasters (*Lachesis*) are discontinuously distributed from southern Nicaragua in Central America to the Atlantic forests of southeastern Brazil in South America (fig. 15.28A). Traditionally one species (*L. muta* sensu lato) with four subspecies (*L. m. muta, L. m. melanocephala, L. m. stenophrys,* and *L. m. rhombeata*) has been recognized (Campbell and Lamar 1989). Recently, Zamu-

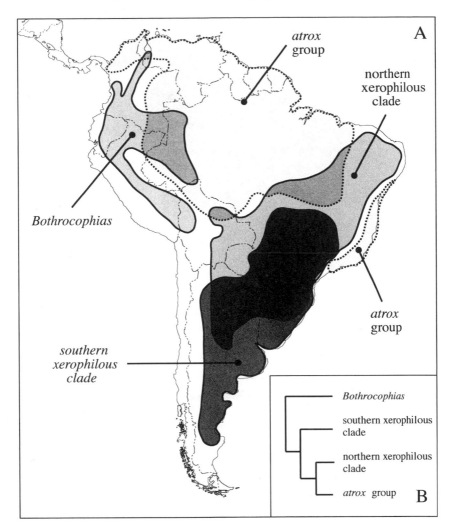

Figure 15.27 (A) Generalized distribution of major clades of South American terrestrial bothropoids. (B) Relationships among the major clades based on fig. 15.2B–C.

dio and Greene (1997), in an analysis of mtDNA variation among these subspecies, concluded that sufficient genetic differences exist between *L. m. stenophrys, L. m. melanocephala,* and the Brazilian forms (*L. m. muta* and *L. m. rhombeata*) to warrant elevation of the former two taxa to full species rank. They also found that bushmasters from Mato Grosso were more closely related to the Atlantic forest form (*L. m. rhombeata*) than to their congeners to the northwest, in the Amazon basin. This result would render *L. m. muta* para-

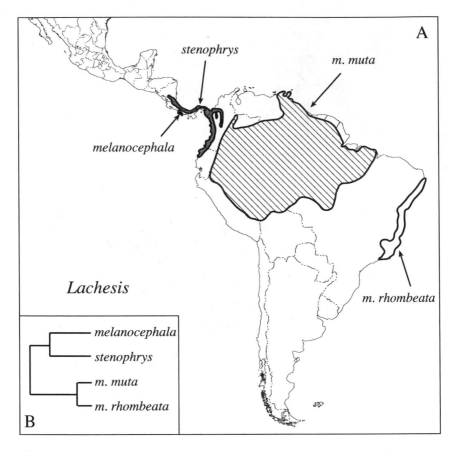

Figure 15.28   (A) Approximate distribution of *Lachesis* (sensu stricto) in Middle America and South America (adapted from Campbell and Lamar 1989; Zamudio and Greene 1997). (B) Relationships among the bushmasters (based on Zamudio and Greene 1997).

phyletic with respect to *L. m. rhombeata,* if these two forms were elevated to specific status. The relationships among the bushmasters, based on Zamudio and Greene (1997), are depicted in figure 15.28B.

Zamudio and Greene (1997) also provided a phylogeographic scenario, illustrated in figure 15.29, for *Lachesis,* based on hypothesized divergence times (estimated from rates of mtDNA evolution), in the context of the geologic and physiographic history of the Neotropics. They suggested that the first vicariant event occurred between the "Central American" bushmasters (*L. melanocephala, L. stenophrys*) and the *cis*-Andean "South American" forms (*L. m. muta, L. m. rhombeata*) in the middle Miocene, concordant with the uplift of the northern Andes (fig. 15.29A). This implies that the ancestral form of the Central American bushmasters was distributed in northwestern South America (trans-Andean) and in Lower Central America (not yet connected to the austral continent). Gene flow must have occurred across a marine portal via fragmented subaerial landmasses (fig. 15.29B). They further suggest a late Miocene to early Pliocene separation between *L. stenophrys* and *L. melanocephala* in Lower Central America, precipitated by the uplift of the Talamancan range. There is still some disagreement about whether Lower Central American (Chorotega Block) in the middle to late Miocene existed as an island arc (e.g., Coates and Obando 1996) or as a single block of subaerial land sutured to the Chortis Block (Donnelly 1989; Escalante 1990). Regardless of which geologic hypothesis is correct, gene flow would have had to occur across a marine portal, between *L. stenophrys* (now isolated from *L. melanocephala*) in Lower Central America and conspecifics in northwestern South America, up to the early to middle Pliocene. The suggestion of Zamudio and Greene (1997) that the differentiation of *L. m. muta* and *L. m. rhombeata* (fig. 15.29D–E) represents an event of recent divergence (< 0.8 mya) driven by Pleistocene glacial cycles is supported by similar events hypothesized for the *Bothrops atrox* group and *Bothriopsis*.

An alternative hypothesis (fig. 15.30) that excludes the assumption of gene flow across the marine portal, isolating Lower Central America from South America, can be constructed. This hypothesis for the evolution of bushmasters is consistent with the phylogeny and estimated divergences of these taxa as suggested by Zamudio and Greene (1997). It requires a *Lachesis* ancestor in South America but absent in Middle America throughout most of the Cenozoic. In this model, the ancestor of modern *Lachesis* was present in South America at least by the late Paleogene. One inherent assumption is the isolation of the ancestor to the Central American bushmasters in the extreme northwestern region of South America (i.e., the northwestern "microcontinent") by the development of a marine embayment to the east of the emergent northern Andes (Hoorn 1993; Hoorn et al. 1995; Lundberg et al. 1998) during the middle Miocene. This latter period is consistent with the time frame estimated by Zamudio and Greene (1997) for the separation of the two major clades of *Lachesis*. Pliocene orogenic events with the continued elevation of the northern Andes (i.e., Cordillera Oriental) and the formation of a xeric climatic barrier to the north in Colombia and Venezuela would have isolated the ancestor of *L. melanocephala* and *L. stenophrys* on the Pacific versant of northern South America, from the Amazonian population(s) (fig. 15.30B–C). These barriers would have prevented contact between ancestral clades of *Lachesis* (Central American versus South American), despite the retreat and contraction of the marine embayment to the north. The later establishment of a continuous, subaerial isthmus between Lower Central America and northwestern South America would have allowed the ancestor of the Central America bushmasters to disperse

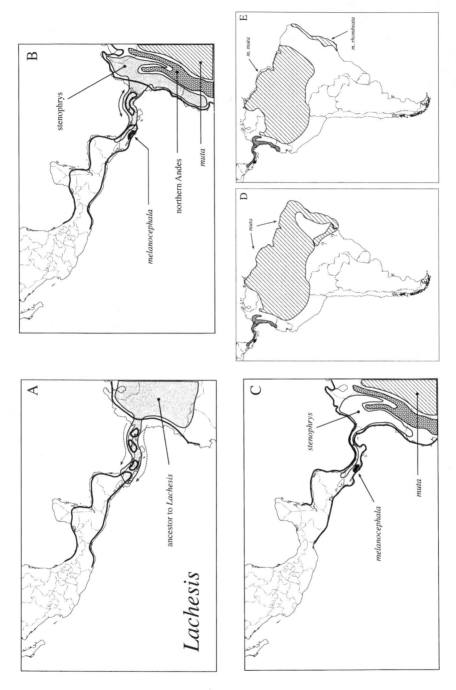

Figure 15.29 (A–E) Author's interpretation of the biogeographic hypothesis for *Lachesis* (sensu stricto; presented in Zamudio and Greene 1997). Dashed arrows (A) indicate probable dispersal events. The solid arrow in (B) indicates possible gene flow. Stippling in (B) and (C) indicates emergence of the northern Andes.

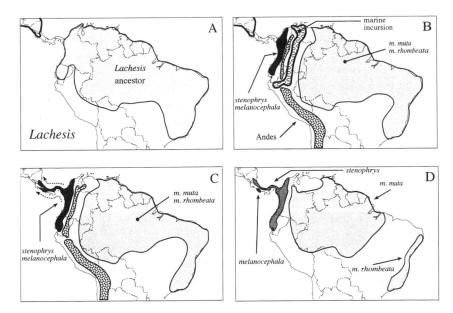

Figure 15.30 (A–D) Alternative vicariant explanation for the biogeographic history of *Lachesis* (sensu stricto), based on the phylogram in fig. 15.28B. The dark solid line around Nuclear Central America indicates subaerial emergent land. Dashed arrows in (C) indicate dispersal north into Lower Central America.

northward into Panama, Costa Rica, and southern Nicaragua, through mesic lowland forest on either side of the Talamancas. The climatic changes that subsequently developed on the Pacific coast of Lower Central America probably isolated the ancestor of *L. melanocephala* in the Golfo Dulce region (xeric climatic barrier to the northwest and southeast; montane barrier to the northeast) and prevented northward range expansion into northwestern Costa Rica (Guanacaste). This would have occurred in the late Pliocene to early Pleistocene. This time frame is consistent with the upper margin (4 mya) of the estimated divergence time between *melanocephala* and *stenophrys* (Zamudio and Greene 1997). Lastly, the South American (Amazonian and Atlantic forest) bushmasters (*L. m. muta, L. m. rhombeata*) diverged most recently during climatic fluctuations in the Pleistocene (fig. 15.30D). Evidence suggests that these two forest regions were at least continuous during mesic episodes, through a northern corridor (Bigarella and de Andrade-Lima 1982), a southeastern corridor through the Mato Grosso (Duellman 1982), or both. The recent separation of the Amazonian and Atlantic coast bushmasters is supported by similar events postulated for the *Bothrops atrox* group and *Bothriopsis bilineata*.

It has been suggested that *Lachesis* is a Middle American group that eventually became restricted to Lower Central American and dispersed southward into South America (Gutberlet and Campbell 2001, 12; Gutberlet and Harvey 2002). Such an interpretation would require *Lachesis* to be a relatively old taxon (supported in part by evidence of a deep phylogenetic divergence among Neotropical pitvipers) that underwent rapid expansion southward into Amazonia at the closure of the Panamanian portal. It is possible that the ancestor of *L. m. muta* and *L. m. rhombeata* dispersed around the northern Andes during mesic episodes and entered the Amazon basin to spread across to the Atlantic forest. However, the lack of fossil or otherwise comparative evidence from other groups of similar distribution is likely to preclude the resolution of this question.

Comments on the Evolution of Component Lineages

Currently, there is support for the monophyly of most bothropoid pitviper genera (Gutberlet and Harvey 2002) and moderate support for the monophyly of all New World pitvipers (Parkinson 1999; Parkinson et al. 2002). The exception to this is *Bothrops* (sensu stricto). The position of *Bothriopsis* as an internal lineage within *Bothrops* (Werman 1992; Kraus et al. 1996; Parkinson 1999; Salomão et al. 1999; Parkinson et al. 2002) will likely result in either the reassignment of *Bothriopsis* to *Bothrops* or the division of the latter genus into two or more genera. It is not clear currently if the xerophilous clades within *Bothrops* (sensu stricto) represent lineages attributable to new generic allocations.

Gutberlet and Campbell (2001) postulated that the Middle American genera—*Atropoides, Bothriechis, Cerrophidion,* and *Porthidium*—may constitute a monophyletic group. To date the phylogenetic evidence suggests that these Middle American lineages are paraphyletic to a monophyletic South American radiation composed of *Bothriopsis, Bothrops,* and *Bothrocophias* (Martins et al. 2001 and references therein; Gutberlet and Harvey 2002; Parkinson et al. 2002). The Middle American lineages diverged during the late Paleogene and Neogene. The proposition here is that these genera may have evolved from ancestral stocks that underwent elevational (gradient) vicariance. An elevational separation of habitat could have allowed for a *Porthidium* (lowland) ancestor to differentiate from an upland ancestor of *Atropoides* and *Cerrophidion*. Continued elevational separation could have then separated *Cerrophidion* (high elevation) from *Atropoides* (generally moderate elevational distribution). The subsequent speciation events within genera likely resulted from habitat vicariance.

The evolution of *Ophryacus* and *Bothriechis* probably involved the early divergence of these arboreal descendents (lost in *O. melanurus*) from a terrestrial ancestor (i.e., *Atropoides-Cerrophidion-Porthidium* ancestor) to one of arboreality. Subsequent to this, *Ophryacus* diverged from *Bothriechis* by a vicariant event at the Isthmus of Tehuantepec (Crother et al. 1992). Within *Bothriechis* to the south, elevational vicariance led to the separation of *B. schlegelii* from the other highland members of this genus.

Outside of the Middle American clades, the phylogenetic history of the pitviper genera becomes particularly uncertain. The relationships of the Middle American genera to *Bothrocophias, Lachesis, Bothrops-Bothriopsis, Crotalus,* and New World *Agkistrodon* are not resolved. Phylogenetic analyses of phenotypic data (Werman 1992; Gutberlet 1998) and mtDNA data (cf. Kraus et al. 1996, Parkinson 1999, and Parkinson et al. 2002) for these genera are incongruent (Werman 1999). This incongruence persists even among recent investigations (Martins et al. 2001 and references therein; Gutberlet and Harvey 2002; Parkinson et al. 2002). Thus, the ancient diversification of the bothropoid pitviper (and New World) genera is open to future investigation. One might speculate that *Lachesis* is an old sister taxon to the Middle American clade that separated in the early Paleogene with the loss of the early intercontinental connection (Savage 1982). *Bothrops* (including *Bothriopsis*) may form a second clade of early divergence. The ancestor of *Bothrops* could have entered South America through the GAARlandia land span at the transition of the Eocene and Oligocene (Iturralde-Vinent and MacPhee 1999) by a short over-water dispersal event from the area of present-day Florida. The subsequent fragmentation of this land span would have isolated the ancestor of *Bothrops-Bothriopsis* in South America from *Lachesis*. Climatic changes in North America would have led to the early divergence of *Crotalus-Sistrurus* from New World *Agkistrodon*. In any event, the divergences among these genera were likely ancient, and therefore the relationships among these groups will be difficult to resolve, given the limitations of the present application of gene sequence data and phenotypic characteristics to address these phylogenetic questions.

Conclusions

The diversity among Middle American pitvipers and the degree of differentiation seen among the many genera of this region suggest that these taxa evolved in relative isolation throughout much of the Cenozoic. This is largely consistent with the hypothesis of Savage (1966, 1982) and the proposition of Gutberlet and Campbell (2001) that the Middle American pitvipers constitute a distinct biotic element in the New World. The pitvipers in this region probably underwent major speciation events from the Miocene forward, when the geotectonic events were extremely kinetic. The differentiation of the major pitviper clades in this region is closely associated with climatic and orogenic changes in the region. The present patterns of distribution and relationship of *Atropoides, Bothriechis, Cerrophidion,* and *Porthidium* were affected by elevational (gradient) vicariance as well as platygeographic vicariance. *Agkistrodon, Crotalus,* and

Lachesis appear to be minor components in the picture of Nuclear Central America.

With the establishment of the isthmian link between present-day Panama and Colombia, limited dispersal took place among pitvipers. *Bothriechis schlegelii, Porthidium nasutum,* and *P. lansbergi* penetrated only a short distance into northwestern South America. Other pitvipers, such as *Crotalus durissus,* apparently underwent an extensive distributional expansion into South America during xeric episodes after the Pliocene. From South America, *Bothrops asper* invaded Middle America, and its range now extends as far north as Tamaulipas in southeastern Mexico. *Bothriopsis punctata* is found primarily in the South American Choco, although this species apparently reaches into extreme southeastern Panama (Campbell and Lamar 1989).

The evolution of pitvipers in South America is closely tied to the development of the Andes and the climatic changes that occurred in this continent from the Neogene forward. Early differentiation likely occurred by the separation of west coast, Andean, and eastern *cis*-Andean ancestral populations. The proposed isolation of northwestern South America as a microcontinent in the Miocene may have played an early role in the establishment of diversity in *Lachesis* and possibly the vicariance of *Bothrocophias* from *Bothrops* and *Bothriopsis.* The present diversity seen in *Bothrops* seems to have been generated by habitat fragmentation to a minor extent in the Amazon basin, but to a larger extent in the relatively xeric regions of the extreme southeastern coastal and inland areas of the continent. The oscillating sectioning of xeric and mesic habitats during the early drying and cooling periods of South America as well as during glacial cycling in more recent times has had the effect of isolation, in a repeated manner, for the ancestral distributions of the xerophilous pitvipers in this region. This mechanism was probably responsible for the high species richness in the xeric southeastern belt as well as the close relationships among Amazonian forest and Atlantic forest inhabitants. In addition, formation of forest refugia in the Pleistocene may have affected diversification of the several closely related species (e.g., the *B. atrox* group).

It is difficult, at present, to ascertain the biogeographic history deep within the New World pitviper radiation. The relationships among the genera of early divergence are not known to any acceptable degree of certainty. It is possible that *Lachesis* is sister to the Middle American clades and that both represent ancient divergences in the New World between Lower Central America and South America. Gutberlet and Campbell (2001) may be correct in their supposition that three basic clades differentiated early in the evolution of New World pitvipers. These include an *Agkistrodon-Crotalus-Sistrurus* clade, a Middle American bothropoid clade, and a South American clade composed of *Bothrocophias, Bothriopsis,* and *Bothrops.* Under this hypothesis, the affinities of *Lachesis* remain an enigma. Because these ideas are not void of speculation, the continued study of the bothropoids and other Neotropical pitvipers will eventually lead to a more refined idea of the biogeographic history for this diverse and widely distributed assemblage of snakes.

Acknowledgments

I am grateful to Brian Crother, Maureen Donnelly, Craig Guyer, Marvalee Wake, and Mary White for providing the opportunity to contribute to this volume. I am especially appreciative of Marvalee Wake and her patience in dealing with the reviews and revisions of this chapter. I thank three anonymous reviewers for important comments and criticisms of the manuscript and especially thank Jonathan Campbell for constructive comments on an earlier, rougher version and for generously providing electronic files of base maps for the Neotropics. I am indebted to Ulrich Kuch for allowing me to incorporate some of his information on pitviper systematics in this chapter. I thank Ronald Gutberlet and Christopher Parkinson for graciously providing drafts of manuscripts on pitviper phylogeny. I thank Jane Heitman, of the Mesa State College Library for her extraordinary effort in obtaining certain literary materials. I also acknowledge the Lathrop Research Fund of the Mesa State College Foundation for financial support for this project.

16

The Herpetofauna of the Rincón Area, Península de Osa, Costa Rica, a Central American Lowland Evergreen Forest Site

Roy W. McDiarmid and Jay M. Savage

Ten to twelve thousand years ago, when humans first came from the north to enter the vast tropical evergreen lowland forests of Mexico and Central America, these forests stretched for what must have seemed an eternity, from the Veracruz lowlands 2,600 km southeastward into northern South America and beyond. Even eight to ten millennia later, when the Spanish conquistadores arrived on the shores of the American continent, these forests, although heavily disturbed in some areas by native Amerindian cultures, presented an awe-inspiring continuum, a great and dark sea in the words of Fernandez de Oviedo y Valdez (1526), over most of the lowlands of Middle America. Until fairly recently, the humid lowland forest areas of the region have been relatively safe from the destructive agricultural and forestry practices that in postconquest times have denuded, "developed," or otherwise laid waste much of the drier and upland habitats of tropical America. Now even these long protected remnants of the most diverse and complicated of ecosystems are threatened with total annihilation, first, in the 1970s and 1980s, by the worldwide drive of the Food and Agriculture Organization of the United Nations to convert these natural wonders into cattle land, and now to ever-increasing human population pressures for land and forest products. At the present rate of cutting and development, within the first two decades of this century nearly all of the Central American humid lowland evergreen forest habitats, the products of millions of years of evolution, seem certain of destruction.

Among the most interesting and diverse of the still-extant tracts of lowland evergreen forest in Central America is the relatively undisturbed region of the Península de Osa on the Pacific side of Costa Rica. The Osa forests are part of the geographic unit so beautifully described by Paul H. Allen (1956) as the rain-forests of the Golfo Dulce. They are unique in being one of two extensive areas (the other is in Chiapas, Mexico, and adjacent Guatemala) on the Pacific versant of Central America with sufficient annual rainfall to support a broad-leaf evergreen forest similar to that found along much of the Atlantic lowlands from Veracruz to Colombia (fig. 16.1). A few favorable situations in Pacific coastal Panama also support forests of this general type, the largest isolated on the Azuero Peninsula. Specifically, the rainforests of the Golfo Dulce show many similarities and relationships to the Chocó forest of Pacific lowland Colombia, and in this regard, they are very different from other Central American forests.

The Golfo Dulce rainforest formerly occupied a continuous region from approximately Bahía Herradura (fig. 16.2) south below an elevation of 500 m along the Pacific coast of Costa Rica to extreme southwestern Panama, somewhat west of Concepción. Although selected areas of flat terrain within the region were cleared for banana culture by the United Fruit Company beginning in 1938, until thirty years ago extensive and continuous stands of virtually undisturbed forest were still common over most of the area. The flood of cattle-raising developments in the 1980s, followed by renewed clearing and population increases in Costa Rica from 1.2 million people in 1960 to 4.0 million in 2001, has changed all this, and the forest is now practically eliminated except on the fairly inaccessible Península de Osa. This chapter provides an introduction to the herpetofauna of the area, especially around Rincón de Osa, a review of current knowledge, and a comparison of this site with other well-sampled lowland forest localities in Lower Central America.

A substantial area of the Osa Peninsula (from the main dividing ridge westward to the sea) was set aside as the Parque Nacional Corcovado (fig. 16.2) by executive degree of President Daniel Oduber Q. of Costa Rica on 31 October 1975. A number of international agencies, including the Nature Conservancy, the World Wildlife Fund, and the Organization for Tropical Studies, aided Costa Rica in establishing this park (Wright 1976). The courageous and enlightened actions of President Oduber in establishing the park cannot be overestimated. To date, the park has been protected and continued under the administration of Presidents Rodrigo Carazo O., Luis Alberto Monge A., Oscar Arias S., and their successors.

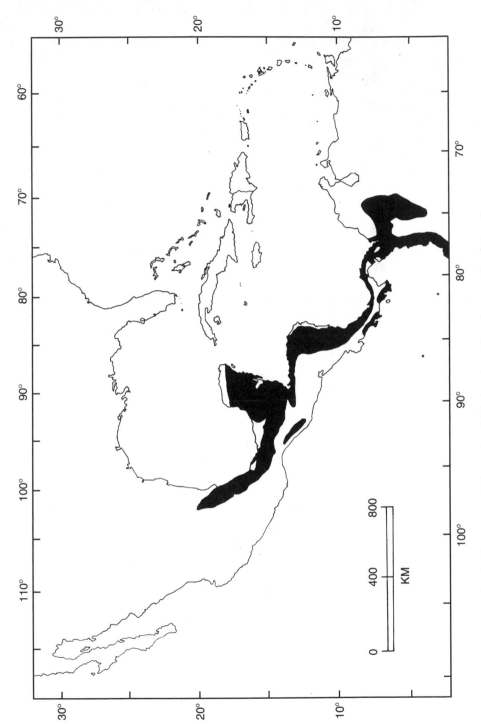

Figure 16.1   Distribution of lowland evergreen rainforests in Middle America and northwestern South America.

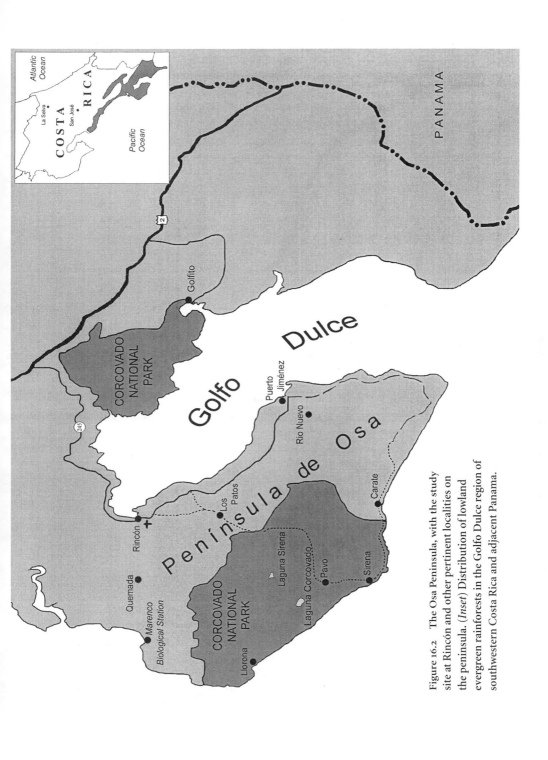

Figure 16.2   The Osa Peninsula, with the study site at Rincón and other pertinent localities on the peninsula. (*Inset*) Distribution of lowland evergreen rainforests in the Golfo Dulce region of southwestern Costa Rica and adjacent Panama.

Sources of Material

MATERIAL BASIS OF THE REPORT

Before 1962, the herpetofauna of the Península de Osa was known from a few specimens taken near Puerto Jiménez (Wettstein 1934) and Playa Blanca. In 1962, Savage and the late Charles F. Walker visited the Rincón area for four days and assembled the first large collection of material from the peninsula, and McDiarmid first worked in the Rincón area for nine days in 1966. Since these initial ventures, Savage carried on fieldwork near Rincón for greater or lesser periods in 1964, 1968, and 1973; McDiarmid worked the area in 1967, 1969, 1971, and for a period of more than 60 days in the summer of 1973. Other material was accumulated through the activities of our associates during various time periods (especially by Norman J. Scott Jr.) and forms a further base for the present study. Samples and dates of collection for the Rincón de Osa material housed in the Costa Rica Expeditions Collection are summarized in appendix 16.1. We estimate that about 5,500 person-hours of collecting were recorded from 1961 to 1973. Even so, sampling was quite irregular by month, with about 65% of it occurring in July and August, and no samples taken during five months (January, April, September, October, and December). The bulk of these materials has been deposited in the collections of the Natural History Museum of Los Angeles County. Some material collected by McDiarmid in 1973 is deposited at the National Museum of Natural History. Since 1973, some specimens have been taken at Rincón by the Organization for Tropical Studies courses and by herpetologists from the Universidad de Costa Rica (UCR), primarily the late Douglas C. Robinson, Federico Bolaños, Federico Muñoz, and students from UCR. This material is deposited in the collections of the Museo de Zoología (UCR). All species recorded from Rincón and vicinity are listed in appendix 16.2.

OTHER SAMPLES FROM THE PENINSULA

The earliest published report on herpetological materials from the Península de Osa was by Wettstein (1934). His samples were taken mostly by Rudolf Zimara and Dr. Otto Koller and forwarded to the Wien Museum. The specimens in this collection are from Puerto Jiménez (Porto Jiménez), or from about 3 km west, near the Río Nuevo, a tributary of the Río Tigre that flows into the western margin of the Golfo Dulce a little north of Puerto Jiménez (fig. 16.2). Among the species from Puerto Jiménez were five lizards and two snakes; a frog and a turtle were taken at Río Nuevo. All are widespread species and, with the exception of the Black Spiny-tailed Iguana, *Ctenosaura similis,* are known from the Rincón area as well (appendix 16.2). The *Ctenosaura* record is confirmed by recently collected specimens at the Universidad de Costa Rica from that locality and from Piedra El Arcón the outer coast of the peninsula. All these localities in southwestern Costa Rica are along the coast, and it seems likely that south of the main portion of its range in dry forest, spiny-tailed iguanas are restricted to open, sandy, or rocky habitats. The scattered southern Costa Rican records and their essentially coastal distribution may reflect transport by humans, inasmuch as the species was an Amerindian food item.

The only other herpetological specimen taken from the peninsula, before 1962, was one frog, *Dendrobates auratus,* formerly in the Museo Nacional de Costa Rica, as noted by E. R. Dunn (unpub. data).

More recently and since the establishment of the Parque Nacional de Corcovado (Boza 1978; Boza and Sevo 1998), amphibians and reptiles have been collected principally in the area around the original park headquarters at Sirena and between Sirena and Llorona, about 16 km northwest of Sirena. These localities are 26 km south-southwest and 28 km southwest of Rincón, respectively. Some material has been collected at other localities within the park as follows (the approximate distances from Sirena and Rincón are indicated in parentheses): Laguna Corcovado (7 km north-northwest of Sirena, 23 km southwest of Rincón), Laguna Sirena (10 km north-northeast of Sirena, 19 km southwest of Rincón), Pavo Forest (4 km northwest of Sirena, 24 km southwest of Rincón), and Los Patos (19 km northeast of Sirena, 10 km south of Rincón). We were unable to track down all these scattered materials or verify all reports and photographs. Many species reported from the park were collected by students and faculty from the University of Texas, especially Gad Perry and Karen Warkentin, and vouchers that exist are in the collections at the Museo de Zoología, Universidad de Costa Rica, and the Texas Memorial Museum, University of Texas. We have included definite records for species from the park in appendix 16.2.

The snake *Conophis lineatus* is included tentatively on this list on the basis of a sight record by a competent naturalist (Janzen, in Scott 1983) in second-growth vegetation at Corcovado National Park. Nevertheless, we are skeptical of this report and recommend that efforts be made to verify its presence by a specimen or photograph.

Among specimens at the Museo de Zoología are representatives of five species (three native and two introduced) from the peninsula that are not represented in other samples from Rincón and Corcovado National Park (appendix 16.2). The hylid frog, *Hyla microcephala,* and the colubrid snake, *Enulius sclateri,* are from the Marenco Biological Station just north of the Corcovado National Park boundary, approximately 25 km north-northwest of Sirena and 20 km west of Rincón. A specimen of *Leptodeira rubricata,* a snake found only in mangroves along the southwestern Pacific coast of Costa Rica and adjacent western Panama, is from near Puerto Jiménez. Two nonnative geckos have recently been collected from the peninsula: *Hemidactylus frenatus* from human habitations in Puerto Jiménez and *Lepidodactylus lugubris* from Quemada, north of Corcovado National Park. We note their occurrence for the sake of completeness but do not include them in our analysis of the native herpetofauna.

Finally, Brian I. Crother and Lisa Aucoin of Southeastern Louisiana University spent 10 days in May 2001 at the Marenco Biological Station with a tropical field ecology course, observing and photographing the biota. Species of amphibians and reptiles that we can identify positively from photographs and notes provided by Crother also have been included in appendix 16.2.

Since 1961, a number of papers mentioning materials from the vicinity of Rincón have appeared. These fall into four categories: (*a*) monographic studies of particular systematic groups that have included reference to examples from the Costa Rica Expeditions Collections and from the Museo de Zoología at the Universidad de Costa Rica; (*b*) accounts of common or interesting species presented in *Costa Rican Natural History,* edited by Janzen (1983); (*c*) ecological or behavioral notes, usually based on incidental studies by students in the field courses of the Organization for Tropical Studies (OTS); and (*d*) ecological research sponsored or supported by OTS. Germane papers among these are cited at appropriate places in this chapter. The report on the herpetofauna of the leaf litter (Scott 1976) and species accounts in Janzen's (1983) book are especially notable.

The Study Area

GEOGRAPHY

The Península de Osa projects into the Pacific from the southwest Costa Rican coast (fig. 16.2) and forms the seaward margin for the Golfo Dulce. The peninsula itself is a low-lying series of ridges (highest elevation, 745 m) that is separated from the mainland to the north by a complicated and extensive swampy zone associated with the Río Sierpe. The axis of the peninsula is about 55 km in length and runs from west to southeast. Except for several areas that support small settlements along the Golfo Dulce shore, much of the 1,200 km^2 of the peninsula is covered by undisturbed evergreen forest.

COLLECTION LOCALITIES

Materials forming the basis of this report are from the area adjacent to Rincón de Osa, a small settlement near the head of the Golfo Dulce (figs. 16.2, 16.3). In

1961 this place was the center for a lumber company, Osa Productos Forestales, that undertook to build roads, survey the forests, and establish housing to accommodate visiting scientists and students, especially those participating in courses presented by the Organization for Tropical Studies. Because plans for utilization of the forests by the lumber company never materialized, most of the areas accessible from Rincón remained relatively undisturbed until the past decade. Our sampling was concentrated along the principal all-weather road, the Carretera al Pacífico, especially within a 5-km radius to the west and south of Rincón (fig. 16.3).

The primary localities and/or particularly significant sites are listed in appendix 16.3. The names of collecting sites were adopted for special areas during our fieldwork. Some of these names are reported in the literature (e.g., Findley and Wilson 1974) and are associated with voucher specimens, so we use them here. The primary localities are indicated on the accompanying map (fig. 16.3) and by reference to standard localities in the list in the appendix. All distances mentioned are straight-line measurements, usually with reference to Rincón.

CLIMATE

The climate of the Península de Osa is typical for lowland hot and moist tropical regions with seasonal effects produced by changes in prevailing winds. The climate, according to the Köppen system, is Afi, meaning that the mean temperature for the coldest month is at least 17.8°C, each month has at least 61 mm of precipitation, and the range of temperatures is less than 12.8°C. Although accurate temperature data are lacking, Holdridge et al. (1971) estimated the mean annual temperature as between 26.4°C and 27.8°C, with a mean of approximately 27.5°C for the area near Rincón. The mean annual precipitation (fig. 16.4) at Rincón totals 3,909 mm and at the airfield 4,576 mm, based on ten years and eight years of records, respectively. A definite dry season occurs in January, February, and March, when monthly precipitation averages well below 200 mm, and the number of rainy days averages 8, 5, and 7 per month, respectively. Rainfall begins to increase in late March or early April and rises dramatically to June; thereafter, average monthly precipitation levels off at 400–500 mm through September. Rain falls an average of 21–22 days per month in May, June, July, and August, with a two- to three-week period of little rainfall usually occurring in late July and/or early August. This very short dry season or *veranillo* often is not noticeable in the monthly totals. Very heavy rains commence in late August and peak in October; these months average 26 rainy days, and October typically has more than 700 mm of precipitation. Rainfall decreases sharply through November and December as the next dry season approaches. Holdridge et al. (1971) noted that the Osa Peninsula was part of the Tropical (Lowland) Wet Forest life zone (Holdridge 1967) with a mean annual biotemperature

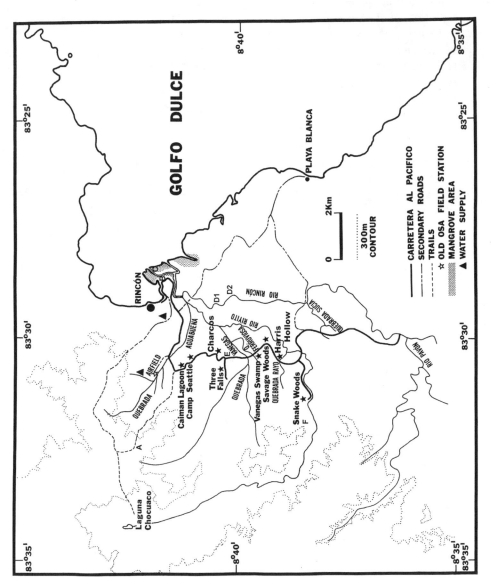

Figure 16.3   The Rincón area, with the major collecting localities for this study.

greater than 24°C and mean annual precipitation between 4,000 and 8,000 mm. Strictly speaking, the area at Rincón (fig. 16.4), with less than 4,000 mm annual precipitation, is in Holdridge's Tropical (Lowland) Moist Forest life zone. This site is in the lee of a steep ridge that reduces the impact of rainstorms coming from the west. Nevertheless, rainfall at Rincón has reached 4,779 mm in a single year (1971).

VEGETATION AND SOILS

Holdridge et al. (1971) described in detail the vegetation and soils at seven sites in the vicinity of Rincón. The following summary is generalized from that report, Sawyer and Lindsey (1971), and our own field observations. This forest may be classified as Tropical (Lowland) Wet Forest. A typical Osa site has a tall (45–55 m), multistrata, evergreen forest with a few canopy species that may be briefly deciduous when flowering in the dry season. The number of tree species is high; at one site on the Holdridge Trail, 103 species are present, whereas Sawyer and Lindsey (1971) gave a value of 503 individual trees per hectare at a site in the general area with 195 tree species. The common canopy forms include several species of *Brosimum* and *Protium, Anacardium excelsum,* and *Terminalia lucida.* The canopy species usually have clear, smooth, light-barked boles up to 30 m tall, with high buttresses. A lower canopy occurs at 30 – 45 m and fills in the spaces between the higher canopy trees. An understory with numerous stilt-rooted palms forms a tall, dense layer at 10 – 20 m. These lower layers commonly include species of *Virola* and *Cryosophila guagara.* A shrub layer consists mostly of dwarf palms reaching a height of about 2 m. The ground layer is minimal; usually only a shallow layer of leaf litter with a few ferns and scattered seedlings is present. Epiphytes, including orchids, bromeliads, and large-leafed herbaceous climbing aroids and cyclanths are common but not conspicuous in the green matrix of the dense forest canopy. Large vines and bush ropes are relatively uncommon, and epiphytic shrubs and strangling trees are rare.

Soils on well-drained sites such as the Holdridge Trail are strongly acidic, red to reddish brown latosols of clay texture. In the flats along the banks of the Río Rincón, the soils are relatively shallow, with leaf litter 2–5 cm deep, an A zone 10–30 cm in vertical extent, and a B zone 40–200 cm but usually 40–70 cm. The soils are deficient in most plant nutrients, although well-drained ridges and uplands tend to have moderate to low amounts of nitrogen, calcium, and manganese.

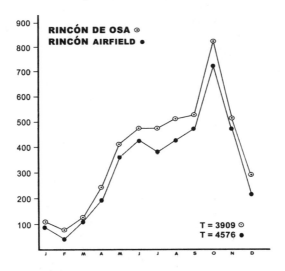

RINCÓN DE OSA ⊙
RINCÓN AIRFIELD ●

T = 3909 ⊙
T = 4576 ●

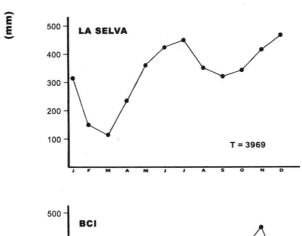

(mm)

LA SELVA

T = 3969

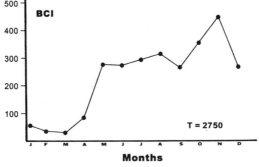

BCI

T = 2750

Months

The Herpetofauna

The following annotated list includes all species of amphibians and reptiles known to occur in the area near Rincón on the Península de Osa (appendix 16.2). The list serves two functions. First, it provides the basic data for discussion of the herpetofauna known from the vicinity of Rincón, and for ecological and biogeographic comparisons with the herpetofaunas of other sites in Lower Central America (e.g., Sasa and Solórzano 1995). Second, it serves as an introduction to the natural history of these animals for researchers who, in visiting this tropical wet lowland forest for the first or a brief time, wish to maximize their efforts, and to students in OTS courses or other field classes who have little familiarity with amphibians and reptiles, or at best only a minimal knowledge of the species. For these reasons, pertinent ecological data based on our observations of the species at Rincón are delineated. Most of our observations and comments apply to the same species in field situations within Corcovado National Park, where current educational and research activities are concentrated.

Comments on size refer to adult lengths expressed in millimeters. Standard length (snout to vent) is used for salamanders, frogs and toads, and lizards, carapace length for turtles, and total length for caecilians, snakes, and crocodilians. Approximate maximum sizes for the caecilian, salamanders, turtles, and crocodilians are included in the text. Size limits for frogs and toads, lizards, and snakes are in table 16.1.

Relative abundances are approximated by the following values: species are considered *rare* when fewer than 5 adult examples have been collected in the study area; *moderately common* species are represented by 5–10 adult specimens; *common* species include those represented by 11–25 adult individuals; and *abundant* species are those represented in our samples by more than 25 adult specimens. These abundance categories do not apply especially well to snakes, which as a group are much less abundant than most other components (e.g., frogs and lizards) of the herpetofauna and perhaps deserve their own relative abundance categories. However, we decided to use the same categories for all groups because that approach more accurately reflects the probability of encountering one species relative to another at a site. A slightly different approach

Figure 16.4 Seasonal distribution of rainfall at three sites in Lower Central America. Values from Rincón are based on 10 years of data, and those from the Rincón Airfield on 8 years of recordings. Total mean annual precipitation (T) for a comparable 10-year period from La Selva Biological Station was higher than the average (3,850 mm) as calculated from long-term records. In contrast, the total from Barro Colorado Island (BCI) was slightly lower.

Table 16.1   Sizes of anurans and squamates from the Osa
Peninsula (in mm)

Size	Taxon		
	Frogs and toads	Lizards	Snakes
Small	<30	<32	<650
Moderate	30–50	32–100	650–1,500
Large	51–80	101–150	1,501–3,500
Very large	>80	>150	>3,500

was taken for species abundance at a dry forest site in northwestern Costa Rica
(see Sasa and Solórzano 1995).

Comments on activity patterns, habitat utilization, food habits, and other
ecological and behavioral traits of species are based primarily on our experience
with specimens from the Rincón area or, in a few instances with rarer species,
on our collective experience with the species in Costa Rica. Most descriptive
terminology we use is generally understood by field biologists. Our definitions
for select terms are as follows: *fossorial* species spend substantial time in soil, often in burrows; *semifossorial* species spend substantial time in and under debris
(e.g., logs, rocks, leaf litter, etc.) on the forest floor; *epigeal* species spend substantial time near or on the surface of the ground; *epiphyllous* forms are most
often found on leaves or stems of the herb or shrub layer; *arboreal* species are
most often encountered on the trunks or limbs of trees from the understory to
the canopy; *insectivorous* species eat insects and other arthropods; *myrmecophagous* species feed primarily on ants.

The broad geographical, elevational, and ecological distribution of each
species throughout its range also is indicated to provide a basis for evaluating
the relationships of the Rincón de Osa herpetofauna to those at other Neotropical lowland evergreen forest sites. Elevational ranges are approximated by the
following terms: *lowland,* sea level to 500 m; *premontane,* 501–1,500 m; *lower
montane,* 1,501–2,500 m; *montane,* 2,501–3,500 m; and *subalpine* (sub-Andean),
2,500–3,000 m. Vegetation mentioned in the range statements follows the
modified Holdridge (1967) system as outlined by Savage (1975).

AMPHIBIA (46)

Gymnophiona (1)

Caeciliidae
Dermophis occidentalis Taylor—Small, to 235 mm; nocturnal; feeds primarily
on earthworms and insect larvae; fossorial, moderately common in wet season

in deep leaf litter, especially at bases of large buttressed trees in forest; known from lowland and premontane evergreen forest on Pacific versant of southern Costa Rica. This caecilian was called *Dermophis parviceps* in earlier lists (e.g., Scott et al. 1983; Savage and Villa 1986; Villa et al. 1988). Savage and Wake (2001) resurrected *Dermophis occidentalis* for the smaller species with high numbers of primary and secondary annuli from Pacific Costa Rica.

Caudata (4)

Plethodontidae

Bolitoglossa colonnea (Dunn)—Moderate size, to 55 mm; nocturnal; insectivorous; epiphyllous on plants along streams; one specimen taken during dry season along Río Riyito; lowland and premontane evergreen forests on both slopes in Costa Rica and western Panama.

Bolitoglossa lignicolor (W. Peters)—Moderate size, to 80 mm; nocturnal; insectivorous; epiphyllous on broad-leaved plants; common in forest, occasional in old second-growth vegetation; lowland evergreen forests of southwestern Costa Rica and adjacent Panama.

Oedipina alleni Taylor—Elongate, moderate size, to 60 mm; nocturnal; insectivorous; semifossorial; moderately common in shallow leaf litter, beneath logs, and in termite tunnels in logs, in forest and along road cuts; lowland and premontane evergreen forests of southwestern Costa Rica and western Panama. This tropical worm salamander had been called *Oedipina parvipes* by most authors (e.g., Scott et al. 1983; Savage and Villa 1986; Villa et al. 1988), until Good and Wake (1997) resurrected *O. alleni* for the lowland form (also see García-París and Wake 2000).

Oedipina pacificensis Taylor—Elongate, moderate size, to 50 mm; nocturnal; insectivorous; semifossorial; abundant in leaf litter, especially near tree buttresses and beneath logs in forest; evergreen forests of southwest Pacific lowlands of Costa Rica and adjacent southwestern Panama. This tropical worm salamander had been called *Oedipina uniformis* by most authors (e.g., Scott et al. 1983; Savage and Villa 1986; Villa et al. 1988), until Good and Wake (1997) resurrected *O. pacificensis* for Pacific lowlands populations.

Anura (41)

Bufonidae

Bufo coniferus Cope—Moderate size; nocturnal; primarily myrmecophagous; primarily terrestrial but sometimes found above ground in low vegetation in forest; rare; lowland and premontane evergreen forests from Nicaragua to northwestern Ecuador.

Bufo haematiticus Cope—Moderate size; nocturnal; myrmecophagous; abundant in forest and along streams at dawn and dusk; juveniles abundant along streams in dry season; males call from stream edges in forest during heavy rains and at dawn in wet season; occasionally found asleep on leaves up to 1 m above ground; known from lowland and premontane evergreen and mixed forests from Honduras to central Ecuador.

Bufo marinus (Linnaeus)—Very large; nocturnal; eats a wide variety of animal prey from small insects to small amphibians and reptiles, birds, and mammals; abundant along roads and around dwellings between Rincón and old Osa Field Station; breeds in temporary shallow ponds and stream backwaters in wet season, primarily in cleared areas and second-growth vegetation; Wassersug (1971) discussed palatability of tadpoles of this and several other species from Rincón area; widespread from Mexico to South America.

Bufo melanochlorus Cope—Moderate size; nocturnal; insectivorous; abundant along forest edge, especially on Carretera al Pacífico during wet season; breeds in small streams during dry season; two males calling from open pond with several *Bufo marinus* in August; widespread in lowland and premontane evergreen forests of Costa Rica.

Centrolenidae

Centrolene prosoblepon (Boettger)—Small; nocturnal; insectivorous; arboreal and riparian; common along small forest streams that flow into Quebrada Aguabuena; males epiphyllous and call from tops of leaves and branches along stream bank; breeding known to occur in June to August; lowland and premontane evergreen forests from eastern Honduras to northwestern Ecuador.

Cochranella albomaculata (Taylor)—Small; nocturnal; insectivorous; arboreal and riparian; common; known in Rincón area only from Quebrada Rayo east of Carretera al Pacífico; males epiphyllous, call from tops of leaves 1–2 m above water; amplexing pairs taken in May and July; lowland and premontane evergreen forests from north central Honduras to western Colombia.

Cochranella granulosa (Taylor)—Small to moderate; nocturnal; insectivorous; common; arboreal and riparian; found along Quebrada Aguabuena and less frequently in its feeder streams; males epiphyllous, call from tops of leaves up to 4 m above water; breeds from June through September; lowland and premontane evergreen forests of eastern Honduras, Nicaragua, Costa Rica, and Panama.

Cochranella spinosa (Taylor)—Small; nocturnal; insectivorous; arboreal and riparian; rare; known from Rincón area from few specimens from water supply stream behind Rincón and single specimen from Quebrada Aguabuena at Carretera al Pacífico; lowland evergreen forests of Costa Rica and Panama, south to northwestern Ecuador.

Hyalinobatrachium colymbiphyllum (Taylor)—Small; nocturnal; insectivorous; arboreal and riparian; abundant especially along Quebrada Aguabuena and several of its feeder streams, also from Three Falls; males call from underside of leaves; breeding occurs from May through September; McDiarmid (1975, 1978) discussed reproductive ecology and behavior of this species and *H. valerioi;* known from lowland and premontane evergreen forests from Costa Rica south to northern and western Colombia.

Hyalinobatrachium pulveratum (W. Peters)—Small; nocturnal; insectivorous; arboreal and riparian; moderately common; known from Quebrada Aguabuena and Río Riyito; males epiphyllous, call from tops of leaves and small branches up to 3 m above water; breeding known to occur in late July and August; lowland evergreen forests from north central Honduras south through Nicaragua, Costa Rica, and Panama to northern Colombia.

Hyalinobatrachium valerioi (Dunn)—Small; nocturnal; insectivorous; arboreal and riparian; abundant, especially along forest streams flowing into Quebrada Aguabuena, also from Quebrada Rayo; males call from undersides of leaves over water; breeding occurs from May through September; McDiarmid and Adler (1974) and McDiarmid (1978, 1983) discussed male behavior of this form at Rincón; lowland and premontane evergreen forests of Costa Rica to western Colombia.

Dendrobatidae

Colostethus flotator (Dunn)—Small; diurnal; insectivorous; abundant; terrestrial, found in forest leaf litter along streams; males call most commonly in morning; tadpoles with umbelliform oral disc, common in slow waters of small streams in forest; known from lowland and premontane evergreen forests from southeastern and southwestern Costa Rica and Panama to western Colombia. Listed as *Colostethus nubicola* by many recent authors (e.g., Savage and Villa 1986; Villa et al. 1988).

Colostethus talamancae (Cope)—Small; diurnal; insectivorous; terrestrial; abundant in leaf litter on forest floor; males call most commonly in afternoon during rain; male found carrying 29 tadpoles on back; males place tadpoles in small puddles of water in depressions on fallen trees or on forest floor; known from lowland evergreen forests of Costa Rica, Panama, and western Colombia.

Dendrobates auratus (Girard)—Moderate; diurnal; myrmecophagous; abundant in forest at most localities; tadpoles taken in wet and dry seasons from small depressions on trunks of fallen trees and from tree hollows and bromeliads; primarily terrestrial, frequently found around bases of large trees or spiny palms that provide ample cover; occasionally arboreal, individuals seen climbing up to 3 m on tree trunks; one tadpole taken in bromeliad in tree 20–25 m above ground (McDiarmid and Foster 1975); occurs in lowland evergreen forests of eastern Nicaragua, Costa Rica, Panama, and extreme northwestern Colombia.

Dendrobates granuliferus Taylor—Small; diurnal; myrmecophagous; terrestrial; abundant, especially along small forest streams; males and females show site fidelity; males usually call from elevated sites late in dry season; Goodman (1971) and Crump (1972, 1983) discussed territorial and mating behavior at Rincón de Osa; adult transporting tadpole was photographed by Michael Fogden (Jacobson and Fogden 1984), and details of larval morphology were described by Van Wijngaarden and Bolaños (1992); restricted to lowland evergreen forest below 100 m in southwestern Costa Rica and adjacent Panama; perhaps also native to southeastern Costa Rica (Myers et al. 1995).

Phyllobates vittatus (Cope)—Small; diurnal; myrmecophagous; abundant; terrestrial, generally restricted to forest along small streams, less common than *Dendrobates granuliferus* or either species of *Colostethus;* males have been observed wrestling; 3, 4, 8, and 36 tadpoles taken from backs of males; tadpoles taken in shallow, intermittent forest stream near field station; known from lowland evergreen forest in southwestern Costa Rica and adjacent western Panama.

Hylidae

Agalychnis callidryas (Cope)—Large; nocturnal; insectivorous; abundant; arboreal and epiphyllous; breeds in temporary ponds in forest along Río Rincón east of mangroves southeast of Rincón and in forest ponds between Quebrada Aguabuena and airfield; eggs placed on leaves over water; found in lowland and premontane evergreen forests from southern Mexico to central Panama and northern Colombia. Warkentin (1995, 1997, 1999a, 1999b, 2000) studied this species intensively at Corcovado National Park.

Agalychnis spurrelli (Boulenger)—Large; nocturnal; insectivorous; arboreal; abundant; occasionally encountered on leaves in forest; incredibly dense breeding aggregations (Scott and Starrett [1974] estimated 13,000 individuals in a breeding aggregation at Caiman Lagoon in August 1970) on vegetation around larger forest ponds; eggs put on leaves over water; known from lowland evergreen forest of southeastern and southwestern Costa Rica and Panama to northwestern Ecuador.

Hyla ebraccata Cope—Small (males) to moderate (females); nocturnal; insectivorous; arboreal; abundant; pond-breeding; males epiphyllous, call from vegetation along edges of ponds; eggs placed on leaves over water; amplexing pairs found in June and August; occurs sporadically in temporary and relatively permanent ponds along airfield and Río Rincón, and frequently in Caiman Lagoon and Vanegas Swamp; known from lowland and premontane evergreen and mixed forests of southern Mexico to Guatemala and Belize, and from northeastern Honduras and Nicaragua to northwestern Colombia.

Hyla rosenbergi Boulenger—Large; nocturnal; insectivorous; arboreal; abundant; found around small, temporary ponds, puddles, and small streams; males terrestrial or epiphyllous, call from ground (often concealed) or from vegetation at water's edge; males territorial, construct mud nests at stream edges and along flooded roadways; males known to call from June through August; eggs and tadpoles found in August on Playa Blanca road between Río Riyito and Río Rincón and in small ponds along Río Rincón behind mangroves; occurs in lowland evergreen forests of southwestern Costa Rica and Panama to northwestern Ecuador.

Phrynohyas venulosa (Laurenti)—Very large; nocturnal; insectivorous; arboreal; rare, single adult male collected as it called from pond near Vanegas Swamp after a heavy rain in March and one juvenile taken near Caiman Lagoon in June; skin secretions highly irritating; elsewhere in its range this large treefrog is an explosive breeder, usually for one night after first torrential rains of wet season and probably behaves similarly on Osa Peninsula; wide ranging in lowland forests from Mexico through tropical South America.

Scinax boulengeri (Cope)—Moderate; nocturnal; insectivorous; abundant; males epiphyllous, usually call from plant stem in head-down direction, often concealed by dense vegetation along edges of shallow ponds adjacent to Quebrada Aguabuena, at scattered localities along lower parts of airfield, and in Vanegas Swamp; known from lowland evergreen forests from central Nicaragua to northwestern Ecuador.

Scinax elaeochroa (Cope)—Moderate; nocturnal; insectivorous; arboreal; common around edges of temporary ponds; males epiphyllous, call from leaves of bushes and tall grass; known from temporary ponds along Río Rincón and lower end of airfield; males heard in June, July, and August; occurs in lowland evergreen forests of eastern Nicaragua, Costa Rica, and adjacent western Panama.

Smilisca phaeota (Cope)—Large; nocturnal; insectivorous; abundant; terrestrial and arboreal; found primarily in temporary ponds and puddles; males call from water's edge or while floating in water in March, June, July, and August; breeding apparently initiated by heavy rains; eggs deposited as surface film; frequently encountered in shallow ponds in disturbed areas along road from Rincón to field station; widespread in lowland and premontane evergreen forests from northeastern Honduras to northwestern Ecuador.

Smilisca sila Duellman and Trueb—Moderate (males) to large (females); nocturnal; insectivorous; terrestrial and arboreal; rare in Rincón area, known only from rocky, steep stream that forms water supply for old Osa Field Station; breeds during dry season; known from lowland, evergreen forests on Pacific slopes of southern Costa Rica, western Panama, eastern Panama, and northern Colombia.

Smilisca sordida (W. Peters)—Moderate (males) to large (females); nocturnal; insectivorous; terrestrial and arboreal; especially abundant during dry season along Quebrada Aguabuena; males call from edges of streams, frequently from rocks or gravel bars, apparently all year; amplexing pairs found in March, June, July, and August; known from lowland and premontane evergreen and mixed forests from northeastern Honduras to western Panama.

Leptodactylidae

Eleutherodactylus crassidigitus Taylor—Moderate; nocturnal, juveniles diurnal in leaf litter; insectivorous; very abundant throughout forest; males epiphyllous, often calling from branches of understory plants up to 1.5 m above ground; females may be active in leaf litter during day; lowland and premontane evergreen forests of southwestern Costa Rica, through Panama to northwestern Colombia. On the Osa Peninsula, this species has been called *Eleutherodactylus longirostris* by many previous authors (e.g., Scott et al. 1983; Savage and Villa 1986).

Eleutherodactylus cruentus (W. Peters)—Small; nocturnal; insectivorous; arboreal; rare, only known specimen from Rincón area was found in an arboreal bromeliad high in forest near the airfield; lowland and premontane evergreen forests from Costa Rica at least to central Panama.

Eleutherodactylus diastema (Cope)—Small; nocturnal; insectivorous and myrmecophagous; epiphyllous; common; males call from leaves 1–2 m above ground in forest; lowland and premontane evergreen forests from eastern Nicaragua to western Ecuador.

Eleutherodactylus fitzingeri (O. Schmidt)—Moderate; nocturnal; insectivorous; epiphyllous; abundant; males call from leaves along streams, trails, and ' forest edge, usually less than 1 m above ground; lowland evergreen and deciduous forests of northeastern Honduras, Nicaragua, Costa Rica, and Panama to western Colombia.

Eleutherodactylus ridens (Cope)—Small, males much smaller than females; nocturnal; insectivorous, small spiders important in diet; common; males epiphyllous, call from leaves in shrub layer and understory up to 3 m above ground; known from lowland and premontane evergreen forests from eastern Honduras to western Ecuador.

Eleutherodactylus rugosus (Cope)—Large; adults nocturnal, juveniles diurnal; insectivorous, eating primarily beetles and occasionally lizards and other frogs; adults in shallow depressions in leaf litter; common on forest floor; known from lowland and premontane evergreen forests of southwestern Costa Rica and adjacent western Panama. Osa populations of this species have been called *Eleutherodactylus biporcatus* by many previous authors (e.g., Scott et al. 1983; Savage and Villa 1986).

Eleutherodactylus stejnegerianus (Cope)—Small; diurnal; insectivorous; abundant in leaf litter, aggregates along streams in dry season; known from evergreen and mixed lowland and premontane forests of Pacific slope Costa Rica, and western Panama. At Rincón, this species was referred to as *Eleuthero-*

dactylus bransfordii by many previous authors (e.g., Scott et al. 1983; Savage and Villa 1986).

Eleutherodactylus taurus Taylor—Large; nocturnal; insectivorous; terrestrial and riparian; abundant; one night a female was seen along a steeply banked section of Quebrada McDiarmid with the legs of a large amblypygid hanging from her mouth; males call from stream banks; restricted to rainforests of Golfo Dulce region in Costa Rica and immediately adjacent southwestern Panama.

Eleutherodactylus vocator Taylor—Small, less than 20 mm; usually nocturnal; insectivorous; common; males epiphyllous, call from leaves very near ground, sometimes in late afternoon; adults often taken in leaf litter during day; known in lowland and premontane evergreen forests of Costa Rica, Panama, and northern Colombia.

Leptodactylus bolivianus Boulenger—Large to very large, females to 88 mm, males to 94 mm; nocturnal; insectivorous; terrestrial; common; generally found around shallow ponds and marshes, often in areas of second-growth vegetation; males call from deep grass and tangles of vegetation near ponds, sometimes during daylight on dark rainy days; known from evergreen and deciduous lowland forests of Pacific Costa Rica, Panama, and most of tropical South America. Some authors (e.g., Frost 1985; Rand and Myers 1990) have used the name *Leptodactylus insularum* Barbour to refer to this taxon.

Leptodactylus melanonotus (Hallowell)—Moderate; nocturnal; terrestrial; rare; known from marshy areas in second-growth vegetation along road from Rincón to airfield and along road to Playa Blanca; tadpoles form schools; widespread in lowlands and premontane zones from Mexico to Ecuador.

Leptodactylus pentadactylus (Laurenti)—Very large (adults to over 150 mm); nocturnal; carnivorous; terrestrial; abundant; common along forest edges and in second-growth vegetation, occasionally found along streams and in forest; tadpoles predaceous; ecology of tadpole of this species (Valerio 1971) and tadpoles of other pond breeders from near Rincón (Heyer et al. 1975); lowlands from Honduras throughout Central America and portions of tropical South America primarily in evergreen and mixed forests.

Leptodactylus poecilochilus (Cope)—Moderate; nocturnal; insectivorous; abundant; terrestrial, found in second-growth vegetation and along roads between Rincón and field station, and near Camp Seattle; males call from concealed sites beneath ground near shallow temporary ponds; lowlands of western Costa Rica south through northern Colombia and Venezuela.

Physalaemus pustulosus (Cope)—Moderate; nocturnal; insectivorous; extremely abundant; terrestrial; commonest in and around temporary, small ponds and puddles in second-growth vegetation or disturbed areas (e.g., around airfield), along roads, and near habitations where it breeds; males heard calling at night from puddle (that filled in a root hollow that formed after a large tree fell in the forest near Quebrada Rayo); eggs placed in foam nests, usually exposed; known from lowlands and foothills from Mexico to northern Colombia and Venezuela.

Microhylidae

Nelsonophryne aterrima (Günther)—Moderate; nocturnal; myrmecophagous; semifossorial (beneath logs) and in leaf litter; rare, collected once in leaf litter near Three Falls; tadpoles described from Panamanian material (Donnelly et al. 1990); found in evergreen lowland and premontane forests from Costa Rica to northwestern Ecuador.

REPTILIA (69)

Testudines (3)

Emydidae

Trachemys ornata (Gray)—Moderate to large, to 600 mm carapace length; diurnal; chiefly herbivorous but opportunistic; aquatic; commonly seen along larger rivers (Río Riyito and Río Rincón); occurs in lowlands on both coasts from Mexico to northern Colombia and northwestern Venezuela.

Kinosternidae

Kinosternon leucostomum (A. Duméril and Bibron)—Small, to 175 mm carapace length; nocturnal; aquatic; opportunistic and omnivorous; rare in Rincón area, known from single specimen found in pond along Gravel Pit Road; lowland and premontane areas in Atlantic drainages of eastern Mexico south to northern Colombia and in Pacific drainages from central Costa Rica to central western Ecuador.

Kinosternon scorpioides (Linnaeus)—Small, to 270 mm carapace length; nocturnal; aquatic; omnivorous; rare, known from near Rincón by single specimen in collections of University of Costa Rica; lowland and premontane areas from Mexico south to Ecuador and northern Argentina.

Squamata—Sauria (22)

Gekkonidae

Lepidoblepharis xanthostigma (Noble)—Small; diurnal; insectivorous but also consumes spiders and mites; terrestrial; common in forest leaf litter, often near streams and occasionally in second-growth areas; lowland evergreen forest from southeastern Nicaragua and Costa Rica to northern Colombia.

Thecadactylus rapicauda (Houttuyn)—Large; nocturnal; insectivorous; arboreal; moderately common on large tree trunks, especially between buttresses and in hollows; eggs deposited in leaf litter at base of trees; young hatch in May;

deciduous and evergreen forests at low elevations along Caribbean versant from Chiapas, Mexico, and on Pacific side of Costa Rica southward, to northwestern Ecuador and through northern South America to Brazil and Bolivia; also Lesser Antilles.

Sphaerodactylus graptolaemus Harris and Kluge—Small; probably insectivorous; rare in Rincón area, known from single specimen collected as it ran across floor of dormitory of old Osa Field Station; lowland evergreen forests of southwestern Costa Rica and adjacent western Panama.

Corytophanidae

Basiliscus basiliscus (Linnaeus)—Very large; diurnal; young primarily insectivorous, adults omnivorous; terrestrial and arboreal; abundant along larger rivers and streams of Rincón area, and relatively uncommon along smaller, forest streams (e.g., Quebrada Crump, Quebrada Rayo, etc.) and in open, forested areas; frequently found asleep at night on vegetation along streams; lowland deciduous and evergreen forests of southwestern Nicaragua, Pacific Costa Rica, and Panama, south on both coasts of central and eastern Panama to northwestern Colombia and northwestern Venezuela.

Corytophanes cristatus (Merrem)—Large; diurnal; insectivorous; abundant; sometimes found on ground but most often seen on vertical stems of understory trees and shrubs, on vines, or on trunks of large trees, where it remains motionless, relying on concealment to escape detection; frequently found asleep in same situations at night; Andrews (1979) reported on diet and field behavior of this lizard at Rincón; occurs in lowland and premontane evergreen forests of Atlantic versant from Mexico to northwestern Colombia and in Golfo Dulce area of Pacific Costa Rica.

Iguanidae

Iguana iguana (Linnaeus)—Very large; diurnal; herbivorous; arboreal; common in Rincón area; may be more abundant than records indicate, especially along large rivers; a few specimens seen and/or collected along Carretera al Pacífico where it parallels Río Rincón south of Rincón, two juveniles near old Osa Field Station, and a few sight records on Río Riyito near Carretera al Pacífico; a wide-ranging species in lowland deciduous and evergreen forests from northern Mexico to Bolivia, Paraguay, and southern Brazil, especially along waterways; also in Lesser Antilles.

Polychrotidae

Dactyloa insignis (Cope)—Large; diurnal; arboreal; rare, known from a single specimen collected from overhanging vegetation along Quebrada Aguabuena; occurs in lowland and premontane evergreen forests of Costa Rica and Panama.

Norops aquaticus (Taylor)—Moderate; diurnal; insectivorous; abundant; terrestrial and epiphyllous; riparian, found on banks, rocks, and tree trunks of heavily shaded streams including upper reaches of Quebrada Aguabuena and other streams; often seeks shelter in water; found only in lowland and premontane evergreen forests of Golfo Dulce area in Costa Rica and adjacent southwestern Panama.

Norops biporcatus (Wiegmann)—Large; diurnal; insectivorous; arboreal; moderately common on large trees in forest near old Osa Field Station; occasionally found at night asleep on branches 1.5–2 m above ground; found in lowland evergreen forests from Mexico to Venezuela on Atlantic and from Costa Rica to Ecuador on Pacific versants.

Norops capito (W. Peters)—Moderate; diurnal; insectivorous, sometimes eats snails; abundant; often found on ground, on stems of small understory shrubs, and on vines; lowland and premontane evergreen forests from southern Mexico to Panama on Atlantic and from southwestern Costa Rica through Panama on Pacific versants.

Norops limifrons (Cope)—Moderate; diurnal; insectivorous but small spiders represent a large proportion of diet; moderately common in old second-growth vegetation along airfield and occasionally along small streams and in forest; one sleeping on leaf at night along stream; lowland deciduous and evergreen forests on Atlantic side from Honduras to Panama and on the Pacific slope from Costa Rica to eastern Panama.

Norops pentaprion (Cope)—Moderate; diurnal; arboreal; moderately common but may be abundant in forest canopy; four taken from upper branches of a large tree cut along south slope north of Carretera al Pacífico between airfield and Rincón; lowland deciduous and evergreen forests from southern Mexico to northern Colombia.

Norops polylepis (W. Peters)—Moderate; diurnal; insectivorous; abundant; probably most abundant lizard in forests on Osa Peninsula; epiphyllous, usually found on small branches or leaves of understory shrubs up to 2 m above forest floor, often terrestrial on forest floor; Andrews (1971) described the ecology of *Norops polylepis* at Rincón de Osa in some detail, and Perry (1996) studied it near Sirena, Corcovado National Park; occurs only in lowland and premontane evergreen forests of Golfo Dulce region of Costa Rica and adjacent western Panama.

Scincidae

Mabuya unimarginata Cope—Moderate; diurnal; insectivorous; terrestrial; rare; usually found in disturbed habitats; wide ranging in lowland, deciduous and evergreen forests from Mexico to at least eastern Panama.

Sphenomorphus cherriei (Cope)—Moderate; diurnal; insectivorous, but isopods represent a large proportion of the diet; terrestrial to semifossorial; moderately common in leaf litter in open forest and under old banana plants near field station; occurs in deciduous and evergreen lowland and premontane forests from southern Mexico to extreme western Panama.

Gymnophthalmidae

Bachia blairi (Dunn)—Moderate; secretive; presumably diurnal; insectivorous; semifossorial; moderately common, known from five specimens collected in leaf-litter accumulations between buttresses of large trees west of old Osa Field Station; McDiarmid and DeWeese (1977) discussed systematic status and natural history of species; lowland evergreen forests of Osa Peninsula and adjacent western Panama.

Leposoma southi Ruthven and Gaige—Moderate; diurnal; insectivorous; terrestrial; moderately common in leaf litter on forest floor near old Osa Field Station; Atlantic and Pacific lowland evergreen forests of Costa Rica and Panama to western Colombia.

Neusticurus apodemus Uzzell—Moderate; diurnal; presumably insectivorous; terrestrial and riparian; rare, single specimen collected among rocks in the upper reaches of the Airfield Water Supply stream; known elsewhere from two localities in lowland evergreen forest of Golfo Dulce area of Costa Rica.

Teiidae

Ameiva festiva (Lichtenstein and von Martens)—Large; diurnal; insectivorous; abundant; terrestrial, most often seen along forest edge, trails, and heavily vegetated roadsides; most common teiid lizard found in the vicinity of Rincón; Hillman (1969) reported on habitat specificity of three species of *Ameiva* at Rincón de Osa; occurs in lowland and premontane evergreen forests from southern Mexico to northern and western Colombia.

Ameiva leptophrys (Cope)—Large; diurnal; insectivorous; abundant and more restricted to forest habitats, especially along forest trails and near small light gaps, than other species of *Ameiva;* known from lowland evergreen forest of Golfo Dulce area of Costa Rica and adjacent western Panama to western Colombia.

Ameiva quadrilineata (Hallowell)—Moderate; diurnal; insectivorous; terrestrial; abundant along beaches, riverbanks, unshaded roads, dry beds of larger streams (e.g., Quebrada Aguabuena), and in other, open, bare-ground areas; lowland evergreen forests of Nicaragua to western Panama on Atlantic and southwestern Costa Rica and western Panama on Pacific sides.

Xantusiidae

Lepidophyma reticulatum Taylor—Moderate; probably nocturnal; insectivorous, may eat fruits and other plant material; moderately common; terrestrial to semifossorial, found in leaf litter, especially between tree buttresses; occurs in lowland and premontane evergreen forests of Pacific slope of Costa Rica.

Squamata—Serpentes (42)

Boidae

Boa constrictor Linnaeus—Very large; crepuscular to nocturnal; feeds on lizards, birds, and mammals; moderately common in Rincón area; known from old second-growth areas around airfield but probably also occurs in primary forest; lowland deciduous, evergreen, and premontane evergreen forests from northern Mexico to Argentina, South America; also Lesser Antilles.

Corallus ruschenbergerii (Cope)—Large, to about 1,650 mm; arboreal; elsewhere feeds on frogs, lizards, birds, and mammals; rare; known from single specimen taken in forest near Osa Field Station; lowland evergreen forests from southwestern Costa Rica through Panama to northern Colombia and northern Venezuela; also Trinidad and Tobago. Listed as *Corallus hortulanus* by Scott et al. (1983) and Savage and Villa (1986), and as *Corallus enhydris* by Villa et al. (1988). Henderson (1997) recognized four species for *Corallus hortulanus* and resurrected *Corallus ruschenbergerii* (Cope, 1876) for the Central American species.

Colubridae

Amastridium veliferum Cope—Moderate; terrestrial to semifossorial; eats frogs; rare, known from two specimens collected in forest, one near airfield and another near end of Water Fall Road; lowland and premontane evergreen forests from southeastern Nicaragua to central Panama on Atlantic versant and in southwestern Costa Rica on Pacific versant.

Chironius carinatus (Linnaeus)—Large, diurnal; terrestrial, sometimes arboreal; feeds mostly on frogs but elsewhere reported to eat salamanders, lizards, birds, and mammals; rare, young specimens from old second-growth areas near airfield; occurs in lowland wet forests from Costa Rica to western Ecuador and Amazonian Brazil; also Trinidad and Lesser Antilles.

Chironius grandisquamis (W. Peters)—Large; diurnal; feeds mostly on frogs but also eats salamanders, lizards, birds, and mammals; moderately common in forest; adults shiny black, juveniles brownish black; adults terrestrial, seen on ground during day and arboreal at night, observed coiled on leaf 4 m above water in Quebrada Rayo, two immatures found at night coiled in branches 1.5 m and 2 m above water on Quebrada McDiarmid; lowland and premontane evergreen forests from Honduras to northwestern Ecuador.

Clelia clelia (Daudin)—Large; crepuscular and nocturnal; feeds on snakes and lizards, occasionally mammals; terrestrial; moderately common; adults black, juveniles bright red with black head and white collar; several specimens found crossing road between Rincón and old Osa Field Station during or after light rain; widespread in lowland deciduous and lowland and premontane evergreen forests from southeastern Mexico to Bolivia.

Coniophanes fissidens (Günther)—Small; diurnal; terrestrial; eats primarily frogs, but arthropods, earthworms, lizards, and rarely snakes are eaten; moderately common; on forest floor near old Osa Field Station; lowland and premontane deciduous and evergreen forests from Mexico to central western Ecuador.

Dendrophidion percarinatum (Cope)—Moderate; diurnal; terrestrial; eats frogs; common; most often found in forest, especially along small streams; occasionally seen swimming; occurs in lowland evergreen forests from Honduras to western Ecuador and northern Venezuela.

Dipsas tenuissima Taylor—Moderate; elongate; nocturnal; arboreal; mollusk eater; very rare only nine known specimens; single example from near Rincón in collection at the Universidad de Costa Rica; restricted to the Golfo Dulce area of Costa Rica and western Panama.

Drymarchon melanurus (Duméril, Bibron, and Duméril)—Large (to 3,000 mm); diurnal; known to feed on mammals (e.g., *Sigmodon hispidus*) at Rincón but also takes birds, snakes, turtles, lizards, amphibians, and fish; moderately common; found in second-growth areas around field station and airfield and along major rivers; disjunct and widespread in deciduous, lowland, and premontane evergreen forests from southern United States to western Ecuador and northern Venezuela.

Erythrolamprus mimus (Cope)—Moderate; rare; known from one specimen taken near old Osa Field Station; probably diurnal; eats mostly snakes, synbranchid eels, and some lizards; occurs in lowland forests from Honduras to western Colombia and Ecuador, and to northwestern Venezuela, eastern Ecuador, and Peru.

Geophis hoffmanni (W. Peters)—Small; earthworm specialist, also eats softbodied insect larvae; fossorial; rare in Rincón area; known from single specimen collected near Quebrada Ferruviosa; occurs in lowland, premontane, and montane forests from Honduras to western Panama.

Imantodes cenchoa (Linnaeus)—Moderate; very elongate; nocturnal; feeds on lizards and frog eggs; arboreal; moderately common in forest and along streams near old Osa Field Station; may forage 5 m or more above ground; in deciduous and evergreen forests in lowlands and in premontane and lower-montane evergreen forests from southern Mexico to southern South America.

Imantodes inornatus Boulenger—Moderate; very elongate; nocturnal; arboreal; forages for sleeping lizards (e.g., small *Basiliscus, Norops* spp.) and frog eggs at night in vegetation along streams and trails; moderately common in forest; from lowland, premontane, and lower-montane evergreen forests of Honduras southward to central western Ecuador.

Leptodeira septentrionalis (Kennicott)—Moderate; nocturnal; arboreal; common in the Rincón area; often found foraging in low bushes and trees around ponds or along streams for frogs or frog eggs (e.g., *Agalychnis* spp., *Hyalinobatrachium* spp.); widespread, in lowland deciduous and evergreen forests and in premontane evergreen forests from southern Texas to western Colombia, Ecuador, and northern Peru.

Leptophis ahaetulla (Linnaeus)—Large; diurnal; arboreal, sometimes terrestrial; feeds on frogs, lizards, small birds, and bird eggs; moderately common near Rincón; most specimens come from second-growth areas and around inhabited areas; widespread in lowland deciduous and evergreen forests from southern Mexico to southwestern Ecuador, and on the Amazonian side to Argentina.

Leptophis riveti Despax—Moderate; diurnal; arboreal; known to eat frogs; rare in Rincón area; known from single specimen taken near old Osa Field Station; occurs in lowland and premontane forests of the Golfo Dulce area of Costa Rica and from central Panama to western Ecuador, Amazonian Peru, and on Trinidad.

Mastigodryas melanolomus (Cope)—Moderate; diurnal; terrestrial; eats primarily lizards, but known to consume variety of small vertebrates; moderately common; most often found in second-growth areas and along sunny banks of larger streams; specimens with red and specimens with greenish white venters sympatric near airfield; wide ranging in lowland deciduous and evergreen forests and premontane evergreen forests from Mexico to northern Colombia.

Ninia maculata (W. Peters)—Small; nocturnal; semifossorial to terrestrial; feeds on soft-bodied prey, mostly earthworms and slugs, and some insect larvae; rare in leaf litter in Rincón area; lowland, premontane, and montane evergreen forests from Nicaragua to eastern Panama.

Ninia sebae (Duméril, Bibron, and Duméril)—Small; body red, head black; nocturnal leaf-litter denizen; feeds primarily on earthworms, slugs, and small land snails; single specimen from near Rincón is in the Universidad de Costa Rica collection; occurs in lowlands and on premontane slopes from Mexico to southwestern Costa Rica.

Nothopsis rugosus Cope—Small; presumably nocturnal; feeds on frogs and salamanders; semifossorial to terrestrial; rare in shallow leaf litter in forest west of old Osa Field Station in May; lowland and premontane evergreen forests from eastern Honduras to Panama on Atlantic slope, and in Golfo Dulce area of Costa Rica, Panama, western Colombia, and northwestern Ecuador on Pacific versant.

Oxybelis aeneus (Wagler)—Moderate; elongate; diurnal; arboreal; eats mostly frogs, lizards, and birds; rare in Rincón area; juvenile found asleep at night on leaf 2 m above ground along Airfield Water Supply trail; widespread in deciduous, mixed, and evergreen lowland and premontane forests from southern Arizona and Mexico south over northern half of South America to northern Bolivia and southern Brazil.

Oxyrhopus petolarius (Linnaeus)—Moderate; primarily nocturnal; eats lizards and small mammals; rare in Rincón area; widespread in lowland and premontane evergreen forests from southeastern Mexico and Costa Rica to western Ecuador, northern South America and south over most of Brazil and the Amazon basin, south to northern Bolivia.

Pseustes poecilonotus (Günther)—Large; diurnal; terrestrial, may venture into trees; carnivorous on birds and mammals; moderately common, especially in dry season; adults taken along Carretera al Pacífico between Rincón and airfield and on Ridge Trail about 2 km west of Rincón; widespread in lowland and premontane forests from Mexico to western Ecuador and in Amazon basin to Bolivia and Brazil.

Rhadinaea decorata (Günther)—Small; diurnal; terrestrial; feeds on salamanders, frogs, small lizards, and other litter organisms; moderately common in and on leaf litter in forest near old Osa Field Station; widespread in lowland and premontane evergreen forests from Mexico to northwestern Ecuador.

Scaphiodontophis annulatus (Duméril, Bibron, and Duméril)—Moderate; diurnal; terrestrial; habits unknown; eats mostly skinks, but also other lizards and sometimes small frogs; rare in Rincón area; lowland and premontane evergreen forests of tropical Mexico south to northern and central Colombia.

Sibon dimidiatus (Günther)—Moderate; elongate; nocturnal; arboreal; feeds on mollusks; rare, four specimens found in vegetation at night 4–4.5 m above ground in forest along trails near old Osa Field Station; lowland and premontane evergreen forests of Atlantic versant from southern Mexico to Nicaragua and on Pacific versant of Guatemala, Honduras, and central and southern Costa Rica.

Sibon nebulatus (Linnaeus)—Moderate; elongate; nocturnal; arboreal; mollusk eater; rare, two specimens taken at night along trails through forest near station; widespread in lowland, deciduous, and evergreen forests from central Mexico through Central America to Colombia, eastern and western Ecuador and northern Venezuela to northern Brazil.

Spilotes pullatus (Linnaeus)—Large; diurnal; arboreal and terrestrial; feeds on mammals and birds and bird eggs; rare in Rincón area; widespread from Mexico through Central America to western Ecuador and Argentina primarily in lowland and premontane evergreen forests.

Stenorrhina degenhardtii (Berthold)—Moderate; nocturnal; terrestrial; feeds on scorpions, centipedes, and other arthropods; rare, one specimen taken near Rincón; low elevations from southern Mexico to northwestern Peru and north central Venezuela in deciduous and evergreen forests.

Tantilla ruficeps (Cope)—Small; nocturnal; semifossorial; habits unknown but known to eat centipedes elsewhere; rare; found in lowland and premontane evergreen forests from eastern Nicaragua, through Costa Rica to western Panama.

Tantilla schistosa (Bocourt)—Small; presumably nocturnal; eats primarily centipedes; rare, in leaf litter in forest west of Osa Field Station; low and moderate elevations from Mexico to central Panama in lowland and premontane evergreen forests.

Tantilla supracincta (W. Peters)—Small; semifossorial; eats primarily centipedes; rare, two adults known from area; one collected under a small log in forest near juncture of Gravel Pit Road and Carretera al Pacífico and another unearthed by bulldozer clearing old second-growth vegetation near old Osa Field Station; known from scattered localities in lowland and premontane evergreen forests from Nicaragua through Costa Rica to central Panama and northwestern Ecuador.

Tripanurgos compressus (Daudin)—Moderate; nocturnal; terrestrial; only eats lizards; rare in Rincón area, single specimen moving on bank of Quebrada Aguabuena about 300 m below field station at entrance to Quebrada Crump; scattered localities in lowland evergreen forest from Golfo Dulce of Costa Rica and adjacent Panama to western Ecuador and Bolivia, Brazil, and Paraguay.

Urotheca fulviceps (Cope)—Small; diurnal and nocturnal; terrestrial to semifossorial; feeds on small leaf litter vertebrates; rare, two specimens, one in leaf litter in forest near old Osa Field Station and on lower slopes (60 m) near Airfield Water Supply in May; known from lowland evergreen forests of Golfo Dulce, central and eastern Panama to northwestern Ecuador.

Urotheca guentheri (Dunn)—Small; presumably diurnal; terrestrial; consumes frogs and salamanders; rare, one specimen found in litter sample west of old Osa Field Station in May; lowland, premontane, and montane evergreen forests of eastern Nicaragua, Costa Rica, and western Panama.

Xenodon rabdocephalus (Wied)—Diurnal; terrestrial; eats mostly toads but also other anurans; heavy bodied and resembles fer-de-lance (*Bothrops asper*); rare, known from single, small specimen collected along Quebrada Aguabuena; widespread in lowland and premontane evergreen and deciduous forests from Mexico through Central America to western Ecuador and Bolivia.

Elapidae

Micrurus alleni K. Schmidt—Moderate; nocturnal and diurnal; terrestrial; feeds primarily on synbranchid eels and other snakes; rare, one specimen found crawling just after dark near Rincón Airfield (reportedly diurnal at other sites in Corcovado National Park); lowland and premontane evergreen forests of Atlantic eastern Honduras, Nicaragua, Costa Rica, and Panama and Golfo Dulce area of southwestern Costa Rica and adjacent Panama.

Viperidae

Bothriechis schlegelii (Berthold)—Moderate; arboreal; moderately common; feeds on bats (e.g., *Carollia castanea*) and probably birds, frogs, and lizards; collected in small trees and bushes along forest trails northwest of airfield, in Savage Woods and Snake Woods, and along Ridge Trail from Rincón to Laguna Chocuaco; lowland and premontane evergreen forests from southern Mexico to Venezuela and northwestern Peru.

Bothrops asper (Garman)—Very large; primarily nocturnal but often moving during day in forest; adults terrestrial, young may be arboreal; eats rodents (e.g., *Sigmodon hispidus*), frogs, and probably lizards; abundant, specimens taken at essentially every major collecting site near Rincón; most frequently encountered at forest edges and in old second-growth vegetation, but also seen swimming in rivers and creeks and coiled along trails and on logs in forest; large individual found at edge of Carretera al Pacífico near Quebrada Aguabuena eating an adult *Leptodactylus pentadactylus;* widespread in evergreen forests from Mexico through Central America to northwestern Ecuador.

Lachesis melanocephala Solórzano and Cerdas—Very large; terrestrial; feeds on mammals; rare, in undisturbed forest on Ridge Trail and in Savage Woods; found in lowland evergreen forests of Golfo Dulce region in Costa Rica and western Panama. The population on the Osa Peninsula, referred to as *Lachesis muta* or *Lachesis muta melanocephala* by previous authors (e.g., Scott et al. 1983; Savage and Villa 1986; Villa et al. 1988), was elevated to species status by Zamudio and Greene (1997).

Porthidium nasutum (Bocourt)—Small; terrestrial; feeds primarily on frogs; moderately common; most frequently encountered in leaf litter in forest west of the old Osa Field Station, in Savage Woods, and in forest around Airfield Water Supply; small juvenile with a yellowish white tail tip found in forest on 15 August; occurs in lowland evergreen forest from southeastern Mexico along Caribbean Central America to Panama and on Pacific side in Golfo Dulce area of Costa Rica, in western Colombia and northwestern Ecuador.

Crocodilia (2)

Alligatoridae

Caiman crocodilus (Linnaeus)—Moderate to large, to 2.7 m, specimens over 2 m are rare; nocturnal; carnivorous; aquatic; abundant; frequents temporary to relatively permanent ponds along Carretera al Pacífico; lowlands from southern Mexico on the Pacific to northwestern Ecuador, and from Honduras on the Atlantic through Central America to central Brazil and Bolivia in the Amazon Basin.

Crocodylidae

Crocodylus acutus (Cuvier)—Gigantic, to 7 m, specimens over 4 m rare; nocturnal; carnivorous; aquatic; moderately common; found in major rivers; four specimens seen in two nights in March 1966 on Río Riyito near Carretera al

Pacífico; reported to have been abundant in Laguna Corcovado on western side of peninsula and to have been hunted commercially in past years; current populations unknown; former mainland distribution from northwestern Mexico south to northern Peru on the Pacific versant and to northern Colombia and Venezuela on the Caribbean side; also southern Florida and Greater Antilles.

MARINE SPECIES

Tosi (1975) listed four species of sea turtles (*Chelonia mydas* [— *Chelonia agassizii* of some authors], *Eretmochelys imbricata*, *Lepidochelys olivacea*, and *Dermochelys coriacea*) from the Osa Peninsula, but we know of no specimens from there. An aerial survey revealed nesting tracks of *Lepidochelys olivacea*, *Chelonia mydas*, and a few *Dermochelys coriacea* at several places along the west coast of the peninsula in October, but none was seen on suitable beaches on the Golfo Dulce side (Richard and Hughes 1972). Other workers (Cornelius 1982; Frazier and Salas 1983) noted that all four forms probably occur in waters off the peninsula and may come ashore, especially on the beaches of the outer coast to nest. Cornelius (1982) commented that *Caretta caretta* is common in Panama and may nest around the Península de Osa. One or a few *Dermochelys coriacea* and *Lepidochelys olivacea* have recently been observed nesting on southern beaches on the outer coast of the peninsula (A. Chaves, pers. comm.) and *Lepidochelys olivacea* on beaches on the mainland south of Golfito. The pelagic or yellow-bellied sea snake, *Pelamis platurus*, occurs along the Pacific coast in warmer waters from southern California to Peru. Snakes often are associated with ocean slicks in deeper water but occasionally wash ashore. Sea snakes have been seen at various places along the outer coast of the Osa Peninsula and taken in the mouth of the Golfo Dulce; specimens have been found on the beach near Llorona and Sirena. Because our treatment deals only with the terrestrial and freshwater components of the herpetofauna, the status of these forms is not considered further.

ERRONEOUSLY RECORDED SPECIES

Several lists of species from the Rincón area have been prepared by OTS course instructors in mimeographed form and are the basis for the summary by Schnell (1971) in the OTS handbook. In addition, in 1973 McDiarmid prepared a list of amphibians and reptiles from the peninsula as part of a report to the Costa Rican government on the need to establish a major national park in this unique region. His summary was incorporated into Tosi's (1975) report and list of species. Unfortunately, these lists sometimes contained names of species that were based on misidentifications or other errors. Here, we review some of those reports to clarify the situation.

The following species appear to have been included in one list or the other simply in error or as geographical probabilities for Rincón: caecilians— *Gymnopis multiplicata;* frogs—*Eleutherodactylus fleischmanni, Leptodactylus labialis* (= *L. fragilis*), *Hyla loquax, Hyla* or *Ololygon* (= *Scinax*) *staufferi, Centrolenella* (= *Hyalinobatrachium*) *fleischmanni, Rana palmipes* (= *R. vaillanti*), *Rana warszewitschii;* turtles—*Chelydra serpentina;* lizards—*Anolis* (= *Dactyloa*) *frenatus, Anolis* (= *Norops*) *humilis, Ameiva undulata, Gymnophthalmus speciosus;* snakes—*Leptophis depressirostris.* Although not currently known from the Rincón area, a few of these species (e.g., *Rana warszewitschii, Chelydra serpentina*) have been recorded from Corcovado National Park (appendix 16.2). Several forms were listed by synonyms or as misidentifications with closely allied species. They include *Eleutherodactylus crassidigitus* as *E. longirostris, Corallus ruschenbergerii* (also called *Corallus hortulanus* by some authors) by its synonym *C. enhydris, Chironius grandisquamis* formerly called *C. fuscus* in Costa Rica, *Leptodeira septentrionalis* as *L. annulata,* and *Sibon dimidiatus* as *S. annulatus.* Other erroneously listed species require additional comment.

The frog *Anotheca spinosa* was included on McDiarmid's 1973 list on the basis of a field identification of a tadpole taken 20–25 m above the ground in a bromeliad. Subsequent examination and comparison to other larvae in the laboratory showed the tadpole to be one of *Dendrobates auratus,* and this was subsequently reported by McDiarmid and Foster (1975).

The treefrog *Hyla microcephala* has been listed several times as occurring at Rincón, but the nearest authenticated records for this species at the time were from lowland evergreen forests near Palmar Sur and Golfito (fig. 16.2). A single tadpole that may be of this species was collected from the Vanegas Swamp, but no adults had ever been taken. For these reasons we refrained from listing the species as a member of the Rincón herpetofauna. Recently, an adult *Hyla microcephala* was collected by personnel from the Universidad de Costa Rica at Marenco Biological Station north of Corcovado National Park. This record suggests that the species may yet be collected near Rincón, but we prefer to wait for verifiable specimens before adding it to the Rincón list. We have included the species in the herpetofauna of the Península de Osa (appendix 16.2).

The treefrog *Hyla rufitela* has been included as part of the Osa Peninsula, on the basis of tadpoles and recent metamorphs. A metamorph that RWM identified in the field as *Hyla rufitela* was pale greenish tan with discrete, dark brown dorsal flecks; in part, his identification was based on previous reports of the species from the Osa Peninsula (Duellman 1970, 2001). JMS became suspicious when no adults of *H. rufitela* turned up in any of the Osa collections. Our reexamination of tadpoles and metamorphs in our collections indicates a misidentification; we are now convinced that all are *Hyla rosenbergi,* a species whose larvae and recent metamorphs resemble those of *Hyla rufitela.* Juveniles of both species are cream with dark brown spots in preservative. We suspect that other reports of *H. rufitela* from the Golfo Dulce region may also be based on mis-identifications. Accordingly, we omit *Hyla rufitela* from our list of amphibians from the Osa Peninsula.

Two turtles, *Chelydra serpentina* and *Kinosternon scorpioides,* have also appeared on various lists as coming from Rincón. We know of no specimens of *Chelydra* from the vicinity of Rincón and have not included it in that fauna. However, this large turtle has been recorded recently in the large lagoon-like habitats in Corcovado National Park (appendix 16.2) and probably occurs in similar habitats elsewhere on the Osa Peninsula. The original report of *Kinosternon scorpioides* from Rincón was based on a misidentified specimen of *Trachemys ornata* at the Universidad de Costa Rica. Nevertheless, other specimens of *Kinosternon scorpioides* have been recorded from near Rincón and Sirena, and we have included this turtle in the Rincón fauna.

EFFECTIVENESS OF SAMPLING

The only published account attempting to evaluate sampling effort at a comparable tropical locality is that of Myers and Rand (1969) for Barro Colorado Island, Panama. Their data indicate that at this intensively studied site approximately 81% of the species in the herpetofauna known at the time had been collected during the first ten years of sampling (1920–30) but that six species of snakes were added to the list between 1940 and 1967. An additional five species—one frog, three lizards, and a snake—have been added since 1969 (Rand and Myers 1990). Although the area near Rincón has not been subject to the same long-term sampling, it seems likely that the present list is equivalent in terms of search effort to that developed for the Panama site by 1931. In other words, between 80% and 85% of the species in the Rincón herpetofauna have been sampled. During the 13 years of sampling at Rincón, species not previously recorded were added nearly every year, including 1973 (see the chronology of sampling in appendix 16.1). Since that time, only three species—a turtle (*Kinosternon scorpioides*) and two snakes (*Dipsas tenuissima* and *Ninia sebae*)—have been added to the Rincón list, and one species, a frog (*Hyla rufitela*), removed from the list.

By the late 1980s, most OTS courses and researchers had shifted their field activities to Sirena in Corcovado National Park. This move resulted in relatively little being added to the Rincón herpetofaunal list. Even so, the snakes *Dipsas tenuissima* and *Ninia sebae* were first collected near Rincón in the 1990s. However, the increased presence at Sirena added substantially to the total species list of amphibians and reptiles for the Península de Osa. Sampling of amphibians and reptiles in the Parque Nacional de Corcovado over the past ten years has turned up one caecilian, one frog, one turtle, one lizard, and three snakes (sighting of a fourth species of snake needs verification) that were not recorded from Rincón. These records, together with the collection since 1990 of four previously unreported species (one frog, one lizard, and two snakes) from elsewhere on the Osa Peninsula (appendix 16.2), have increased the peninsular diversity even more. Adding the five species of marine reptiles to this compilation raises the number of native species known from the Osa Peninsula to 133 (135 if the two introduced lizards are included). We are confident that additional fieldwork will add terrestrial and freshwater species to the list of amphibians and reptiles known from the peninsula. On the basis of our knowledge of the fauna of the Golfo Dulce region and experience in sampling lowland wet forest herpetofaunas, we expect that the herpetofauna of the Rincón area will reach 135 species and estimate the herpetofauna of the Península de Osa to be about 149 terrestrial and freshwater species.

Our estimate is reinforced by close examination of the distributional records of species known from elsewhere in the Golfo Dulce forests but as yet not taken near Rincón. A number of conspicuous species that have high population densities in other sections of the Golfo Dulce forests were eliminated from consideration as possible members of the Rincón herpetofauna because they have never been seen elsewhere on the peninsula. As near as we can determine, possible additions other than the seven species from Corcovado National Park and the four from elsewhere on the peninsula (appendix 16.2) include one caecilian, two frogs, three lizards, and five snakes for a total of 22 taxa. Both frog species breed in temporary ponds but neither has been seen nor heard near Rincón and probably do not occur there. Several relatively conspicuous snakes may be added to the Rincón list in the future. Two rarely collected snakes, *Hydromorphus concolor* and *Ungaliophis panamensis,* first reported on the peninsula (at Sirena) in the 1990s, may also eventually appear.

Ecological Comparisons

As previously pointed out by Savage (1966, 1982), the Golfo Dulce forest region of Costa Rica and immediately adjacent southwestern Panama forms an isolated Pacific lowland evergreen forest region. Its closest biotic affinities are with the evergreen forests that, before the last 30 years, extended more or less continuously along the Atlantic lowlands from central Veracruz, Mexico, to Colombia. The Golfo Dulce forest (fig. 16.2) is separated from direct contact with the eastern forests by the Cordillera de Talamanca–Chiriquí axis to the northeast, the lowland semideciduous and deciduous forest areas of western Central America to the north, and the semideciduous and deciduous forests and savannas of western and central Panama to the east. These latter areas also isolate the Golfo Dulce forests and its herpetofauna from the evergreen rainforests of western Colombia and Ecuador. In this section, we provide a comparative evaluation of the differences between the Rincón herpetofauna and those from other lowland evergreen sites.

The best-known herpetofauna of a local area within the Isthmian region is that of Barro Colorado Island, Panama Province, Panama. This fauna has been sampled and studied from about 1923, when the island was made a wildlife preserve and research station, onward (Myers and Rand 1969; Rand and Myers 1990). A second intensively studied local herpetofauna occurs at La Selva Biological Station, Heredia Province, Costa Rica. This site, originally under the ownership of Dr. Leslie R. Holdridge, now serves as a major field research station for the Organization for Tropical Studies (Clark 1990; McDade and Hartshorn 1994). This site has been the locus of a detailed analysis of the ecologic roles of the species of amphibians and reptiles in the forest community (Lieberman 1986); the herpetofauna was reviewed by Guyer (1990, 1994a, 1994b) and Donnelly (1994a, 1994b). No other areas of Central American or Mexican lowland evergreen forest provide comparable data based on long-term and repetitive sampling. For this reason, the following discussion concentrates on an ecological comparison of the herpetofaunas at La Selva, Rincón, and Barro Colorado Island (BCI).

ENVIRONMENTS AT COMPARATIVE SITES

La Selva

La Selva research station lies on the west bank of the Río Puerto Viejo, a tributary of the Río Sarapiquí (Río San Juan drainage), 5 km southwest of the town of Puerto Viejo in the Atlantic lowlands of Costa Rica (fig. 16.2), at an elevation between 35 and 137 m. La Selva Biological Station has a total area of about 15.3 km² (McDade and Hartshorn 1994). The climate in the Köppen system is Afi, the same as that of Rincón; the mean annual temperature is 26.2°C; and precipitation averages 3,850 mm per year, with one to two relatively dry months (occurring in January through April) with less than 200 mm of rainfall (Sanford et al. 1994). Rainfall increases from March through July (averages 445 mm), then drops off during August through October to less than 350 mm per month, and finally rises again to above 400 mm per month in November and December (fig. 16.4). La Selva lies in the Tropical (Lowland) Wet Forest life zone (Holdridge 1967) and is slightly drier than most sites near Rincón de Osa. Holdridge et al. (1971) described the vegetation and soil at a typical site within the undisturbed forest at La Selva; at that site, the forest was of three tree strata with the canopy at 50 m. Of approximately 113 tree species at La Selva, 57 species represented by 404 individual trees per hectare occurred at the study site (Sawyer and Lindsay 1971). Hartshorn and Hammel (1994) reviewed the vegetation types and floristics of the site. Soils on this and other well-drained sites are strongly acid, dark yellowish brown clay latosols that become reddish brown at a depth of 1.5 m; concentrations of magnesium, calcium, potassium, and phosphorus are very low, with concentrations of sodium, manganese, and nitrogen only slightly higher. Sollins et al. (1994) reviewed the soils and soil processes at La Selva.

Barro Colorado Island

The island, previously a hilltop, was formed in 1914 as the result of the building of the Panama Canal and the damming of the Río Chagres in 1912 to create Gatun Lake. Barro Colorado Island lies in the rather low rolling country of central Panama, north of the continental divide on the Atlantic versant; it is approximately 14.5 km² in extent and lies at an elevation of 164 m above sea level and 138 m above Gatun Lake. The climate in the Köppen system is Am; the mean annual temperature is 26.9°C, and precipitation averages 2,750 mm per year, with four dry months (January through April) with less than 100 mm of rainfall. Rainfall increases markedly in April and remains at about 280 mm per month through August; after a slight decrease in amount in September, heavy rains return and peak in November at about 440 mm, just before falling to slightly under 300 mm for December (fig. 16.4). The island is in the Tropical (Lowland) Moist Forest life zone (Holdridge 1967). Croat (1978) listed 365 tree species from the island, of which 211 attain heights of 10 m or more. Bennett (1963) described the vegetation on the island, with the best-developed forest consisting of two tree strata with the canopy at around 30 m in height. Soils at sites with well-developed forest are slightly acid to alkaline, alluvial clay, red to deep-red in color, and somewhat brownish in the uppermost 5 cm. This site has been the focus of intensive and long-term study (see Leigh et al. 1982; Leigh 1999).

SPECIES COMPOSITION AND DIVERSITY

The composition and diversity of the herpetofaunas at the three sites are shown in appendix 16.2. The number of species and percentage that each amphibian and reptile group contributes to the herpetofauna at each site and to the total combined herpetofauna ($N = 215$) are shown in table 16.2. With two exceptions, the herpetofaunas at the three sites are similar in total numbers of species by group. One notable exception is the lower number of frog species recorded from Barro Colorado Island (33) compared with those recorded from La Selva (45) and Rincón de Osa (41). This difference may be directly related to the amount of available water (rainfall and habitats) on Barro Colorado Island, which has lower mean annual precipitation and a more marked dry season

Table 16.2 Herpetofauna among three sites in Lower Central America, combined, and at Santa Cecilia, Ecuador

| | Site | | | | | | | | | | | | | |
| | Rincón | | | La Selva | | | BCI | | | Combined | | Santa Cecilia | |
Taxon	N	%S	%T	N	%S	%T	N	%S	%T	N	%T	N	%T
Amphibia	46	40.0	21.4	49	35.5	22.8	36	33.3	16.7	82	38.1	86	49.7
Gymnophiona	1	0.9	0.5	1	0.7	0.5	1	0.9	0.5	3	1.4	3	1.7
Caudata	4	3.5	1.9	3	2.2	1.4	2	1.9	0.9	8	3.7	2	1.2
Anura	41	35.7	19.1	45	32.6	20.9	33	30.5	15.3	71	33.0	81	46.8
Reptilia	69	60.0	32.1	89	64.5	41.4	72	66.7	33.5	133	61.9	87	50.3
Testudines	3	2.6	1.4	5	3.6	2.3	5	4.6	2.3	7	3.3	6	3.5
Amphisbaenia	0	0.0	0.0	0	0.0	0.0	1	0.9	0.5	1	0.5	1	0.6
Sauria	22	19.1	10.2	25	18.1	11.6	22	20.3	10.2	43	20.0	27	15.6
Serpentes	42	36.5	19.5	57	41.3	26.5	42	38.9	19.5	80	37.2	51	29.5
Crocodilia	2	1.7	0.9	2	1.5	0.9	2	1.9	0.9	2	0.9	2	1.1
Totals	115	100	53.5	138	100	64.2	108	100	50.2	215	100	173	100

Source: Donnelly 1994b; Guyer 1994b; Rand and Myers 1990; Duellman 1978.

Note: N, number of species; %S, percentage of contribution of each taxonomic category to the herpetofauna at each site; %T, percentage of total herpetofauna.

(January to April) than the other sites (fig. 16.4). Because many species of amphibians depend on mesic microhabitats and adequate water for breeding, and because many forest species cannot survive long periods of drought, the differences in the amount of rainfall and its seasonal patterns almost certainly are responsible for some of the differences in anuran diversity. In addition to fewer suitable pond habitats than are found at Rincón and La Selva, BCI also lacks permanent streams and is missing many frogs associated with that habitat. One group that reflects this difference is the family Centrolenidae. Three species are known from Barro Colorado Island, but six species occur less than 10 km away on the Río Frijoles, a permanent stream (A. S. Rand, pers. comm.). Similar trends in comparative diversity occur among those amphibian species with direct development, namely, plethodontid salamanders and, in the La Selva to BCI comparison, frogs of the genus *Eleutherodactylus*. Finally, the insular nature of BCI also makes recolonization or invasion by rare species less likely than at the other two sites.

The other exception is the snake fauna, of which 57 species have been recorded from La Selva, 15 more than at BCI and at Rincón. We suggest that part of the difference in snake diversity between La Selva and the Osa Peninsula is the result of inadequate sampling at Rincón de Osa. Support for our assertion can be found by comparing the species list for Corcovado and the other sites to that of Rincón (appendix 16.2); some of the six species known only from elsewhere on the peninsula almost certainly will be taken near Rincón with additional sampling. Between La Selva and BCI, the differences are more likely real. Because fewer species of frogs and toads are part of the Barro Colorado Island herpetofauna and many snake species are frog specialists, the proportion of snake species at BCI (38.9%) is similar to that at La Selva (41.3%). A similar effect is seen in the proportion of reptiles in the total fauna at Barro Colorado Island. In contrast, the lizard faunas are similar in species number and percentages among the three sites. On the basis of these comparisons and our familiarity with herpetofaunas at other Neotropical sites, we expect that a typical herpetofauna from a lowland wet evergreen forest site in Lower Central America will approximate the group composition shown in table 16.2. A taxonomic breakdown of the groups that contribute more than 10% to the total fauna is shown in table 16.3.

Some interesting patterns emerge from comparisons within faunal groups among sites. Leptodactylid species diversity is higher than hylid diversity and contributes the most to frog diversity (34%–37%) at each site; it also is the highest contributor (11%–13%) among amphibians to the total diversity. Three other family groups (Bufonidae, Centrolenidae, Dendrobatidae) generally make up 4%–17% of the frog faunas and 1%–6% of the total herpetofauna. Two notable exceptions to the similarity among faunas in this comparison include the relatively high centrolenid diversity at Rincón (17.1% of frogs and 6.1% of

Table 16.3 Contribution by family of the major components (frogs, lizards, and snakes) of the herpetofaunas of Rincón de Osa, La Selva, and Barro Colorado Island

	Site								
	Rincón			La Selva			BCI		
Taxon	N	%G	%H	N	%G	%H	N	%G	%H
Amphibia	46	—	40.0	49	—	35.5	36	—	33.3
Anura	41	89.1	35.7	45	91.8	32.6	33	91.7	30.6
Bufonidae	4	9.8	3.5	3	6.7	2.2	3	9.1	2.8
Centrolenidae	7	17.1	6.1	7	15.6	5.1	3	9.1	2.8
Dendrobatidae	5	12.2	4.3	2	4.4	1.4	3	9.1	2.8
Hylidae	10	24.4	8.7	13	28.9	9.4	10	30.3	9.3
Leptodactylidae	14	34.1	12.2	16	35.6	11.6	12	36.4	11.1
Microhylidae	1	2.4	0.9	1	2.2	0.7	0	0	0
Ranidae	0	0	0	3	6.7	2.2	2	6.0	1.9
Reptilia	69	—	60.0	89	—	64.5	72	—	66.7
Sauria	22	31.9	19.1	25	28.1	18.1	22	30.6	20.4
Anguidae	0	0	0	3	12.0	2.2	0	0	0
Gekkonidae	3	13.6	2.6	4	16.0	2.9	4	18.2	3.7

	Rincón			La Selva			BCI		
	N	%G	%H	N	%G	%H	N	%G	%H
Corytophanidae	2	9.1	1.7	3	12.0	2.2	2	9.1	1.9
Iguanidae	1	4.5	0.9	1	4.0	0.7	1	4.5	0.9
Polychrotidae	7	31.8	6.1	9	36.0	6.5	10	45.5	9.3
Scincidae	2	9.1	1.7	2	8.0	1.4	1	4.5	0.9
Gymnophthalmidae	3	13.6	2.6	0	0	0	1	4.5	0.9
Teiidae	3	13.6	2.6	2	8.0	1.4	2	9.1	1.9
Xantusiidae	1	4.5	0.9	1	4.0	0.7	1	4.5	0.9
Serpentes	42	60.9	36.5	57	65.9	41.3	42	58.3	38.9
Anomalepididae	0	0	0	0	0	0	2	4.8	1.9
Boidae	2	4.8	1.7	2	3.5	1.4	3	7.1	2.8
Tropidophiidae	0	0	0	1	1.8	0.7	0	0	0
Colubridae	35	83.3	30.4	47	82.5	34.1	33	78.6	30.6
Elapidae	1	2.4	0.9	3	5.3	2.2	2	4.8	1.9
Viperidae	4	9.5	3.5	4	7.0	2.9	2	4.8	1.9

Note: N, number of species; %G, percentage of contribution of species in each category to the next higher taxonomic group; %H, percentage of contribution of species in each category to the total number of species at each site (Rincón, 115; La Selva, 138; BCI, 108).

the fauna) and La Selva (15.6% and 5.1%, respectively), and the low dendrobatid diversity at La Selva (4.4% and 1.4%, respectively). The high centrolenid diversity, at least at the Osa Peninsula, may reflect more thorough sampling for centrolenids, because one of us (RWM) focused considerable field time on this group (McDiarmid 1975, 1978), and at BCI, because of the lack of suitable breeding habitats. The low dendrobatid diversity at La Selva is surprising; these diurnal frogs usually are obvious components of a fauna, and we would expect any additional species (e.g., *Dendrobates auratus*) that occurs there to have been recorded previously.

Among lizards, species of polychrotids are the most diverse at all three sites and make up about 32%–46% of the lizard fauna and 6%–9% of the total fauna. The gekkonids are the next most diverse group (approximately 14%–18%) at each site, although at Rincón they are equaled by the gymnophthalmids and teiids. The low diversity of teiids (2 species; 8% of the lizards and 1.4% of the total fauna) and the absence of gymnophthalmids at La Selva, and the higher diversity of these two groups on the Osa Peninsula (3 species in each family, representing a combined 27.2% of lizards and 5.4% of the total fauna) are notable. We are unable to account for the low teiid–gymnophthalmid diversity at La Selva; one teiid, *Ameiva quadrilineata*, may now be extirpated (Guyer 1994). Other species of teiids and gymnophthalmids occur in the northeastern part of Costa Rica (Savage and Villa 1986) and eventually may be recorded from La Selva. The high anguid diversity at La Selva may explain the low teiid–gymnophthalmid diversity because some anguid species are ecologically similar (arboreal and leaf-litter species) to species of teiids and gymnophthalmids and functionally may operate in place of these groups at La Selva. Anguids are unknown at Rincón and BCI, but *Coloptychon* is known from near Golfito and probably occurs near Rincón on the Osa Peninsula.

The snake faunas among the three sites are similar (table 16.3). Colubrids make up approximately 79%–83% of snakes and 30%–34% of the total faunas at each site. Boids, elapids, and viperids each usually account for approximately 2%–10% of the snakes and 1%–3.5% of the totals. The low diversity of elapids from the Osa Peninsula (2.4% of snakes and 0.9% of the total fauna) may be a sampling artifact. At least one, possibly two, other species of *Micrurus* are known from the Golfo Dulce area, and one has been taken in the Corcovado National Park (appendix 16.2).

Duellman (1966, 711–12) compared the numbers of species of amphibians and reptiles from six areas of lowland evergreen forest in Mexico and Central America. His comparisons were based on a survey of samples from relatively large geographic units (e.g., Limón Province, Costa Rica) of substantially greater area than the three intensively studied sites considered here. Generally speaking, species diversity for the large areas would be expected to be higher than for a specific site within one area, because more habitats occur in the for-

Table 16.4   Number of species of amphibians and reptiles
recorded from each of six areas of lowland evergreen forest in
Mexico and Central America

Area	No. species	Percentage
Southern Veracruz, Mexico	93	38.9
El Peten, Guatemala	84	35.1
Bonanza area, Nicaragua	113	47.3
Limón Province, Costa Rica	127	53.1
Caribbean Canal Zone, Panama	125	52.3
Golfo Dulce area, Costa Rica	94	39.3

Source: Duellman 1966.
Note: The percentage is the proportion of the combined fauna of
239 species found in each region times 100. The current Golfo Dulce
herpetofauna is estimated to be approximately 156 species.

mer. Nevertheless, the comparisons are instructive although somewhat out of
date because more sampling has occurred in the intervening years, particularly
in the Osa portion of the Golfo Dulce region. The total number of species represented in the six areas is 239. The number of species known from each area in
1966 and the percentages of the combined fauna that they represent are shown
in table 16.4.

Aside from the Golfo Dulce area, which is now known (Savage 2002) to have
a herpetofauna of about 156 species (excluding marine forms and *Conophis*),
these data are remarkably similar to ours when the proportions of total available species are compared by area. Thus, the percentage for Limón Province
(53.1%, table 16.4) and that for La Selva (64.2%) and Barro Colorado Island
(50.2%) from table 16.2 (%T) tend to confirm that the area of highest species
richness in the Atlantic versant herpetofauna is in northeastern Costa Rica. The
highest species richness in lowland evergreen forests west and north of the
Darien, however, seems to be in the Golfo Dulce forests. Our data suggest that
although the greatest number of species at a single intensively collected site
within these lowland evergreen forests is at La Selva, it is probable that further
work will reveal that the highest single-site richness is at a Golfo Dulce forest site
and likely on the Osa Peninsula.

Additional data on the overall herpetofaunal similarities at the three sites
provide another kind of comparison. Table 16.5 presents values for the coefficient of faunal resemblance (Simpson 1960) for pair-wise (irrespective of
species occurrence at other sites) and restricted (species known only from the
two sites) pair-wise comparisons. Of the 43 (20%) species in common to all
three sites (the percentage of the total herpetofauna for three sites is given in

Table 16.5   Herpetofaunas from three tropical lowland sites in Central America: Rincón de Osa, Puntarenas Province, Costa Rica; La Selva Biological Station, Heredia Province, Costa Rica; and Barro Colorado Island, Panama Province, Panama

Site	Rincón	La Selva	BCI
Rincón	**115**	70 (60.9)	58 (52.8)
La Selva	27 (23.5)	**138**	62 (58.3)
BCI	14 (13.0)	20 (18.5)	**108**

Note: Values in boldface are total species diversity; values above the diagonal are total species in common between sites and the coefficient of similarity (in parentheses); values below the diagonal are number of species restricted to the two sites and Simpson's coefficient of similarity (in parentheses).

parentheses), 11 (5.1%) are amphibians (all anurans) and 32 (14.9%) are reptiles, as follows: 1 (0.5%) turtle, 9 (4.2%) lizards, 20 (9.3%) snakes, and 2 (0.9%) crocodilians. Of the 27 (12.6%) species that occur only at La Selva–Rincón, 1 (0.5%) is a salamander, 11 (5.1%) are frogs, 3 (1.4%) are lizards, and 12 (5.6%) are snakes; 14 (6.5%) species in common to Osa Peninsula–Barro Colorado Island include 7 (3.2%) frogs, 1 (0.5%) turtle, 3 (1.4%) lizards, and 3 (1.4%) snakes; the 20 (9.3%) species shared uniquely by Barro Colorado Island–La Selva include 8 (3.7%) frogs, 3 (1.4%) turtles, 2 (0.9%) lizards, and 7 (3.3%) snakes.

The actual number (58) of species shared between the Rincón and Barro Colorado Island samples is lower than that shared between La Selva and the Osa Peninsula (70). The coefficients of faunal resemblance of pair-wise comparisons (table 16.5) indicate a strong similarity among all three sites (also see Donnelly 1994a; Guyer 1994a). The Rincón fauna is more similar to that at La Selva than to the BCI fauna; BCI fauna is more similar to La Selva fauna than to that at Rincón. Restricted pair-wise comparisons (table 16.5) indicate greatest similarity between Rincón and La Selva, followed by La Selva and BCI, with Rincón and BCI being the least similar. The faunal similarity between Rincón and La Selva is even stronger than that indicated by the coefficient comparisons because several species pairs (i.e., species whose closest relative occurs at the other site) are known from Rincón and La Selva, respectively, but not from Barro Colorado Island. Among these are *Oedipina pacificensis* and *O. gracilis*, *Dendrobates granuliferus* and *D. pumilio*, *Phyllobates vittatus* and *P. lugubris*, *Eleutherodactylus taurus* and *E. ranoides*, *Sphaerodactylus graptolaemus* and *S. homolepis*, and *Urotheca fulviceps* and *U. pachyura* (appendix 16.2). Interestingly, 48 species known from La Selva do not occur at the other sites. By contrast, 31 species from Rincón and 31 from Barro Colorado Island are not known from the other two sites. The difference between La Selva and Barro Colorado Island probably reflects the greater species richness at the former locale, and the presence of unique species of southern affinities at the latter. The difference between La Selva and Rincón probably reflects the greater intensity of sampling at the former, particularly among snakes.

Santa Cecilia (Napo Province, Ecuador) is the only other Neotropical evergreen lowland forest site where sampling effort has been comparable to that expended at the three Lower Central American sites just discussed. This upper Amazonian locality was sampled from 1966 to 1973 by field teams from the University of Kansas. The herpetofauna at Santa Cecilia, not including species taken at other nearby localities along the Río Aguarico, comprises 86 species of amphibians and 87 species of reptiles, for an astounding total of 173 species. The general composition of this herpetofauna, including the percentage of each group in the total fauna, is presented in table 16.2.

Comparisons with the three Central American sites indicate that differences between Santa Cecilia and those areas lie in the larger amphibian fauna at the Ecuadorian locale. At Santa Cecilia, amphibians are much more diverse (86 species) and make up 49.7% of the fauna, compared with 33.3%–40.0% at the three Central American sites. Although the anurans at Santa Cecilia make up 47% of the total fauna, the values are 33%, 36%, and 31% for La Selva, Rincón, and Barro Colorado Island, respectively. Among frogs, hylids are more diverse than leptodactylids at Santa Cecilia (less diverse in Central America) and represent 21.4% and 14.4% of the fauna, respectively. For hylids, this is more than twice the contribution to the fauna than in Central America (8.7%–9.4%). The species diversity and percentage contribution for the other frog families are similar to those of the Central American faunas. Lizard diversity is about the same, with 27 species contributing 15.6% of the fauna. At Santa Cecilia, gymnophthalmids (9 species, 5.2%) are more diverse than polychrotids (6 species, 3.5%), whereas at the Central American sites, polychrotids (7–10 species, 6.1%–9.3%) are more diverse than gymnophthalmids (0–3 species, 0%–2.6%). Diversities of other lizard families that are in common are comparable. Interestingly, lizards seem to show more family-level endemism between areas than do amphibian and other reptile groups. Snake diversity (51 species) is lower than that recorded for La Selva (57) but higher than that at Rincón (42) and Barro Colorado Island (42) (table 16.2). The proportion of snakes within the total fauna is lower (29.5%) than for that at La Selva (41.3%), Rincón (36.5%), or Barro Colorado Island (38.9%). Colubrids (36 species) contribute less to snake diversity at Santa Cecilia (70.6% of the snake fauna) than in Central America (33–47 species, 78.6%–83.3%), whereas boids (5 species, 9.8%, at Santa Cecilia vs. 2–3 species, 3.5%–7.1%, at the Central American sites) and elapids (5 species, 9.8% versus 1–3 species, 2.4%–5.3%) have higher diversity.

These differences and similarities, in part, may be related to the amount and pattern (or lack thereof) of precipitation at Santa Cecilia. This locality is just north of the equator at 340 m elevation and normally receives between 4,200 and 4,300 mm of rainfall per year. There is no definite dry season, but the amount of rainfall is erratic, with some months having in excess of 500 mm and some less than 300 mm. The higher humidity in that area seemingly favors higher frog-species richness compared with that at Barro Colorado Island, which averages much less annual rainfall (2,758 mm) and has a definite dry season. La Selva also averages less annual rainfall (3,850 mm) and has a short dry season, and Rincón (4,576 mm of rainfall/year) has a definite dry season.

Only 4 frogs, 1 turtle, 2 lizards, 11 snakes, and a caiman occur both at Santa Cecilia and at one or more of the Central American sites; suffice it to say that Amazonian and Central American herpetofaunas are very different (also see Donnelly 1994a; Guyer 1994a). Of the species found at Santa Cecilia, 153 (88%) of 173 do not occur at any of the three Central American sites. Or to put it another way, the coefficients of similarity between Santa Cecilia and La Selva, Rincón, and Barro Colorado are 8%, 12%, and 14%, respectively, and essentially reflect the increasing geographic distances of each from Santa Cecilia.

Biogeographic Considerations

The Península de Osa is part of the recently uplifted Isthmian Link that reunited the North and South American continents after their separation during most of the Cenozoic by the marine waters of the Panamanian portal. Not surprisingly, its herpetofauna consists primarily of a mixture of (*a*) autochthonous Central American taxa and (*b*) South American groups that apparently dispersed into Lower Central America through the Isthmian connection. In terms of its herpetofauna the peninsula is part of the Golfo Dulce area of endemism (Savage 1982) that extends along the Pacific versant from central Costa Rica to extreme southwestern Panama. A number of forms that occur elsewhere in the Golfo Dulce region and as yet are not reported for the Península de Osa probably occur there. For purposes of this section, only taxa known from the peninsula are included, but the description of historical events and conclusions are applicable to the region as a whole.

At the present time, the Golfo Dulce rainforests form an island of evergreen vegetation separated from contact with similar forest stands by a series of substantial barriers. The first of these is the main mountain axis (2,500–3,800 m) of Lower Central America that borders the Golfo Dulce region on the north and northeast. On the northwest, the region abuts an area of deciduous forest that forms the southern boundary of an extensive subhumid to arid region extending through the Pacific lowlands from Costa Rica to northern Mexico. On the southeast, a lowland, dry forest barrier in Pacific western Panama borders the

Golfo Dulce region. Farther to the east on the Pacific slope of Panama, evergreen forest alternates with subhumid savanna or dry forest habitats to form a filter barrier to exchanges between the rainforests of northwestern South America and areas to the west, including the Golfo Dulce rainforests.

Three major source units have contributed to the herpetofauna of the Golfo Dulce area following the system developed by JMS (Savage 1966, 1982):

1. Old Northern Element (Central American component)—a component of Laurasia affinities that has been disjunct from more northern components of the Old Northern Element during most of later Tertiary and Quaternary and has evolved in situ in tropical Mesoamerica; its isolation was produced by the development of a temperate, semiarid to arid climatic barrier to the north;
2. South American Element—stocks that evolved in situ in South America through its isolation during most of the Cenozoic; and
3. Middle American Element—groups anciently related to South American ones that evolved in situ in tropical North America during the Cenozoic.

For purposes of this account, we recognize 83 major lineages of amphibians and reptiles (exclusive of marine species, introduced forms, and *Conophis*) as represented on the Osa Peninsula as a whole. In our view, major lineages are equivalent to genera, subgenera (in the case of the genus *Eleutherodactylus*), and species groups (for the polyphyletic genera *Bufo* and *Hyla*). In the following comparisons, we indicate the percentage of the total major lineages in the herpetofauna belonging to each of the three biogeographic units described, followed (in parentheses) by the percentage of total species in the fauna belonging to each unit: Central American component of the Old Northern Element 28% (23%), Middle American Element 36% (40%), and South American 36% (37%). Thus, the herpetofauna was assembled primarily from autochthonous Mesoamerican taxa, 64% (63%), with a strong South American representation.

The present distributions and ecologic valence of these lineages and the individual species ally the herpetofauna of the Golfo Dulce region most closely with those from the lowland evergreen forests of the Atlantic versant of Lower Central America (eastern Nicaragua to eastern Panama) and the Pacific versant of Panama. These faunas consist of a base core of Mesoamerican taxa that was enriched by the dispersal of South American groups, primarily from the lowland evergreen forest region of northwestern South America, across the completed Isthmian Link beginning in the Pliocene. Most of these South American groups today occur no farther north than the Golfo Dulce region on the Pacific versant of Central America, while many range northward to various latitudes between Costa Rica and Mexico in the Atlantic slope evergreen forests.

The major steps in the development of the herpetofauna of the area may be summarized as follows:

1. Dispersal event I. Initially the region formed part of a Central American peninsula that gradually emerged from north to south during mid- to late Tertiary; by early Miocene (20 mya), representatives of the Old Northern (Central American component) and Middle American elements were doubtless widespread throughout Central America, and rainforest vegetation was dominant and continuous in distribution; by this time, at the latest, the Golfo Dulce region had a rainforest herpetofauna composed of these elements.

2. Vicariance I. Orogenic activity beginning at about this time led to the fragmentation of rainforest areas into Atlantic and Pacific lowland segments, and associated climatic changes brought about the replacement of lowland rainforest and associated herpetofaunas along the Pacific versant from Costa Rica northward.

3. Dispersal II. With the relinking of Central and South America by the Panamanian Isthmus (approximately 3 mya), representatives of the South American Element dispersed northward into rainforest areas but were prevented from ranging north of the Golfo Dulce region on the Pacific versant by the now fully developed lowland dry forest and thorn woodland zone that extends from northern Mexico to Costa Rica.

4. Vicariance II. Drying trends during the last one to two million years fragmented the formerly continuous lowland rainforest of Pacific Costa Rica and Panama and effectively isolated the Golfo Dulce region when deciduous forest developed to the southwest in western Panama.

From the herpetofaunal perspective, the Golfo Dulce region is an area of endemism among the lowland evergreen forest environments of Central America. Currently, 12 species are recognized Golfo Dulce region endemics: *Oscaecilia osae, Oedipina alleni, Oedipina pacificensis, Dendrobates granuliferus* (Myers et al. 1995 reported the occurrence of a population of *D. granuliferus* from the southeastern Atlantic lowlands of Costa Rica; however, there remains a question about whether these animals were accidentally introduced by the Dirección de Fauna Silvestre through the release of confiscated frogs), *Phyllobates vittatus, Eleutherodactylus rugosus, Eleutherodactylus taurus, Sphaerodactylus graptolaemus, Norops aquaticus, Bachia blairi, Neusticurus apodemus,* and *Lachesis melanocephala.* At least 7 other forms—*Bufo melanochlorus, Hyalinobatrachium valerioi, Colostethus flotator, Eleutherodactylus cruentus, Micrurus alleni, Bothrops asper,* and *Porthidium nasutum*—have distinctive Golfo Dulce populations that may well be regarded as separate species by future workers. Of these 19 taxa, 11 are of Mesoamerican and 8 of South American origins. This suggests that Mesoamerican ancestors were in the Golfo Dulce region somewhat earlier than the South American groups, indicating slightly different effects and timing of isolation on speciation events. Currently, it is not possible to determine which of the two vicariance events described was principally responsible for the development of the endemic taxa in this herpetofauna, because detailed studies of relationships have not been made for most of them. Mesoamerican taxa may have been derived from ancestral populations isolated from their congeners by the rise of the Talamanca-Baru cordillera to the east and northeast and the development of the deciduous forest environments to the north. However, the fact that both Mesoamerican and South American groups are involved requires that final isolation of the region from southern less-humid habitats occurred after South American lineages were present (i.e., sometime after the Panamanian portal was closed).

Postscript

During the years since our most concentrated fieldwork in the Rincón area, many changes have occurred on the Península de Osa. Most of the forests in which our work was accomplished have been cut, the old Osa Field Station near the airfield has been demolished, and subsistence farming has replaced much of the once imposing natural scene. Nevertheless, Corcovado National Park remains essentially intact. A road now connects Rincón around the head of the Golfo Dulce with the Carretera Interamericana and continues south through Puerto Jiménez to the tip of the peninsula.

Corcovado National Park faced a major crisis in its early days, when the Park Service had to remove the numerous squatters who had established themselves within park boundaries. The squatters did not want to leave so only through payments totaling $1.7 million were they bought out. Unfortunately, many resettled on the eastern portion of the peninsula outside the park limits.

Another crisis threatened the park in 1985. The new road had brought a flood of eager gold miners to the peninsula, many of whom were also local subsistence farmers and/or former squatters in the park, but most were refugees from the mainland, where cessation of banana operations had left many unemployed. Major placer mining companies had established themselves in the forest "preserves" around the park so that the only way left for individual miners to operate was by illegal gold panning and mining within the park itself. As the crisis developed, some 1,400 miners and many of their hangers-on were in the park seeking gold. Ultimately the miners were evicted at considerable cost, both financially and politically, to everyone involved, but even today illegal mining and logging take place in the park because of understaffing of the Park Service (Wallace 1992).

The 1990s saw a resurgence of ecotourism on the Osa Peninsula, and today a steady stream of visitors can be seen making their way south along the road to the tip of the peninsula, where they can hike into Corcovado National Park. Between Puerto Jiménez and Carate there are now several privately owned nature lodges set amid large tracts of the remaining forest. The volume of visitors to Corcovado National Park has led the Park Service to give serious consideration to limiting the number of visitors entering the park. Threats to the integrity of the Osa forests have not ceased. In the mid-1990s, serious consideration was given to building a causeway across the Golfo Dulce from Golfito to Puerto Jiménez, a project since abandoned. A proposed chip mill, although based on harvesting plantations of nonnative trees, also set off alarms. Environmentalists saw seed dispersal of nonnative species, chemical pollution of the gulf, and other effects as affecting Corcovado National Park and, together with the local community, have resisted this development.

Fortunately, the Costa Rican people remain committed to the preservation of much of the majesty and profound beauty of the Osa forests through continuing protection within the boundaries of Corcovado National Park. The Park Service has increased efforts to integrate the local community adjacent to the park boundaries more fully into park programs and operations for mutual benefit, especially along the vulnerable northeast margin and around Puerto Jiménez. The economy of the region is now substantially supported by ecotourism involving local Costa Ricans. These developments give promise for long-term stability in maintaining the remnants of the Golfo Dulce rainforests for posterity.

We were privileged to work in the Osa rainforest at a time when it covered the entire peninsula and when few others had the opportunity to experience its magnificence. How can we ever forget seeing our first harpy eagle and king vulture, hearing the peccaries gnashing their teeth in the woods and the howlers howling in the trees 50 m above, or sighting a 2-m-long, black-headed bushmaster barely visible against the leaf litter on the forest floor. But most of all we recall the nights spent in the great forest with good companions searching for glassfrogs along a tropical stream.

Appendix 16.1

Table 16A.1 History of collections: dates and collectors of amphibians and reptiles from the vicinity of Rincón de Osa, Osa Peninsula, Puntarenas Province, Costa Rica

Year	Dates	Collector(s)	Materials
1961	Summer	Lewis Steinmetz	1 *Leptodactylus bolivianus*
1962	13–18 August	Los Angeles County Museum field party: R. Casebeer, C. McLaughlin, A. Schoenherr, and F. Truxal	general collection
	17–20 August	J. M. Savage and C. F. Walker	general collection
1964	16–19 June	D. Huckaby, J. M. Savage, N. J. Scott, and C. F. Walker	general collection
1965	1–4 March	D. H. Janzen	a few specimens
1966	2–10 March	R. W. McDiarmid in OTS course	general collection
	8–13 November	D. P. Paulson	general collection
1967	7–21 March	D. P. Paulson with OTS course	general collection
	21–22 March	R. W. McDiarmid	general collection
	9 May	D. P. Paulson	2 centrolenids
	22–29 July	A. Starrett with OTS course	frogs and tadpoles
	1–11 August	R. W. McDiarmid with OTS course	general collection
	11 August	W. T. O'Day	tadpoles
1968	27 February–14 March	N. J. Scott with OTS course	general collection
	22 July–6 August	J. M. Savage with OTS course	general collection
	7–13 August	N. J. Scott with OTS course	general collection
	20–29 November	I. Straughan	frogs
1969	5–10 March	N. J. Scott with OTS course	leaf-litter fauna
	14–22 July	N. J. Scott	general collection
	27 July–10 August	R. W. McDiarmid with OTS course	general collection
1970	4 March	N. J. Scott with OTS course	leaf-litter fauna
	30 July–10 August	A. Starrett with OTS course	tadpoles
	8 August	N. J. Scott with OTS course	leaf-litter fauna
1971	2–18 May	R. W. McDiarmid with OTS course	general collection

(*continued*)

Table 16A.1   (continued)

Year	Dates	Collector(s)	Materials
1973	17–20 March	J. M. Savage and R. T. Harris with OTS course	frogs
	9–22 June	R. W. McDiarmid, W. R. Heyer, and J. J. Talbot	general collection
	30 June–11 July	J. M. Savage and GORE field parties	general collection
	7 July–31 August	R. W. McDiarmid and students	general collection

Note: Field activities after 1973 are summarized in the text.

Appendix 16.2

This appendix lists all 135 species of amphibians (49) and reptiles (86) reported from the Osa Peninsula, Puntarenas Province, Costa Rica. Records of species from the vicinity of Rincón and the rest of the Osa Peninsula are listed separately. Records for sea turtles and sea snakes are based on sightings along the coast, inside and outside of Corcovado National Park. They are included for completeness, as are records for two introduced geckos. For comparison, all species of amphibians and reptiles known from La Selva Biological Station, Costa Rica, and Barro Colorado Island (BCI), Panama, are included.

Table 16A.2   Herpetofaunal composition and species richness

	Site			
Taxon	Rincón	Osa	La Selva	BCI
Amphibia	(46)	(47)	(49)	(36)[a]
Gymnophiona	(1)	(2)	(1)	(1)
Caeciliidae				
Dermophis occidentalis	+	SI,MA		
Gymnopis multiplicata			+	
Oscaecilia ochrocephala				+
Oscaecilia osae		SI,PT		

Taxon	Site			
	Rincón	Osa	La Selva	BCI
Caudata	(4)	(4)	(3)	(2)
Plethodontidae				
Bolitoglossa colonnea	+	MA	+	
Bolitoglossa lignicolor	+	SI,LC,MA		
Oedipina alleni	+	SI,LC		
Oedipina complex				+
Oedipina cyclocauda			+	
Oedipina gracilis			+	
Oedipina pacificensis	+	SI,MA		
Oedipina parvipes				+
Anura	(41)	(41)	(45)	(33)
Bufonidae	(4)	(4)	(1)	(1)
Bufo coniferus	+	MA	+	
Bufo granulosus				+[b]
Bufo haematiticus	+	SI,MA	+	
Bufo marinus	+	SI,MA	+	+
Bufo melanochlorus	+	SI,MA		
Bufo cf. *typhonius*				+
Centrolenidae	(7)	(6)	(7)	(3)
Centrolene prosoblepon	+	LL	+	+
Cochranella albomaculata	+	MA	+	
Cochranella granulosa	+	SI,MA	+	
Cochranella spinosa	+		+	+
Hyalinobatrachium colymbiphyllum	+	MA		
Hyalinobatrachium fleischmanni			+	+
Hyalinobatrachium pulveratum	+	SI	+	
Hyalinobatrachium valerioi	+	SI,MA	+	
Dendrobatidae	(5)	(5)	(2)	(3)
Colostethus flotator	+	SI,LL		+
Colostethus talamancae	+	SI,LL,MA		
Dendrobates auratus	+	SI,MA		+
Dendrobates granuliferus	+	SI,MA		
Dendrobates pumilio			+	
Minyobates minutus				+[b]
Phyllobates lugubris			+	
Phyllobates vittatus	+	SI,MA		
Hylidae	(10)	(11)	(13)	(10)
Agalychnis calcarifer			+	+
Agalychnis callidryas	+	SI,MA	+	+

(*continued*)

Taxon	Site			
	Rincón	Osa	La Selva	BCI
Agalychnis saltator			+	
Agalychnis spurrelli	+	SI,LC,MA		+
Hyla ebraccata	+	SI,MA	+	
Hyla loquax			+	
Hyla microcephala		MA		+
Hyla phlebodes			+	+
Hyla rosenbergi	+	SI,MA		
Hyla rufitela			+	+
Phrynohyas venulosa	+	SI,MA		+
Scinax boulengeri	+	SI,MA	+	+
Scinax elaeochroa	+	SI,MA	+	
Smilisca baudinii			+	
Smilisca phaeota	+	SI,MA	+	+
Smilisca puma			+	
Smilisca sila	+	SI,MA		+
Smilisca sordida	+	SI,MA	+	
Leptodactylidae	(14)	(14)	(16)	(12)
Eleutherodactylus altae			+	
Eleutherodactylus bransfordii			+	
Eleutherodactylus bufoniformis				+
Eleutherodactylus caryophyllaceus			+	
Eleutherodactylus cerasinus			+	+
Eleutherodactylus crassidigitus	+	SI,MA	+	+
Eleutherodactylus cruentus	+	MA	+	
Eleutherodactylus diastema	+	SI,MA	+	+
Eleutherodactylus fitzingeri	+	SI,MA	+	+
Eleutherodactylus gaigeae				+
Eleutherodactylus megacephalus			+	+
Eleutherodactylus mimus			+	
Eleutherodactylus noblei			+	
Eleutherodactylus ranoides			+	
Eleutherodactylus ridens	+	SI,MA	+	+
Eleutherodactylus rugosus	+	SI,MA		
Eleutherodactylus stejnegerianus	+	SI,MA		
Eleutherodactylus taeniatus				+
Eleutherodactylus talamancae			+	
Eleutherodactylus taurus	+	SI,MA		
Eleutherodactylus vocator	+	SI,MA		
Leptodactylus bolivianus[b]	+	SI,MA		+
Leptodactylus melanonotus	+	MA	+	

Taxon	Site			
	Rincón	Osa	La Selva	BCI
Leptodactylus pentadactylus	+	SI,MA	+	+
Leptodactylus poecilochilus	+	RN*,MA		
Physalaemus pustulosus	+	MA		+
Microhylidae	(1)	(0)	(1)	(0)
Gastrophryne pictiventris			+	
Nelsonophryne aterrima	+			
Ranidae	(0)	(1)	(3)	(2)
Rana vaillanti			+	+
Rana taylori			+	
Rana warszewitschii		PT,LC,MA	+	+[c]
Reptilia	(69)	(75)	(89)	(72)
Testudines	(3)	(8)	(5)	(5)
Chelydridae	(0)	(1)	(1)	(1)
Chelydra serpentina		LC,LS	+	+
Kinosternidae	(2)	(2)	(2)	(1)
Kinosternon angustipons			+	
Kinosternon leucostomum	+	SI,RN*	+	+
Kinosternon scorpioides	+	SI		
Emydidae	(1)	(1)	(2)	(3)
Rhinoclemmys annulata			+	+
Rhinoclemmys funerea			+	+
Trachemys ornata	+	SI,MA		+
Cheloniidae	(0)	(3)	(0)	(0)
Chelonia mydas		marine		
Eretmochelys imbricata		marine		
Lepidochelys olivacea		marine		
Dermochelyidae	(0)	(1)	(0)	(0)
Dermochelys coriacea		marine		
Squamata—Sauria	(22)	(25)	(25)	(22)
Amphisbaenidae	(0)	(0)	(0)	(1)
Amphisbaea fuliginosa				+
Anguidae	(0)	(0)	(3)	(0)
Celestus hylaius			+	
Diploglossus bilobatus			+	
Diploglossus monotropis			+	
Gekkonidae	(3)	(6)	(4)	(4)
Gonatodes albogularis		MA		+
Hemidactylus frenatus[d]		PJ		
Lepidoblepharis sanctaemartae				+

(*continued*)

Taxon	Site			
	Rincón	Osa	La Selva	BCI
Lepidoblepharis xanthostigma	+	SI,MA	+	
Lepidodactylus lugubris[d]		QU		
Sphaerodactylus graptolaemus	+	SI,MA		
Sphaerodactylus homolepis			+	
Sphaerodactylus lineolatus				+
Sphaerodactylus millipunctatus			+	
Thecadactylus rapicauda	+	SI,MA	+	+
Corytophanidae	(2)	(2)	(3)	(2)
Basiliscus basiliscus	+	SI,MA,PJ*		+
Basiliscus plumifrons			+	
Basiliscus vittatus			+[e]	
Corytophanes cristatus	+	SI,LC,MA	+	+
Iguanidae	(1)	(2)	(1)	(1)
Ctenosaura similis		SI,PJ*,PJ,MA		
Iguana iguana	+	SI,MA,LC,PJ*	+	+
Polychrotidae	(7)	(6)	(9)	(10)
Dactyloa frenatus				+
Dactyloa insignis	+			
Norops aquaticus	+	SI,PA,MA		
Norops auratus				+
Norops biporcatus	+	SI,MA	+	+
Norops capito	+	SI,PA,MA	+	+
Norops carpenteri			+	
Norops humilis			+	
Norops lemurinus			+	
Norops limifrons	+	SI,LC,MA	+	+
Norops lionotus				+
Norops oxylophus			+	
Norops pentaprion	+	SI,MA,PJ*	+	+
Norops polylepis	+	SI,LC,MA,PJ*		
Norops vittigerus				+
Norops sp.				+
Polychrus gutturosus			+	+
Scincidae	(2)	(2)	(2)	(1)
Mabuya unimarginata	+	SI,MA	+	+
Sphenomorphus cherriei	+	SI,PA,MA	+	
Gymnophthalmidae	(3)	(3)	(0)	(1)
Bachia blairi	+	SI,MA		
Leposoma southi	+	SI,MA		+
Neusticurus apodemus	+	MA		

Taxon	Site			
	Rincón	Osa	La Selva	BCI
Teiidae	(3)	(3)	(2)	(2)
Ameiva festiva	+	SI	+	+
Ameiva leptophrys	+	SI,LC,MA		+
Ameiva quadrilineata	+	SI,PA,MA	+[e]	
Xantusiidae	(1)	(1)	(1)	(1)
Lepidophyma flavimaculatum			+	+
Lepidophyma reticulatum	+	SI,MA		
Squamata—Serpentes	(42)	(44)	(57)	(42)
Anomalepididae	(0)	(0)	(0)	(2)
Anomalepis mexicanus				+
Liotyphlops albirostris				+
Boidae	(2)	(2)	(2)	(3)
Boa constrictor	+	SI,MA	+	+
Corallus annulatus			+	+
Corallus ruschenbergerii	+	SI,MA		
Epicrates cenchria				+
Tropidophiidae	(0)	(1)	(1)	(0)
Ungaliophis panamensis		SI	+	
Colubridae	(35)	(34)	(47)	(33)
Amastridium veliferum	+	SI,MA	+	+
Chironius carinatus	+	SI		+
Chironius grandisquamis	+	SI,MA	+	+
Clelia clelia	+	SI,MA	+	
Coniophanes fissidens	+	SI,MA	+	+
Conophis lineatus		SI[f]		
Dendrophidion percarinatum	+	SI	+	+
Dendrophidion vinitor			+	
Dipsas tenuissma	+	SI		
Dipsas variegata				+
Drymarchon melanurus	+	PA	+	+
Drymobius margaritiferus			+	
Drymobius melanotropis			+	
Drymobius rhombifer			+	
Enulius flavitorques				+
Enulius sclateri		MA	+	+
Erythrolamprus bizona				+
Erythrolamprus mimus	+	PA	+	
Geophis hoffmanni	+	SI	+	+
Hydromorphus concolor		SI,MA	+	

(*continued*)

Taxon	Site			
	Rincón	Osa	La Selva	BCI
Imantodes cenchoa	+	SI,LC,MA	+	+
Imantodes gemmistratus				+
Imantodes inornatus	+	SI	+	
Lampropeltis triangulum			+	
Leptodeira annulata			+	+
Leptodeira rubricata		PJ		
Leptodeira septentrionalis	+	SI,MA	+	+
Leptophis ahaetulla	+	SI,LC,MA,PJ*	+	+
Leptophis depressirostris			+	
Leptophis nebulosus			+	
Leptophis riveti	+			
Liophis epinephalus			+	+
Mastigodryas melanolomus	+	SI,MA	+	+
Ninia maculata	+	MA	+	+
Ninia sebae	+		+	
Nothopsis rugosus	+		+	
Oxybelis aeneus	+	SI,MA	+	+
Oxybelis brevirostris			+	
Oxybelis fulgidus			+	
Oxyrhopus petolarius	+	SI,MA	+	+
Pseudoboa neuwiedii				+
Pseustes poecilonotus	+	SI,MA	+	+
Rhadinaea decorata	+	SI,MA	+	+
Scaphiodontophis annulatus	+	SI,MA	+	
Sibon annulatus			+	
Sibon dimidiatus	+			
Sibon longifrenis			+	
Sibon nebulatus	+	SI,MA	+	
Siphlophis cervinus				+
Spilotes pullatus	+	SI,MA,PJ*	+	+
Stenorrhina degenhardtii	+			+
Tantilla albiceps				+
Tantilla melanocephala				+
Tantilla reticulata			+	
Tantilla ruficeps	+	MA	+	
Tantilla schistosa	+	SI		
Tantilla supracincta	+	PA	+	
Tretanorhinus nigroluteus			+	
Trimetopon barbouri				+
Trimetopon pliolepis			+	

Taxon	Site			
	Rincón	Osa	La Selva	BCI
Tripanurgos compressus	+	SI		
Urotheca decipiens		SI	+	
Urotheca euryzona			+	+
Urotheca fulviceps	+			+
Urotheca guentheri	+	PA	+	
Xenodon rhabdocephalus	+	SI	+	+
Elapidae	(1)	(3)	(3)	(2)
Micrurus alleni	+	SI,MA	+	
Micrurus mipartitus			+	+
Micrurus nigrocinctus		SI,MA	+	+
Pelamis platurus		marine		
Viperidae	(4)	(4)	(4)	(2)
Bothriechis schlegelii	+	SI,MA	+	+
Bothrops asper	+	SI,LC,MA	+	+
Lachesis melanocephala	+	SI		
Lachesis stenophrys			+	
Porthidium nasutum	+	MA	+	
Crocodilia	(2)	(2)	(2)	(2)
Alligatoridae	(1)		(1)	(1)
Caiman crocodilus	+	SI	+	+
Crocodylidae	(1)		(1)	(1)
Crocodylus acutus	+	LC,MA	+	+
Total species	(115)	(126)	(138)	(108)

Note: Numbers in parentheses indicate the number of species in each taxonomic category for a particular site. Most records of species at localities other than Rincón are within Corcovado National Park from the vicinity of the former headquarters at Sirena (SI). Other records from within the park are noted as follows: LC, Laguna Corcovado; LS, Laguna Sirena; LL, Llorona; PA, Pavo Forest; PT, Los Patos. Specimens from localities outside the park were reported by Wettstein (1934) from Puerto Jiménez (PJ) and Río Nuevo (RN); others have been recorded from Marenco Biological Station (MA), Puerto Jiménez (PJ), Quemada (QU), and marine waters and beaches along the outer coast (marine). Species recorded by Wettstein (1934) are indicated by an asterisk (*).

[a] Campbell 1999 listed 52 species of amphibians from Barro Colorado Island, but this value included species from the adjacent mainland (Rand and Myers 1990).

[b] Some authors have used *Leptodactylus insularum* for this species. We agree that the Central American species is probably not *L. bolivianus* but prefer to maintain that name until definitive evidence about assignment of names is published.

[c] Species formerly known from Barro Colorado Island but now extirpated (Myers and Rand 1969).

[d] Nonnative species introduced on the Península de Osa.

[e] Species formerly known from La Selva but now possibly extirpated (Guyer 1994a).

[f] Species identification questionable; see text.

Appendix 16.3   Collecting Localities and Other Important Sites on the Osa Peninsula

Airfield Water Supply: Water-collecting site on a small, steep tributary of Quebrada Aguabuena, north of the airfield and 3 km west of Rincón; most specimens were taken along the lower reaches of the stream (60–80 m), but some were from near the top at the ridge (300 m) drained by the water supply stream.

Caiman Lagoon: Small lake created by road-building activities in the late 1950s or early 1960s, east of Carretera al Pacífico; another large pond (70 × 50 m) with the same name is west of road; both are about 3 km southwest of Rincón, 35 m.

Camp Seattle: Abandoned lumber camp, 3 km southwest of Rincón, just east of Carretera al Pacífico, 30 m.

Carretera al Pacífico: Road running southwest from Rincón to Río Rincón near its juncture with Río Pavón; formerly it extended a little farther, but bridges over the Río Riyito and Río Rincón have been washed out.

Charcos: Abandoned village site, east of Carretera al Pacífico and north of junction with Water Fall Road, 4 km southwest of Rincón, 20 m.

Gravel Pit Road: Side road off Carretera al Pacífico going west to Snake Woods, north of Río Riyito, 60 m.

Harris Hollow: Heavily forested depression just east of Carretera al Pacífico, 6 km southwest of Rincón, between Savage Woods and Quebrada Rayo, 10 m.

Holdridge Trail: Trail beginning at Carretera al Pacífico near Caiman Lagoon and running northwest onto ridge (400–440 m); connects with Ridge Trail to Laguna Chocuaco; shown but not labeled in figure 16.3.

Laguna Chocuaco: Lake 10 km west of Rincón, 200 m.

Mangrove Area: Mangrove forest 1–2 km southeast of Rincón on Golfo Dulce shore, near the mouth of Río Rincón; Carretera al Pacífico parallels mangroves for a short distance east of Rincón, 0–3 m.

Old Osa Field Station: Field building erected for Tropical Science Center, just south of Rincón Airfield and north of Quebrada Aguabuena, 40 m.

Playa Blanca: Settlement on eastern shore of peninsula, 9 km southeast of Rincón, 3 m.

Puerto Jiménez: Major settlement on Golfo Dulce on southeastern shore of peninsula, 9.5 km southeast of Playa Blanca, 7 m (fig. 16.2).

Quebrada Aguabuena: Tributary of Río Riyito, draining slopes of a small valley south and west of Rincón; samples were taken over most of the length of the stream, which is reduced to intermittent trickles in dry season and may rise 1.5 m in wet season; most material is from about 2 km southwest of Rincón near Carretera al Pacífico at 30 m and immediately south of the old Osa Field Station, 3 km west of Rincón at 40 m.

Quebrada Crump: Small forest stream flowing eastward into Quebrada Aguabuena, about 300 m southeast of the old Osa Field Station, 40 m (Crump Creek of Findley and Wilson 1974); shown but not labeled in figure 16.3.

Quebrada Ferruviosa: Tributary of Río Riyito that crosses Carretera al Pacífico about 5.5 km southwest of Rincón, 20 m.

Quebrada McDiarmid: Small forest stream flowing eastward to join Quebrada Aguabuena just south of the old Osa Field Station, 40 m (McDiarmid Creek of Findley and Wilson 1974); shown but not labeled in figure 16.3.

Quebrada Rayo: Small tributary of Río Riyito that crosses Carretera al Pacífico about 6.5 km southwest of Rincón, 40 m.

Quebrada Vanegas: Small tributary of Río Riyito that crosses Carretera al Pacífico, 5 km southwest of Rincón, 35 m (= Quebrada Banegas on some maps).

Ridge Trail: Trail that begins at the terminus of a secondary road south of Rincón and extends north and west onto ridge (> 400 m) and eventually reaches Laguna Chocuaco; shown but not labeled in figure 16.3.

Rincón Airfield: Dirt and gravel landing strip suitable for small planes, 2.5 km west of Rincón, 0.5 km northeast and parallel to Quebrada Aguabuena, 40 m.

Rincón de Osa: Settlement and former headquarters for Osa Productos Forestales near western head of Golfo Dulce, 4 m.

Rincón Water Supply: Small, steep creek immediately west of Rincón, 15–100 m.

Río Nuevo: Small settlement, 3 km southwest of Puerto Jiménez, 10 m (fig. 16.2).

Río Pavón: Major tributary of Río Rincón, which it joins near Carretera al Pacífico, 11.5 km south-southwest of Rincón, 60 m.

Río Rincón: Primary river-draining region that exits to Golfo Dulce via northern and eastern branch around large island, 1.5–2 km east of Rincón.

Río Riyito: Major tributary of Río Rincón that crosses Carretera al Pacífico, 7 km southwest of Rincón, 40 m.

Savage Woods: Primary forest on river flat of Río Riyito, 5.5 km southwest of Rincón, east of Carretera al Pacífico, 20 m.

Snake Woods: Primary forest at end of Gravel Pit Road just north of Río Riyito, 8 km southwest of Rincón and 2 km west of Carretera al Pacífico, 60 m.

Three Falls: Waterfalls in small stream crossing Water Fall Road, 1 km west of Carretera al Pacífico, 30 m.

Vanegas Swamp: A permanent swamp, 5.5 km southwest of Rincón, 20 m.

Waterfall Road: Side road going west from Carretera al Pacífico to Three Falls, 4 km southwest of Rincón, 30 m; shown but not labeled in figure 16.3.

The sites studied by Holdridge et al. (1971) for their vegetation analysis are as follows:

A. Holdridge Trail, 420 m.

B. Mora Swamp, 1.5 km southeast of Rincón, near northern Boca de Río Rincón, just east of Carretera al Pacífico, 1.5–2 m.

C. Mangrove swamp, 2 km southeast of Rincón, near Boca de Río Rincón, on an island between two mouths of the Río Rincón, 0–1 m.

D1. Eastern bank of Río Rincón, 3 km south of Rincón, 9 m.

D2. Eastern bank of Río Rincón, 3.5 km south of Rincón, 8 m.

E. 3.5 km south of Rincón Airfield, 300 m.

F. 6 km south of Rincón Airfield, 200 m north of Río Riyito, 100–200 m.

Acknowledgments

We thank the many friends and colleagues who shared with us the excitement of working in the Osa forests near Rincón; primary among these are Ron and Miriam Heyer, Norm Scott, Andy and Kathy Shumaker, Andy Starrett, Jim Talbot, and the late Charles Walker. Discussions with Craig Guyer and Mo Donnelly helped clarify our thinking about the Osa fauna. The late Doug Robinson and Federico Bolaños shared their expertise with us and provided access to material from the Osa Peninsula in the Museo de Zoología, Universidad de Costa Rica. Karen Warkentin and Gad Perry, former graduate students at the University of Texas, and Doug Wechsler, then at the University of Washington and now at the Academy of Natural Sciences of Philadelphia, provided us with lists of species they collected, observed, and/or photographed at Corcovado National Park. David Cannatella and Jess Rosales made available specimen records from Corcovado National Park in collections of the Texas Memorial Museum. Brian Crother and the late Lisa Aucoin furnished notes and photographs of specimens they captured at Marenco Biological Station. Bill Lamar and Alejandro Solórzano shared their knowledge of the Marenco fauna. Anny Chaves, formerly of the Ostional Marine Turtle Center, provided information on sea turtles. The contributions of these colleagues are especially appreciated for helping us to present a full coverage of the Osa herpetofauna.

In 1973, the new management of Osa Productos Forestales, which owned 100,000 acres (approximately 41,000 ha) of forest, began promoting land sales to wealthy North Americans; they drew up plans to extend the road across the peninsula, to build retirement communities on the western beaches, and to dredge the Laguna Corcovado to make a large marina. We decided that this development and the ever-increasing clearing and burning near Rincón by squatters made it imperative that we mount a specially intensive collecting program before all of the forest was gone. Our concern was heightened when we learned that the company was considering sale of the remainder of its holdings to a Japanese company that also was negotiating with the Costa Rican government to clear-cut most of the peninsula. Participants who joined us in this effort (sometimes referred to as GORE I and II) included Doug Allen, Jim and Rosemary DeWeese, Carl Lieb, Andy and Kathy Shumaker, Holly Starrett, Jim Talbot, Ian Straughan, Wayne Van Devender, Dave and Marvalee Wake, and George Zug. We especially thank them for their efforts in making the Osa herpetofauna better known. We also want to acknowledge the many students and faculty who shared the Osa experience with us during our several stints as instructors of courses given by the Organization for Tropical Studies at Rincón; particularly appreciated were interactions with Marty Crump, Jim DeWeese, Ron Harris, Dennis Paulson, Allan Schoenherr, Wade Sherbrooke, Wayne Van Devender, and Don Wilson.

The support of the following organizations is gratefully acknowledged: Osa Productos Forestales and the Tropical Science Center invited us to work near Rincón, and the former provided radio communication to the outside world when emergency aid was needed. The Organization for Tropical Studies gave logistical support and other assistance through many years to enhance our effectiveness in the field. Fieldwork was supported in part by a Faculty Release-Time Award from the University of South Florida (RWM) and a John Simon Guggenheim Fellowship and National Science Foundation grants (G. 40747, G. B. 9200081) (JMS).

Several persons assisted us in preparing the final text of this chapter. Mercedes Foster read early drafts of some sections and helped with the numerous recalculations of similarity coefficients. Fiona Wilkinson and Kinard Boone (USGS Patuxent Wildlife Research Center) prepared the final maps and figures; early drafts of figures 16.3 and 16.4 were prepared by Ron Harris. Stan Rand and two anonymous reviewers made suggestions and shared their experiences with the faunas, which improved our thinking. Finally, the editors of this volume, especially Craig Guyer, facilitated inclusion of our contribution. To all of these, we extend our special thanks.

17

The Iwokrama Herpetofauna: An Exploration of Diversity in a Guyanan Rainforest

Maureen A. Donnelly, Megan H. Chen, and Graham G. Watkins

The Iwokrama Forest is a 360,000-ha tract of lowland tropical rainforest in central Guyana. The forest is administered by the Iwokrama International Centre for Rain Forest Conservation and Development, whose goal is to conserve and use tropical rainforest to benefit the people of Guyana and the international community. The Academy of Natural Sciences of Philadelphia was contracted in 1996 to survey the vertebrate fauna of the Iwokrama Forest, and our knowledge of the herpetofauna is based on collections made during the survey, during an Amerindian ranger-training course in 1999, and additional observations made by G. G. Watkins between 1997 and 1999. We relate the Iwokrama fauna to other faunas known from northern South America, and consider patterns of abundance for amphibians and reptiles of the Iwokrama Forest.

Guyana is located on the northeastern shoulder of South America and is the westernmost of the three "Guianas" (Guyana, Suriname, and French Guiana). The three Guianas, part of Venezuela (Bolívar and Amazonas states), and northern Brazil make up the Guianan region of northern South America (sensu Hoogmoed 1979a; see Gorzula and Señaris 1998 for a different definition). This part of northern South America has a long geologic history because the Guiana Shield and the Brazilian Shield form the foundation of the South American continent. The Guiana Shield and Western Africa are part of the same ancient Gondwanan geological formation, and neither has undergone major geological change since their separation (Fjeldså and Lovett 1997).

The climate of Guyana is classified as a Tropical Rainforest Climate (Af) in the Köppen system (van Kekem et al. 1996), with two rainy seasons (May to August and December to January). The annual precipitation is 2–3 m, and annual rainfall patterns are driven by the northern and southern movements of the Intertropical Convergence Zone (ter Steege 1993; van Kekem et al. 1996; ter Steege et al. 1996). The human population in Guyana is concentrated along the coast, and approximately 80% of the country (~21.5 million ha) is covered by primary forest (ter Steege et al. 1996). Most of Guyana is classified as tropical wet forest in the Holdridge life zone system (Holdridge et al. 1978). The forests of Guyana are part of a block of "*cis*-Andean" forests that accounts for approximately 80% of the tropical forests in South America (Lynch 1979).

The Iwokrama Forest is about 300 km south of Georgetown in Administrative Region No. 8 (fig. 17.1). The forest lies between 4°0′ and 5°0′ N and 58°5′ and 59°5′ W. The old-growth areas of Iwokrama are typical of lowland, evergreen forests found in much of northern and central Guyana. The trees that form the dense canopy are generally 20–30 m high. There are differences in the forest types across the Iwokrama Reserve. Most of the forest has no particular tree species that dominates (Kerr 1993), although there are patches with a single dominant or co-dominant species elsewhere in central Guyana (ter Steege 1993). The soils of Iwokrama are dominated by brown sands, laterite, and white sands that form during in situ erosion of sandstone substrates or deposition of sediments (from sandstone of the Pacaraimas; see Clarke et al. 1999). The Iwokrama Forest includes and is bordered by rivers and creeks; the Essequibo River (the largest river in Guyana) forms the eastern boundary of the reserve, and the Siparuni River forms the northwest border. The major physical features of the forest are the centrally located Iwokrama Mountains, rising to 1,000 m, and the Pakatau Hills, rising to about 300 m in the northeast. The Georgetown–Lethem Road runs through the Iwokrama Forest, and the Rupununi Savanna borders the forest on the south. Several field camps were established near the road, and others were established along rivers before the start of the vertebrate survey. We sampled amphibians and reptiles in the vicinity of these field camps.

Botanists and forest ecologists at the Iwokrama Centre have classified several forest types (http://www.iwokrama.org/ResearchProjects/ZoningPeerReview-Doc_part_b.pdf), and six characterize our study sites (table 17.1). The Mabura Hill study site (between 4°40′ and 5°20′ N and 58°26′ and 58°54′ W) is close to Iwokrama, and some of the generalities from Mabura Hill probably apply to Iwokrama. Quantitative forest studies in Guyana began in the 1930s (Davis and Richards 1933, 1934). Tropenbos started research in the Mabura Hill region in 1989, and the Biodiversity of the Guianas Program based at the Smithsonian Institution continues to study Guyana's botanical diversity. The Mabura Hill site has two basic forest types: mixed rainforest and dry evergreen forest (van Kekem et al. 1996), and it has poor to moderate tree species richness (ter Steege

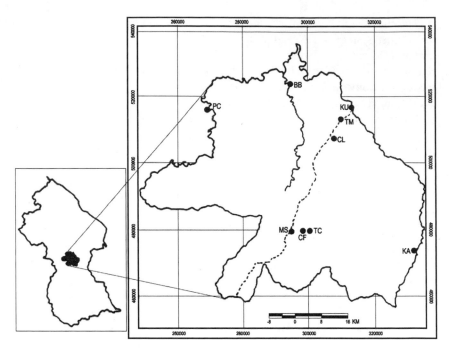

Figure 17.1   The location of Iwokrama within Guyana as well as the locations of the collecting localities or camps. Camps are indicated as follows: BB, Burro Burro; CF, Cowfly; CL, Cutline; KA, Kabocali; KU, Kurupukari; MS, Muri Scrub; PC, Pakatau Creek; TC, Third Camp; TM, Three Mile (see table 17.1).

1993). Seven species of trees account for 68% of all individuals and 76% of total basal area (ter Steege 1993). The major timber species in Guyana (Greenheart, *Chlorocardium rodiei*) accounts for only 0.5%–1.5% of total standing stock yet represents 45% of harvested timber (ter Steege 1993). Greenheart occurs as a codominant in several parts of Iwokrama (table 17.1), and its protection in portions of the Iwokrama Reserve will be important for conservation of this heavily harvested species. Ek (1997) described 11 forest types at Mabura Hill, and the forest types vary across soil types (brown vs. white sand) and drainage type (well drained vs. poorly drained). It has been suggested that in central Guyanan forests, water availability is more important than nutrient availability in explaining patterns of tree distribution (ter Steege et al. 1996). Characterization of forest types in Mabura Hill, and presumably in other sites in Guyana, is difficult because tree species diversity is high and density is low. In general, the acidic soils of the Mabura Hill region are poor in nutrients and are typical of deeply weathered soils in the humid tropics (van Kekem et al. 1996).

The Iwokrama Herpetological Collections

COLLECTION 1: THE PRE-HERPETOLOGICAL SURVEY

Between January and May 1997, amphibians and reptiles were collected opportunistically at nine sites (Burro Burro, Kurupukari, Kabocali, Maipuri, Muri
Scrub, Pakatau, Third Camp, Tiger Creek, and Three Mile; see table 17.1) by
field teams that did not include herpetologists. No attempts were made to standardize sampling activity, and a list of species was the only information obtained from the first collecting effort. Amphibians and reptiles were also collected from pitfall traps that were arrayed to trap mammals along drift fences at
Three Mile camp. The mammalogists collected and preserved mammals, amphibians, and reptiles removed from the traps. Two 50-m drift fences, made
from black plastic sheeting (1 m high), were established perpendicular to the
main trail. Each fence was coupled with 19-liter plastic pitfall traps sunk into the
ground every 5 m (Corn 1994). The pitfall traps were open continuously during
the stay at Three Mile and were checked at least once a day. For the most part,
the only information available for specimens obtained during the first collecting effort is locality data.

COLLECTION 2: THE HERPETOLOGICAL SURVEY

The herpetology portion of the vertebrate survey was conducted from May to
July 1997. Our field crew included Cynthia Watson (University of Guyana),
Zacharias Norman (Wowetta Village), Dexter Torres (Wowetta Village), Ron
Allicock (Surama Village), Paulette Allicock (Surama Village), Daniel Allicock
(Surama Village), Cynthia Jacobs (Kwatamang Village), and Matthew Baber
(Florida International University).

During the herpetological survey, we collected two data sets for amphibians
and reptiles. Our observational data set included all sightings made during the
herpetological survey. Our specimen data set was a subset of the observational
data set because we did not collect every individual we encountered. Animals
prepared as specimens were euthanized. Tissues (muscle and liver) were taken
from a subset of specimens and stored in 95% ethanol. Specimens were fixed in
10% formalin and stored in 70% ethanol. Tissues are stored at Florida International University. Specimens were deposited in collections at the University of
Guyana Centre for Biodiversity, the United States National Museum, the American Museum of Natural History, and at Florida International University. During the herpetological survey we worked extensively in eight field camps (table
17.1). Three camps (Three Mile, Cutline, and Muri Scrub) were located on the
Georgetown–Lethem Road. Three camps (Muri Scrub, Cowfly, and Third
Camp) were part of the Iwokrama Mountain Trail; the Muri Scrub camp was

Table 17.1 Collecting sites sampled during the Iwokrama survey

Camp	Longitude/ latitude	Elevation (m)	Forest type	Collection dates	No. days in camp	Sampling method				Total samples
						DVES	PLOTS	NVES	OP	
Muri Scrub	4°25.20 N 58°50.96 W	80	MS/MMCT	9–14 May	5	2	25	11	6	44
Cowfly	4°20.00 N 58°49.00 W	120	MMCT/MLS	14–20 May	6	2	28	8	6	44
Third Camp	4°20.00 N 58°48.00 W	224	GBKW/MLS	20–27 May	7	6	20	4	8	38
Three Mile	4°37.98 N 58°42.87 W	102	GBKW	28 May–8 June	11	17	36	25	11	89
Burro Burro	4°43.86 N 58°51.04 W	83	MKSW/MMCT	9–20 June	11	14	50	26	9	99
Kabocali	4°17.10 N 58°30.56 W	101	MMCT/GSS/ GBKW	24 June–3 July	9	21	36	27	8	92
Cutline	4°35.00 N 58°44.85 W	70	GBKW	5–11 July	6	14	20	11	3	48

Pakatau Creek	4°45.00 N 59°01.00 W	85	MLS	13–24 July	11	22	40	24	9	97
Maipuri	4°45.25 N 58°35.28 W	n.a.	n.a.	pre–May 1997						
Tiger Creek	4°31.47 N 58°33.62 W	n.a.	n.a.	pre–May 1997						
Kurupukari	4°43.91 N 58°59.00 W	70	MMCT	1997 and 1999						

Note: The first eight sites were sampled during the 1997 herpetological survey; the last three sites were sampled before and after the 1997 survey. The forest types are as follows: MS, muri scrub; MMCT, mora-manicole-crabwood-tyrsil; MLS, mixed low-stature forest on high hills; GBKW, mixed Greenheart–black kakaralli–wamara; MKSW, manicole–kokerite–soft wallaba; GSS, mixed greenheart–sand baromalli–soft wallaba. Muri scrub = *Humiria balsamifera*; manicole = *Euterpe oleracea* or *E. precatoria*; crabwood = *Carapa guianensis*; tyrsil = *Pentaclethra macroloba*; greenheart = *Chlorocardium rodiei*; black kakaralli = *Eschweilera sagotiana*; wamara = *Swartzia leiocalycina*; kokorite = *Attalea maripa*; soft wallaba = *Epeura falcata*; sand baromalli = *Catostemma fragrans*. The sampling methods are as follows: DVES, daytime visual encounter survey; PLOTS, leaf litter plots; NVES, nighttime visual encounter survey; OP, opportunistic collecting. n.a., missing data.

closest to the road, and Third Camp was closest to the Iwokrama Mountains. Three camps (Burro Burro, Kabocali, and Pakatau) were located along rivers (fig. 17.1).

We sampled amphibians and reptiles using four methods at all sites: daytime visual encounter surveys (Crump and Scott 1994), nighttime visual encounter surveys, litter plots (Scott 1976; Jaeger and Inger 1994), and opportunistic collecting (see Chen 1998). The visual encounter surveys were conducted (1) along trails in forest, (2) along the Georgetown–Lethem Road, (3) in ponds, or (4) along watercourses (sampled from boats). Trails were either established before May 1997 or were established the first few days we were in each camp. We sampled 5 × 5-m and 8 × 8-m litter plots in forest at all sites because initially we wanted to determine if abundance and richness were related to plot size. Plots were selected haphazardly (on relatively flat terrain) and were separated from each other by at least 25 m. Plot perimeters were demarcated, and we removed all litter and coarse woody debris by hand (or tool) while we searched for amphibians and reptiles. We also collected animals opportunistically in or near camp. We did not keep track of the time spent collecting opportunistically, so as a conservative estimate we assumed that each person in camp spent one hour per day engaged in this activity. During our stay at Three Mile camp, we also used drift fences with pitfall traps to sample amphibians and reptiles (these were left from the mammal survey at Three Mile camp). During the Three Mile portion of the herpetological survey, pitfall traps were open continuously, and traps were typically checked in the morning and in the afternoon.

The time spent in each camp was not equivalent, and the collecting activities varied among camps (table 17.1). The difference in collecting activities was in part related to the different amount of time spent in each camp (table 17.1) and to differences in processing times (i.e., processing time decreased as we spent more time in each camp and reached collection limits). Collection 2 provided us with a species list and allowed us to describe relative abundance of amphibians and reptiles.

COLLECTION 3: THE RANGER TRAINING COURSE

In 1999, as part of a series of workshops designed to teach wildlife biology to local Amerindian rangers working in the Iwokrama Forest, we collected amphibians and reptiles at Kurupukari base camp using the four methods described earlier in the chapter. We also carried out nighttime visual encounter surveys at Three Mile camp, and along streams and rivers near Kurupukari. We made collections during 16–26 August 1999. This collection is important because concentrated efforts to sample amphibians and reptiles at Kurupukari were not possible during the vertebrate surveys of 1997.

The Iwokrama Herpetofauna

The Iwokrama herpetofauna includes 128 species, of which 20 have not yet been assigned to particular species because they have not been conclusively identified (e.g., juvenile *Paleosuchus* sp.), are undescribed, or are part of problematic taxa (appendix 17.1). Like other South American sites, the amphibian fauna is dominated by frogs and toads and the reptile fauna by squamates (lizards and snakes; see table 17.2). Lynch (1979) and Duellman (1999) noted that most South American amphibian faunas are dominated by leptodactylids and hylids. However, the amphibian fauna of Iwokrama includes more hylids than leptodactylids (table 17.2). Additionally, in the Iwokrama Forest, Colubridae is the most diverse squamate family, followed by Teiidae (including Gymnophthalmidae), Gekkonidae, Polychrotidae, and Boidae (table 17.2).

During the first collecting effort (collection 1), 107 specimens of 52 species were collected. The second collection included 99 species, of which 45 were new to the list (1,874 adult specimens, 41 unidentified juvenile frogs, 9 lots of anuran eggs, and 8 lots of tadpoles). The third collection included 217 specimens of 52 species (10 were new for Iwokrama) and 1 lot of juvenile frogs. The representation of different amphibian and reptile groups differed among collections (table 17.3). In the first collection, squamate reptiles accounted for more than half of the species obtained (mostly snakes), whereas amphibians were more important in the second and third collections. We attribute this to an emphasis on night collecting during collecting efforts 2 and 3. Only collection 2 includes representatives of all higher taxa (table 17.3). Of the 121 species obtained as specimens during the three collecting efforts (7 species from Iwokrama are known from observations made between the vertebrate survey and the ranger training course), 63 are represented in only one collection and 32 of these are known from single specimens (table 17.4). Thirty-eight species are represented in two collections, and 20 taxa are represented in all three collections (table 17.5). The species common to all collections are either from the Amazonian basin, widespread, or lowland endemics of the Guianan region of South America (Hoogmoed 1979a). Aside from new species yet to be described, all taxa from Iwokrama are known from the Guianan region of northern South America or are widespread throughout Amazonia (see appendix 17.1). As collecting activities increase in northern South America, our knowledge of the Amazonian fauna will increase and range extensions will be common (Heyer 1988; Rodriguez and Cadle 1990; Zimmerman and Rodrigues 1990).

We collected three taxa from Iwokrama that have not been reported previously from Guyana. *Anolis trachyderma* previously was known from western Amazonia and Brazil (Avila-Pires 1995). Elsewhere, *A. trachyderma* is known as a forest lizard that uses the ground and low vegetation. *A. trachyderma* was collected at Pakatau in Iwokrama during a nighttime visual encounter survey.

Table 17.2   Number of species in each family

Taxon	No. species	Percentage of class	Percentage of group
Amphibia			
Caecilians			
Caeciliidae	3	5.8	75.0
Rhinatrematidae	1	1.9	25.0
Anurans			
"Allophrynidae"	1	1.9	2.1
Bufonidae	6	11.5	12.4
Centrolenidae	3	5.8	6.3
Dendrobatidae	2	3.8	4.2
Hylidae	24	46.2	50.0
Leptodactylidae	9	17.3	18.7
Microhylidae	1	1.9	2.1
Pipidae	1	1.9	2.1
Ranidae	1	1.9	2.1
Reptilia			
Turtles			
Bataguridae	1	1.3	14.2
Chelidae	2	2.6	28.6
Pelomedusidae	2	2.6	28.6
Testudinidae	2	2.6	28.6
Squamates			
Amphisbaenidae	1	1.3	4.2
Iguanidae	1	1.3	4.2
Polychrotidae	5	6.6	20.8
Tropiduridae	3	3.9	12.5
Gekkonidae	5	6.6	20.8
Teiidae (includes Gymnophthalmidae)	8	10.6	33.3
Scincidae	1	1.3	4.2
Typhlopidae	1	1.3	2.4
Anilidae	1	1.3	2.4
Boidae	5	6.6	12.2
Colubridae	27	35.6	65.9
Viperidae	3	3.9	7.3
Elapidae	4	5.3	9.8
Crocodilians			
Alligatoridae	4	5.3	100.0

Note: The third column indicates the percentage of each family within a class (Amphibia or "Reptilia"); the fourth column indicates the percentage of representation for each group (e.g., caecilians, turtles, snakes).

Table 17.3   Number of species collected during each of three collecting efforts made in the Iwokrama Forest (percentages in parentheses)

Taxon	Collection 1	Collection 2	Collection 3
Amphibia	19 (36.5)	47 (47.5)	27 (51.9)
Testudinata	0	3 (3.0)	0
Squamata	32 (61.5)	46 (46.5)	22 (42.3)
Crocodilia	1 (2.0)	3 (3.0)	3 (5.8)
Total	52	99	52

Table 17.4   Amphibian and reptile taxa of Iwokrama represented by unique specimens or observations

Amphibians
 Caecilian sp. 1
 Caecilian sp. 2
 Caecilian sp. 3
 Epicrionops niger
 Centrolenid sp. 3
 Colostethus sp.
 Osteocephalus sp. 1
 Phyllomedusa tomopterna

Turtles
 Geochelone denticulata
 Rhinoclemmys punctularia

Lizards
 Anolis sp.
 Anolis trachyderma
 Bachia flavescens
 Cercosaura ocellata
 Pseudogonatodes guianensis

Snakes
 Atractus flammigerus
 Bothriopsis bilineata
 Chironius sp.
 Corallus caninus
 Drymoluber dichrous
 Hydrops triangularis
 Lachesis muta
 Micrurus circinalis
 Micrurus hemprichii
 Micrurus lemniscatus
 Micrurus psyches
 Oxyrhopus petola
 Pseudoboa coronata
 Pseudoboa neuwiedii
 Pseustes sulphureus
 Tantilla melanocephala

Crocodilians
 Caiman crocodilus

Hydrops martii callostictus is known from Peru (Peters and Orejas-Miranda 1970). *Scinax proboscideus* was listed as occurring east of the Essequibo River (Hoogmoed 1979a) and in the northeastern part of Amazonia-Guiana (Duellman 1999). Our collections of this species from Iwokrama extend the range of the species westward. Two other species (*Chironius* sp. and *Paleosuchus* sp.) may represent known taxa, but they have not been conclusively identified (appendix 17.1). The taxonomy of tropical American frogs in the genus *Physalae-*

Table 17.5   The 20 species of amphibians and reptiles represented in the three collections made in the Iwokrama Forest

Amphibians—11 species	Lizards—5 species
Allophryne ruthveni (LE)	*Tropidurus umbra* (AM)
Bufo guttatus (AP)	*Thecadactylus rapicauda* (W)
Bufo marinus (W)	*Ameiva ameiva* (W)
Bufo margaritifer (W)	*Kentropyx calcarata* (AM)
Hyla boans (W)	*Mabuya bistriata* (AM)
Hyla crepitans (W)	
Osteocephalus taurinus (AM)	Snakes—4 species
Scinax boesemani (LE)	*Corallus hortulanus* (AM)
Leptodactylus bolivianus (W)	*Eunectes murinus* (AM)
Leptodactylus knudseni (AM)	*Helicops angulatus* (AM)
Leptodactylus mystaceus (AM)	*Bothrops atrox* (AM)

Note: Distribution patterns are in parentheses (sensu Hoogmoed 1979a): LE, lowland endemic; AP, peripheral Amazonian (known from the western and northern margins of the basin); AM, Amazonian (widespread through Guianan region of South America); W, widespread (from Mexico through Central America over *cis*-Andean tropical South America). Distribution patterns for three taxa— *Leptodactylus knudseni* (from Frost 2001), *Mabuya bistriata* (from Avila-Pires 1995), and *Corallus hortulanus* (McDiarmid et al. 1999)—were placed into one of Hoogmoed's categories.

mus are under study, and we await the results of that work before we associate the Iwokrama frogs with a scientific name. Two taxa on our faunal list (*Osteocephalus* sp. and *Anolis* sp.) are represented by juveniles that we cannot accurately assign to species.

SPECIES ACCUMULATION PATTERNS WITHIN IWOKRAMA

We observed more than half of the total species in the first 15 days of sampling but continued to add species as we changed camps (fig. 17.2A). The shape of our species accumulation curve is similar to that presented in Rodriguez and Cadle (1990, fig. 22.1). We encountered more species at Iwokrama in 78 days than Rodriguez and Cadle did in their preliminary study of the herpetofauna of Cocha Cashu, Peru. The steepness of the "discovery curve" for Iwokrama resembles those shown for Peruvian sites (Duellman and Mendelson 1995). The Iwokrama herpetofauna includes three species that were seen more than 200 times (fig. 17.2B). The number of observations decreases quickly as less common species are considered, creating a distribution that resembles the shape of importance curves for amphibians and reptiles from Peru (Duellman and Mendelson 1995).

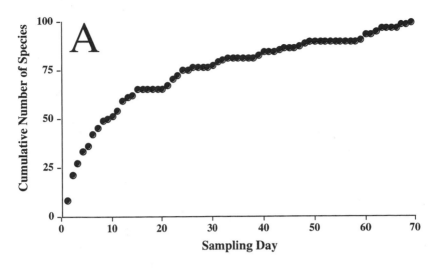

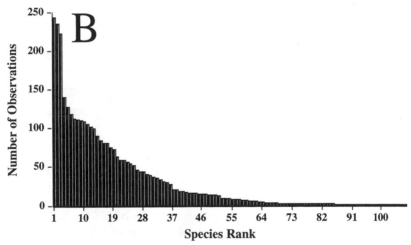

Figure 17.2 (A) Species accumulation based on observations made during collection 2. (B) Species importance curve; the first species (in rank) was seen the greatest number of times.

We examined patterns of species accumulation using all four methods across the eight camps, based on cataloged specimens and observations of all organisms encountered during sampling activities for collection 2. Similar patterns of species accumulation were obtained for specimen-based and observation-based data (fig. 17.3), but the observation-based data typically included one or two more species than did the specimen-based data. Species were accumulated most rapidly at Pakatau Creek (fig. 17.3). Except for relatively "flat" species accumulation curves at Third Camp and Cutline (fig. 17.3), continued accumulation of new taxa was expected if camps were sampled for longer time periods.

Comparison of Sites in Northern South America

Knowledge of South American herpetofaunas is increasing rapidly (see Duellman 1999), but it remains fragmentary because many sites have not been explored (see Heyer 1988) and most sites have not been sampled extensively. The collections from Iwokrama represent the first herpetological collections made from the central part of Guyana. Duellman (1999) argued that the anuran fauna of the Guianan region of northern South America is a subset of a widespread Amazonian fauna, whereas Hoogmoed (1979a) argued that Guiana is a distinct region. Iwokrama is bordered on the east by the Essequibo River, and Hoogmoed (1979a) suggested that the Essequibo–Río Branco Depression may act as a barrier for the east–west dispersal of some forest species and as a conduit for northern movement of aquatic Amazonian species (e.g., *Melanosuchus niger* and *Chelus fimbriatus*). If the Essequibo acts as a barrier, it might be expected that the Iwokrama Forest would have a herpetofauna that differs from those of lowland forest sites in Suriname and French Guiana. The Iwokrama herpetofauna might resemble those from forest sites in the Venezuelan lowlands to the west.

COEFFICIENTS OF BIOGEOGRAPHIC SIMILARITY

To examine patterns of similarity among South American herpetofaunas, we used coefficients of biogeographic similarity because this approach has been employed frequently by other researchers. We calculated coefficients of similarity (= coefficient of biogeographic resemblance [CBR] of Duellman [1990], where CBR = $2C/(N_1 + N_2)$ and C = number of species in common between sites 1 and 2, N_1 = number of species at site 1, and N_2 = number of species at site 2) for five sites in northern South America (appendix 17.2, table 17A.1), for 10 northern South American frog faunas (appendix 17.2, table 17A.2), and for 13 northern South American snake faunas (appendix 17.2, table 17A.3). We calculated an average CBR value for each published table and calculated average CBR values for tables we generated (table 17.6). Because we selected sites in northern South America and excluded sites from Amazonia, we expected that the CBR values we obtained would be higher than CBR values that include several sites from different parts of the continent.

The lowest average CBR value we obtained was that for Duellman's (1999) analysis of anuran faunas across regions of South America. Many of the regions

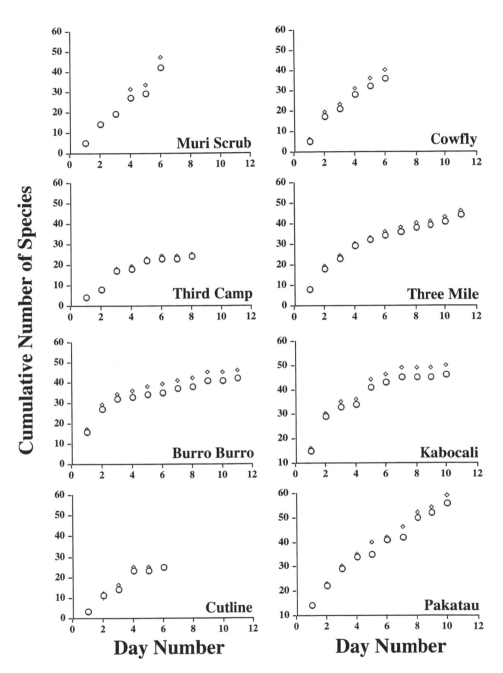

Figure 17.3 Species accumulation curves for eight camps, based on the 1997 herpetological survey (collection 2). Circles indicate collection data; diamonds indicate survey data.

Table 17.6 Average coefficients of biogeographic resemblance for different groups of amphibians and reptiles

CBR score	Source
Herpetofauna	
.270	Duellman 1990
.364	This study, table 17A.1
Amphibians	
.029	Duellman 1999
.309	Duellman 1990
.332	This study, table 17A.1
Frogs	
.220	Duellman 1990
.281	Duellman 1997; all sites
.301	Duellman 1997; northern South American sites
.332	This study, table 17A.1
.356	This study, table 17A.2
.428	Duellman 1997; savanna sites
.493	Hoogmoed 1979a; rainforest frogs
.514	Duellman and Mendelson 1995
.620	Hoogmoed 1979a; savanna frogs
Reptiles	
.315	Duellman 1990
.388	This study, table 17A.1
Lizards	
.496	Duellman and Mendelson 1995
.394	This study; data not included in appendix 17.2
.637	Hoogmoed 1979a; rainforest lizards
.812	Hoogmoed 1979a; savanna lizards
Snakes	
.360	Duellman 1990
.384	This study, table 17A.1
.552	This study, table 17A.3
.567	Hoogmoed 1979a; open formation snakes
.695	Hoogmoed 1979a; rainforest snakes

Note: The three values in the herpetofauna group are based on all amphibians and reptiles.

considered by Duellman have distinctive amphibian faunas that share no species; the large number of zeros in his data matrix result in a low average. We excluded the zero values and recalculated a mean CBR value for Duellman's data for South American regions. The resulting average CBR value for Duellman's data (.06) indicates that there is very little overlap of anuran species across regions of South America that share species. Generally, frog faunas are less similar than are lizard and snake faunas (table 17.6). Hoogmoed (1979a) suggested that the sedentary habits of frogs and their reliance on moisture for reproduction limit their ability to disperse broadly, and this results in limited distributions. The data that have accumulated since 1979 lend support to Hoogmoed's assertion.

The savannas of northern South America have similar frog faunas, as do rainforests of Peru (table 17.6). The lowest values obtained for frogs included studies that spanned broad geographical regions or habitat types. Hoogmoed (1979a) compared four faunal regions in northern South America (eastern Guiana, western Guiana, Brazilian Guiana, and Belém) by using CBR scores and noted that the anuran fauna of eastern Guiana (east of the Essequibo River) was more similar to that in Brazilian Guiana than to the fauna in western Guiana. Duellman (1997) included six savanna sites (four in Venezuela and two in Suriname) in his study. Hoogmoed (1979a) suggested that when the Essequibo–Río Branco Depression flooded, it could serve as a migration route for savanna frogs; if so, this feature might explain the relatively high similarity we report here (table 17.6). The ten forest sites from Guiana that we compare (appendix 17.2, table 17A.2) are more similar to each other than are Duellman's (1997) forest sites from South America. We focused on lowland forests, and Duellman's study included one highland site. Given the high level of endemicity in the Guianan Highlands (Duellman 1999; see McDiarmid and Donnelly, chap. 18 of this volume), it is not surprising that inclusion of a highland site would lower overall similarities for a given CBR table.

The reptile faunas of Madre de Dios, Peru, are more similar to each other than are the reptile faunas of northern South America (Morales and McDiarmid 1996). Perhaps the forests of Madre de Dios are more uniform than those in northern South America; the northern part of the continent includes several savannas, and this mix of habitat types might reduce similarity values. We expect that forests next to savannas might have moderately high species richness because of high diversity in ecotonal regions. Hoogmoed (1979a) presented data for rainforest and savanna lizards and found that savanna lizards are more similar to each other than are forest lizards (table 17.6). Although the flooding of the Essequibo–Río Branco Depression might inhibit savanna lizard movements in the rainy season, the drying of the depression could serve as a corridor in the dry season (Hoogmoed 1979a). Snake assemblages in South American rainforests are among the richest known (Dixon 1979; Cadle and

Greene 1993), and we found that rainforest snake assemblages are more similar to each other than are snake assemblages in open formations (table 17.6).

When Hoogmoed (1979a) reviewed the biogeography of Guianan amphibians and reptiles, he noted that reptiles tend to have larger geographic distributions than amphibians do. He also noted that reptiles have greater mobility than amphibians. Calculation of similarity indices helps us understand patterns, and we suggest that use of CBR-type values is most informative when comparisons are made across similar habitat types in a region (e.g., rainforests in central Amazonia) rather than across regions.

The Iwokrama Herpetofauna: Patterns of Abundance

Several biologists have examined rarity in a conservation context (Rabinowitz 1981; Rabinowitz et al. 1986; Kunin and Gaston 1993, 1997; Bevill and Louda 1999; Maina and Howe 2000), but few have explored patterns of abundance. Hanski et al. (1993) demonstrated that species with extensive distributions tend to be more locally abundant than species with restricted distributions. Hanski et al. (1993) suggested that this pattern reflects a sampling artifact because rare species are more difficult to detect than locally abundant species, and this is certainly true for tropical amphibians and reptiles. Hanski et al. (1993) suggested that geographical distributions are relatively well known, but this is not the case for Neotropical taxa (see Heyer 1988).

Morales and McDiarmid (1996) defined abundance on the basis of the percentage of all observations: taxa were considered to be abundant if they accounted for more than 3.8% of all observations, common if they accounted for 1.0%–3.79% of all observations, and uncommon if they accounted for less than 1% of all observations. We used their categories and added a fourth. If a taxon was collected once, we called it rare. Although numeric abundance is easily quantified and understood, distributional extent and detectability also contributed to observed patterns of abundance. For example, if a taxon was widespread throughout Iwokrama, it could be considered "spatially" abundant. If a taxon was collected using every sampling method (i.e., detectable), it would be more "abundant" than a taxon collected using a single method.

The three collections (specimen data only) differ in the distribution of taxa across abundance categories (table 17.7). Approximately half of the species were uncommon or rare, and only one taxon (*Bufo margaritifer*) was numerically important in all three collections (table 17.8). Four or five of the species in each collection were only abundant in that collection, and this may represent temporal differences among species in their phenology or activity.

To document numeric abundance of the Iwokrama herpetofauna, we used the data collected during the herpetological survey in 1997 (collection 2) to de-

Table 17.7   Representation of abundance categories in the three collections
of Iwokrama

Category	Collection 1	Collection 2	Collection 3
No. species collected	52	95	49
Abundant	8 (15.4)	10 (10.5)	10 (20.4)
Common	23 (26.9)	23 (24.2)	15 (30.6)
Uncommon	30 (57.7)	38 (40.0)	5 (10.2)
Rare		24 (25.3)	19 (38.8)

Source: Morales and McDiarmid 1996.
Note: The numbers refer to cataloged specimens. Percentages are in parentheses.

Table 17.8   The ten most abundant species according to the three collections

Collection 1	Collection 2	Collection 3
Dendrophryniscus minutus	*Bufo* sp. cf. *typhonius*	*Hyla geographica*
Bufo margaritifer	*Dendrophryniscus minutus*	*Physalaemus* sp.*
Atelopus sp.*	*Hyla geographica*	*Hyla minuta**
*Bothrops atrox**	*Hyla granosa**	*Bufo margaritifer*
*Bufo marinus**	*Osteocephalus taurinus*	*Hyla boans*
*Corallus hortulanus**	*Osteocephalus* sp. 3	*Leptodactylus knudseni*
Leptodactylus knudseni	*Gonatodes humeralis*	*Scinax ruber**
Osteocephalus sp. 3	*Leptodactylus mystaceus**	*Gonatodes humeralis*
*Chironius scurrulus**	*Leptodactylus petersi**	*Scinax nebulosus**
Hyla boans	*Scinax boesemani**	*Osteocephalus taurinus*

Note: The species are ranked in order of abundance. An asterisk indicates that a species is
abundant in only one collection.

termine the 10 most abundant species collected and the 10 species observed
most frequently to document numeric abundance of the Iwokrama herpetofauna. We used the observation-based data from collection 2 to determine if
taxa were widespread (found in 7–8 camps), common (found in 4–6 camps),
uncommon (found in 2–3 camps), or restricted (found in one camp; see appendix 17.3).

We defined detectability on the basis of the number of methods that were
effective in detecting a particular taxon. The majority of the Iwokrama amphibians and reptiles were captured using a single method (appendix 17.4).
Nighttime visual encounter surveys and opportunistic collecting obtained

Table 17.9 Four measures of abundance observed for the Iwokrama herpetofauna

Specimens[a]	Observations[b]	Detectable[c]	Widespread[d]
Bufo margaritifer	*Bufo margaritifer*	*Bothrops atrox*	*Bufo margaritifer*
Dendrophryniscus minutus	*Hyla boans*	*Bufo margaritifer*	*Gonatodes humeralis*
Hyla geographica	*Dendrophryniscus minutus*	*Dendrophryniscus minutus*	*Kentropyx calcarata*
Hyla granosa	*Hyla geographica*	*Leptodactylus knudseni*	*Bothrops atrox*
Osteocephalus taurinus	*Leptodactylus petersi*	*Leptodactylus mystaceus*	*Dendrophryniscus minutus*
Osteocephalus sp. 3	*Leptodactylus knudseni*		*Hyla boans*
Gonatodes humeralis	*Osteocephalus* sp. 3		*Hyla geographica*
Leptodactylus mystaceus	*Hyla calcarata*		*Hyla granosa*
Leptodactylus petersi	*Leptodactylus mystaceus*		*Leposoma percarinatum*
Scinax boesemani	*Hyla granosa*		*Leptodactylus knudseni*
			Leptodactylus rhodomystax
			Osteocephalus taurinus
			Phyllomedusa bicolor
			Physalaemus sp.

[a]These are the 10 most abundant specimens in collections gathered during the herpetological survey in 1997.
[b]These 10 species were observed most frequently during the 1997 survey.
[c]These species were captured using all four standard methods.
[d]These species were collected in seven or eight camps.

most of the least detectable taxa. Litter plots were not effective in Iwokrama (in terms of number of individuals obtained per person hour), but two species known from single specimens were collected using the litter-plot method. Daytime visual encounter surveys did not obtain unique taxa.

When we used all four criteria, we found that only two toads, *Bufo margaritifer* and *Dendrophryniscus minutus,* were abundant (table 17.9). Four additional anuran taxa met three of the four criteria, suggesting that only six anurans are abundant in the Iwokrama Forest. Our finding that the assemblage is dominated by a few abundant species and several uncommon or rare ones is consistent with results obtained in other studies (Maina and Howe 2000).

ABUNDANCE PATTERNS ACROSS IWOKRAMA

We examined patterns of abundance for the eight camps that we sampled in 1997 (collection 2) to see if the general pattern (a few abundant species or several uncommon and rare ones) occurred at a smaller spatial scale. Table 17.10 lists, on the basis of the observation data, the ten most abundant species found in each of the eight camps. We used the observation-based data because they give the most complete picture of what we encountered during our sampling activities. In the majority of camps, frogs were the animals most frequently encountered. We characterized the fauna of each camp using the aforementioned criterion of Morales and McDiarmid (1996). We found that approximately half of the species collected at each camp were either abundant or common (fig. 17.4). The herpetofaunal assemblage of a given site includes several taxa that are abundant or common, but the species that form local assemblages vary across Iwokrama (table 17.10). Our data demonstrate that descriptions of abundance patterns for taxa depend on spatial scale. The proportion of abundance types for Kurupukari (collection 3) more closely resembles that of the camps of 1997 than that of collection 1 (cf. table 17.7 and fig. 17.4). The composition of the local fauna varied considerably among sites, and future studies in Iwokrama will hopefully explain this interesting spatial variation.

Conclusions

Our preliminary study of the amphibians and reptiles of Iwokrama begins to fill a gap in the knowledge of northern South American herpetofaunas (Heyer 1988). The site has a moderately rich herpetofauna, and the known diversity will increase with additional study. On the basis of the Mabura Hill studies, we know that forest type is associated with soil type and drainage, and that soils in this part of Guyana are nutrient poor. These forests may have low productivity, and this may affect species richness. Emmons (1984) noted that the density of nonflying mammals is lowest in forests on poor, sandy soils. Eisenberg (1979)

Table 17.10   The 10 most abundant species observed at each of the camps studied during the 1997 herpetology survey (collection 2)

Muri Scrub (67.7%)
Scinax boesemani
Osteocephalus taurinus
Hyla granosa
Hyla minuscula
Leptodactylus rhodomystax
Hyla minuta
Hyla crepitans
Dendrophryniscus minutus
Leptodactylus knudseni
Leptodactylus mystaceus

Cowfly (68.7%)
Osteocephalus taurinus
Bufo guttatus
Leptodactylus knudseni
Gonotodes sp.
Hyla fasciata
Centrolenid sp. 1
Hyla geographica
Phyllomedusa bicolor
Atelopus sp.
Bufo margaritifer

Third Camp (85.7%)
Neusticurus rudis
Atelopus sp.
Osteocephalus sp. 1
Hyla granosa
Bufo margaritifer
Iphisa elegans
Osteocephalus taurinus

Kentropyx calcarata
Gonatodes humeralis
Gonotodes sp.

Three Mile (71.5%)
Hyla boans
Bufo marinus
Dendrophryniscus minutus
Hyla minuta
Hyla geographica
Bufo margaritifer
Ameiva ameiva
Hyla leucophyllata
Physalaemus sp.
Leptodactylus knudseni

Burro Burro (64.7%)
Hyla calcarata
Dendrophryniscus minutus
Leptodactylus petersi
Hyla boans
Leptodactylus mystaceus
Caiman crocodilus
Pipa pipa
Phyllomedusa bicolor
Hyla granosa
Hyla geographica

Kabocali (58.3%)
Dendrophryniscus minutus
Osteocephalus sp. 2
Bufo margaritifer

Hyla geographica
Chiasmocleis sp.
Caiman crocodilus
Bufo guttatus
Leptodactylus knudseni
Kentropyx calcarata
Leptodactylus bolivianus

Cutline (73.1%)
Ameiva ameiva
Epipedobates femoralis
Kentropyx calcarata
Leptodactylus knudseni
Mabuya nigropunctata
Uranoscodon superciliosus
Scinax nebulosus
Coleodactylus
  septentrionalis
Gonatodes humeralis
Leposoma percarinatum

Pakatau (66.7%)
Bufo margaritifer
Leptodactylus petersi
Stefania evansi
Gonatodes humeralis
Leptodactylus mystaceus
Hyla boans
Leptodactylus bolivanus
Osteocephalus sp. 2
Caiman crocodilus
Hyla geographica

Note: The number next to a camp name indicates the percentage of all observations accounted for by the 10 most abundant species.

reported the same pattern for primates. Emmons (1984) suggested that low species richness of mammals occurred when taxa disappeared from forests on nutrient-poor soils. Future studies across forests with different soil types would be useful in understanding how productivity affects diversity.

Information on the herpetofauna of Iwokrama is important because it can help us understand the relationship between the western part of the Guianan

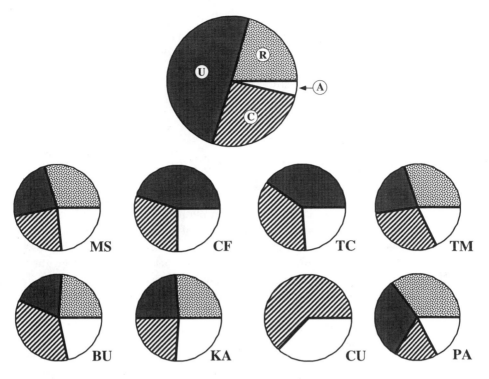

Figure 17.4   All observations made during the herpetological survey in 1997 (*top*) and the distribution of abundance categories for eight camps (*bottom*). The four abundance categories follow the modified scheme of Morales and McDiarmid (1996): A, abundant; C, common; R, rare; U, uncommon. Camps are abbreviated as follows: MS, Muri Scrub; CF, Cowfly; TC, Third Camp; TM, Three Mile; BU, Burro Burro; KA, Kabocali; CU, Cutline; PA, Pakatau Creek.

region (Venezuela) and the eastern part (Suriname and French Guiana). Phytogeographically, the Guianan region is distinct (Berry, Huber, et al. 1995), and four provinces are recognized: Eastern, Central, and Western Guyana provinces, and Pantepui. Guyana lies in the Eastern and Central phytogeographical provinces, and it might be expected that the herpetofauna of Iwokrama represents a mixture of species from these regions. The collections from Iwokrama include several species new to science and extend the ranges of other species. Part of the difficulty in making comparisons among South American forest sites is associated with the confused taxonomy of some widespread species (e.g., *Bufo typhonius, Scinax ruber, Hyla minuta*) that are probably the southern equivalents of the North American *Rana "pipiens"* (Lynch 1979; Heyer 1988). Resolution of the systematics and taxonomy of widespread species will help us understand the relationships among South American sites and will certainly provide insights about the relationship of Guiana with Amazonia. The forest in Iwokrama is heterogeneous, and the herpetofauna varies substantially across the Reserve. The factors responsible for the faunal differences within the forest are complex and worthy of future study. We look forward to future fieldwork in the Iwokrama Forest to understand better the herpetofauna of this lowland forest.

Appendix 17.1   Amphibians and Reptiles of Iwokrama

Taxa with an asterisk are those unique to one of the three collections. Taxa with a plus sign are those that have been seen in Iwokrama in addition to those represented in collections. Distributional patterns of Hoogmoed (1979a) are abbreviated as follows: AB, Amazon basin species that occur along the southern edge of Guianan region; AM, Amazonian species that occur throughout the greater part of the Guianan region; AP, peripheral Amazonian species, distributed along the northern and western margin of the Amazon basin; EA, species reaching the eastern limit of distribution on the Guiana Shield; LE, lowland species endemic to Guianan South America; SB, species from Brazil that reach Guiana; and W, species that are widespread (from Mexico or Central America over entire *cis*-Andean tropical South America). Data are from Hoogmoed (1979a) except for that relating to *Atelopus spumarius, Leptodactylus knudseni, Leptodactylus pallidirostris, Leptodactylus petersi* (Frost 2001); *Anolis nitens* (Peters and Donoso-Barros 1970); *Tropidurus hispidus, Mabuya nigropunctata, Anolis trachyderma, Coleodactylus septentrionalis, Tupinambis teguixin* (Avila-Pires 1995); *Corallus hortulanus* (McDiarmid et al. 1999); *Atractus flammigerus* (Hoogmoed 1980); *Hydrops marti callostictus* (Peters and Orejas-Miranda 1970); and *Micrurus circinalis* (Roze 1996).

AMPHIBIANS
Caeciliidae
    Caecilian sp. 1*
    Caecilian sp. 2*
    Caecilian sp. 3*
Rhinatrematidae
    Epicrionops niger* (LE)
"Allophrynidae"
    Allophryne ruthveni (LE)
Bufonidae
    Atelopus sp.
    Atelopus spumarius* (AM)

    Bufo guttatus (AP)
    Bufo marinus (W)
    Bufo margaritifer (W)
    Dendrophryniscus minutus (AM)
Centrolenidae
    Centrolenid sp. 1*
    Centrolenid sp. 2*
    Centrolenid sp. 3*
Dendrobatidae
    Colostethus sp.*
    Epipedobates femoralis* (AM)

Hylidae
  Hyla boans (W)
  *Hyla calcarata** (AM)
  Hyla crepitans (W)
  *Hyla fasciata** (AM)
  Hyla geographica (AM)
  Hyla granosa (AM)
  Hyla leucophyllata (AM)
  *Hyla marmorata** (AP)
  Hyla minuscula (LE)
  Hyla minuta (AM)
  Osteocephalus sp. 1
  Osteocephalus sp. 2
  Osteocephalus sp. 3*
  Osteocephalus sp. 4*
  Osteocephalus taurinus (AM)
  Phyllomedusa bicolor (AP)
  *Phyllomedusa hypochondrialis**
    (AM)
  *Phyllomedusa tomopterna** (AM)
  Phyllomedusa vaillanti (AM)
  Scinax boesemani (LE)
  *Scinax proboscideus** (LE)
  Scinax ruber (W)
  Scinax nebulosus (AM)
  *Stefania evansi** (LE)
Leptodactylidae
  Adenomera andreae (AM)
  *Ceratophrys cornuta** (AM)
  Leptodactylus bolivianus (W)
  Leptodactylus knudseni (AM)
  Leptodactylus mystaceus (AM)
  *Leptodactylus pallidirostris** (AM)
  Leptodactylus petersi (AM)
  Leptodactylus rhodomystax (AP)
  Physalaemus sp.
Microhylidae
  Chiasmocleis sp.
Pipidae
  Pipa pipa (AM)
Ranidae
  *Rana palmipes** (W)

REPTILES
Bataguridae
  *Rhinoclemmys punctularia** (EA)
Chelidae
  Chelus fimbriatus+ (AB)
  *Platemys platycephala** (AM)
Pelomedusidae
  Podocnemis expansa+ (AM)
  Podocnemis unifilis+ (AM)
Testudinidae
  Geochelone carbonaria+ (AM)
  *Geochelone denticulata** (AM)
Amphisbaenidae
  Amphisbaena fuliginosa (AM)
Iguanidae
  Iguana iguana+ (W)
Polychrotidae
  Anolis fuscoauratus (AP)
  *Anolis nitens** (LE)
  *Anolis ortonii** (AM)
  Anolis sp. 1*
  *Anolis trachyderma** (AM)
Tropiduridae
  Tropidurus hispidus+ (AM)
  Tropidurus umbra (AM)
  Uranoscodon superciliosa (LE)
Gekkonidae
  *Coleodactylus septentrionalis**
    (AM)
  Gonatodes humeralis (AM)
  Gonatodes sp.*
  *Pseudogonatodes guianensis** (AP)
  Thecadactylus rapicauda (W)
Teiidae
  Ameiva ameiva (W)
  *Bachia flavescens** (LE)
  *Cercosaura ocellata** (AM)
  *Iphisa elegans** (AM)
  Kentropyx calcarata (AM)
  Leposoma percarinatum (LE)
  Neusticurus rudis (LE)
  Tupinambis teguixin (AM)

Scincidae
  Mabuya nigropunctata (AM)
Typhlopidae
  Typhlops reticulatus* (AM)
Anilidae
  Anilius scytale (AM)
Boidae
  Boa constrictor (W)
  Corallus caninus* (AM)
  Corallus hortulanus (AM)
  Epicrates cenchria (W)
  Eunectes murinus* (AM)
Colubridae
  Atractus flammigerus* (AM)
  Chironius sp.*
  Chironius fuscus* (W)
  Chironius scurrulus (AB)
  Dipsas catesbyi* (AM)
  Dipsas pavonina (AM)
  Dipsas variegata* (W)
  Drymoluber dichrous* (AP)
  Erythrolamprus aesculapii* (AM)
  Helicops angulatus (AM)
  Hydrops martii callostictus* (AM)
  Hydrops triangularis* (AM)
  Imantodes cenchoa (W)
  Leptodeira annulata* (W)

Leptophis ahaetulla* (W)
Liophis reginae* (W)
Liophis typhlus (AM)
Oxybelis aeneus (W)
Oxybelis fulgidus* (W)
Oxyrhopus petola* (W)
Oxyrhopus trigeminus (SB)
Pseudoboa coronata* (AP)
Pseudoboa neuwiedii* (EA)
Pseustes poecilonotus (W)
Pseustes sulphureus* (AM)
Tantilla melanocephala* (W)
Tripanurgos compressus (W)
Viperidae
  Bothriopsis bilineata* (AM)
  Bothrops atrox (AM)
  Lachesis muta* (W)
Elapidae
  Micrurus circinalis* (LE)
  Micrurus hemprichii* (AM)
  Micrurus lemniscatus* (AP)
  Micrurus psyches* (AP)
Alligatoridae
  Caiman crocodilus* (W)
  Melanosuchus niger+ (AB)
  Paleosuchus palpebrosus (AM)
  Paleosuchus sp.*

Appendix 17.2

Table 17A.1 Comparison of northern South American herpetofaunas

Site	IWOK	GUY	GUIANA	VENEZ	FG
IWOK	128	.437	.363	.431	.328
GUY	58	137	.432	.471	.235
GUIANA	95	115	395	.494	.251
VENEZ	75	84	152	220	.202
FG	32	24	58	29	67

Note: The five sites are abbreviated as follows: IWOK, Iwokrama Forest; GUY, Guyana (species list provided by the U.S. National Museum, 1998); GUIANA, herpetofauna of the Guianan region (Hoogmoed 1979a); VENEZ, Venezuelan Guayana (Gorzula and Señaris 1998); FG, Petit Saut, French Guiana (Hoogmoed and Avila-Pires 1991). Numbers along the diagonal are the number of species known from each site; numbers above the diagonal are the coefficients of biogeographic resemblance (Duellman 1990); numbers below the diagonal are the number of species shared by pairs of sites.

Table 17A.2 Coefficients of biogeographic resemblance for frogs (431 species) at 10 sites

Site	IWOK	GUY	VEN	SUR	FG1	FG2	FG3	FG4	GUI	AG
IWOK	48	.442	.289	.400	.424	.475	.424	.477	.296	.204
GUY	23	61	.405	.237	.265	.519	.286	.376	.411	.257
VEN	21	31	92	.226	.202	.321	.217	.242	.392	.277
SUR	16	11	14	32	.290	.333	.290	.333	.270	.154
FG1	18	13	13	10	37	.523	.838	.416	.302	.193
FG2	28	34	26	21	28	70	.486	.600	.529	.336
FG3	18	14	14	10	31	26	37	.494	.302	.199
FG4	21	19	16	12	16	33	19	40	.356	.215
GUI	32	47	51	27	31	63	31	37	168	.440
AG	36	47	55	26	33	34	37	37	104	305

Note: Numbers along the diagonal are the number of species known from each site; numbers below the diagonal are the number of shared species at each pair of sites; numbers above the diagonal are the coefficients of biogeographic resemblance. The sites are abbreviated as follows: IWOK, Iwokrama Forest (data from this study); GUY, Guyana (species list provided by the U.S. National Museum, 1998); VEN, Venezuelan Guayana (Gorzula and Señaris 1998); SUR, Suriname (Hoogmoed 1979b; Goin 1971); FG1, Petit Saut, French Guiana (Hoogmoed and Avila-Pires 1991); FG2, French Guiana (Lescure 1976); FG3, Petit Saut (Duellman 1997); FG4, Oyapock (Duellman 1997); GUI, Guiana (Duellman and Hoogmoed 1992; Hoogmoed 1979a, 1979b); AG, Amazonia-Guiana (Duellman 1999).

Table 17A.3 Coefficients of biogeographic resemblance for snakes (178 species) at 13 sites

Site	IWOK	GUY	BG1	GUY2	VEN	VG	SUR	FG1	FG2	FG3	FG4	GUI1	GUI2
IWOK	48	.537	.523	.547	.529	.610	.561	.231	.552	.615	.615	.437	.421
GUY	22	41	.455	.589	.569	.542	.579	.192	.552	.598	.581	.448	.444
BG1	23	20	47	.574	.407	.468	.567	.138	.514	.488	.553	.511	.475
GUY2	26	28	29	54	.644	.702	.756	.215	.653	.631	.692	.567	.576
VEN	27	29	22	37	61	.710	.612	.167	.623	.613	.599	.567	.576
VG	36	32	29	46	49	77	.680	.159	.682	.654	.693	.724	.705
SUR	32	33	34	48	41	51	73	.238	.759	.711	.805	.699	.641
FG1	6	5	4	7	6	7	10	11	.192	.230	.253	.153	.128
FG2	37	37	36	48	48	58	63	10	93	.828	.864	.752	.717
FG3	36	35	30	41	42	50	53	8	70	76	.868	.670	.650
FG4	36	34	34	45	41	53	60	11	73	66	76	.718	.660
GUI1	38	39	46	53	55	76	72	11	85	70	75	133	.882
GUI2	38	38	42	53	55	73	65	9	80	67	68	116	130

Note: Numbers along the diagonal are the number of species known from each site; numbers below the diagonal are the number of shared species at each pair of sites; numbers above the diagonal are the coefficients of biogeographic resemblance. The sites are abbreviated as follows: IWOK, Iwokrama Forest (data from this study); GUY, Guyana (species list provided by the U.S. National Museum, 1998); BG1, British Guiana (Hoogmoed 1982); GUY2, Guayanan region of northern South America (Hoogmoed 1982); VEN, Venezuelan Guayana (Gorzula and Señaris 1998); VG, Venezuelan Guayana (Hoogmoed 1982); SUR, Suriname (Hoogmoed 1980, 1982); FG1, Petit Saut, French Guiana (Hoogmoed and Avila-Pires 1991); FG2, French Guiana (Chippaux 1986); FG3, French Guiana (Gasc and Rodrigues 1980); FG4, French Guiana (Hoogmoed 1982); GUI1, Guianan region of South America (Hoogmoed 1982); GUI2, Guianan region of northern South America (Hoogmoed 1979a).

Appendix 17.3

Table 17A.4   Patterns of occurrence of amphibians and reptiles based on 1997 survey data

Widespread	Common	Uncommon	Restricted
8 camps	**6 camps**	**3 camps**	**1 camp**
Bufo margaritifer	*Ameiva ameiva*	*Hyla minuscula*	*Anolis trachyderma*
Gonatodes humeralis	*Bufo guttatus*	*Leptodactylus pallidirostris*	*Atelopus spumarius*
Kentropyx calcarata	*Bufo marinus*	*Scinax proboscideus*	*Bothropsis bilineatus*
	Caiman crocodilus	*Tupinambis teguixin*	*Caecilian* sp. 1
7 camps	*Coleodactylus septentrionalis*		*Centolenid* sp. 1
Bothrops atrox	*Leptodactylus mystaceus*	**2 camps**	*Centrolenid* sp. 2
Dendrophryniscus minutus	*Mabuya nigropunctata*	*Allophryne ruhveni*	*Centrolenid* sp. 3
Hyla boans	*Tripanurgos compressus*	*Anilius scytale*	*Ceratophrys cornuta*
Hyla geographica	*Uranoscodon superciliosa*	*Anolis nitens*	*Cercosaura ocellata*
Hyla granosa		*Atelopus* sp.	*Chironius scurrulus*
Leposoma percarinatum	**5 camps**	*Chiasmocleis* sp.	*Colostethus* sp.
Leptodactylus knudseni	*Leptodactylus petersi*	*Chironius fuscus*	*Corallus caninus*
Leptodactylus. rhodomystax	*Paleosuchus palpebrosus*	*Dipsas catesbyi*	*Epicrates cenchira*
Osteocephalus taurinus	*Tropidurus umbra*	*Gonatodes* sp.	*Epicrionops niger*
Phyllomedusa bicolor		*Helicops angulatus*	*Hydrops martii callostictus*
Physalaemus sp.		*Hyla crepitans*	*Hydrop triangularis*

4 camps

Anolis fuscoauratus
Corallus hortulanus
Epipedobates femoralis
Geochelone denticulata
Hyla calcarata
Hyla leucophyllata
Leptodactylus bolivianus
Neusticurus rudis
Osteocephalus sp. 3
Phyllomedusa vaillanti
Pipa pipa
Platemys platycephala
Scinax boesemani
Scinax nebulosa
Thecadactylus rapicauda

Hyla fasciata
Hyla minuta
Iguana iguana
Leptodeira annulata
Liophis reginae
Liophis typhlus
Osteocephalus sp. 2
Oxybelis aeneus
Oxybelis fulgidus
Pseustes poecilonotus
Typhlops reticulatus

Iphisa elegans
Melanosuchus niger
Micrurus circinalis
Micrurus psyches
Osteocephalus sp. 1
Oxyrhopus trigeminus
Phyllomedusa hypochondrialis
Pseudoboa coronata
Pseustes sulphureus
Rana palmipes
Rhinoclemmys punctularia
Scinax ruber
Stefania evansi
Tantilla melanocephala

Appendix 17.4

Table 17A.5   Detectability of amphibians and reptiles at Iwokrama

Detectable	Conspicuous	Inconspicuous	Scarce
Bufo margaritifer	*Atelopus* sp.	*Adenomera andreae*	*Caecilian* sp. 1
Dendrophryniscus minutus	*Bufo marinus*	*Hyla boans*	*Allophryne ruthveni*
Leptodactylus knudseni	*Epipedobates femoralis*	*Hyla calcarata*	*Atelopus spumarius*
Leptodactylus mystaceus	*Hyla crepitans*	*Hyla geographica*	*Bufo guttatus*
Bothrops atrox	*Hyla granosa*	*Hyla minuscula*	*Centrolenid* sp. 1
	Hyla leucophyllata	*Leptodactylus pallidirostris*	*Centrolenid* sp. 2
	Hyla minuta	*Osteocephalus* sp. 3	*Centrolenid* sp. 3
	Leptodactylus petersi	*Osteocephalus* sp. 4	*Ceratophrys cornuta*
	Leptodactylus rhodomystax	*Phyllomedusa bicolor*	*Colostethus* sp.
	Osteocephalus taurinus	*Rana palmipes*	*Hyla fasciata*
	Phyllomedusa vaillanti	*Scinax boesemani*	*Leptodactylus bolivianus*
	Physalaemus sp.	*Geochelone denticulata*	*Osteocephalus* sp. 1
	Chiasmocleis sp.	*Platemys platycephala*	*Phyllomedusa hypochondrialis*
	Anolis fuscoauratus	*Ameiva ameiva*	*Pipa pipa*
	Anolis nitens	*Gonatodes* sp.	*Scinax proboscideus*
	Coleodactylus septentrionalis	*Mabuya nigropunctata*	*Scinax ruber*
	Gonatodes humeralis	*Tupanimbis teguixin*	*Scinax nebulosus*
	Kentropyx calcarata	*Anilius scytale*	*Stefania evansi*
	Leposoma percarinatus	*Leptodeira annulata*	*Rhinoclemmys punctularia*
	Neusticurus rudis	*Liophis typhlus*	*Anolis trachyderma*
	Tropidurus umbra	*Tripanurgos compressus*	*Cercosaura ocellata*
	Thecadactylus rapicauda	*Typhlops reticulatus*	*Iguana iguana*
	Uranoscodon superciliosus	*Paleosuchus palpebrosus*	*Iphisa elegans*

Boa constrictor
Bothropsis bilineatus
Chironius fuscus
Chironius scurrulus
Corallus caninus
Corallus hortulanus
Dipsas catesbyi
Epicrates cenchira
Eunectes murinus
Helicops angulatus
Hydrops marti callostictus
Hydrops trianuglaris
Liophis reginea
Micrurus circinalis
Micrurus psyches
Oxybelis aeneus
Oxybelis fulgidus
Oxyrhopus trigeminus
Pseudoboa coronata
Pseustes poecilonotus
Pseustes sulphureus
Tantilla melanocephala
Caiman crocodilus
Melanosuchus niger

Note: Detectable taxa were collected with four standard methods, conspicuous taxa were collected with three of the four methods, inconspicuous taxa were collected with two methods, and scarce taxa were collected with a single standard method.

Acknowledgments

We thank our field companions for their contributions to the Iwokrama collections. Our fieldwork was supported by a GEN/UNDP Assistance Grant to the Iwokrama International Rainforest Program (project no. GUY/92/G31). Fieldwork in Guyana for MHC was supported by a Florida International University Provost's Award (571244850) to MAD. Ralph Saporito helped in all phases of the identification of the Iwokrama collection. We thank the curators of amphibians and reptiles at the United States National Museum (USNM) for the list of species from Guyana. We thank Robert Reynolds, Steve Gotte, and Ronald Crombie (USNM) for courtesies in Washington, D.C. We thank W. Ron Heyer and Roy W. McDiarmid (USNM) for help in identifying problematic taxa. Kirsten Nicholson identified the anoles. The Smithsonian Institution provided funds to MAD to help defray travel expenses. We thank Linda S. Ford, Thomas Trombone, and Darrel R. Frost for assistance at the American Museum of Natural History (AMNH). MHC received a Collection Study Grant from the AMNH and a fellowship from the Tropical Biology Program at Florida International University. We thank the Miami Herp Group (D. Bickford, K. Nicholson, J. Watling, K. Hines, and C. Ugarte), Harald Beck, William E. Duellman, Marinus S. Hoogmoed, Teresa Avila-Pires, and Robert Reynolds for constructive comments on the manuscript. This is contribution 78 to the program in Tropical Biology at Florida International University.

18

The Herpetofauna of the Guayana Highlands: Amphibians and Reptiles of the Lost World

Roy W. McDiarmid and Maureen A. Donnelly

South America has an extremely diverse herpetofauna that includes about one-third of all living amphibian and one-fifth of all living reptile species (Duellman 1979b, 1999; Uetz 2000). The rate of discovery and description of amphibian and reptile species, unlike that of birds and mammals, continues to increase. Many of these new taxa are from South America (Bauer 1998; Glaw and Köhler 1998), a continent for which 60% of the species of amphibians have been described in the past 40 years (Duellman 1999). In contrast to other continents, South America has only moderate topographic complexity. Lowlands cover about half of the continent and are associated primarily with the major drainage systems of the Orinoco, Amazon, and Paraná rivers. The highland regions are generally discrete and occur in the northeast (Guayana Shield), the southeast (Brazilian Shield), and along the western margin of the continent (Andes). Narrow coastal lowlands fringe these upland areas in the east and west. Even though the tropical lowland regions and their wet forests contribute substantially to herpetofaunal species diversity, species diversity among amphibians is highest in montane regions, where the numbers of endemics are high (usually above 75%; Duellman 1999). Comparable data for species of reptiles are not available, but typically, species diversity of reptiles in montane regions lags behind that of amphibians (e.g., amphibians make up 64% of the Andean herpetofauna; Duellman 1979c). In contrast, reptile diversity in lowland Neotropical regions may be 20%–30% higher than that of amphibians in any particular area (e.g., Hoogmoed 1979a; Duellman 1990).

In 1979, Hoogmoed reviewed the Guianan herpetofauna and published a list of species of amphibians and reptiles known from the region. Hoogmoed pointed out the relatively high degree of endemicity in highland components of the fauna. He evaluated the distributional patterns of the endemics in light of various hypotheses that had been used to explain distributions of other organisms from the same region. The appearance of that seminal publication, which coincided with early work by McDiarmid on the eastern tepuis, provided a starting point for our efforts and those of others to document the herpetofauna of the Guayana Highlands. Recently, Gorzula and Señaris (1998) published an account of the species of amphibians and reptiles collected by Gorzula from 1974 to 1990, largely in the Río Caroní watershed in Bolívar state. Although restricted to the Venezuelan Guayana, that treatment made another important contribution to our understanding of the distributions and species diversities of the amphibians and reptiles of the region. Gorzula and Señaris (1998, 1) expressed the hope that "these raw data could provide both a stimulus and a starting point for others to attempt a more definitive account of the herpetofauna of this region." Their statement, combined with the natural appeal of the region, our research interests, and past and on-going field investigations, served as the stimulus for this review. In this chapter we briefly describe the physical and biological characteristics of the Guayana Highlands, review the fieldwork and publications that have appeared since Hoogmoed's pioneering contribution, compile a list of the highland herpetofauna by massif/tepui, identify patterns reflected in the distributions of the component species, and offer some explanation for those patterns.

The Guayana Region

A brief comparison of the three highland areas of South America will help to put our discussion of the Guayana Highlands in perspective. Each has a complex and contrasting history, provides an array of different habitats, and harbors a diverse and highly endemic fauna. The series of meridionally oriented Andean mountain chains that extend along the entire western edge of the continent (Tierra del Fuego in the south to Panama in the north) forms the most obvious and most extensive highland area. The Andes include some of the highest mountains in the world and exert a primary influence on the climates of western South America. Compared with the other two upland areas, the Andes are relatively young. Lundberg et al. (1998) gave a detailed reconstruction of the Andean orogeny through its 90-million-year history and focused on the complex geologic history of the continent relative to its major river systems. They showed unequivocally that the current patterns of west-to-east flow of the Amazon and Orinoco rivers resulted from the final uplift of the Mérida Andes and Eastern Cordillera of Colombia approximately 8 million years ago.

Two immense and older upland areas, the Brazilian and Guayana shields, occupy the eastern parts of South America and are essentially Precambrian in age (Simpson 1979). These areas, derived from the western section of Gondwanaland, contain some of the oldest terrestrial habitats on the continent. The Brazilian Shield covers much of the continent south of the Amazon and north and east of the Paraná. Although more extensive, the Brazilian Shield tends to be lower, supports drier habitats, and lacks the distinctive, isolated mountains characteristic of the Guayana region. Together with the Andes, the Guayana and Brazilian shields contribute noticeably to the geographic layout of the continent (Lundberg et al. 1998) and have affected the composition and distribution of its biota both historically and ecologically.

The Guayana Shield occupies a vast area that extends approximately 1,500 km in an east-to-west direction from the coast of Suriname to southwestern Venezuela and adjacent Colombia and covers much of the southern portions of Venezuela, Guyana, and Suriname as well as portions of French Guiana and extreme northern Brazil. Extensive lowland forests and savannas, as defined by the Orinoco, Negro, and Amazon rivers, separate this region from the Andean highlands to the north and west and from the northern portions of the Brazilian Shield to the south (Lundberg et al. 1998). Various authors have used the terms *Guiana, Guyana,* and *Guayana* and their derivatives to refer to the entire region or to parts of it, and three countries in the region have had or currently use some form of the word in their names. In this chapter, we follow an author's usage when referring to his or her work (see Berry, Holst, et al. 1995). Otherwise, we use *Guayana* because it has a broader and less political meaning.

The Guianan region was broadly delimited by Hoogmoed (1979a) as the area south of the Río Orinoco, east of the Cassiquiare Canal and the Río Negro, north of the Río Amazonas, and west and south of the Atlantic Ocean. From a phytogeographical perspective, the Guayana region encompasses the area from central French Guiana, Suriname, and Guyana, westward through the northern portions of the Brazilian states of Pará, Roraima, and Amazonas, most of the Venezuelan states of Bolívar and Amazonas, and parts of the southeastern Colombian departments of Guainía, Vichada, and Vaupés (Huber 1994; Berry, Huber, et al. 1995). Contrary to treatments by certain authors (e.g., Maguire 1979), that of Berry, Huber, et al. (1995) specifically excluded the Cordillera de Macarena located in Meta at the base of the Colombian Andes, and we concur. Topographically, the Guayana region is marked by extensive uplands (500–1,500 m), isolated highlands (> 1,500 m), and peripheral lowlands (< 500 m) that occur along the rivers and on the coastal plain. This region is mostly characterized by nutrient-poor soils and a flora of notable species richness (approximately 15,000 species), high endemism, and diversity of growth forms. The most significant geographic feature of the Guayana region is the presence of mid- to high-elevation areas that make up its physiographic core. These highlands, collectively called the Guayana Highlands or Pantepui (fig. 18.1), consist of loosely clustered groups of isolated mountains that range

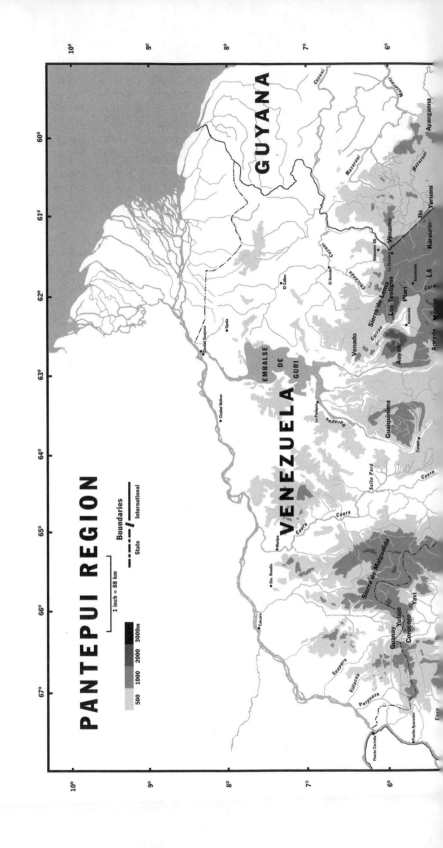

PANTEPUI REGION

1 inch = 88 km

Boundaries

— — State   —·—·— International

500 1000 2000 3000m

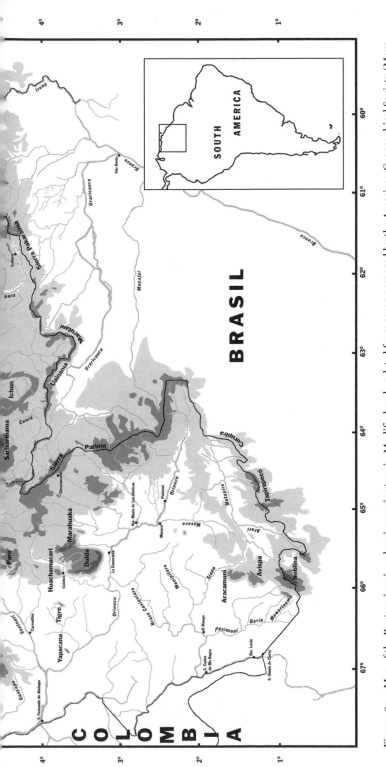

Figure 18.1 Map of the Pantepui region showing major tepuis. Modified and updated from maps prepared by the American Geographical Society (Mayr and Phelps 1967), Missouri Botanical Garden (Steyermark et al. 1995a), and International Travel Maps (Healy 1994).

from approximately 1,300 to more than 3,000 m in elevation and harbor a diverse and highly endemic herpetofauna.

Pantepui: The Guayana Highlands

HISTORY

Mayr and Phelps (1955) first used the term *pantepui* in the title of a paper they gave in 1954. In a subsequent paper, Mayr and Phelps (1967, 275) explicitly proposed *Pantepui* as an arbitrary name to refer collectively to "the sandstone tabletop mountains in the Venezuelan Territorio Amazonas and Estado Bolívar and in the adjacent border regions of Brasil and Guyana." Although Mayr and Phelps (1967) specifically stated that Cerro el Negro (a granite mountain of 1,200 m elevation in northern Amazonas) and Cerro de la Neblina (600 km to the south on the Brazilian border) were not tabletop mountains, they included them in their treatment of the Pantepui mountains on faunistic grounds.

Maguire (1970, 1972, 1979) used *Guayana* in a general sense to refer to the region overlain by the Roraima Formation (see the section titled "Geology") and its outwash sediments, but restricted Guayana as a floristic province to those parts of the Guayana Shield that included tepuis; he considered the Tafelberg in Suriname and the Cordillera (Sierra) de Macarena in Colombia to be the easternmost and westernmost sites, respectively, in that province.

Müller (1973) used *Pantepui* and *Roraima* as names for two dispersal centers for terrestrial vertebrates in the region, and Brown (1975, 1987) used *Pantepui* as a name for a center of endemism of lowland forest butterflies in the southeastern part of Bolívar state. Relatively few workers have used Müller's classification, and like Huber (1987), we find it difficult to incorporate Brown's concepts, derived from forest butterflies, into a broader biogeographic pattern for Pantepui.

Steyermark (1982, 200) expanded Mayr and Phelps's definition of Pantepui to include "not only the sandstone tabletop mountains in the Venezuelan Territorio Amazonas and Estado Bolívar and in the adjacent border regions of Brazil and Guyana, . . . but also the Gran Sabana at the base of the eastern Venezuelan tepuis, the edaphic lowland sand savannas and igneous 'laja' formations of the western part of the Territorio Federal Amazonas, and the extreme northeastern sector drained by the Río Venamo and tributaries." In so doing, he included lowlands in his notion of Pantepui, in contrast to Mayr and Phelps (1967), who specifically defined the term to accommodate only the mountains. Steyermark (1982) argued that because the flora of the Guayana Highlands appears to be unified but also differentiated, it was preferable to maintain it as a single large unit but recognize eastern and western subdivisions. In this sense, he followed the lead of Mayr and Phelps (1967), who divided Pantepui into eastern and western subdivisions separated primarily by the Río

Figure 18.2   Schematic representation of a typical tepui with forested slopes and surrounding savannas (modified from a figure in Means 1995).

Caroní. Mayr and Phelps (1967) also noted that the Pantepui avifauna was not
evenly distributed, despite the relatively high vagility of birds, and that species
differed between the two divisions.

Huber (1987) recommended use of the term *Pantepui* to refer to the complex
of mountains in the Guayana region of southern Venezuela, northeastern Guyana, southern Suriname, and northern Brazil whose major components were
derived from the sandstone of the Roraima Group that lies atop the Guayana
Shield. The majority of these mountains (fig. 18.2) are relatively isolated, have
summits that today are above 1,200 m elevation, and share a common geological history. Huber's concept of Pantepui was more than a geographically based
entity; he reasoned that because Pantepui has common geological, geomor-
phological, chemical, and biological attributes, it is distinct from the surrounding lowland ecosystems. In this chapter, we use Huber's concept of Pantepui.

PHYSIOGRAPHY

The Guayana region has been divided into subregions that differ among authors (e.g., Maguire 1979; Huber 1987, 1995a). In the most recent phytogeographical treatment, Berry, Huber, et al. (1995) recognized four provinces: Pantepui, and Eastern, Central, and Western Guayana. The Pantepui Province
comprises the central highlands, whereas the latter three include lowlands with
some scattered upland regions. Huber (1987) initially subdivided the Pantepui
Province into five geographically defined sectors, but subsequently Berry, Huber, et al. (1995) modified that scheme and recognized four districts defined primarily by geographical criteria (major drainages) and floristics. We follow their

arrangement, with only minor modifications, because it has a strong geographical component and treats lowlands and highlands separately (appendix 18.1).

The Eastern Pantepui District includes those highlands that occur east of the Río Caroní, a major drainage that flows north into the lower Río Orinoco (fig. 18.1). Most of these tepuis are in the Caroní basin, but those in western Guyana are drained by the Mazaruni and Essequibo rivers. Maguire (1979) indicated that the Guyana tepuis might deserve separate treatment when they are better known. Tepuis in Brazil are in the drainage of the Río Branco, a tributary of the Amazon. Water from the summit of Roraima flows into all three drainages. We recognize three subdistricts, Roraima, Los Testigos, and Chimantá, for the eastern, north central, and western tepuis, respectively. We separated the Los Testigos group from the Chimantá group because of its geographically intermediate position. Berry, Huber, et al. (1995) also commented on the transitional nature of these tepuis.

The Western Pantepui District is the most extensive and includes sandstone and granite mountains whose summits reach between 1,300 and 2,350 m elevation; it is bounded by the Río Caura, Río Orinoco, and Río Ventuari systems (fig. 18.1). The summits (between 1,800 and 2,350 m) of the granitic Sierra Maigualida and other high mountains along the eastern edge of the district are mostly forested and make up the Maigualida Subdistrict. The Yutajé Subdistrict includes Guanay, Yaví, Corocoro, and Yutajé tepuis, the summits of which lie at elevations from 1,800 to 2,300 m and have a high diversity of shrubland (M. A. Donnelly, pers. obs.). Cerro Yaví and Cerro Yutajé are relatively dry tepuis with little or no peat on their summits. The Cuao-Sipapo Subdistrict includes several small sandstone tepuis and granitic mountaintops that reach elevations between 1,400 and 2,000 m. Tepuis in this subdistrict are poorly explored and have wet meadows, high-tepui shrub, and open rock habitats.

The Central Pantepui District (the Jaua-Duida District of Berry, Huber, et al. 1995) includes a number of large sandstone massifs in southwestern Bolívar and east-central Amazonas states, Venezuela (fig. 18.1). The summits of most of these tepuis are between 1,500 and 2,800 m. We recognize four subdistricts rather than three (Berry, Huber, et al. 1995), tentatively placing Cerro Guaiquinima in its own subdistrict. Cerro Guaiquinima is a centrally located, well-defined large tepui that lies in the Paragua drainage well to the west (> 100 km) of the Chimantá-Auyán massifs of the Eastern Pantepui District. Berry, Huber, et al. (1995, 173, 175) placed Guaiquinima, along with Cerro Yapacana, the Tafelberg (a Suriname outlier), and other uplands with summits above 1,500 m elevation, in the Central Guayana Province. However, the summit of Cerro Guaiquinima lies at 1,650 m (Huber 1995a) and thus, by their own definition, should be in Pantepui; accordingly, we place Cerro Guaiquinima in its own subdistrict within the Central Pantepui District. The other three subdistricts are Jaua-Sarisariñama, in the upper Paragua and Caura river drainages (we also place the

Sierra Marutaní here); Asisa, which includes the poorly explored Parú Massif at approximately 2,200 m elevation, and the lower (up to 1,300 m), forested Parima uplands in the Río Ventuari system; and Duida-Marahuaka in the upper Río Orinoco drainage. The latter subdistrict includes the moderately well-explored Cerro Duida and an adjacent tepui, Cerro Marahuaka; together with Cerro Huachamacari to the north, these tepuis have summits that lie between 1,900 and 2,800 m, are somewhat isolated, and have several distinctive vegetation types. We have included the much lower (1,300 m) Cerro Yapacana here as well, because of geographic location.

The Southern Pantepui District includes some high sandstone tepuis and granitic mountains along the southern border of Venezuela (fig. 18.1). For the most part, these highlands, whose summits range from 1,600 to 3,014 m, are drained by tributaries of the Río Negro. This region is complex geologically and consists of both quartzitic and igneous units (Berry, Huber, et al. 1995). Extensive shrublands and broad-leaved, high-tepui meadows occur on Neblina, which has been well explored compared with the other highlands in this district. Although poorly known as a whole, this region has the highest number of endemic plant taxa of the entire Guayana region.

The major tepuis and other uplands and highlands that make up Pantepui of the Guayana region are listed in appendix 18.1; most are shown on the map in figure 18.1. If the maximum elevation of a tepui was less that 1,500 m (i.e., an upland, not a highland), we usually added it to our list. Thus, some uplands whose highest elevations do not exceed 1,500 m are included (e.g., Cerro Venado in the Chimantá Subdistrict). When known, the approximate elevational range for each tepui is included in appendix 18.1 (data from Huber 1987, 1995a; T. Hollowell, pers. comm.; available maps).

GEOLOGY

The Guayana Shield is made up of igneous and metamorphic rock (granites, gneisses, schists) of Proterozoic age. A thick layer of sedimentary rock (sandstone and quartzite) called the Roraima Group (also Series or Formation) lies atop this crystalline basement. Younger intrusive rocks (diabases and granites) have penetrated the basement and sedimentary layers, and their sills are scattered throughout the formation. The granitic basement apparently formed during four orogenic events between 0.8 and 3.6 billion years ago. Geologic evidence argues that the multiple layers of sand that were heavily compressed and cemented together to form the Roraima Group, which must have been thousands of meters deep in some areas, were deposited atop the granite between 1.5 and 1.8 billion years ago (Ghosh 1985). The absence of fossil material also supports a Precambrian age. Ghosh (1985) noted that the original Roraima sediments could have had a minimal surface area of approximately 250,000 km^2.

Extensive ripple marks and cross-bedded sections in freshly exposed sandstone surfaces on the summits of several eastern tepuis attest to several periods of deposition (in a shallow sea or large inland lake [S. Ghosh, pers. comm.]; Huber 1995a), probably beginning in the east and later filling in western and southern basins. Whatever the case, the Roraima sandstone has been in place much longer than most terrestrial life. Huber (1995a), Maguire (1970), Schubert and Briceño (1987), and Steyermark (1986) summarized the geologic history of the region, and detailed treatments were written by Gansser (1974) and Ghosh (1977, 1985).

Reid (1974) delineated four strata of the Roraima Formation whose distinctive thicknesses and compositions are easily seen on Mount Roraima. Rock in each stratum differs in color and hardness, evidently reflecting compositional differences. Strata on eastern tepuis are often level-bedded (e.g., Roraima) or gently sloped (e.g., Auyán), whereas those on western tepuis, where visible (e.g., Duida), are largely deformed. The four stratigraphic patterns described for the eastern tepuis are not evident on many western, central, and southern tepuis that we have visited. These differences may reflect variations in patterns of deposition in eastern versus western tepuis. If so, then the eastern strata are likely older than western and southern ones, and the older granitic mountains of Parima and Maigualida may have separated the different sedimentation basins (Huber 1995a). Given the extreme age of these events, it seems unlikely that any separation would be reflected in present-day patterns of herpetofaunal diversity. If, however, the distinct rocky habitats characteristic of the summits of the eastern tepuis are a consequence of differential patterns of deposition and subsequent erosion of the sandstone, then the occupation of these habitats should be manifest in the evolutionary history of the herpetofauna.

The initial uplift of the granitic base and roofing sandstone of the Guayana Highlands probably occurred more than 2 billion years ago (Schubert and Briceño 1987). Three other periods of uplift have been reported, one each in the Mesozoic, the Paleocene, and the later Tertiary. Recent history has been marked by intense erosion and minor eustatic changes (Simpson 1979). A schematic model that describes the erosional surfaces seen at different elevations on tepuis is available (Schubert and Briceño 1987). The flat-topped summits that characterize many tepuis (fig. 18.3) apparently are the result of horizontally layered sandstone and quartzite strata. Most summits occur at elevations between 2,000 and 2,600 m, with several in the eastern chain being higher. Occasionally, the flat summits are surmounted by higher conical peaks that rise to over 3,000 m, as occurs in the Sierra de la Neblina. Most tepuis have sculpted, vertical walls, or escarpments, that drop 300–700 m, and sometimes more than 1,000 m, to the surrounding lowlands; some (e.g., Auyán) have a series of vertical drops that form step-like sides (fig. 18.4). Several larger massifs (e.g., Auyán, Jaua, Duida) have sizable streams that may flow continuously; others (e.g., Roraima,

Figure 18.3 Aerial view of Aparamán-tepui looking west from Murisipán-tepui, in the Los Testigos Massif. Murochiopán is the small tepui in the foreground between Murisipán and Aparamán, and Padapué-tepui of the Los Hermanos group is to the left and beyond Aparamán. The northwest extent of Auyán-tepui appears in the distance.

Figure 18.4   **The step-like escarpments of Auyán-tepui as seen from the south.**

Kukenán, Ptari, Yaví) accumulate rainwater in shallow depressions, some of which may be of considerable expanse, and drain through a few nonpermanent streams that plunge as intermittent waterfalls over the escarpment. The lower slopes and the talus that forms at the base of these cliffs often include very large blocks of stone and rocky debris that have separated and fallen from the escarpment. The slopes often are steep and covered by a dense mossy forest that is bathed by moisture from dense clouds that form along their cliffs. Summits are often deeply cut by streams that follow rock crevices and contact zones between strata (e.g., Auyán-tepui). Presumably, valleys between adjacent tepuis within some massifs (e.g., Chimantá) were formed by rivers that began as small water courses on their summits.

The principal consequences of this long erosional history are the flat-topped mountains with vertical pink cliffs and cascading waterfalls. This type of highland is what Pemón Amerindians of southeastern Venezuela call a *tepui*. Appropriately, this term has been added as a suffix to many specifically named mountains in the region. Tepuis have provided the settings for some exciting science and science fiction (e.g., *The Lost World* by Arthur Conan Doyle, 1912; see fig. 18.5) and likely will continue to do so for decades to come.

SOILS

The soils of the Guayana Highlands are moderately diverse, highly acidic, and generally low in nutrients. Poor soil quality is the result of low mineral content

Figure 18.5 (*Right*) View of the western flank of Roraima with Kukenán in the background. (*Below*) The Eastern Tepui Chain from the Gran Savanna. From the left, the eastern tepuis are Tramen, Ilú, and Karaurín, with the pinnacle of Wadakapiapué lying in the gap (*left of center*), followed (*at right*) by Yuruaní, Kukenán, and Roraima.

of the parent rock as well as the long and extreme weathering to which they have been subjected. The following comments are based on Huber (1995a) and our observations. Many tepui summits have relatively little soil, and summits of many eastern tepuis are largely bare sandstone. Here, plants grow wherever they can get a start, often in shallow sand derived from weathered rock, and some plants respond to nutrient limitation by trapping insects or nutrients (e.g., sundews and bromeliads; Huber 1995a). On other tepuis (e.g., Duida), deep (up to 2 m) layers of decomposing organic materials (peat) form histosols atop rocky substrates. These boggy areas are generally open and devoid of trees, provide easy helicopter access, and frequently are used for campsites (e.g., Neblina). Entisols are characteristic of the extensive savannas that surround many tepuis and are made up of fine- to coarse-grained white sands. These soils are often high in iron and aluminum and are known for their poor quality (Huber 1995a). The soils (e.g., oxisols) that support forests in the region are composed of white sand with varying amounts of clay.

CLIMATE

The climate of the Guayana Highlands environment is the result of the interplay between the trade winds and the annual oscillations of the Intertropical Convergence. Trade winds carry cool moisture-laden air onto the warm continent, and the eastern tepuis are among the first major upland areas contacted in Guayana. As a result, mean annual rainfall decreases from east to west and south to north. The Sierra de la Maigualida may cause a local rain shadow on its western slopes and on tepuis to the west (e.g., Cerro Yaví); similar effects are seen at sites in the southern Gran Sabana. The mean annual temperature in the lowlands (0–500 m) is always greater than 24°C, and the annual rainfall varies from approximately 2,000 mm in the north to 4,200 mm in the upper Caroní, Paragua, and Caura river basins. A distinct dry season occurs from December through February in northern areas and is almost imperceptible in December and January to the south. Mean annual temperatures at upland elevations (500–1,200 m) range from 18° to 24°C, and mean annual rainfall is 2,000 mm or higher (data are sparse for the region). Mean annual temperatures at sites above 1,500 m elevation (highlands) are cool, between 12° and 18°C, and the mean annual precipitation is high (2,500–3,500 mm). Dense clouds and prolonged mist are common and provide additional moisture at highland sites. Precipitation is less common from December through February, when insolation is high. Because rainy weather makes helicopter access difficult, most work has been done on the summits of higher tepuis from December through March. The highest tepuis (summits > 2,350 m) probably have a mean annual temperature below 10°C. Frost is unknown on tepuis, but freezing temperatures occasionally occur; one of us (RWM) found ice in a metal drinking cup outside his tent one January morning on Roraima. A detailed treatment of the climate of the Guayana Highlands is provided by Huber (1995a).

VEGETATION

The Guayana Highlands are characterized by high floristic diversity and high ecological diversity in both physiognomy of the plant communities and their distributions. The vegetation relevant to our analyses includes zones of montane (800 –1,500 m elevation) forest on tepui slopes and upper-montane (cloud) forests that develop near or along bases of vertical cliffs at approximately 1,500 – 2,000 m. Epiphytic mosses, ferns, lichens, and orchids are common in the upper-montane forests. On some tepui slopes (e.g., Roraima, Auyán, Uei), forests have been destroyed by locally set fires (Means 1995). The summit vegetation (2,000 –3,000 m) includes low-growing tepui forest, tepui scrub, and high mountain meadows and grasslands. Leaves of most trees and shrubs are small and sclerophyllous, whereas those of most herbs are thick and coriaceous (Huber 1995c; R. W. McDiarmid, pers. obs.; see Steyermark et al. 1995a for vegetation map).

Low, evergreen forests grow in sheltered areas on some tepui summits, usually on peat overlying sandstone or on mineral soils generated from weathering of diabasic intrusions (Huber 1995c). On the eastern tepuis, such evergreen forests are often low (tree layer of 6 –12 m) and species poor, and the undergrowth is dominated by large rosette-forming herbs (e.g., bromeliads of the genus *Brocchinia*). Montane forest on the summits of other tepuis (e.g., Sarisariñama) are taller (15 –25 m) and have a well-developed, species-rich understory. Low (7 –10 m), evergreen upper-montane forests cover the central and southern sections of Cerro Duida between 1,500 and 2,200 m, whereas a low, evergreen high-tepui forest occurs in occasional depressions on the summit of Marahuaka above 2,600 m. Low (8 –15 m), evergreen upper-montane forest occurs between 1,500 and 2,000 m on Cerro de la Neblina, and a taller, evergreen, montane forest type occurs at lower elevations (800 –1,500 m) on Neblina's slopes.

Shrublands are important components of the Guayana environments. Most of the shrubs are less than 5 m tall and have woody stems and leaves that are distributed along the stem or concentrated as a terminal rosette. The makeup and distribution of shrubland habitats probably reflect local edaphic conditions more than do those of either forest or grassland. Shrublands (1–3 m tall) occur on the summits of Auyán and Chimantá massifs, sometimes on peat soils and sometimes on rocky slopes and outcrops. Areas of well-developed shrublands occur on rocky soils in the Jaua-Sarisariñama Massif, and dense sclerophyllous scrub (usually < 4 m) is found on the summits of Guanay, Yutajé, and Duida. A peculiar shrubland grows on shallow organic soils between 1,600 and

2,000 m elevation on Cerro de la Neblina. This "Neblinaria scrub" is composed primarily of a single species (*Bonnetia maguireorum*, Theaceae), which forms dense stands of plants 2–3 m tall. The plants have terminal rosettes of leathery leaves that can hold considerable water. Fire has been shown to play a role in maintenance of these shrublands (Givnish et al. 1986; Means 1995).

Grass-dominated meadows (savannas) are common in the lowlands but less so on tepuis. Highland meadows comprise nongramineous plants with a variety of growth forms. Broad-leaved meadows vary floristically and physiognomically on different tepuis. In some places (e.g., between 2,400 and 2,750 m on Roraima), the meadows consist of isolated clumps of mixed species on exposed rocky surfaces. Meadows grow in dense patches on dead peat on Auyán (1,600–1,900 m) and on Guaiquinima (1,200–1,600 m). Species of *Stegolepis* (Rapateaceae) are often found in these meadows. Eastern tepui (e.g., Ptari, Ilú) meadows are dominated by species in the Xyridaceae and Ericaulaceae. Tubiform bromeliads (*Brocchinia*) and tubiform pitcher plants (*Heliamphora*, Sarraceniaceae) are also found in some tepui meadows. High-tepui grasslands are rare and have been found in waterlogged or flooded sites on Auyán, Chimantá, Sierra de Maigualida, and Cerro Marahuaka (Huber 1995c).

Pioneer plant formations occur throughout the Guayana Shield on granite outcrops at low elevations and on exposed sandstone in the highlands. Early colonizers include the cyanobacteria *Stigonema panniforme*, which gives rise to the black color of the rock surfaces, many lichens, a few mosses, and several kinds of vascular plants. These plants often grow in small, elongate vegetation islands (up to 300 m²) and form over depressions in which sand and organic materials have accumulated (Michelangeli 2000).

The Pantepui Herpetofauna

CREATING A LIST

Current knowledge of the Pantepui biota comes from publications and materials collected by a series of expeditions, the history of which we summarize in appendix 18.5. We compiled a species list of amphibians and reptiles known to be above 1,500 m on tepuis, on the basis of previously published summaries (e.g., Hoogmoed 1979a; Duellman 1999; Gorzula and Señaris 1998), other published literature (e.g., Donnelly and Myers 1991; Avila-Pires 1995; Myers and Donnelly 1996, 1997, 2001; Barrio 1998; Galán 2000), and known specimens in museum collections. Reports of the occurrence of widespread, lowland species on tepuis that were based on specimens unaccompanied by specific locality and/or elevation data (e.g., Roze's [1958a, 1958b] papers reporting on material collected by others from Auyán and Chimantá tepuis) were evaluated individually. Elevation and/or precise locality data were available for a few specimens in museum catalog records; if these records indicated an occurrence above

1,500 m, the species was included in our list. In other instances where precise data were not available, a decision to include or exclude a specific record was based on other considerations (e.g., the distribution of the species elsewhere and our experience with it in the Guayana region). Although we did exclude a few records (e.g., *Anolis auratus*, *Anilius scytale*) as likely having been collected from a locality near the tepui but in the surrounding lowlands and below 1,500 m, we made such decisions cautiously.

Hoogmoed (1979a, appendix) gave an elevational range for each amphibian and reptile species from Guayana and assigned each to one of 12 general categories of distribution. Of the 55 species of frogs reported from localities above 1,000 m, 18 were known only from localities above 1,000 m and were called highland endemics by Hoogmoed. He listed 9 of 38 species of reptiles from above 1,000 m in the same category. Many (perhaps most) of the highland species that Hoogmoed listed as "sp." (i.e., unidentified) in his 1979 compilation have now been identified and/or described (e.g., for *Stefania* sp. A–C, see Duellman and Hoogmoed 1984; Señaris et al. 1996).

Duellman (1999, table 5.2) reported 76 species (74 in his appendix 5.1) of anurans from the Guayana Highlands, including several taxa known primarily from upland sites near La Escalera in Venezuela, the north slope of Roraima in Guyana, and possibly elsewhere (see Duellman and Hoogmoed 1992; Duellman 1997). The following species have not been found on the summit of any tepui and are not included in our list: *Oreophrynella macconnelli*, *Cochranella helenae*, *C. oyampiensis*, *Hyalinobatrachium iaspidiensis*, *H. ostracodermoides*, *Colostethus parimae*, *C. parkerae*, *Hyla kanaima*, *H. lemai*, *H. loveridgei*, *H. ornatissima*, *H. pulidoi*, *H. roraima*, *H. warreni*, *Scinax danae*, *S. exigua*, *Stefania roraimae*, *S. scalae*, *S. woodleyi*, *Tepuihyla galani*, *T. rodriguezi*, *T. talbergae*, *Eleutherodactylus pulvinatus*, *Leptodactylus sabanensis*, and *Otophryne robusta*. Duellman (1999, appendix 5.1) also included a few more-widespread forms that occur at localities above and below 1,000 m. We assume (no elevational data were given) that Duellman's count of 76 species was comparable to Hoogmoed's general value of 55 species. However, Duellman's (1999, appendix 5.1) Guayana Highlands list does not include many of the 37 species that Hoogmoed (1979a) reported from the highlands (above 1,000 m) and from localities below 1,000 m. Duellman listed many of those 37 species in his Amazon basin–Guiana lowlands category, rather than in both categories, as would have been appropriate based on their elevational range. Five species that Duellman (1999) omitted from his Guayana Highlands list (*Hyla boans*, *H. minuta*, *Osteocephalus taurinus*, *Leptodactylus longirostris*, and *L. rugosus*) had been reported from the top of Cerro Guaiquinima (Donnelly and Myers 1991; Mägdefrau et al. 1991; Schlüter 1994). Thus, Duellman's Guayana Highlands amphibian fauna is not the same as Hoogmoed's highland endemic group. Duellman's count(s) represents about a 30% increase in the total known anuran diversity for the Guayana Highlands in the 20-year period since Hoogmoed's (1979a) treatment.

We included records of some species that had been collected from uplands (1,000–1,500 m) rather than highlands (> 1,500 m), when such sites had been treated as tepuis (see previous discussion of physiography; also Huber 1995a). Accordingly, species collected at moderate elevations but from the tops of Guaiquinima (e.g., *Hyla boans* at 780 m and *Hyla crepitans* at 930 m; Schlüter 1994) and Cerro Yapacana (*Minyobates steyermarki* described from a single specimen collected at 1,200 m near the summit) were added to the list, even though these and other species records from similar upland and slope elevations (but not on tepui tops) were not included. Likewise, species collected at localities of moderate to high elevations in the Sierras Maigualida, Marutaní, and Tapirapecó, primarily granitic mountains within the Guayana Highlands, were included. Finally, we added RWM's unpublished records of taxa encountered during fieldwork on the tepuis and those of others with which we were familiar.

We had a more difficult time deciding how to treat species reports from localities at intermediate elevations (up to 1,500 m) on tepui slopes, given our adoption of 1,500 m elevation as a workable though completely arbitrary lower limit for a highland to qualify as a tepui. An often-used approach is to include any species that has been taken at a locality above some minimal elevation (i.e., the lowest elevation used in defining the highlands) as a member of the highland assemblage (e.g., Chapman 1931). Such an approach certainly facilitates species assignments and in some studies may be appropriate. On the other hand, when the decision to add a species is based on specimen availability (e.g., Chapman 1931) rather than some biologically meaningful elevational limit, it deserves scrutiny. In instances in which a species occurrence can be documented by numerous samples taken along elevational transects, one may be able to detect natural breaks or discontinuities in species distributions by using statistical approaches. Such a statistically defined limit could then be applied to the distribution of the same species in comparable sites with some confidence. Given the inadequacy of sampling on tepui slopes, such an approach currently is not possible. Ultimately, we excluded species that had been reported from localities below but near 1,500 m, if the bulk of other evidence indicated that the species occurs primarily at lower elevations. On the other hand, we always included a primarily lowland species if it had been collected from a locality above 1,500 m on a tepui slope.

Most troublesome were reports of species collected from the "base" or "foot" of Roraima (e.g., certain species collected by Quelch and McConnell and described in Boulenger 1900a, 1900b). The "base" of Roraima is somewhat variable, and accessible localities above 1,500 m occur in this area (see Chapman 1931). However, most reports of specimens from the base of Roraima use 3,500 ft. (approximately 1,067 m) as the elevation, and we did not include those records in our compilation. Many of the same species, as well as several others, have been collected from montane forest sites on the north slope of Roraima (August–September 1971) at camps reportedly located between approximately 3,000 ft. (camp 5, approximately 915 m) and 6,700 ft. (camp 9, approximately 2,040 m), and possibly higher (Warren 1973). Unfortunately, Warren did not list all species collected or clearly indicate which species came from which camp or at which elevation. As a result, we were forced to determine specific locations from his narrative, which is somewhat problematic. Certain frogs collected by Warren and his associates were described by Duellman and Hoogmoed (1992) and listed as coming from localities between 1,430 and 1,480 m. Two years after Warren's trip (October–November 1973), Michael Tamessar of Georgetown, Guyana, accompanied a British climbing expedition to Roraima's summit (MacInnes 1976). They made use of the camps and trails prepared by Warren's group (appendix 18.5). Tamessar collected the holotype of *Stefania roraimae* at 1,402 m (Duellman and Hoogmoed 1984) and a paratype of *Hyla roraima* and another specimen of *Hyla kanaima* at 1,430 m, reportedly at the same place that Warren had collected these two species (Duellman and Hoogmoed 1992). Warren reported elevations for his camps (1–4) along the Waruma at above 2,000 ft. (2,100–2,300 ft., 640–700 m), but current maps and recent GPS readings indicate that the entire stretch of the river where the trails and camps were located is actually between 1,600 and 1,700 ft. (488–520 m; C. Milensky, pers. comm.). On the basis of these reports, comparisons of elevations given by Warren with those on topographic maps (not available in 1971), and recent measures with GPS devices, it seems likely that Warren's elevations were too high, perhaps by as much as 150 m. Confusion about elevations and the locations at which specimens had been collected, and knowledge that other specimens from the same region collected by Warren and others (e.g., Smithsonian Guyana Expedition, 2001) have been neither identified nor reported in the literature, led to the decision to exclude records of species from the north slope of Roraima from our list. Future work will show which, if any, of these species are members of the Pantepui herpetofauna.

Our final list of amphibians and reptiles from the Guayana Highlands includes 159 species, of which 61% are amphibians and 39% are reptiles (appendix 18.3). A recent collection from Mount Ayanganna in Guyana (Royal Ontario Museum Ayanganna Expedition [2000], Biological Diversity of the Guianas Program) contains some species on our list, others not currently treated in our compilation and known previously only from the slopes of Roraima, as well as undescribed species of *Oreophrynella*, *Stefania*, *Arthrosaura*, and possibly others (A. Lathrop and R. MacCulloch, pers. comm.). We have not included those records in our list but refer to species when appropriate. The pattern of species accumulation over time indicates that herpetofaunal species of Pantepui remain to be discovered (fig. 18.6). We are convinced that additional fieldwork at higher elevations on this and other tepuis in Guyana and

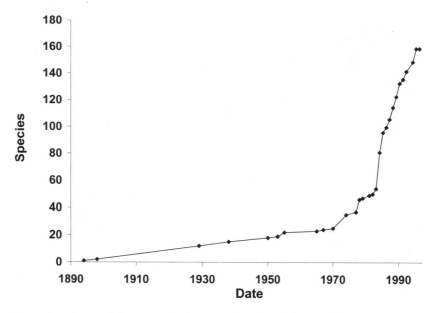

Figure 18.6   Accumulation curve for the 159 species of amphibians and reptiles from Pantepui.

Brazil (e.g., Uei, Wei-assipu, Wokomung), as well as on the slopes and summits of many of the poorly sampled Venezuelan tepuis, will increase the Pantepui herpetofauna to more than 200 species.

From our lists of amphibians and reptiles (appendix 18.3) and tepuis (appendix 18.1), we compiled a list of amphibians and reptiles for each tepui (appendix 18.4). Forty of 72 tepuis (approximately 55%) have some herpetological records, but certain tepuis are better known than others. We arranged tepuis and massifs into 12 groups (subdistricts), which we combined into four regional clusters (districts), generally following the phytogeographical system devised for the Guayana region (Berry, Huber, et al. 1995). For each tepui, we included select information about its geography, physical and biological features, and the date of the first herpetological sample. We indicated which species were endemic, listed pertinent general and herpetological references, and provided other relevant comments.

FACTORS AFFECTING SPECIES DISTRIBUTIONS

Assumptions

Both historical and contemporary factors affect the distributions of amphibians and reptiles of the Lost World. Aside from inferences derived from geological and geographical attributes of tepuis, historical influences await elucidation of phylogenetic relationships among tepui taxa. Given the absence of historical information, we used multivariate statistical methods to determine which ecologic and geographic factors are associated with the observed patterns of amphibian and reptile diversity on tepuis. We assumed that tepuis are habitat islands and, therefore, that island size, elevation, slope (an indication of relative accessibility), distance from neighboring "islands," ecological complexity, and other such factors interact to determine diversity. Further, we assumed that all tepuis were of approximately the same age and older than any significant component of their summit biota. Finally, we assumed that any large-scale environmental change would have affected the evolutionary history of the biota on each tepui in more or less the same fashion and at about the same time(s).

Abiotic Factors

We derived data on abiotic (physical) features of the tepuis from Huber (1995a) or other published sources. Tepui/massif size was based on measures of summit and slope areas. Tepuis with large slope areas often have more gradual rises (i.e., slope habitat) and potentially easier access to their summits. Those with a smaller slope–summit area relationship often have steeper slopes with prominent vertical cliffs; slope habitats suitable for amphibians and reptiles often are lacking, and access by lowland species is difficult. Slope area is not known for all tepuis (i.e., some tepuis were too close together to have individual slope areas); in those instances we used slope values that were provided by Huber (1995a) for a massif rather than an individual tepui. Slope or summit area estimates were not available for three southern sierras (Marutaní, Maigualida, and Tamacuari), and they were excluded from the multivariate analyses. We used the maximum elevation recorded for a tepui as an index of the degree of isolation from the surrounding lowlands and of habitat diversity on the mountain. In our experience, flat-topped tepuis (e.g., Roraima) generally have fewer habitats than those with distinct peaks (e.g., Neblina).

Sampling effort was not equal among adjacent or connected tepuis. For this and other reasons (e.g., geographical proximity and lack of slope data for individual tepui [Huber 1995a]), we combined neighboring tepuis into single massifs for analysis. For example, ten tepuis make up the Chimantá Massif (appendix 18.1; Huber 1995a, fig. 1.27), and some are interconnected; each of the Chimantá tepuis has been sampled disproportionately (Huber 1992), and one (Agparamán) has no reported records of amphibians and reptiles (Gorzula and Señaris 1998). For our analyses, all Chimantá tepuis were treated as a single massif. In some instances, a massif (e.g., Guaiquinima) consists of a single tepui; in others, a massif (e.g., Eastern Tepui Chain) includes several discrete tepuis. Some massifs contain tepuis each with a different but closely related species, whereas others show little or no inter-tepui differentiation (appendix 18.4).

Chimantá Massif is interesting because only 6 of the 17 species recorded from the massif are known from single tepuis, whereas 7 others occur on four or more tepuis within the Chimantá Massif. Given the disproportionate sampling and our combining of data for massifs rather than individual tepuis, we felt compelled to include some measure of this geographical reality in our analysis. In the absence of knowledge of the timing of fragmentation of a massif into discrete tepuis (e.g., seemingly a more recent event within the Chimantá Massif than within the Eastern Tepui Chain), and of the responses of different taxa to such a vicariant event(s), we used the number of tepuis with species of amphibians and reptiles that make up each massif as a variable in our analyses.

As an estimator of geographic isolation, we measured the nearest neighbor distance for each tepui by using the map of Mayr and Phelps (1967; modified and reproduced here as fig. 18.1). For each massif we calculated the nearest-neighbor distance as the average of the nearest-neighbor distance for each contained tepui.

Biotic Factors

The number and kinds of animals at any site are influenced by their interactions with each other and with other biological components of the environment. We used two approaches to incorporate some independent measures of the biological complexity of each tepui into our analyses. We used the number of vegetation types documented from the slopes and summit of tepuis (from the vegetation map in Steyermark et al. 1995a; see appendix 18.2) as an indicator of habitat diversity. We gathered distributional data for several vascular plant taxa from the accounts in the published volumes of the *Flora of the Venezuelan Guayana* (Steyermark et al. 1995b, 1997, 1998, 1999) with the idea that plant diversity and distribution patterns might reflect factors that have contributed to the diversity and distribution patterns of the amphibians and reptiles. We included a taxon in our plant data set when its occurrence on a specific massif was noted; we did not include species whose ranges were described in general terms (e.g., "found on tepuis in Bolívar state"). Four values were scored for each massif: total number of primitive vascular plant species, total number of endemic primitive vascular plant species, total number of angiosperm plant species (as taken from the alphabetical listing of families [A–L] in the published volumes), and total number of endemic angiosperm species (families A–L).

Montane herpetofaunas are aggregates of species with different elevational ranges and distributions. We reasoned that tepuis have faunas composed of different assortments of taxa, reflecting their relative degrees of temporal and spatial isolation, and that the patterns of distributions of their component species result from such isolation. Hoogmoed (1979a) recognized the importance of these patterns and provided an elevational range for each species; he also assigned species to specific distributional categories (e.g., lowland endemic, widespread, disjunct Amazonian, highland endemic, and so forth). In our analysis we augmented Hoogmoed's data with our own and with those from the literature and determined the elevational and distributional ranges for each taxon (see table 18A.1 in appendix 18.3).

Tepui herpetofaunas often include species with narrow elevational ranges (i.e., known from a single locality on the summit or from a narrow range of slope elevations), species with moderate elevational ranges (i.e., they occur on the slopes and on the summit), and a few species that occur in the surrounding lowlands, on the slopes, and on the summit. To characterize each tepui fauna on the basis of the elevational ranges of its component species, we generated an elevational range (ER) for each species by subtracting the minimum from the maximum known elevation (table 18A.1) and then assigned each species to one of four categories: (1) 0–500 m (the elevational range is 0 when a taxon is known only from a single locality); (2) 501–1,000 m; (3) 1,001–1,600 m; and (4) 1,601–2,300 m. We calculated a value for each tepui by summing the number of species in each category, which we weighted by multiplying each by 1, 2, 10, and 100, respectively, or, SUMER = $ER_1(1) + ER_2(2) + ER_3(10) + ER_4(100)$, where ER_1 is the number of taxa known from the massif in category 1, ER_2 is the number of taxa in category 2, ER_3 is the number of taxa in category 3, and ER_4 is the number of taxa in category 4. We chose our weighting factors to reflect the observation that available geographic range increases as a power of elevational range. The SUMER values range from 1 (Angasima, Upuigma, and Yapacana) to 728 (Auyán). The SUMER values reflect observed differences among massifs with similar species diversities but different compositions of species from the four ER groups. For example, Cerro Guaiquinima, Duida, and Auyán massifs have faunas of similar size (22, 23, and 24 species, respectively), but they have very different SUMER values (Guaiquinima = 520; Duida = 150; Auyán = 728). The herpetofaunas of the Cerro Guaiquinima and Auyán massifs include several taxa with wide elevational ranges, whereas that of the Duida Massif includes more taxa with narrow elevational ranges. The SUMER value likely will change with more complete sampling, but we believe that similar differences will remain.

Species in the Pantepui herpetofauna vary from forms that are narrowly restricted in their distributions (i.e., known from a single locality) to forms that are more widespread (i.e., occur over much of the continent). In an attempt to capture these differences, we assigned each species to one of the following distributional pattern (DP; table 18A.1) categories: (1) known from a single tepui (highly restricted); (2) known from localities on two or more tepuis (moderately restricted); (3) broadly distributed across, but still restricted to, the Guayana region (Guayanan endemic); and (4) wide ranging in South America (widespread). From these data we generated the following weighted distributional range score for each massif: SUMDP $= DP_1(1) + DP_2(2) + DP_3(10) + DP_4(100)$, where DP_x values represent the number of species in that particular category on each tepui or massif. Again, our weighting factors reflect the increase in surface area covered by each successive category. Values ranged from 1 (Yapacana) to 1,336 (Guaiquinima) and were 226 (Neblina), 457 (Auyán), 1,336 (Guaiquinima), and 324 (Duida) for the tepuis with high species diversity (27, 24, 23, and 22, respectively). SUMDP values (226–457) for Neblina, Auyán, and Duida massifs are more tightly clustered than are their SUMER values (60–728), reflecting the relative differences among their respective herpetofaunas as revealed by the elevational and distributional distribution data.

Sampling Effort

We incorporated a measure of sampling effort into the analysis by scoring each tepui or massif with a value derived from our estimate of the total number of sampling days devoted to herpetological exploration on each tepui. These data were taken from the literature and from our field notes. Although crude in some instances (e.g., the high Duida value was influenced by the estimate of three months spent by Tate and his associates on Duida although we have little knowledge of the actual time spent collecting amphibians and reptiles), the values (1–131 days) provide an index to actual sampling effort on each tepui/massif. The temporal bias (most sampling has been done during the dry season) often has a major impact on results obtained at lowland sites, but few data are available to gauge its impact on tepui summits.

ANALYSES

To understand the sources of variance in the total number of species among massifs, we used Principal Component Analysis (PCA; SAS Institute 2000) to reduce 13 independent variables (table 18.1) to a smaller set of orthogonal (uncorrelated) factors. We deleted three massifs (Tamacuari, Marutaní, and Maigualida) from this analysis because of missing data (Tamacuari and Maigualida lack slope area estimates, and Marutaní lacks a summit area estimate). We used variance maximizing (varimax) rotation of the original variable space. The varimax rotation maximizes the variability of the factor while minimizing the variance around the factor (Statsoft 1994).

The PCA identified three factors with Eigenvalues greater than 1.0 (factor I $= 8.03$, accounting for 61.8% of variation; factor II $= 2.28$, 17.5%; factor III $= 1.06$, 8.1%). Several variables loaded heavily on factor I. The highest values (table 18.2), in decreasing order, were the number of endemic primitive vascular plant species, the number of primitive vascular plant species, the number of endemic angiosperm plant species, the number of angiosperm plant species,

Table 18.1   Variables used in the PCA

LOC	TOT	REP	AMP	TEP	ELE	SUM	SLOP	NN	FT	DC	VPT	VPE	PPT	PPE	ER	DR
ETC	9	1	8	4	2,810	70	320	383	1	50	135	56	112	19	9	11
PTA	2	0	2	1	2,400	2.5	28	290	2	2	94	28	71	10	2	2
LT	5	1	4	3	2,400	12	116	317	1	3	33	13	46	10	6	8
AUY	24	11	13	3	2,450	667.7	715	272	4	54	165	76	83	17	728	457
APR	2	0	2	1	2,500	6	210	264	1	6	17	9	15	5	2	3
CHI	17	11	6	9	2,650	615	915	268	4	81	242	129	156	28	426	340
ANG	1	0	1	1	2,250	2	32	270	1	3	7	5	0	0	1	2
UPU	1	0	1	1	2,100	0.6	13	281	1	1	5	0	0	0	1	2
GUQ	22	13	9	1	1,650	1,096	410	260	3	28	111	40	41	4	520	1,336
MAR	1	1	0	1	1,500	n.a.	740	278	2	1	20	4	17	0	101	101
COC	1	0	1	1	1,500	526	400	249	2	1	8	5	4	0	100	100
JAU	18	6	12	2	2,350	1,170	770	259	4	29	113	63	53	7	236	327
MAI	2	0	2	1	2,400	440	n.a.	303	2	7	39	26	22	0	11	11
YAV	7	4	3	1	2,300	5.6	70	331	2	9	19	13	6	3	7	10
YUT	10	5	5	2	2,400	275	143	361	3	17	108	61	28	4	127	212

(continued)

Table 18.1 (continued)

LOC	TOT	REP	AMP	TEP	ELE	SUM	SLOP	NN	FT	DC	VPT	VPE	PPT	PPE	ER	DR
GUY	8	3	5	1	2,080	165	113	367	1	11	36	22	5	3	233	225
YAP	1	0	1	1	1,300	10.5	38	421	1	7	28	13	5	0	1	1
DUI	23	9	14	3	2,800	1,219	1,100	346	7	131	220	135	187	28	150	324
TAM	11	4	7	1	2,340	0.01	n.a.	466	1	8	8	1	0	0	137	228
NEB	27	9	18	2	3,014	473	1,515	538	5	69	199	118	180	23	246	226

Note: The 13 variables used in the analysis are those in cols. 5–17. Column headings are abbreviated as follows: LOC, massif or tepui; TOT, total number of amphibian and reptile species known from each massif; REP, number of reptile species known from each massif; AMP, number of amphibian species known from each massif; TEP, number of tepuis in each massif known to have amphibians and reptiles; ELE, maximum elevation (m); SUM, summit area (m²); SLOP, slope area (m²); NN, average nearest-neighbor distance between each massif (km); FT, number of forest/vegetation types from each massif; DC, days spent collecting; VPT, total number of vascular plants known from each massif; VPE, number of endemic vascular plants known from each massif; PPT, total number of primitive plants known from each massif; PPE, number of endemic primitive plants known from each massif; ER, summed elevation range variable; DP, summed distribution variable. Missing data are indicated by n.a. (not available). Massifs or tepuis used in the analysis are abbreviated as follows: ETC, Eastern Tepui Chain; PTA, Ptari; LT, Los Testigos; AUY, Auyán; APR, Aprada; CHI, Chimantá; ANG, Angasima-tepui; UPU, Upuigma-tepui; GUQ, Cerro Guaiquinima; MAR, Sierra Marutani; COC, Cerro Guanacoco; JAU, Jaua; MAI, Sierra de Maiqualida; YAV, Cerro Yavi; YUT, Yutajé; GUY, Cerro Guanay; YAP, Cerro Yapacana; DUI, Duida-Marahuaka; TAM, Cerro Tamacuari; NEB, Neblina-Aracamuni.

the number of tepuis in a massif known to support amphibians and reptiles, the maximum elevation of a massif, the number of collecting days, the slope area, and the number of forest types. Three variables, distributional patterns (SUMDP), summit area, and elevational range (SUMER) loaded heavily (and negatively) on factor II (table 18.2). Average nearest-neighbor distance was the only variable that loaded heavily on factor III.

We used the rotated PCA factor scores in three multiple regression analyses to describe the sources of variance in total number of amphibians and reptiles known from tepui massifs, total number of reptiles alone, and total number of amphibians alone. For all amphibians and reptiles (i.e., total number of species), the multiple regression was statistically significant ($F_{3,13} = 43.93$, $p < .0001$), and the three factors explained 88.9% of the variance in observed species richness. All three factors were significant in the effect tests (factor I: $F = 42.58$, $p < .0001$; factor II: $F = 80.02$, $p < .0001$; factor III: $F = 9.19$, $p = .0096$). For reptiles, the multiple regression was statistically significant ($F_{3,13} = 29.84$, $p < .0001$), and the three factors explained 84.4% of the variance in the number of reptile species among massifs. The effect tests showed that factor I ($F = 20.68$, $p = .0005$) and factor II ($F = 68.84$, $p < .0001$) were significant, but factor III ($F = 0.0021$, $p = .9638$) was not. For amphibians, the multiple regression model was statistically significant ($F_{3,13} = 25.03$, $p < .0001$), and the

Table 18.2 Factor-loading scores (varimax normalized) for variables used in the PCA

Original variable	Factor I	Factor II	Factor III
No. tepuis with amphibians and reptiles	.828*	−.055	−.411
Maximum elevation (m)	.821*	.173	.167
Summit area (m²)	.270	−.896*	.021
Slope area (m²)	.681*	−.527	.362
Nearest-neighbor distance (km)	.233	.180	.860*
No. forest types	.603*	−.648	.296
Days spent collecting	.790*	−.480	.208
No. vascular plant species	.847*	−.472	.115
No. vascular plant endemics	.852*	−.447	.217
No. primitive vascular plants	.901*	−.269	.273
No. primitive vascular endemics	.956*	−.199	.107
SUMER	.249	−.796*	−.228
SUMDP	−.048	−.903*	−.121

Note: An asterisk indicates that the original variable loads heavily on the PCA factor, and the loadings are similar to correlation coefficients. The abbreviations SUMER and SUMDP indicate the loadings for the summed elevational range variable and the summed distribution pattern variable, respectively.

three factors explained 81.8% of the variance in the number of amphibian species among massifs. All three PCA factors were significant in the effect tests (factor I: $F = 27.59$, $p = .0002$; factor II: $F = 30.53$, $p < .0001$; factor III: $F = 16.97$, $p = .0012$).

Factor I is an index of habitat diversity. The positive relationship between amphibian and reptile diversity and the four measures of plant diversity, the number of forest types, and probably the slope area together show that habitat diversity has a positive influence on species richness. One might argue that factors affecting plant species distributions and diversity are likely to influence amphibian and reptile diversity in a similar manner. Even thought certain plants (e.g., those with wind-dispersed spores and seeds) may move easily among tepuis and between tepuis and other montane habitats, we believe that plant diversity itself positively influences amphibian and reptile species diversity. The varied and unusual growth forms of many tepui plants (e.g., *Brocchinia*, Bromeliaceae) are important contributors to the life history and ecology of many amphibians and reptiles on tepuis. The abrupt, insular nature of tepuis (elevated rocky mesas rising from a sea of lowland forest) is revealed in the high number of endemic taxa of amphibians and reptiles characteristic of Pantepui environments and dependent on two other variables that loaded heavily on factor I—the number of discrete tepuis in a massif and their maximum elevations. As each increases, the chance of adding different habitats rises. These variables have a positive affect on herpetofaunal diversity. It is interesting that factor-loading scores for endemic plants are higher in both categories than for non-endemic plants and that the highest values are for primitive vascular species rather than for angiosperms.

The relatively well-sampled massifs with high factor II scores have large summit areas and faunas that include several geographically and elevationally widespread taxa. Most of these massifs (e.g., Guaiquinima, Auyán, Jaua, Duida) also have large slope areas. Accordingly, we interpret factor II as an index of "accessibility" because the variables that loaded heavily on this factor were generally associated with surface area. We suggest that the probability of a lowland species of amphibian or reptile becoming established on a tepui summit is inversely related to summit elevation and extent of vertical cliffs, and directly related to summit and slope areas.

High factor III scores are associated with variables that measure distance among massifs and, hence, isolation. The relative degree of isolation of certain massifs (e.g., Duida, Neblina) explains some of the species diversity and endemism characteristic of these massifs (appendix 18.4).

Results from the multivariate analyses indicate that the biotic and abiotic factors we examined explain a substantial amount of the variance seen in the number of species known from the massifs included in our analysis. For two of the larger tepuis (Auyán and Guaiquinima), species richness is higher than might be expected based on habitat diversity alone (see factor II scores). An examination of the species lists from these two tepuis (appendix 18.4) confirms the presence of several lowland species in their faunas. Although Auyán is a high tepui (up to 2,450 m) with impressive cliffs (Angel Falls is on the north side) and has a slope area that is only slightly larger than its summit (table 18.1), a series of step-like escarpments (fig. 18.4) on the southern and eastern sides support forest and savanna-type habitats and provide relatively easy access to the top (R. W. McDiarmid and M. A. Donnelly, pers. obs.). Auyán also is relatively well sampled and has a multibranched river complex on its summit (see Huber 1995a, fig. 1.25; R. W. McDiarmid and M. A. Donnelly, pers. obs.) that provides additional habitats uncommon on many other high tepuis. In contrast, Guaiquinima is a low tepui whose summit ranges from 730 m in the south to 1,650 m in the northeast. Although its surface area is larger than that of Auyán (1,096 vs. 667 km²), its slope area is only approximately 40% of that of the summit and consists of a series of circular piedmont escarpments. Dense forest alternating with scrub and herbaceous vegetation covers much of the summit plains (Huber 1995a), a feature that presumably provides relatively easy access for lowland species. Two major streams drain Guaiquinima's summit to the south, and like Auyán, this tepui has been relatively well sampled. This combination of unusual traits may help explain why these two massifs differ from the general trends described by the principal components analysis.

Origin of the Herpetofauna

PATTERNS OF ENDEMISM AND DIVERSITY

Hoogmoed (1979a) was the first to comment on the high levels of endemism characteristic of the herpetofauna of the Guayana region. Of the 55 species of amphibians then known from highland localities (> 1,000 m), he noted that 18 (33%) occurred exclusively above 1,000 m and referred to them as highland endemics. Comparable data for reptiles identified 9 of 38 species (24%) as highland endemics. Hoogmoed remarked that most of the sampling of amphibians and reptiles on tepuis since 1894 had been done by explorers and scientists other than herpetologists (see appendix 18.5), and he speculated that the scanty information at the time was a reflection of inadequate sampling rather than geographically limited species distributions. Twenty years later, Duellman (1999) noted that 71 of 76 species of amphibians that he recorded from the Guayana Highlands were endemic and that this represented the highest endemism value (93%) for amphibians in South America. Although Duellman did not specify his criteria, we assume that his Guayana Highlands species were those occurring at localities above 1,000 m.

Our data support assertions of high endemism on tepuis. We define endemics relatively rigidly, however, restricting the term to those species known only from a single tepui (i.e., they are highly restricted). We recorded 159 species (appendix 18.3) from tepuis in the Guayana Highlands. Of those, 109 are known from the slope or summit of only a single tepui (i.e., 68.5% of the Pantepui herpetofauna is highly restricted). Levels of endemism differ between amphibians and reptiles ($G_{1df} = 8.76$, $p = .003$); there are more highly restricted species of amphibians (75 of 97 known species, 77.3%) than reptiles (34 of 62 known species, 54.8%). The observation that the percentage of highly restricted reptiles is considerably lower than that for amphibians suggests that amphibians tend to have smaller ranges than do reptiles in the Pantepui and that moderate- to wide-ranging reptiles are more likely to invade tepui habitats than are moderate- to wide-ranging amphibians. Even though more species of amphibians compared with reptiles occur in the Guayana Highlands, reptile species outnumber amphibian species throughout the Venezuelan portion of Guayana (Péfaur and Rivero 2000). We classified 13 other species that are known from more than one tepui within one of five massifs as moderately restricted: *Epipedobates rufulus*, *Stefania ginesi*, *Anadia* sp. A, *Arthrosaura* sp. A, *Atractus steyermarki*, and *Thamnodynastes chimanta* (Chimantá); *Oreophrynella nigra* and *Riolama leucosticta* (Eastern Tepui Chain); *Colostethus shrevei*, *Stefania goini*, *Atractus riveroi* (Duida); *Arthrosaura testigensis* (Los Testigos); and *Prionodactylus phelpsorum* (Jaua). We also considered 9 species that occur on more than one massif to be moderately restricted: *Stefania satelles* (Angasima, Aprada, Los Testigos, and Upuigma); *Tepuihyla edelcae* (Auyán, Chimantá, and Los Testigos); *Otophryne steyermarki* (Chimantá, Jaua, and Eastern Tepui Chain [Mount Ayanganna; A. Lathrop and R. MacCulloch, pers. comm.]); *Eleutherodactylus cantitans*, *E. yaviensis*, and *Prionodactylus goeleti* (Yaví and Yutajé); *Arthrosaura synaptolepis* (Neblina and Tamacuari); *Arthrosaura tyleri* (Duida and Jaua); and *Plica pansticta* (Guanay and Yutajé). If we relax our definition of endemism to include the moderately restricted forms as well (i.e., those 22 species restricted to tepuis and/or massifs), then we would treat 131 Pantepui species as highland endemics. That is 82.4% of the entire Guayana Highlands herpetofauna as we have defined it and represents 87.6% of the amphibians and 74.2% of the reptiles. As was pointed out earlier, amphibians in tropical highland faunas (e.g., tropical parts of the Andes) may outnumber reptiles between 1.5:1 to 2:1, and endemic species of amphibians may outnumber endemic reptiles even more. As expected, the Pantepui amphibian-to-reptile ratio (1.56:1) is similar to that in portions of the Andes, and the inclusive endemicity values also are higher (1.85:1).

A brief review of appendix 18.3 shows that species of centrolenids, *Colostethus*, and *Eleutherodactylus* are especially numerous and together make up more than half of the anuran fauna. These three taxa contribute significantly to the species diversity of frog faunas in tropical South America, and their numerical contributions are relatively similar to those reported for the same groups in the Andes (Duellman 1999). Inclusion of the endemic radiations of bufonids (7 species), *Stefania* (9 species), *Tepuihyla* (4 species), and certain hylids (8 species) accounts for more than 80% of the fauna. Other noteworthy and distinctive components of the amphibian fauna are manifest in species of *Adelophryne* (2), *Dischidiodactylus* (2), and *Otophryne* (1). Among reptiles, species of gymnophthalmids are dominant and represent 66% of the lizard fauna, and a few of them (e.g., *Adercosaurus* and *Riolama*) are endemic to the region. Other genera are more widely distributed (e.g., *Anadia, Arthrosaura, Neusticurus,* and *Prionodactylus*), but they include Pantepui species. Another noteworthy tepui clade is represented by the three species of *Phenacosaurus*, all of which have been described in the last decade. It will be interesting to learn how the Pantepui clade is related to the Andean *Phenacosaurus*. Among snakes, a few species of *Atractus* and *Thamnodynastes* are noteworthy for their limited distributions, but these genera also are widespread in South America.

Given the long evolutionary history of Pantepui and its biota, one might expect to find unusual life-history traits, interesting behaviors, peculiar morphological characteristics, and other such distinguishing features that serve to differentiate taxa of the Guayana Highlands from relatives found in other habitats and regions on the continent. Reproductive modes are diverse among South American anurans, and examination of reproductive patterns in tepui anurans might explain some of the observed patterns. As currently understood, several species of tepui bufonids (in contrast to most other South American bufonids), species of the Guayanan endemic *Stefania* (Hylidae), those of the wide-ranging genus *Eleutherodactylus,* and likely the two species of *Dischidiodactylus* (Leptodactylidae) all have some form of independently derived direct development. The Pantepui species of *Adelophryne* may also fit this pattern. The presence of relatively permanent streams on several tepui summits has provided interesting evolutionary challenges manifest in the larvae of some frogs. Species of centrolenids, *Colostethus,* certain hylids, and *Otophryne* generally breed in or along streams, and the tadpoles of some of these taxa are modified for specific tepui habitats. For example, the larvae of the stream-breeding hylids have a complex oral disc with many rows of labial teeth that facilitate their feeding in highenergy rocky streams (Ayarzagüena and Señaris 1993; R. W. McDiarmid and M. A. Donnelly, pers. obs.), and the tadpoles of *Otophryne* have curious modifications of the jaw sheaths and an elongate spiracular tube that purportedly are associated with feeding beneath the fine Roraima sands on the bottoms of shallows and seeps of small streams in the Guayana uplands (Wassersug and Pyburn 1987). The rock-climbing habits and peculiar tumbling behavior of species of *Oreophrynella* seemingly are unique among South American amphibians (McDiarmid and Gorzula 1989). Although less is known about the natural history of most tepui reptiles, a few observations are notable. Certain arboreal snakes (species of *Leptophis*) have been observed foraging across the tops of and searching among the leaves of several tepui plants (e.g., *Brocchinia,* Bromeliaceae; *Neblinaria celiae,* Theaceae); the overlapping leaves and water-holding capacities of these plants provide ample retreats for amphibians and foraging areas for frog-eating snakes. Likewise, wet meadows on the summits of some tepuis (e.g., Duida) often support populations of large earthworms, thereby providing good foraging habitat for worm-eating snakes (e.g., *Atractus;* B. Means, pers. comm.; R. W. McDiarmid, pers. obs.).

BIOGEOGRAPHIC HYPOTHESES

In many ways, amphibians and reptiles are better subjects for biogeographical study than are other terrestrial vertebrates. They are usually less vagile than birds and most mammals, and may be closely tied to specific habitats. Given the age and degree of isolation of Pantepui habitats, we expected the herpetofauna to reflect some generalized biogeographic patterns not duplicated in younger faunas at comparable elevations (e.g., Andes).

In their analysis of the origin of the bird fauna of the Venezuelan highlands, Mayr and Phelps (1967) predicted a greater similarity between western tepui and Andean faunas than between the eastern tepui and Andean faunas, and found higher endemism and species diversity in the eastern than in the western avifaunas. Mayr and Phelps (1967, 289) attributed this to "greater ecological opportunities" in the tepuis and highlands around the Gran Sabana, as compared with those of the more isolated western tepuis. They also noted that the eastern tepuis were more closely aggregated, which might have provided more opportunities to birds than were available with the more dispersed western tepuis. Finally, they hypothesized that the more clumped eastern tepuis may have represented a sort of reservoir for highland elements, whereas the western tepuis were more like isolated towers in a vast area of lowlands. Accordingly, they posited that the eastern tepuis were more suitable for the development and maintenance of a highland fauna.

Mayr and Phelps (1967, 289) examined five hypotheses to explain the origin of the Pantepui bird fauna: Distance Dispersal, Cool Climate, Habitat Shift, Specialized Habitat, and Plateau. Hoogmoed (1979a) assessed four of these (excluding the Specialized Habitat hypothesis) together with a Mountain Bridge hypothesis. No geological evidence supports the notion of a physical connection between the tepuis and the Andes; hence, the Mountain Bridge hypothesis is not considered here. The Specialized Habitat idea may explain the presence of some birds (swifts and swallows along cliffs and in canyons) but seems to have little relevance to amphibians and reptiles, with the possible exception of species of rock-dwelling lizards of the genera *Plica* and *Tropidurus,* which must have been derived independently from lowland ancestors (Frost et al. 2001). The other four hypotheses deserve more attention.

Distance Dispersal

The Distance Dispersal hypothesis was considered important relative to the origin of the Pantepui bird fauna, accounting for 50% of the 96 species treated by Mayr and Phelps (1967). Some Andean bird species are active colonizers; they or their ancestors presumably crossed the valleys of the Orinoco and Negro rivers to become established on some of the western tepuis, from whence they "island hopped" across Pantepui to the eastern massifs. Mayr and Phelps (1967) and others (e.g., Cook 1974) presented evidence that supported this notion and indicated that eastern tepuis had a higher diversity than western ones, a relationship contrary to predictions derived from measures of distance to source. Mayr and Phelps (1967) argued that the eastern tepuis provided greater ecological opportunities for bird colonists and thereby retained more species (i.e., fewer went extinct), even though the distance from the Andes was greater. Although this is an interesting idea and may well explain some diversity patterns exemplified by tepui birds, no evidence on species turnover rates was presented. Immigration rates may be high, as indicated by reports of several accidental and chance occurrences among Neblina birds (Willard et al. 1991).

Mindful of the inappropriateness of this idea for amphibians and reptiles (i.e., they do not island hop by flying long distances), we examined the presence of highly and moderately restricted taxa (appendix 18.3) across the four Pantepui districts (table 18.3) for comparison with the bird data. A similar pattern emerges from our data. The eastern tepuis as a group have an equivalent or higher number of amphibian (except for the central tepuis, which have one more amphibian) and reptile genera, the highest number of amphibian species, and the second highest number of reptile species (equivalent in number to the central tepuis), when compared with any of the more western districts. Tepuis in the Western District (nearest straight-line distance to an Andean source) have the lowest number of endemic amphibians (genera and species). Reptile genera are more numerous than amphibian genera in the Western District. Additionally, the number of species of reptiles (8) is equivalent in the Western and Southern districts, and this value is about half that of the Central and Eastern districts (15 species). However, if one were to sum the numbers for the western tepuis (Western, Central, and Southern districts), those tepuis would have about twice the number of endemic species as found in the eastern tepuis (58/28 for amphibians, 31/15 for reptiles; table 18.3). These results are in sharp contrast to those reported by Mayr and Phelps (1967) for birds.

In considering the similarities and differences between the amphibian/ reptile and bird comparisons, we considered the discrepancy of samples between summits and slopes, and the elevational limits of tepui samples for the bird data. Mayr and Phelps (1967, table 1) were cognizant of these sampling

Table 18.3　Occurrence of endemic (highly and moderately restricted) taxa of amphibians and reptiles in the four Pantepui districts

	Pantepui District							
	Eastern		Western		Central		Southern	
	A	R	A	R	A	R	A	R
Genera	10 (4)	10 (2)	4 (0)	6 (2)	11 (6)	10 (1)	8 (4)	6 (2)
Species	28	15	8	8	26	15	24	8

Note: A, amphibians; R, reptiles. Values for genera include Pantepui endemics/clades in parentheses.

issues but performed the analysis because they considered the Pantepui bird fauna to be reasonably well known. They (Mayr and Phelps 1967, 283) commented further that future fieldwork undoubtedly would turn up new records, new subspecies, and possibly even species, but that those additions would not modify their picture of this fauna to any major extent. Two points relative to Mayr and Phelps's views deserve comment. First, Roraima has the longest history of exploration and the greatest bird diversity (76, 78, or 84 species [Mayr and Phelps 1967; Cook 1974; Willard et al. 1991, respectively]). Many of these collections were made on the slopes of Roraima in forest, a large part of which on the southern and eastern sides was destroyed by fire in 1926 (Tate 1932). The forest apparently is extensive and continuous on the northern side of Roraima in Guyana. On the basis of our experience with Roraima and its herpetofauna, we question the validity of conclusions derived from comparisons between sites at which sampling over time and space differed substantially. In our experience, vertical cliffs of many tepuis separate slope from summit herpetofaunas with little or no overlap in species. We do not know if this is true for birds, but habitats differ considerably on the slopes and summit of Roraima. Second, Mayr and Phelps (1967) reported 38 different Pantepui bird species from Neblina and considered the degree of exploration to be good (in their five-position scale: very poor, poor, fair, good, very good) for both the summit and slopes. Cook (1974) stated that Neblina had only 36 species, less than half the number predicted by his regression of species number against elevation. Cook attributed this peculiarity to Neblina's isolation and noted its high proportion of endemic taxa (19%) compared with other tepuis (0%–9%). Willard et al. (1991) increased the number of montane (above 750 m) species from Neblina to 65, a value more in line with Cook's expectation. These values (very poor to good) represent an informed guess; they are therefore arbitrary and may mislead some workers who are looking for data about a fauna. Had the value "good" been deemed sufficient to those trying to make an informed decision about the value of more bird work on Neblina, we might not have learned that the previous estimate was only 60% of the currently reported diversity. To the best of our knowledge, the only tepui that Mayr and Phelps (1967) cited as having been re-sampled was Cerro Duida. Over a three-month period in 1928–29, personnel from the Tyler Duida Expedition (American Museum of Natural History) collected 52 different species from localities above 1,000 m. In November 1950, an ornithological team from the Phelps Collection secured 38 species, none of which was new to Duida. Accordingly, Duida was rated as "very good" by Mayr and Phelps (1967). In their characterization of the tepui avifauna, Willard et al. (1991) reported 64 species from Duida, an increase of approximately 23% above the diversity suggested in the analysis of Mayr and Phelps.

Cool Climate

Chapman (1931) proposed the Cool Climate hypothesis for birds, and Tate (1939) proposed it for Pantepui mammals. Their arguments were based on the idea that cool temperatures during glacial periods shifted subtropical (montane and upper-montane) habitats down mountains and thereby formed habitat connections between tepuis that previously had been separated by warmer lowland environments. Mayr and Phelps (1967) argued that temperature depressions as great as those proposed by Chapman and Tate (up to 9°C) would have resulted in considerable loss of the lowland fauna, for which there are no convincing data. They also pointed out that virtually no evidence exists in support of a habitat connection between the Andes and Pantepui during glacial periods, and that such would have resulted in a more or less uniform fauna throughout Pantepui. Hoogmoed (1979a) proposed a Modified Cool Climate hypothesis under which subtropical habitats would have been somewhat depressed (with a temperature depression of perhaps 3°C), thereby permitting some upland and highland organisms to occur at sites from which they previously had been absent.

The high level of endemism among amphibian and reptile species seems to argue against the Cool Climate hypothesis. Only a few species occur on more than one tepui (i.e., moderately restricted, in table 18A.1), and most of those 22 species occur on neighboring tepuis. We argue that such a distribution pattern is more likely the result of fragmentation of a continuous population on a plateau than depression of habitats to the lowlands and dispersal to adjacent isolates. Additional sampling of suitable habitat at lower elevations and between tepuis should shed light on this issue.

A few species occur on the slopes of some tepuis and at lower elevations between tepuis. Although samples from low or moderate elevations between tepuis are rare, the detection of species in these areas that previously were known only from the slopes or summits of adjacent tepuis would indicate wider distributions that may have been associated with temperature depressions. Likely candidates for consideration are *Colostethus shrevei,* some species of *Stefania* and *Tepuihyla, Otophryne,* possibly some species of *Eleutherodactylus* (e.g., *E. cantitans* and *E. yaviensis*), *Arthrosaura tyleri,* perhaps some species of *Prionodactylus,* and *Atractus riveroi.*

Habitat Shift

The Habitat Shift hypothesis was used to explain the origins of more than 36% of the Pantepui avifauna (Mayr and Phelps 1967), and Hoogmoed (1979a, 256) advanced this idea for amphibians and reptiles by stating that it "serves to explain the distribution and occurrence of the majority of the taxa living at higher elevations." Basically, this concept assumes that elements of the Pantepui fauna were derived from lowland ancestors that adjusted to intermediate and highland habitats and subsequently differentiated into distinct species. Among amphibians, many of the endemic species of centrolenids, *Colostethus,* and certain *Eleutherodactylus* seem to fit this pattern. With reptiles, we suggest that *Arthrosaura, Neusticurus, Prionodactylus, Tropidurus, Dipsas, Leptophis, Liophis,* and *Thamnodynastes* also fit this pattern.

Presumably all upland taxa at some point in their evolution can be traced to lowland ancestors. Given the age of the Guayana Shield and its associated Roraima Formation, it seems safe to say that in Guayana some type of uplands has been available for colonization from lowland stocks for a very long period. Such colonization probably has been ongoing for the past 100 million years and certainly for much of the Cenozoic. Only two taxa of endemic Pantepui amphibians seem likely to predate this period, and they have arisen from Gondwanaland stocks (also see Rachowiecki 1988, 48–49); these are bufonid frogs of the genera *Oreophrynella, Metaphryniscus,* and their relatives, and microhylid frogs of the genus *Otophryne.* Graybeal (1997) showed that the Pantepui bufonids are most closely allied with African genera other than *Bufo;* we predict that a similar pattern will be found with *Otophryne* (but see Wild 1995). Other amphibian species on tepuis presumably were derived from South American ancestors that were lowland species or that reached Pantepui through the lowlands. If these derivations occurred before the last major uplift (later Tertiary) and the subsequent major erosional events that have configured Pantepui today, then some elements of the fauna likely were distributed more or less continuously on a large plateau or on large blocks of a plateau, and fragmentation of that plateau and its fauna (vicariance) may account for many of the Pantepui endemics seen today (see section titled "Plateau Hypothesis"). Thus, ignoring the temporal component with regard to derivations of highland forms may make the Habitat Shift hypothesis more encompassing than it really is.

Second, Hoogmoed (1979a) suggested that the presence of widespread lowland forms in the herpetofauna of some tepuis supports the Habitat Shift hypothesis. It is true that lowland species are conspicuous components of the herpetofaunas on some tepuis (e.g., Guaiquinima). Mere presence, however, does not provide evidence to support the hypothesis; rather, differentiation (i.e., actual habitat shifts) rather than wide ecological tolerances must be demonstrated. That both Guayanan endemic and widespread forms (appendix 18.4) occur on the tepuis supports the notion that invasions of the highlands are continuing. If highland forms are being derived independently from some widely distributed lowland species, then it should be possible to detect local geographic differentiation within certain taxa (e.g., *Hyalinobatrachium orientale, Hyla benitezi, H. minuta, H. sibleszi, Leptodactylus rugosus, Arthrosaura tyleri, Neusticurus racenisi, N. rudis, Anolis nitens,* and *Tropidurus hispidus*).

Plateau Hypothesis

The last hypothesis asserts that the Pantepui herpetofauna consists of remnants of a herpetofauna that occupied a plateau that has been dissected by uplift and erosion into discrete tepuis. Chapman (1931) wrote about the tepuis as remnants of a more extensive tableland and the fauna as being more extensive than it is today. Tate (1938), in contrast, talked about in situ adaptations to changing environments associated with a slowly rising plateau. That might seem to support the Habitat Shift hypothesis, but he also noted that little change had occurred among the species after the dissection of the plateau into its numerous faunal islands. Tate maintained that the absence of a species from a mountain could only be accounted for by some inimical condition that caused it to die out (i.e., extirpation). Mayr and Phelps (1967) argued that considerable evidence contradicted the Plateau hypothesis, primarily the ancient age of the plateau, which they argued was dissected into its component tepuis long before most of the bird fauna evolved. They suggested that the irregular distributions of species, and their different levels of differentiation from nearest relatives and among themselves, argued against the notion that the present Pantepui bird fauna is a remnant of an old, formerly uniform plateau fauna. Although those arguments may be appropriate in assessing the origin of a vagile Pantepui avifauna, they do not seem appropriate for amphibians and reptiles. The observation that few genera and only a small number of species of birds and mammals are endemic to the Pantepui fauna contrasts markedly with the situation among amphibians and reptiles.

Hoogmoed (1979a) noted that a modified version of the Plateau hypothesis explains the distribution of the hylid genus *Stefania*. We believe that several other genera can be added to that list and argue that events proposed by the Plateau hypothesis have contributed significantly to the herpetofaunal diversity of Pantepui. The complex patterns of differentiation and distribution of certain species of hylid frogs on adjacent tepuis within the Chimantá Massif and of species of *Oreophrynella* and *Riolama* in the Eastern Tepui Chain lend support to the Plateau hypothesis. Likewise, erosional fragmentation of massifs into separate, isolated tepuis results in an increase in the number of isolated, related populations and, in terms of species diversity, approximates patterns seen among islands of some archipelagos. The difference here is that vicariance likely has played a more important role in total herpetofaunal diversity than has dispersal. Recolonizations after local extinctions may be less probable because access to some tepuis is difficult compared with other montane environments. More thorough sampling of the Pantepui herpetofauna and development of phylogenetic hypotheses about the relationships among the endemic taxa are required to gauge the importance of various processes in the origin of the Pantepui herpetofauna.

DISTRIBUTION PATTERNS

We suggest that if species known only from the summits of tepuis are found on more than one tepui, this is evidence that these tepuis were probably connected in the past—that is, that the tepuis were part of a larger physical unit (Pantepui Plateau) that was subsequently fragmented. Individual species known from lowland, intermediate, and upland sites could represent colonization from the lowlands upward or from the summit downward. The former seems more likely because lowland species often have wider geographic ranges and may have greater ecological tolerance than upland forms. Distinguishing between these two alternatives, however, is difficult. If upland populations on separate tepuis are more closely related to each other than they are to lowland populations, then vicariance explains their origins better than independent upland derivations from lowland ancestors. Unfortunately, our knowledge of the phylogenetic relationships of tepui taxa is nonexistent; hence, we lack a basis for making historical arguments. Although several species from tepuis have been described, no phylogenetic treatment includes all known species for any Pantepui lineage.

Only 33 species are known from more than a single tepui (excluding the 17 species that are treated as lowland taxa and have only been collected on the summit of a single tepui: *Colostethus brunneus, Hyalinobatrachium orientale, Hyla boans, H. crepitans, Leptodactylus longirostris, Osteocephalus taurinus, Pseudopaludicola llanera, Ameiva ameiva, Anolis fuscoauratus, Mabuya bistriata, Chironius exoletus, C. fuscus, Imantodes lentiferus, Leptophis ahaetulla, Leptotyphlops albifrons, Thamnodynastes pallidus,* and *Bothriopsis taeniata*). Fourteen of the 33 species are known to have narrow elevational ranges (elevational range < 500 m; see "Biotic Factors" for definition of elevational range used in this chapter): *Oreophrynella nigra, Epipedobates rufulus* (listed as *Dendrobates rufu-*

lus in Duellman 1999), *Eleutherodactylus cantitans*, *E. yaviensis*, *Stefania goini*, *S. satelles*, *Otophryne steyermarki*, *Anadia* sp. A, *Arthrosaura* sp. A, *A. synaptolepis*, *A. testigensis*, *Prionodactylus goeleti*, *Riolama leucosticta*, and *Atractus riveroi*. Six species have moderate elevational ranges (501–1,000 m): *Stefania ginesi*, *Tepuihyla edelcae*, *Arthrosaura tyleri*, *Prionodactylus phelpsorum*, *Atractus steyermarki*, and *Thamnodynastes chimanta*. Six species, *Colostethus shrevei*, *Hyla benitezi*, *H. sibleszi*, *Neusticurus racenisi*, *Tropidurus hispidus*, and *T. panstictus*, have moderately wide elevational ranges (1,001–1,600 m), and seven, *Hyalinobatrachium taylori*, *Hyla minuta*, *Leptodactylus rugosus*, *Anolis nitens* ssp., *Neusticurus rudis*, *Leptodeira annulata*, and *Mastigodryas boddaerti*, have extremely wide elevational ranges (> 1,600 m).

Six of the 14 species with narrow elevational ranges (*Eleutherodactylus cantitans*, *E. yaviensis*, *Stefania satelles*, *Otophryne steyermarki*, *Arthrosaura synaptolepis*, and *Prionodactylus goeleti*) occur on more than one massif. Populations of *E. yaviensis* and *P. goeleti* are known from Cerro Yaví and from Yutajé and Corocoro of the Yutajé Massif, whereas *E. cantitans* is known from Yaví and Yutajé. *Otophryne steyermarki* has been taken on Jaua and Chimantá massifs and recently from Mount Ayanganna, in the Eastern Tepui Chain, and *A. synaptolepis* is known from Neblina and Tamacuari. *Stefania satelles* is known from the Angasima, Aprada, Upuigma, and Los Testigos massifs but, interestingly, not from the Chimantá Massif (see fig. 18.1). Specimens of *S. satelles* from Los Testigos differ from those of other localities (see comments under "Kamarkawarai-tepui" in appendix 18.4 and in Señaris et al. 1996, 37). This suggests that the Los Testigos frogs (on two tepuis) may have been separated from those on the other tepuis at an earlier time. Molecular studies could clarify relationships among these populations. Because these six species are known only from tepui summits, we argue that their distribution patterns are reflective of history, and these highlands at one time must have been parts of larger units that subsequently fragmented.

Each of the eight remaining species of amphibians and reptiles with narrow elevational ranges is found on more than one tepui but always within a single massif (*Oreophrynella nigra* and *Riolama leucosticta* in the Eastern Tepui Chain; *Epipedobates rufulus*, *Anadia* sp. A, and *Arthrosaura* sp. A on the Chimantá Massif; *Stefania goini* and *Atractus riveroi* on the Duida Massif; and *Arthrosaura testigensis* from the Los Testigos Massif). We suggest that the presence of a species on more than one tepui within a massif indicates a past physical connection between tepuis.

The six species with moderate elevational ranges (501–1,000 m) exhibit similar patterns. *Tepuihyla edelcae* has been recorded from eight of the nine tepuis that have been sampled in the Chimantá Massif, as well as from the Auyán (two tepuis) and Los Testigos (one tepui) massifs. The distribution of this frog, together with those of the frog *Stefania ginesi* (known from seven of nine Chimantá tepuis) and the snakes *Thamnodynastes chimanta* (six Chimantá tepuis) and *Atractus steyermarki* (two Chimantá tepuis), provides ample evidence that most elements of the Chimantá Massif share a common history. With few exceptions, absences of particular species from individual tepuis within the Chimantá Massif probably reflect inadequate sampling. In a few instances, lack of suitable habitat may explain an absence (e.g., *A. steyermarki*). It seems likely that the Chimantá Massif was at one time a single sandstone block with its own fauna that was divided into separate entities by erosional forces associated with streams draining in different directions. That relatively little differentiation is evident among the populations on the tepuis of the Chimantá Massif suggests a more recent separation than has occurred among tepuis in other massifs (e.g., Eastern Tepui Chain). Whether the Chimantá Massif was connected to the Auyán and Los Testigos massifs, as is suggested by their common faunal components, remains to be determined. Given their physical proximity, it is likely that these massifs were once part of a single unit. Genetic studies would be useful in determining the relationships among the populations of *Tepuihyla edelcae* and among the populations of *Stefania ginesi* and its presumed close relatives, *S. satelles* and *S. schuberti*, deciphering the sequences of taxon differentiation and elucidating the geographic history of the region.

The other two species with moderate elevational ranges and distributions on more than one tepui are *Arthrosaura tyleri* (Duida and Jaua) and *Prionodactylus phelpsorum* (Sarisariñama and Jaua, according to our treatment). On the basis of our knowledge of other species in these two genera, we suspect that the Pantepui taxa are more widely distributed at intermediate elevations than currently reported (e.g., *A. tyleri*) and may be closely related to other taxa endemic to the Guayana Highlands. Increased sampling at intermediate elevations and phylogenetic analyses of these lizards and their relatives should provide much needed biogeographic insight.

Six of the 13 species with moderately wide (1,001–1,600 m) or very wide (> 1,600 m) elevational ranges are ecologically associated with streams: *Hyalinobatrachium taylori, Hyla benitezi, H. sibleszi, Leptodactylus rugosus* (see Heyer 1995), *Neusticurus racenisi,* and *N. rudis.* We expect that if these species, all of which are Guayanan endemics (appendix 18.3), occur on other tepuis, they will be found only on tepuis with persistent streams and relatively easy access. Hoogmoed (1979a) considered five of the species to be endemic to the lowlands of Guayana; *Hyalinobatrachium taylori* was not described until after his study, but it would also fit the category. These taxa could have moved from the surrounding lowlands to the summits of tepuis or moved from the summits to the lowlands. Hoogmoed (1979a) suggested *Neusticurus, Stefania,* and *Otophryne* as examples of organisms that originated in the highlands and subsequently invaded the lowlands. Donnelly and Myers (1991) referred to *Neusticurus racenisi* as a tepui species but later (Myers and Donnelly 1997) suggested that it was likely a lowland form that had moved upward. On the basis of the lowland distributions of other species of *Neusticurus,* the latter seems a more reasonable explanation. Nearly all of the lizard and snake species with wide or very wide elevational ranges that have been collected on more than one tepui are lowland species, suggesting that they have successfully moved from the lowlands onto the summits of certain tepuis. We are not surprised that amphibians and reptiles have been able to disperse upward into these amazing mountains. None of the "lowland invaders" is known from tepuis with sheer escarpments (e.g., Roraima); rather, they occur on mountains with low elevations (e.g., Guaiquinima, Tamacuari) or on those with extensive, less abruptly rising slope areas (e.g., Auyán-tepui).

Conclusions

The increase in knowledge about the Pantepui herpetofauna over the past 20 years has been remarkable. We now have a working list of the species of amphibians and reptiles known from the Guayana Highlands and some knowledge of the distributions of species among tepuis. Although appreciation of the herpetofauna of several areas clearly would benefit from additional fieldwork, our compilation provides a baseline inventory of taxa from some tepuis (e.g., Roraima, Auyán), and future work should concentrate on tepuis and regions in which collections are either scarce or nonexistent. Primary targets include Ayanganna, Karaurín, Uei, Wei-assipu, and Wokomung in the Eastern Tepui Chain; Soropán and Carrao in the Ptari Massif; Aparamán and others in the Los Testigos Massif; Araopán in the Aprada Massif; any tepui above 1,800 m in the Cuao-Sipapo and Parú massifs; and Avispa, Aracamuni, and other uplands between Neblina and Duida. We are not suggesting that work on other tepuis (e.g., Duida, Marahuaka, Jaua, Sarisariñama) is complete but rather that information from other sites is more important at this time. In addition to the summits of select tepuis and specific summit habitats (e.g., bogs), slope habitats of most tepuis, where accessible, need to be explored.

Throughout this analysis we have stressed the need for detailed phylogenetic studies of Pantepui taxa to hypothesize about the evolutionary history of the herpetofauna. Molecular data would provide an additional and potentially informative perspective, and we recommend that tissues be sampled at every opportunity. Future tepui explorations should include scientists from several disciplines (e.g., malacology, arachnology, botany) with specific goals to collect tissues and produce phylogenetic hypotheses of groups with ages and vagility similar to those of amphibians and reptiles. Such phylogenetic studies would permit cross-group comparisons and potentially produce patterns of geographic congruence. Results also should clarify the relative importance of the Plateau and Habitat Shift hypotheses with regard to understanding the origin of the Pantepui herpetofauna.

An unresolved issue that we encountered during our analyses of distributional patterns among elements of the Pantepui herpetofauna was the potential importance of extinctions and our ability to detect them. Two aspects of this problem deserve mention. First, it seems likely that extinction of some highland populations of amphibians and reptiles occurred as a consequence of differential uplift and erosion of the Pantepui Plateau and the habitat changes that resulted from these processes. Guaiquinima is a large tepui and presumably was part of the Pantepui Plateau, so therefore it must have been much higher than it is today (ranges from 730 to 1,650 m). If true, its geographic location (fig. 18.1) suggests that some Pantepui endemics (e.g., species of *Oreophrynella* or a related bufonid, *Stefania*, *Tepuihyla*) may have occurred there. The fauna today (appendix 18.4) has none of these taxa. Instead, it is dominated by wide-ranging, lowland forms (e.g., species of *Hyla*, *Leptodactylus*, *Anolis*) and a few endemic taxa that likely were derived from lowland ancestors (e.g., species of *Eleutherodactylus*, *Liophis*, *Philodryas*). Whether these lowland species contributed to extirpations through biological interactions (e.g., competition, predation) or whether abiotic factors (e.g., increasing aridity, warmer temperatures) and habitat changes were the principal causes of species loss is immaterial. What seems likely is that extinctions of highland populations must have occurred, and with few chances of recolonization, their faunas gradually changed.

Second, the relatively small area and isolated nature of most tepuis suggest that extinction rates may have been higher there than elsewhere. Even though we are beginning to understand which species occur on which tepuis (appendix 18.4) and may actually know the fauna of certain better-sampled summits (e.g., Roraima, Auyán), the history of herpetofaunal sampling in the Guayana Highlands indicates that our knowledge is incomplete (fig. 18.6). We lack repeated samples from most tepuis over adequate time periods and know relatively little about population sizes. In our experience, certain frogs (e.g., species of *Oreophrynella*, *Eleutherodactylus*, *Tepuihyla*) are moderately abundant and, if present, may be relatively easy to detect directly or through their vocalizations. With adequate sampling on targeted tepuis, statements about actual absence, as compared with lack of detection, may be possible for some species. For other, less-abundant, nonvocalizing species, documentation of an absence is more difficult. Such a reality leads to another question. How likely is the absence of a species in an area where it is expected to occur attributable to local extinction? If one accepts our premise that a tepui is relatively difficult to colonize from another highland area, then our ability to detect historical extinctions of amphibian and reptile species on tepuis may be higher than for other groups. Although hypotheses of phylogenetic relationships of Pantepui endemics are wanting, we remain optimistic that future studies will show congruence among area cladograms and thereby provide insight about past distributions and likely extinctions. Such notions almost certainly will furnish direction about where sampling needs to be done, and we remain confident that repeated sampling of tepuis eventually will allow us to make definitive statements about the origins of their herpetofaunas.

In summary, through the efforts of many herpetologists and countless other field biologists, we currently have a much better idea of the composition and diversity of the Pantepui herpetofauna. Many tepuis are unexplored herpetologically, and we advocate an expanded program of exploration and field research. New ideas and biogeographical predictions made during this study need testing, and we look forward to refinements that will come with additional data for species that live in the Lost World.

Appendix 18.1   Tepuis of the Guayana Highlands

Major tepuis and other uplands and highlands that make up Pantepui of the Guayana region (also see fig. 18.1) are listed in this appendix. The approximate elevational range (m), when known, is included. Tepuis from which herpetological specimens have been collected are indicated with an asterisk (*). Data are from Huber (1987, 1995a), T. Hollowell (pers. comm.), and field notes (RWM).

Eastern Pantepui District
    Roraima Subdistrict
        Eastern Tepui Chain
            Ilú/ Tramen, 1,000 – 2,400/2,700 m*
            Karaurín, 1,000 – 2,500 m
            Kukenán, 1,000 – 2,600 m*
            Roraima, 1,000 – 2,723 m*
            Uei, 1,000 – 2,150 m
            Wadakapiapué, 1,000 – 2,000 m
            Yuruaní, 1,000 – 2,400 m*
            Ayanganna, 550 – 2,043 m
            Wei-assipu, 1,000 – 2,772 m
            Wokomung, 550 – 1,650? m
    Los Testigos Subdistrict
        Isolates
            Cerro Venamo, ? – 1,600 m
            Sierra de Lema, ? – 1,650 m
        Ptari Massif
            Carrao, 1,200 ± 2,200 m
            Sororopán, 1,200 – 2,050 m
            Ptarí, 1,200 – 2,400 m*

Los Testigos Massif
 Aparamán, 400–2,200 m
 Kamarkawarai, 400–2,400 m*
 Murisipán, 400–2,400 m*
 Tereke-yurén, 400–1,900 m*
Chimantá Subdistrict
 Isolates
  Cerro Venado, 400–1,300 m
  Kurún-tepui, 400–1,300 m
  Angasima, 500–2,250 m*
  Upuigma, 500–2,100 m*
 Auyán Massif
  Auyán, 400–2,400 m*
  Cerro El Sol, 400–1,750 m*
  Cerro La Luna, 400–1,650 m*
  Uaipán, 400–1,950 m
 Aprada Massif
  Aprada, 400–2,500 m*
  Araopán, 400–2,450 m
 Chimantá Massif
  Abacapá, 500–2,400 m*
  Acopán, 500–2,200 m*
  Agparamán, 500–2,400 m
  Amurí, 500–2,200 m*
  Apacará, 500–2,450 m*
  Chimantá, 500–2,550 m*
  Churí, 500–2,500 m*
  Murey (Eruoda), 500–2,650 m*
  Tirepón, 500–2,600 m*
  Toronó, 500–2,500 m*
Western Pantepui District
 Maigualida Subdistrict
  Sierra de Maigualida (Cerro Yudi, Serranía Uasadi, and others),
   ?–2,400 m*
 Yutajé Subdistrict
  Isolates
   Cerro Camani, 100–1,800 m
   Cerro Guanay, 100–1,500/2,080 m*
   Cerro Yaví, 100–2,300 m*
  Yutajé Massif
   Cerro Corocoro, 100–1,500/2,400 m*
   Serranía Yutajé, 100–1,300/2,140 m*

Cuao-Sipapo Subdistrict
    Isolates
        Cerro Moriche, 100 to ± 1,250 m
        Cerro Ovana, ? to ± 1,800 m
    Cuao-Sipapo Massif
        Cerro Autana, 100–1,375 m
        Cerro Cuao, 100–2,000 m
        Cerro Sipapo, 100 to ± 1,800 m
Central Pantepui District
    Guaiquinima Subdistrict
        Cerro Guaiquinima, 300–1,650 m *
    Jaua-Sarisariñama Subdistrict
    Isolates
        Cerro Guanacoco, 300 to ± 1,500 m *
        Cerro Ichún, 500 to ± 1,500? m
        Sierra Marutaní, 500 to ± 1,500 m *
    Jaua Massif
        Cerro Jaua, 300–1,300/2,250 m *
        Cerro Sarisariñama, 300–1,250/2,350 m *
    Asisa Subdistrict
    Parú Massif
        Cerro Asisa, 100 to ± 2,200 m
        Cerro Euaja, 100 to ± 2,000 m
        Cerro Parú, 100–900/2,200 m
    Parima uplands
        Sierra Parima, ?–750/1,300 m
    Duida-Marahuaka Subdistrict
    Isolate
        Cerro Yapacana, 100–1,300 m *
    Duida-Marahuaka Massif
        Cerro Duida, 100–2,358 m *
        Cerro Huachamacari, 100–1,900 m *
        Cerro Marahuaka, 100 to ± 2,800 m *
Southern Pantepui District
    Isolates
        Cerro Aratityope, ?–1,700 m
        Sierra Unturán, ? to ± 1,600 m
    Sierra Tapirapeco
        Cerro Tamacuari, ?–2,340 m *
        Sierra Urucusiro, ?–?
        Sierra Curupira, ?–?
    Neblina-Aracamuni Massif

Cerro Aracamuni, 400–1,200/1,500 m*
Cerro Avispa, 400–1,500 m
Cerro de la Neblina, 400–1,800/3,045 m*

Appendix 18.2   Pantepui Plant Formations

Pantepui plant formations are listed in this appendix; their occurrence on tepuis with amphibians and reptiles is listed in appendix 18.4. Numbers of species are from the vegetation map that accompanies volume 1 of *Flora of the Venezuelan Guayana* (Steyermark et al. 1995a).

Upland and Highland Forests
Low evergreen upper-montane forests (eastern tepuis, Auyán, Chimantá)—22
Low evergreen high-tepui forests (Auyán, Chimantá)—23
Medium evergreen montane forest (Guaiquinima type)—26
Medium evergreen montane forest (Guanacoco, Jaua, Sarisariñama, Marutaní type)—27
Medium evergreen upper-montane forest (Jaua, Sarisariñama)—28
Low to medium evergreen upper-montane forest (Sierra de Maigualida)—30
Low evergreen upper-montane forest (Yaví, Corocoro, and Yutajé summits)—37
Low evergreen montane forest (Cerro Yapacana summit)—41
Tall evergreen lower-montane forest (Duida, Marahuaka, and Huachamacari slopes)—44
Medium evergreen montane forest (Duida-Marahuaka Massif)—45
Low evergreen upper-montane forest (Duida, Marahuaka, and Huachamacari)—46
Low evergreen high-tepui forest (Marahuaka summit)—47
Medium to tall evergreen montane forest (Aracamuni and Neblina, upper slopes)—53
Low evergreen upper-montane forest (Neblina, Tapirapecó)—54

Shrublands
Tall upland scrub on rock (Guaiquinima uplands type)—57
Tall upland scrub on rock (Marutaní type)—58
Low tepui-summit scrub and meadows on peat and rock (Auyán, Chimantá summits)—59
Tall upland scrub on rock (Sarisariñama uplands type)—60

Low high-tepui scrub on peat and rock (Jaua)—61
Low to tall tepui scrub on rock (Guanay, Corocoro, Yutajé)—62
Low to tall upland and high-tepui scrub (Duida and Huachamacari summits)—68
Low high-tepui scrub on peat (Neblina)—69

Herbaceous Formations
Broad-leaved, shrubby upland meadows on peat (Guaiquinima)—89
Tubiform, shrubby highland meadows on rock and peat (Guanacoco, Jaua, Sarisariñama)—90
Broad-leaved, shrubby upland and highland meadows on rock and peat (Sierra de Maigualida, Yaví, Corocoro, Yutajé)—91
Broad-leaved, shrubby highland meadows on peat (Duida type)—93
Broad-leaved, shrubby high-tepui meadows on peat and rock (Marahuaka)—94
Broad-leaved, shrubby highland meadows on peat (Huachamacari type)—95
Broad-leaved, shrubby upland meadows on peat (Aracamuni)—97
Broad-leaved, shrubby upland and high-tepui meadows on peat (Neblina)—98

Pioneer Formations
Pioneer vegetation on sandstone summits (Ilú, Yuruaní, Kukenán, Roraima, Los Testigos, Ptari, eastern Auyán, Aprada, Murey, and Tirepón tepuis in the Chimantá Massif, others?)—102

Appendix 18.3   Herpetofauna of the Guayana Highlands

The 159 species of amphibians and reptiles known from the Guayana Highlands are listed in table 18A.1 at the end of this appendix. Data on elevational range (minimum elevation–maximum elevation, or elevation at type locality) were taken from the literature and from field notes (RWM). Distribution patterns are as follows: highly restricted (HR) = known only from a single tepui; moderately restricted (MR) = known from two or more tepuis or massifs; Guayanan endemic (GE) = known only from localities in the Guayana region (see text for definition of Guayana); widespread (WS) = also known from areas of South America outside of Guayana. Numbers in parentheses represent the number of species in that group.

Our treatment of the distribution and identification of certain tepui taxa requires some explanation. We have confirmed records of *Otophryne steyermarki* from Apacará-tepui and Jaua (appendix 18.4) and reliable reports from Mount Ayanganna; Gorzula (1985) and Campbell and Clarke (1998) also reported a population (males heard calling but none collected) from Acopán-tepui, near the type locality in the Chimantá Massif. Gorzula and Señaris (1998) mentioned a specimen of *O. steyermarki* that had been collected from the talus slope (no elevation given) of Roraima during a La Salle expedition, but they did not give a museum number. We have not examined that specimen and defer from including it in our Roraima list. The type specimen of *Otophryne robusta* was collected at the base of Roraima, and other specimens have been taken on the slopes (see "Remarks" for Roraima, appendix 18.4). We are concerned that the specimen reported by Gorzula and Señaris (1998) may be *O. robusta* and not *O. steyermarki.* However, both *O. robusta* and *O. steyermarki* have been collected on Mount Ayanganna (A. Lathrop and R. MacCulloch, pers. comm.), and they may both occur on the slopes of Roraima at elevations compelling their inclusion in that fauna.

Considerable confusion exists both with the identification of specimens and the application of names for certain Pantepui lizards of the genus *Anolis.* The proposed use of the generic name *Norops* is a third issue that adds to the confusion. Although we agree philosophically with many of the points to partition the genus *Anolis* (Guyer and Savage 1986; Savage and Guyer 1989), in this treatment we follow the lead of other students of the Pantepui herpetofauna and use the name *Anolis* rather than *Norops.*

Part of the confusion involves the distinctiveness of *Anolis eewi,* a species Roze (1958b) described from the summit of the Toronó-tepui (Chimantá Massif). Roze compared the unique type to specimens of *Anolis fuscoauratus kugleri,* at the time the only other anole from Chimantá but from lower elevations; he did not compare the Toronó specimen to *A. nitens* (Wagler 1830), which Roze (1958a) had previously reported from Auyán-tepui to the north. Most other authors (e.g., Peters and Donoso-Barros 1970) recognized *A. eewi* and *A. nitens* as distinct species. In a key paper on geographic variation in this group of anoles, Vanzolini and Williams (1970) asserted that only a single species was involved (they examined the holotype of *A. eewi*) and used the name *Anolis chrysolepis* Dumeríl and Bibron 1837 for these lizards (they assigned the few Pantepui anoles available to them to the subspecies *A. c. planiceps*). They opted for *chrysolepis* rather than *nitens* (the older name) because the type of *nitens* is lost and the locality ("America") is imprecise. Vanzolini and Williams noted that the original description was inadequate and that the name *nitens* had been used primarily for a color morph of *chrysolepis* group animals (Shreve 1947). Vanzolini and Williams (1970, 84) commented further that the type of *nitens* "is probably, but hardly with certainty, applicable to some member of the *chryso-*

lepis group." Setting aside the propriety of the name *chrysolepis* (see below), most workers have considered anoles in this group from Pantepui to be one species. However, Mägdefrau et al. (1991) referred specimens from Guaiquinima to both *A. c. planiceps* (camp 5, at 1,520 m) and *A. c. scypheus* (camp 4, at 980 m). Specimens from Guaiquinima (1,030 m; Donnelly and Myers 1991) and Pico Tamacuari (1,270 m; Myers and Donnelly 1997) were referred to *A. c. planiceps*. Gorzula and Señaris (1998) used ratios of tibia to body length and argued that *Anolis eewi* was distinct from *A. nitens;* their sample of seven specimens (two from an upland south of Marutaní, five from the Chimantá Massif, and two from 1,650 m on Cerro Guanay) supposedly showed that *A. eewi* has shorter legs than *A. nitens* (*A. chrysolepis* data from Vanzolini and Williams 1970). Proportional data of this kind that are gathered by different people from few individuals of widely spaced samples are subject to considerable variation, and the resulting analyses are fraught with problems. Also, a plot of tibia length versus snout-vent length for several specimens of *A. nitens* (Avila-Pires 1995, fig. 22) seems to encompass the differences reported by Gorzula and Señaris (1998). Accordingly, we believe it is prudent to treat all of these populations as representative of a single species, pending detailed study of available material. What this taxon should be called is a second problem.

After Vanzolini and Williams (1970), most authors used *Anolis chrysolepis* for these lizards. Hoogmoed (1973) did as well, but correctly pointed out that stability would be best served if Vanzolini and Williams would request that the International Commission of Zoological Nomenclature use its plenary powers to suppress *nitens* Wagler 1830 in favor of *chrysolepis* Duméril and Bibron 1837. Apparently no request was made. Subsequently, various authors (e.g., Savage and Guyer 1991; Avila-Pires 1995) argued for retention of the name *nitens*. Gorzula and Señaris (1998) also called some of the Pantepui anoles *A. n. nitens*. Myers and Donnelly (1997) raised some concerns about the suitability of applying Wagler's *nitens*, which was described as greenish-, to brownish-colored anoles. Although we agree with the concern about coloration and recognize the disquiet caused by Wagler's (1830) inadequate description, some populations of this lizard in the Kartabo area of Guyana were described as greenish above (Beebe 1944, 197).

In our opinion, the lack of adequate samples from the Guayana Highlands has hindered previous analyses and makes current determinations difficult. Until results from a detailed study of material from lowland and highland sites throughout Pantepui are available, we prefer to consider all populations as representative of a single species; in this regard, we are in agreement with most authors. In this chapter we ignore assignment of populations to subspecies and, for reasons of priority, use the name *Anolis nitens* ssp. for these lizards.

Another taxonomic problem exists with gymnophthalmid lizards of the genus *Prionodactylus*. An understanding of the distributions of Pantepui species of these lizards is confounded by conflicting taxonomic opinions. Lancini (1968) described *Euspondylus phelpsi* from a single specimen purportedly collected during the Phelps Expedition of 1967 from the top of Cerro Jaua. Steyermark and Brewer-Carías (1976, 180) pointed out that the expedition never visited Cerro Jaua; rather, personnel collected only on the western summit of Sarisariñama. We therefore assume, as did Gorzula and Señaris (1998), that the type of *E. phelpsi* also was from Sarisariñama, not Jaua. That was confirmed by Phelps (1977), who changed his specimen data from Jaua to Sarisariñama (Myers and Donnelly 2001 n. 26). About 30 years later, Myers and Donnelly (1996) described *Euspondylus goeleti* from six specimens taken at 2,150 m on Cerro Yaví, and they pointed out that the name that Lancini had proposed for the Sarisariñama lizard, *E. phelpsi,* needed to be changed to *Euspondylus phelpsorum* to meet the requirements of the *International Code of Zoological Nomenclature.* In their description of *Euspondylus goeleti,* Myers and Donnelly compared the Yaví specimens to Lancini's description of *E. phelpsorum* and concluded that the two were distinct. Gorzula and Señaris (1998) described *Prionodactylus nigroventris* on the basis of three specimens from 1,650 m on Cerro Guanay and considered all three Pantepui taxa to be in the genus *Prionodactylus.* Gorzula and Señaris distinguished *P. nigroviridis* and *P. phelpsorum* on the basis of ventral coloration and the number of temporal scales. They also suggested that *E. goeleti* from Cerro Yaví was the same as *P. phelpsorum.* Gorzula and Señaris reported five other specimens of what they called *P. phelpsorum* from Cerro Corocoro, Serranía de Yutajé, and Cerro Yaví (specimen[s] from Corocoro and Yutajé listed with same catalog number; also see comment in Myers and Donnelly 2001, 60). According to their interpretation, *P. phelpsorum* occurs on Sarisariñama and Jaua (Jaua Massif, Jaua-Sarisariñama Subdistrict of Central Pantepui District) and on Yutajé, Yaví, and Corocoro (Yutajé Subdistrict of Western Pantepui District; fig. 18.1). Mijares-Urrutia (2000) examined specimens of *P. goeleti* and reviewed the evidence relative to recognition of one or two species. Although the type of *P. phelpsorum* is lost, Mijares-Urrutia concluded that two species were represented, and he resurrected *P. goeleti* from synonymy. Myers and Donnelly (2001) reported another juvenile from Cerro Yutajé that they called *P. goeleti,* thereby independently rejecting the conclusion of Gorzula and Señaris (1998). Because the holotype of *P. phelpsorum* is lost and other specimens from the Jaua-Sarisariñama area are not available for study, further comment on the status of these two species is pointless. If *P. goeleti* proves to be conspecific with *P. phelpsorum,* then the distribution pattern described above suggests a link between the Yutajé Massif, Cerro Yaví, and the Jaua Massif. If the taxa are distinct, then *P. phelpsorum* is restricted to the Jaua Massif and *P. goeleti* links the Yutajé Massif (Yutajé and Corocoro) to Cerro Yaví. At this time we prefer to use the name *P. goeleti* for the species on Yaví, Yutajé, and possibly Corocoro, and restrict *P. phelpsorum* to those lizards from the Jaua Massif.

Confusion besets the systematics of snakes in the genus *Thamnodynastes,* and determining the number of Pantepui species is somewhat problematic. We include six species in appendix 18.3, one of which (*T. pallidus*) is a widespread, lowland species that gets onto a few tepuis at lower elevations (1,270 m on Tamacuari [Myers and Donnelly 1997]; 1,370 m on Cerro Guaiquinima [Gorzula and Ayarzagüena 1995; Gorzula and Señaris 1998]). The other five are small species known from specific tepuis. Myers and Donnelly (1996) described *T. yavi* from three specimens taken at 2,150 m on the summit of Cerro Yaví in 1995 and *T. duida* from a single male collected in 1929 at 2,015 m on the south end of the summit of Cerro Duida. That same year, Gorzula and Ayarzagüena (1995) described *T. marahuaquensis* from a single female taken at 2,500 m on Tepui Marahuaka Norte (Cerro Marahuaka) and *T. corocoroensis* also from a single female collected at 2,150 m on Tepui Corocoro (Cerro Corocoro). Gorzula and Señaris (1998) considered *T. yavi* Myers and Donnelly to be a junior synonym of *T. corocoroensis* Gorzula and Ayarzagüena. Setting aside the problem of name priority (publication date of the Gorzula and Ayarzagüena paper is controversial; Myers and Donnelly 1997, 63 n. 19; 2001, 78–79), Myers and Donnelly (2001) concluded that *T. yavi* and *T. corocoroensis* probably were distinct species but that *T. duida* and *T. marahuaquensis* were more problematic. We do not know why Gorzula and Señaris (1998) did not include *T. marahuaquensis* in their compilation of taxa known from the Venezuelan Guayana. Until more material from Duida and Marahuaka becomes available, we recommend the species be treated as separate.

Three species of reptiles on our list (appendices 18.3 and 18.4) are known from parts of shed skins: Gymnophthalmid sp. B, Tirepón-tepui; *Chironius* sp., Sarisariñama; and Colubrid sp., Auyán-tepui. Two of these sheds were collected by McDiarmid, who has examined all three. Although we think the snake sheds are not of new species, that of the lizard represents the only reptile record from Tirepón-tepui; we therefore scored Gymnophthalmid sp. B as endemic. Detailed comparisons may yet show that it is a shed from one of the other gymnophthalmid species known from the Chimantá Massif (*Anadia* sp. A, *Arthrosaura* sp. A., *Neusticurus rudis*). Additionally, two other lizard taxa, Gymnophthalmid sp. A (*Arthrosaura?*, Cerro Yaví) and *Riolama* sp. C (Cerro Duida), are based on partially digested specimens taken from the stomachs of snakes (Myers and Donnelly 1996). A tail of Gymnophthalmid sp. A was in the stomach of a female *Thamnodynastes yavi* (296 mm total length [TL]), and a specimen of *Riolama* sp. C was taken from the stomach of the male holotype of *T. duida* (455 mm TL). These species "records" demonstrate that even relatively well-explored mountains harbor additional species and that snakes of the genus *Thamnodynastes* are effective lizard collectors.

According to our best estimates, only approximately 61% of the 97 species of amphibians and approximately 71% of the 62 species of reptiles recorded from the Guayana Highlands can be assigned to currently known species. Interestingly, only about 39% of the 18 species of amphibians that Hoogmoed (1979a) listed as highland endemics were known at the time, whereas nearly all (8 of 9 species) of reptiles were known. Although most of the unknown species recorded by Hoogmoed more than 20 years ago have been identified or described since then, the current proportion of undescribed to described taxa is significant and reflects both the high levels of endemism and the incredible rate of discovery of new taxa in the Guayana Highlands. Much remains to be done.

Table 18A.1   Amphibia and Reptilia of the Guayana Highlands

	Elevational range (m)	Distribution pattern
Amphibia		
Anura (97)		
Bufonidae (7)		
Bufonid sp. (Neblina)	2,000–2,400	HR
Metaphryniscus sosai	2,600–2,800	HR
Oreophrynella cryptica	1,750	HR
Oreophrynella huberi	1,700	HR
Oreophrynella nigra	2,300–2,600	MR
Oreophrynella quelchii	2,600–2,800	HR
Oreophrynella vasquezi	2,500	HR
Centrolenidae (12)		
Centrolene gorzulae	1,850	HR
Cochranella duidaeana	2,140	HR
Cochranella riveroi	1,600	HR
Cochranella sp. (Neblina)	1,390–2,100	HR
Hyalinobatrachium auyantepuianum	1,850	HR
Hyalinobatrachium crurifasciatum	1,160–1,200	HR
Hyalinobatrachium eccentricum	1,700	HR
Hyalinobatrachium orientale	100–1,650	WS
Hyalinobatrachium cf. *orientale*	1,800–1,900	HR
Hyalinobatrachium taylori	200–1,900	GE
Hyalinobatrachium sp. A (Guaiquinima)[a]	1,400	HR
Hyalinobatrachium sp. B (Jaua)	1,750–1,800	HR
Dendrobatidae (14)		
Colostethus ayarzaguenai	1,600[b]	HR
Colostethus brunneus	0–1,524	WS
Colostethus guanayensis	1,650	HR
Colostethus murisipanensis	2,350	HR
Colostethus praderioi	1,800–1,950	HR
Colostethus roraima	2,700	HR
Colostethus shrevei	350–1,829	MR
Colostethus tamacuarensis	1,160–1,200	HR
Colostethus tepuyensis[c]	1,600–1,850	HR

	Elevational range (m)	Distribution pattern
Colostethus undulatus	1,750	HR
Colostethus sp. A (Jaua)	1,750–1,800	HR
Colostethus sp. B (Neblina)	140–1,250	HR
Epipedobates rufulus	2,100–2,600	MR
Minyobates steyermarki	1,200	HR
Hylidae (28)		
Hyla aromatica	1,700	HR
Hyla benitezi	800–1,801	GE
Hyla boans	0–1,216	WS
Hyla crepitans	0–1,420	WS
Hyla inparquesi	2,600	HR
Hyla minuta	0–1,800	WS
Hyla sibleszi	900–2,100	GE
Osteocephalus taurinus	10–1,250	WS
Stefania ginesi	1,850–2,600	MR
Stefania goini	1,402–1,700	MR
Stefania oculosa	1,600	HR
Stefania percristata	1,600	HR
Stefania riae	1,400	HR
Stefania riveroi	2,300	HR
Stefania satelles	2,000–2,500	MR
Stefania schuberti	1,750–1,970	HR
Stefania tamacuarina	1,270	HR
Tepuihyla aecii	2,150	HR
Tepuihyla edelcae	1,630–2,600	MR
Tepuihyla luteolabris	2,550	HR
Tepuihyla rimarum	2,400	HR
Hylid sp. A (Maigualida) [d]	2,100	HR
Hylid sp. B (Auyán) [e]	1,600	HR
Hylid sp. C (Neblina)	1,450–2,100	HR
Hylid sp. D (Neblina)	1,450–1,880	HR
Hylid sp. E (Neblina)	1,250	HR
Hylid sp. F (Duida)	1,850	HR
Hylid sp. G (Jaua)	1,750–1,800	HR
Leptodactylidae (35)		
Adelophryne sp. A (Neblina)	1,390–1,515	HR
Adelophryne sp. B (Neblina)	1,730–2,100	HR
Dischidiodactylus colonnelloi	2,550	HR
Dischidiodactylus duidensis	1,402	HR
Eleutherodactylus avius	1,160–1,460	HR
Eleutherodactylus cantitans	1,750–2,150	MR

(*continued*)

	Elevational range (m)	Distribution pattern
Eleutherodactylus cavernibardus	1,160–1,200	HR
Eleutherodactylus memorans	1,160–1,270	HR
Eleutherodactylus pruinatus	2,150	HR
Eleutherodactylus yaviensis[f]	1,700–2,150	MR
Eleutherodactylus sp. A (Guanay)[g]	1,650	HR
Eleutherodactylus sp. B (Aprada)[h]	2,500	HR
Eleutherodactylus sp. C (Auyán)[h]	1,860	HR
Eleutherodactylus sp. D (Murisipán)[h]	2,350	HR
Eleutherodactylus sp. E (Yuruaní)[h]	2,300	HR
Eleutherodactylus sp. F (Jaua)	1,750–1,800	HR
Eleutherodactylus sp. G (Sarisariñama)	1,380–1,420	HR
Eleutherodactylus sp. H (Ptari)	2,350	HR
Eleutherodactylus sp. I (Murey)	2,350	HR
Eleutherodactylus sp. J (Roraima)	2,600–2,620	HR
Eleutherodactylus sp. K (Duida)	1,850	HR
Eleutherodactylus sp. L (Neblina)	1,730–2,100	HR
Eleutherodactylus sp. M (Neblina)	1,390–2,100	HR
Eleutherodactylus sp. N (Neblina)	1,390–1,515	HR
Eleutherodactylus sp. O (Neblina)	1,730–1,850	HR
Eleutherodactylus sp. P (Neblina)	1,390–1,850	HR
Eleutherodactylus sp. Q (Neblina)	1,250	HR
Eleutherodactylus sp. R (Neblina)	1,390–1,850	HR
Eleutherodactylus sp. S (Guaiquinima)[i]	780–1,520	HR
Eleutherodactylus sp. T (Marahuaka)	2,450	HR
Leptodactylus longirostris	0–1,520	WS
Leptodactylus rugosus	90–1,720	GE
Leptodactylus sp. (Neblina)	2,085–2,100	HR
Leptodactylid sp. (Neblina)	2,085–2,100	HR
Pseudopaludicola llanera	100–1,220	WS
Microhylidae (1)		
Otophryne steyermarki	1,800–2,150	MR
Reptilia (62)		
Squamata (Lizards) (35)		
Gymnophthalmidae (23)		
Adercosaurus vixadnexus	1,700	HR
Anadia sp. A (Chimantá)[j]	2,100–2,600	MR
Anadia sp. B (Auyán)[k]	1,970	HR
Arthrosaura synaptolepis	1,200–1,450	MR
Arthrosaura testigensis	1,800–2,300	MR
Arthrosaura tyleri	1,402–2,164	MR
Arthrosaura sp. A (Chimantá)[l]	1,950–2,250	MR
Arthrosaura sp. B (Auyán)[m]	2,100	HR

	Elevational range (m)	Distribution pattern
Neusticurus racenisi	134–1,234	GE
Neusticurus rudis	0–2,100	GE
Neusticurus tatei	1,402	HR
Neusticurus sp. A (Guaquinima)[n]	1,030–1,520	HR
Neusticurus sp. B (Marutaní)	1,200	HR
Neusticurus sp. C (Jaua)[o]	1,600	HR
Prionodactylus goeleti	1,700–2,150	MR
Prionodactylus nigroventris	1,650	HR
Prionodactylus phelpsorum	1,380–1,917	MR
Riolama leucosticta	2,500–2,700	MR
Riolama sp. A (Neblina)	2,085–2,100	HR
Riolama sp. B (Neblina)	1,730–2,100	HR
Riolama sp. C (Duida)[p]	2,015	HR
Gymnophthalmid sp. A (Yaví)[q]	2,150	HR
Gymnophthalmid sp. B (Tirepón)	2,450	HR
Polychrotidae (5)		
Anolis nitens ssp.[r]	0–2,200	WS
Anolis fuscoauratus	0–1,030	WS
Phenacosaurus bellipeniculus	2,150	HR
Phenacosaurus carlostoddi	2,200	HR
Phenacosaurus neblininus	1,690–2,100	HR
Scincidae (2)		
Mabuya bistriata	0–1,800	WS
Mabuya sp. (Neblina)	140–1,880	HR
Teiidae (1)		
Ameiva ameiva	0–1,180	WS
Tropiduridae (4)		
Plica lumaria	780–1,380	HR
Plica pansticta	180–1,220	MR
Tropidurus bogerti	1,600–2,080	HR
Tropidurus hispidus	20–1,420	WS
Squamata (Snakes) (27)		
Leptotyphlopidae (1)		
Leptotyphlops albifrons	0–1,380	WS
Colubridae (25)		
Atractus duidensis	2,050–2,150	HR
Atractus riveroi	1,300–1,800	MR
Atractus steyermarki	1,430–2,250	MR
Atractus sp. (Auyán)[s]	2,100	HR
Chironius exoletus	200–1,402	WS
Chironius fuscus	0–2,283	WS
Chironius sp. (Sarisariñama)	1,380–1,420	HR

(*continued*)

Table 18A.1   (continued)

	Elevational range (m)	Distribution pattern
Dipsas cf. *indica*	1,515	HR
Dipsas sp. (Neblina)	1,250	HR
Imantodes lentiferus	100–1,030	WS
Leptodeira annulata	0–2,150	WS
Leptophis ahaetulla	0–1,850	WS
Leptophis sp. (Neblina)	1,820	HR
Liophis ingeri	1,900	HR
Liophis torrenicola	1,030–1,180	HR
Liophis trebbaui	1,020–1,938	HR
Mastigodryas boddaerti	0–2,200	WS
Philodryas cordata	1,030–1,520	HR
Thamnodynastes chimanta	1,920–2,600	MR
Thamnodynastes corocoroensis	2,150	HR
Thamnodynastes duida	2,015	HR
Thamnodynastes marahuaquensis	2,500	HR
Thamnodynastes pallidus	0–1,370	WS
Thamnodynastes yavi	2,150	HR
Colubrid sp. (Auyán)	1,940	HR
Viperidae (1)		
Bothriopsis taeniata	0–2,000	WS

Source: Data on elevational range were taken from the literature and from RWM's field notes.

Note: In this list of the 159 species known to occur on tepuis in the Guayana Highlands, the numbers in parentheses represent the numbers of species in the particular group. The elevational range represents the minimum elevation–maximum elevation or elevation at a type/single locality. Distribution patterns are as follows: HR, highly restricted, known only from a single tepui; MR, moderately restricted, known from two or more tepuis or massifs; GE, Guayanan endemic, known only from localities in the Guayana region; WS, widespread, also known from areas of South America outside of Guayana.

[a] *Hyalinobatrachium* sp. of Gorzula and Señaris 1998, 32.

[b] Elevation derived from other source (see remarks concerning Jaua in appendix 18.4).

[c] Includes the *Colostethus* sp. of Myers 1997, 3 (C. Myers, pers. comm.).

[d] Hylid sp. A of Gorzula and Señaris 1998, 36.

[e] Hylid sp. B of Gorzula and Señaris 1998, 37.

[f] Includes Eleutherodactylinae series b (Corocoro) of Gorzula and Señaris 1998, 55, according to Myers and Donnelly 2001, 39.

[g] Eleutherodactylinae series a (Corocoro) of Gorzula and Señaris 1998, 55.

[h] Eleutherodactylinae series b (Corocoro) of Gorzula and Señaris 1998, 55.

[i] *Eleutherodactylus* spp. of Mägdefrau et al. 1991, 16, and Schlüter 1994, 79, 83.

[j] *Anadia* sp. A of Gorzula and Señaris 1998, 114.

[k] *Anadia* sp. B of Gorzula and Señaris 1998, 115.

[l] *Arthrosaura* sp. A of Gorzula and Señaris 1998, 124.

[m] Refers to an undescribed species of *Arthrosaura* mentioned in Myers 1997, 4.

[n] *Neusticurus* sp. of Donnelly and Myers 1991, 38; *Neusticurus rudis* of Mägdefrau 1991, 21, and Gorzula and Señaris 1998, 127–28.

Table 18A.1 (continued)

[o] *Neusticurus* cf. *rudis* of Señaris et al. 1996, 24.

[p] *Riolama* sp. of Myers and Donnelly 1996, 52. Molina and Señaris (2001) described *Riolama uzzelli* from two sites on Cerro Marahuaka (1,850 and 2,600 m). We suspect that future work will show that the Duida and Marahuaka specimens represent a single species.

[q] *Arthrosaura* (?) species of Myers and Donnelly 1996, 22.

[r] Includes *Anolis eewi, Anolis nitens nitens,* and *Anolis chrysolepis planiceps* of the authors.

[s] Refers to a specimen of *Atractus* mentioned in Myers 1997, 4.

Appendix 18.4 Tepuis of the Guayana Highlands and Their Herpetofaunas

The major tepuis of the Guayana Highlands from which species of amphibians and/or reptiles are known are listed by district and, where appropriate, by massifs within districts. Arrangement is in a general east-to-west and north-to-south sequence. Tepuis within each of the four districts are listed in alphabetical order by subdistrict, massif, or individually, whichever is appropriate. Unless otherwise specified, much of the tepui data come from Huber (1995a). Endemic species are indicated with an asterisk (*). Pertinent general and herpetological references are listed, as are various clarifying remarks.

Eastern Pantepui District

RORAIMA SUBDISTRICT

Eastern Tepui Chain

Ilú-tepui and Tramen-tepui

Venezuela (Bolívar); latitude × longitude: 05°27′ N 61°03′ W and 05°40′ N 62°37′ W (U.S. Board on Geographic Names [USBGN] 1961); maximum elevation: 2,700 m; summit area: 5.63 km^2; slope area: not available, 300 km^2 for all tepuis in Eastern Tepui Chain except Uei-tepui; massif: Eastern Tepui Chain; number of vegetation types for massif: 1—pioneer vegetation on sandstone summits; date of first herpetological collection-exploration: 1977.

Herpetofauna: 2 species, 1 endemic: Amphibia (1)—*Oreophrynella vasquezi**; Reptilia (1) *Riolama leucosticta.*

Pertinent literature: Gorzula and Señaris 1998; Señaris et al. 1994.

Remarks: McDiarmid accompanied Charles Brewer-Carías here and made collections one day in January 1977. Gorzula visited the tepuis at three different sites; one day in April 1984 and another two in June 1985. The three collection sites had the following coordinates: 05°25′ N 60°58′ W, 05°26′ N 60°59′ W, and 05°25′ N 60°59′ W (Gorzula and Señaris 1998). *Riolama leucosticta* is endemic to the Eastern Tepui Chain.

Kukenán

Venezuela (Bolívar); latitude × longitude: 05°13′ N 60°51′ W (Paynter 1982; US-BGN 1961, spelled Cuquenán); maximum elevation: 2,650 m; summit area: 20.63 km²; slope area: not available, 300 km² for all tepuis in Eastern Tepui Chain except Uei-tepui; massif: Eastern Tepui Chain; number of vegetation types for massif: 1—pioneer vegetation on sandstone summits; date of first herpetological collection-exploration: 1977.

Herpetofauna: 2 species, 0 endemics: Amphibia (1)—*Oreophrynella nigra;* Reptilia (1)—*Riolama leucosticta.*

Pertinent literature: General: Brewer-Carías 1978b; Mayr and Phelps 1967. Herpetological: Gorzula and Señaris 1998; McDiarmid and Gorzula 1989; Señaris et al. 1994.

Remarks: McDiarmid collected on this tepui for three days in January 1977 with Charles Brewer-Carías and others, and Gorzula visited it for one day in April 1984 and for one day in April 1985. Gorzula and Señaris (1998) listed coordinates for the two collection sites as 05°17′ N 60°48′ W and 05°11′ N 60°49′ W. This tepui has also been called Kukenán-, Mataui-, and Matawi-tepui (Huber 1995a), Kukenám and Monte Kukenám (Brewer-Carías 1978a, 1978b), Cuquenán (USBGN 1961), Cuquenám (Mayr and Phelps 1967), Mt. Kukenaam (McConnell 1916, map), and Mount Kukenaam (Warren 1973). The two recorded species, *Riolama leucosticta* and *Oreophrynella nigra,* are endemic to the Eastern Tepui Chain. Mayr and Phelps (1967) listed the elevation at 2,680 m, and Brewer-Carías (1978b) reported 2,600 m and coordinates of 05°12′ N and 60°45′ W for the site of McDiarmid's collections.

Roraima

Venezuela (Bolívar), Brazil (Roraima), and Guyana (Cuyuni-Mazaruni); 05°12′ N 60°44′ W (Paynter 1982; Stephens and Traylor 1985; Paynter and Traylor 1991; USBGN 1961); maximum elevation: 2,810 m; summit area: 34.38 km²; slope area: not available, 300 km² for all tepuis in Eastern Tepui Chain except Uei-tepui; massif: Eastern Tepui Chain; number of vegetation types for massif: 1—pioneer vegetation on sandstone summits; date of first herpetological collection-exploration: 1894.

Herpetofauna: 5 species, 4 endemics: Amphibia (5)—*Colostethus praderioi*, Colostethus roraima*, Eleutherodactylus* sp. J*, *Oreophrynella quelchii**; Reptiles (1)—*Riolama leucosticta.*

Pertinent literature: General: Brewer-Carías 1978a; Clementi 1920; Im Thurn 1885; Mayr and Phelps 1967; Quelch 1921; Tate 1928, 1930; Warren 1973. Herpetological: Boulenger 1895a, 1895b, 1900a, 1900b; Gorzula and Señaris 1998; La Marca 1996; McDiarmid and Gorzula 1989.

Remarks: This was the first tepui to be explored (appendix 18.5), and because of the moderately easy access to the summit, it probably is one of the best known. Several herpetologists have worked on Roraima. McDiarmid spent three days with Charles Brewer-Carías on the summit in January 1977, five days in February 1979, and one day in 1988. Even so, new species are still being found on its top. Also known in the literature as Roraima-tepui (Huber 1995a), Monte Roraima (Brewer-Carías 1978a), Mt. Roraima (McConnell 1916, map), and Mount Roraima (Warren 1973). The single collection cited by Gorzula and Señaris (1998) was made at 2,750 m at 05°11′ N 60°45′ W. USBGN (1961) and Stephens and Traylor (1985) gave Mazaruni-Potaro as the administrative unit in Guyana, but this has recently changed to Cuyuni-Mazaruni (*Times Atlas of the World* 1999). Huber (1995a) listed the maximum elevation for the Venezuelan part of Roraima as 2,723 m; the highest part of Roraima may be on the northern end of the tepui and in Guyana. Other elevations reported are 9,094 ft. (~2,771 m; Warren 1973) and 2,810 m (Paynter 1982; *Merriam-Webster's Geographical Dictionary* 1997). The type locality of *Colostethus praderioi* (La Marca 1996) apparently is in the third and highest quebrada crossed by the summit trail; we include it here because the given elevations were 1,800 and 1,950 m. Several examples of *Otophryne robusta* were reported by M. J. Praderio (pers. comm. in La Marca 1996, 37); we have not examined those specimens and accordingly have not included *Otophryne robusta* in the Roraima compilation. Gorzula and Señaris (1998) mentioned a specimen of *O. steyermarki,* purportedly from the talus slope of Roraima (no elevation given), but we are reluctant to accept this record without verification. Future work may confirm their presence (see appendix 18.3). The report of *Tepuihyla edelcae* from Roraima (Gorzula and Señaris 1998, 255; Galán 2000) apparently is an error; no specimens of this frog are listed in the species account by Gorzula and Señaris, and none is known to us. Specimens of *Bothriopsis taeniata,* previously known from Pantepui only from the holotype of *Bothrops lichenosa* from Chimantá (Roze 1958b), have been collected along the summit trail at the base of Roraima (Lancini 1978; Peter McIntyre, pers. comm.). A few species collected by McConnell and Quelch (appendix 18.5) and described by Boulenger (1900a, 1900b) were listed as coming from the base of Roraima; included are *Oreophrynella macconnelli, Eleutherodactylus marmoratus, Otophryne robusta,* and *Neusticurus rudis.* These and a few other species have been collected from the north slopes in Guyana (appendix 18.5); some were listed by Warren (1973) and others described by Duellman and Hoogmoed (1984, 1992) and Hoogmoed and Lescure (1984). Most, perhaps all, of these records are from localities below 1,500 m and have not been incorporated into this study (see the section titled "Creating a List"). Excluded were *Adelophryne gutturosa, Hyla kanaima, H. roraima,*

H. warreni, *Stefania roraimae*, *Oreophrynella macconnelli*, *Eleutherodactylus marmoratus*, *Otophryne robusta*, and *Neusticurus rudis*. Vegetation on the slopes is low evergreen, upper-montane forest. A lake (Lago Gladys) occurs on the northern end of the summit at about 2,700 m in Guyana (photographs in Brewer-Carías 1978a, 59; Huber 1995c, pl. 27). As currently understood, *Riolama leucosticta* is endemic to the Eastern Tepui Chain.

Yuruaní-tepui

Venezuela (Bolívar); latitude × longitude: 05°16′ N 60°51′ W (USBGN 1961); maximum elevation: 2,400 m; summit area: 4.38 km²; slope area: not available, 300 km² for all tepuis in Eastern Tepui Chain except Uei-tepui; massif: Eastern Tepui Chain; number of vegetation types for massif: 1—pioneer vegetation on sandstone summits; date of first herpetological collection-exploration: 1977.

Herpetofauna: 3 species, 2 endemics: Amphibia (3)—*Eleutherodactylus* sp. E*, *Oreophrynella nigra*, *Stefania riveroi**.

Pertinent literature: Gorzula and Señaris 1998; Mägdefrau and Mägdefrau 1994; Señaris et al. 1994, 1996.

Remarks: McDiarmid collected here for one day on the summit in January 1977 with Charles Brewer-Carías. Yuruaní has also been called Iwara-Karima (Brewer-Carías 1978b) and Mt. Iwalkarima (McConnell 1916, map; Warren 1973). Gorzula and Señaris (1998) listed the coordinates for a collection made over two days in April 1984 at 2,300 m as: 05°19′ N 60°51′ W. We are unable to verify if the *Eleutherodactylus* sp. E (a specimen in series B reported by Gorzula and Señaris 1998, 55) is the same as that represented in the photograph in Mägdefrau and Mägdefrau (1994, 100). We list only a single species from Yuruaní-tepui, pending clarification and additional material. *Oreophrynella nigra* is known from here and Kukenán to the south; such a distribution suggests a former connection between the tepuis or a more extensive distribution of the species.

LOS TESTIGOS SUBDISTRICT

Ptari Massif

Ptari-tepui

Venezuela (Bolívar); latitude × longitude: 05°46′ N, 61°46′ W (Paynter 1982), 05°47′ N 61°47′ W (USBGN 1961); maximum elevation: 2,400 m; summit area: 1.25 km²; slope area: not available, 28 km² for Carrao-tepui and Ptari-tepui; massif: Ptari; number of vegetation types for massif: 2—pioneer vegetation on sandstone summits; low evergreen, upper-montane forests on slopes; date of first herpetological collection-exploration: 1978.

Herpetofauna: 2 species, 2 endemics: Amphibia (2)—*Eleutherodactylus* sp. H*, *Tepuihyla rimarum**.

Pertinent literature: General: Brewer-Carías 1978b; Mayr and Phelps 1967. Herpetological: Ayarzagüena et al. 1992a, 1992b.

Remarks: McDiarmid accompanied Charles Brewer-Carías and collected one day in February 1978. Brewer-Carías (1978b) reported the collecting locality as 05°47' N 61°47' W and 2,400 m. Gorzula visited the summit on a November day in 1984 and collected 16 amphibians at approximately 2,400 m at a site with coordinates of 05°47' N 61°47' W. Mayr and Phelps (1967) listed the elevation as 2,620 m, but the summit had not been visited at that time.

Los Testigos Massif

Kamarkawarai-tepui

Venezuela (Bolívar); latitude × longitude: 05°53' N 61°59' W (Gorzula and Señaris 1998); maximum elevation: 2,400 m; summit area: 5.00 km²; slope area: not available, 88 km² for Los Testigos Massif except Aparamán-tepui; massif: Los Testigos; number of vegetation types for massif: 1—pioneer vegetation on sandstone summits; date of first herpetological collection-exploration: 1986.

Herpetofauna: 1 species, 0 endemics: Amphibia (1)—*Stefania satelles.*

Pertinent literature: Gorzula and Señaris 1998; Señaris et al. 1996.

Remarks: Señaris et al. (1996, 37) did not include the Los Testigos specimens in their description of *Stefania satelles,* commenting that the specimens had much larger tympana than specimens from other localities. Gorzula and Señaris (1998, 47) later commented on their initial reluctance to treat those frogs as *S. satelles,* and then included them in that species. Collectors spent one day in January 1986 on this tepui.

Murisipán-tepui

Venezuela (Bolívar); latitude × longitude: 05°53' N 62°05' W (estimated from Huber 1995a, fig. 1.19); maximum elevation: 2,350 m; summit area: 5.00 km²; slope area: not available, 88 km² for Los Testigos Massif except Aparamán-tepui; massif: Los Testigos; number of vegetation types for massif: 1—pioneer vegetation on sandstone summits; date of first herpetological collection-exploration: 1988.

Herpetofauna: 4 species, 2 endemics: Amphibia (3)—*Colostethus murisipanensis**, *Eleutherodactylus* sp. D*, *Stefania satelles;* Reptilia (1)—*Arthrosaura testigensis.*

Pertinent literature: Gorzula and Señaris 1998; La Marca 1996.

Remarks: Also known as Murosipán, according to La Marca (1996). Confusion exists about the exact location of collecting sites on this tepui. Gorzula and

Señaris (1998, 257) listed coordinates of 05°53' N 62°03' W for a collection made by Gorzula and others at a site on "Terekyurén (Murisipán)" at 2,350 m. According to Huber (1995a, fig. 1.19), this locality has to be on Murisipán, because the maximum elevation of Tereke-yurén is 1,900 m. Therefore, we have decided that the amphibians reported by Gorzula and Señaris (1998, 257) as collected on 2–3 March 1988 from Terekyurén were actually collected on Murisipán. La Marca (1996) also listed coordinates 05°53' N 62°04' W and 2,350 m for the holotype of *Colostethus murisipanensis* that Gorzula collected on Murisipán on 2 March 1988. If our deductions about localities are true, only the westernmost Aparamán, of the tepuis making up the Los Testigos Massif, has not been sampled. USBGN (1961) listed the coordinates for Murisipán at 05°51' N 62°00' W, which is clearly incorrect according to Huber's map. *Arthrosaura testigensis* is endemic to the Los Testigos Massif.

Tereke-yurén-tepui

Venezuela (Bolívar); latitude × longitude: 05°53' N 62°02' W (estimated from Huber 1995a, fig. 1.19; Gorzula and Señaris 1998); maximum elevation: 1,900 m; summit area: 0.63 km²; slope area: not available, 88 km² for Los Testigos Massif except Aparamán-tepui; massif: Los Testigos; number of vegetation types for massif: 1—pioneer vegetation on sandstone summits; date of first herpetological collection-exploration: 1988.

Herpetofauna: 2 species, 0 endemics: Amphibia (1)—*Tepuihyla edelcae;* Reptilia (1)—*Arthrosaura testigensis.*

Pertinent literature: Gorzula and Señaris 1998; Señaris et al. 1996.

Remarks: According to Gorzula and Señaris (1998, 117), the holotype of *Arthrosaura testigensis* was collected at 1,800 m at 05°52' N 62°03' W on the south side of the massif on 15 January 1986, as was a single specimen of *Tepuihyla edelcae.* See remarks for Murisipán-tepui regarding the location of reported localities.

CHIMANTÁ SUBDISTRICT

Auyán Massif

Auyán-tepui

Venezuela (Bolívar); latitude × longitude: 05°55' N 62°32' W (Paynter 1982; USBGN 1961); maximum elevation: 2,450 m; summit area: 667.7 km² excluding Uaipán; slope area: not available, 715 km² for Auyán-tepui, Cerro La Luna, Cerro El Sol; massif: Auyán; number of vegetation types for massif: 4—pioneer vegetation on sandstone summits; low tepui-summit scrub and meadows, on peat and rock; low evergreen, high-tepui forests; low evergreen, uppermontane forests on slopes; date of first herpetological collection-exploration: 1956.

Herpetofauna: 23 species, 12 endemics: Amphibia (12)—*Centrolene gorzulae**, *Colostethus tepuyensis**, *Eleutherodactylus* sp. C*, *Hyalinobatrachium auyantepuianum**, *Hyalinobatrachium* cf. *orientale**, *Hyalinobatrachium taylori, Hyla sibleszi,* Hylid sp. B*, *Leptodactylus rugosus, Oreophrynella cryptica**, *Stefania schuberti**, *Tepuihyla edelcae;* Reptilia (11)—*Anadia* sp. B*, *Anolis nitens* ssp., *Arthrosaura* sp. B*, *Neusticurus rudis, Tropidurus bogerti**, *Atractus* sp.*, *Chironius fuscus,* Colubrid sp., *Leptodeira annulata, Liophis trebbaui, Mastigodryas boddaerti.*

Pertinent literature: General: Brewer-Carías 1978b; Buchsbaum 1969; Barrowclough et al. 1997; Jirak et al. 1968; Mayr and Phelps 1967; Tate 1938. Herpetological: Ayarzagüena 1992; Ayarzagüena et al. 1992a, 1992b; Ayarzagüena and Señaris 1996; Gorzula and Señaris 1998; La Marca 1996; Myers 1997; Roze 1958a; Señaris 1993; Señaris and Ayarzagüena 1993; Señaris et al. 1996.

Remarks: Gorzula and Señaris (1998) listed the following coordinates for their nine collecting sites on this tepui: 05°48′ N 62°32′ W, 05°54′ N 62°38′ W, 05°56′ N 62°33′ W, 05°58′ N 62°29′ W, 06°00′ N 62°35′ W, 06°01′ N 62°26′ W, 06°01′ N 62°37′ W, 06°02′ N 62°40′ W, and 06°03′ N 62°35′ W. Myers (1997) briefly reviewed the history of herpetological investigations on Auyán-tepui and referred to some specimens collected during the American Museum of Natural History–Terramar Expedition of 1994. Among species collected at or near five helicopter-supported camps (camp I, 05°51′ N 62°32′ W, 1,700 m; camp II, 05°54′ N 62°29′ W, 1,750 m; camp III, 05°53′ N 62°38′ W, 1,850 m; camp IV, 05°58′ N 62°33′ W, 1,700 m; camp V, 05°46′ N 62°32′ W, 2,200 m) were specimens of an undescribed *Arthrosaura* (sp. B); a snake of the genus *Atractus* had been collected previously. Even though the specimen of *Atractus* sp. has not been compared with other forms (Myers 1997), we treat it and the *Arthrosaura* sp. B as endemic forms in our analysis. Several lowland species, including *Anolis auratus, Ameiva ameiva, Cnemidophorus lemniscatus, Tropidurus hispidus, Leptotyphlops albifrons, Chironius carinatus, Oxybelis aeneus, Liophis lineatus,* and *Erythrolamprus aesculapii* were reported by Roze (1958a) from Auyán-tepui but without specific locality or elevation. Specimens of these species reported by Roze and in the collections of the American Museum of Natural History have no additional data pertinent to elevation. Accordingly, we assume they were collected near the base of the tepui and exclude them from our treatment. The unidentified colubrid is known from parts of a shed skin and has not been counted as endemic. Ayarzagüena and Señaris (1996) transferred their species *Centrolenella auyantepuiana* to the genus *Hyalinobatrachium* (also see Myers and Donnelly 1997, 71, note added in proof).

Cerro el Sol

Venezuela (Bolívar); latitude × longitude: 06°06′ N 62°32′ W (Gorzula and Señaris 1998); maximum elevation: 1,750 m; summit area: 0.60 km²; slope area: not available, 715 km² for Auyán-tepui, Cerro La Luna, Cerro El Sol; massif:

Auyán; number of vegetation types for massif: 4—pioneer vegetation on sandstone summits; low tepui-summit scrub and meadows, on peat and rock; low evergreen, high-tepui forests; low evergreen, upper-montane forests on slopes; date of first herpetological collection-exploration: 1987.

Herpetofauna: 1 species, 1 endemic: Amphibia (1)—*Oreophrynella huberi**.

Pertinent literature: Diego-Aransay and Gorzula 1987.

Remarks: Three specimens of *Oreophrynella huberi* reportedly were captured in high, dense herbaceous vegetation at 1,700 m on the summit; two among the several scattered males heard calling during the day were collected on 7 May 1987. Cerro El Sol is the name that has been applied to Uei-tepui in the Eastern Tepui Chain by some authors (e.g., Huber 1995a, 36); a locality with this name is listed as Uei-tepuí in USBGN (1961) with coordinates 05°01′ N 60°37′ W.

Cerro La Luna

Venezuela (Bolívar); latitude × longitude: 06°05′ N 62°31′ W (Gorzula and Señaris 1998); maximum elevation: 1,650 m; summit area: 0.20 km²; slope area: not available, 715 km² for Auyán-tepui, Cerro La Luna, Cerro El Sol; massif: Auyán; number of vegetation types for massif: 4—pioneer vegetation on sandstone summits; low tepui-summit scrub and meadows, on peat and rock; low evergreen, high-tepui forests; low evergreen, upper-montane forests on slopes; date of first herpetological collection-exploration: 1986.

Herpetofauna: 1 species, 0 endemics: Amphibia (1)—*Tepuihyla edelcae.*

Pertinent literature: Ayarzagüena et al. 1992a, 1992b.

Remarks: Cerros La Luna and El Sol exist as two tower-like tepuis at the end of a long, forested ridge extending off the north end of Auyán-tepui (Huber 1995a, fig. 1.25).

Aprada Massif

Aprada-tepui

Venezuela (Bolívar); latitude × longitude: 05°26′ N 62°25′ W (Paynter 1982; USBGN 1961); maximum elevation: 2,500 m; summit area: 4.37 km²; slope area: not available; 210 km² for Aprada Massif; massif: Aprada; number of vegetation types for massif: 1—pioneer vegetation on sandstone summits; date of first herpetological collection-exploration: 1984.

Herpetofauna: 2 species, 1 endemic: Amphibia (2)—*Eleutherodactylus* sp. B*, *Stefania satelles.*

Pertinent literature: General: Brewer-Carías 1978b; Mayr and Phelps 1967. Herpetological: Gorzula and Señaris 1998; Señaris et al. 1996.

Remarks: McDiarmid accompanied Charles Brewer-Carías and collected here in 1978; Brewer-Carías (1978b) reported coordinates as 05°26′ N 62°26′ W at 2,450 m elevation. Gorzula and Señaris (1998) listed two other localities for this tepui: 05°24′ N 62°27′ W at 2,500 m and 05°27′ N 62°23′ W at 2,050 m. Mayr and Phelps (1967) gave 2,400 m as the elevation, but this must have been an estimate.

Chimantá Massif

Abacapá-tepui

Venezuela (Bolívar); latitude × longitude: 05°13′ N 62°15′ W (USBGN 1961); maximum elevation: 2,400 m; summit area: 28.13 km²; slope area: not available, 915 km² for Chimantá Massif; massif: Chimantá; number of vegetation types for massif: 4—pioneer vegetation on sandstone summits; low tepui-summit scrub and meadows on peat and rock; low evergreen, high-tepui forests; and low evergreen, upper-montane forests on slopes and in valleys between tepuis; date of first herpetological collection-exploration: 1984.

Herpetofauna: 6 species, 1 endemics: Amphibia (2)—*Stefania ginesi, Tepuihyla edelcae;* Reptilia (4)—*Anadia* sp. A, *Arthrosaura* sp. A, *Anolis nitens* ssp., *Phenacosaurus carlostoddi**.

Pertinent literature: General: Huber 1992. Herpetological: Gorzula 1992; Gorzula and Señaris 1998; Señaris et al. 1996; Williams et al. 1996.

Remarks: Gorzula and Señaris (1998) listed 05°12′ N 62°19′ W as coordinates for their collecting site. We use the name *Anolis nitens* ssp., rather than *Anolis eewi*, contrary to the treatment in Gorzula and Señaris (1998); see appendix 18.3 for explanation. *Stefania ginesi, Anadia* sp. A, and *Arthrosaura* sp. A are endemic to the Chimantá Massif.

Acopán-tepui

Venezuela (Bolívar); latitude × longitude: 05°12′ N 62°04′ W (USBGN 1961); maximum elevation: 2,200 m; summit area: 92.50 km²; slope area: not available, 915 km² for Chimantá Massif; massif: Chimantá; number of vegetation types for massif: 4—pioneer vegetation on sandstone summits; low tepui-summit scrub and meadows on peat and rock; low evergreen, high-tepui forests; and low evergreen, upper-montane forests on slopes and in valleys between tepuis; date of first herpetological collection-exploration: 1984.

Herpetofauna: 6 species, 0 endemics: Amphibia (2)—*Stefania ginesi, Tepuihyla edelcae;* Reptilia (4)—*Arthrosaura* sp A, *Neusticurus rudis, Leptodeira annulata, Thamnodynastes chimanta.*

Pertinent literature: General: Huber 1992; Mayr and Phelps 1967. Herpetological: Barreát et al. 1986; Gorzula 1985, 1992; Gorzula and Señaris 1998; Señaris et al. 1996.

Remarks: Paynter (1982) gave coordinates for this locality as 05°12′ N 62°14′ W. Gorzula and Señaris (1998) had four coordinates for collection sites on this tepui: 05°11′ N 62°02′ W, 05°13′ N 62°05′ W, 05°10′ N 61°59′ W, and 05°12′ N

62°05′ W. This site was listed as Akopan-tepui by Huber (1992) and Señaris et al. (1996). Three (*Stefania ginesi, Arthrosaura* sp. A, and *Thamnodynastes chimanta*) of the six species known from this tepui are endemic to the Chimantá Massif. Gorzula (1985) and Campbell and Clarke (1998) also reported hearing *Otophryne steyermarki* calling from this tepui.

Amurí-tepui

Venezuela (Bolívar); latitude × longitude: 05°10′ N 62°07′ W (USBGN 1961); maximum elevation: 2,200 m; summit area: 36.88 km²; slope area: not available, 915 km² for Chimantá Massif; massif: Chimantá; number of vegetation types for massif: 4—pioneer vegetation on sandstone summits; low tepui-summit scrub and meadows, on peat and rock; low evergreen, high-tepui forests; and low evergreen, upper-montane forests on slopes and in valleys between tepuis; date of first herpetological collection-exploration: 1983.

Herpetofauna: 7 species, 0 endemics: Amphibia (3)—*Epipedobates rufulus, Stefania ginesi, Tepuihyla edelcae;* Reptilia (4)—*Anadia* sp. A, *Neusticurus rudis, Leptodeira annulata, Thamnodynastes chimanta.*

Pertinent literature: General: Huber 1992. Herpetological: Barreát et al. 1986; Gorzula 1988, 1992; Gorzula and Señaris 1998; Señaris et al. 1996.

Remarks: Gorzula and Señaris (1998) listed 05°12′ N 62°06′ W, 05°08′ N 62°07′ W, and 05°08′ N 62°08′ W as collection sites on this tepui. *Epipedobates rufulus, Stefania ginesi, Anadia* sp. A, and *Thamnodynastes chimanta* are endemic to the Chimantá Massif.

Apacará-tepui

Venezuela (Bolívar); latitude × longitude: 05°18′ N 62°13′ W (USBGN 1961); maximum elevation: 2,450 m; summit area: 173.12 km²; slope area: not available, 915 km² for Chimantá Massif; massif: Chimantá; number of vegetation types for massif: 4—pioneer vegetation on sandstone summits; low tepui-summit scrub and meadows on peat and rock; low evergreen, high-tepui forests; and low evergreen, upper-montane forests on slopes and in valleys between tepuis; date of first herpetological collection-exploration: 1983.

Herpetofauna: 5 species, 0 endemic: Amphibia (2)—*Otophryne steyermarki, Tepuihyla edelcae;* Reptilia (3)—*Anolis nitens* ssp., *Leptodeira annulata, Thamnodynastes chimanta.*

Pertinent literature: General: Huber 1992. Herpetological: Barreát et al. 1986; Campbell and Clarke 1998; Gorzula 1992; Gorzula and Señaris 1998; Rivero 1967; Roze 1958b.

Remarks: Huber (1992) and Gorzula and Señaris (1998) listed five collection sites with the following coordinates: 05°20′ N 62°12′ W, 05°25′ N 62°11′ W, 05°19′ N 62°10′ W, 05°19′ N 62°12′ W, and 05°17′ N 62°16′ W for this tepui. A sixth site (camp XIX; Huber 1992) lies between this and the adjacent Murey-tepui in the headwaters of the Río Apakará near 05°22' N 62°08' W. *Thamnodynastes chimanta* is endemic to the Chimantá Massif. We use the name *Anolis nitens* ssp., rather than *Anolis eewi* (contra Gorzula and Señaris 1998; see appendix 18.3).

Chimantá-tepui

Venezuela (Bolívar); latitude × longitude: 05°18' N 62°10' W (Paynter 1982; USBGN 1961); maximum elevation: 2,550 m; summit area: 93.75 km²; slope area: not available, 915 km² for Chimantá Massif; massif: Chimantá; number of vegetation types for massif: 4—pioneer vegetation on sandstone summits; low tepui-summit scrub and meadows on peat and rock; low evergreen, high-tepui forests; and low evergreen, upper-montane forests on slopes and in valleys between tepuis; date of first herpetological collection-exploration: 1953.

Herpetofauna: 11 species, 1 endemic: Amphibia (3)—*Hyla sibleszi, Stefania ginesi, Tepuihyla edelcae;* Reptilia (8)—*Anolis nitens* ssp., *Arthrosaura* sp. A, *Neusticurus rudis, Atractus steyermarki, Leptodeira annulata, Liophis ingeri*, Thamnodynastes chimanta, Bothriopsis taeniata.*

Pertinent literature: General: Huber 1992; Mayr and Phelps 1967. Herpetological: Barreát et al. 1986; Gorzula 1992; Gorzula and Señaris 1998; Roze 1958b; Señaris et al. 1996.

Remarks: Gorzula and Señaris (1998) listed three collecting localities on Chimantá proper: 05°16' N 62°09' W, 05°18' N 62°03' W, and 05°22' N 62°08' W. We use the name *Anolis nitens* ssp., rather than *Anolis eewi* (contra Gorzula and Señaris 1998; see appendix 18.3). *Stefania ginesi, Arthrosaura* sp. A, *Atractus steyermarki,* and *Thamnodynastes chimanta* are endemic to the Chimantá Massif.

Churi-tepui

Venezuela (Bolívar); latitude × longitude: 05°13' N 61°54' W (Paynter 1982; USBGN 1961); maximum elevation: 2,500 m; summit area: 47.50 km²; slope area: not available, 915 km² for Chimantá Massif; massif: Chimantá; number of vegetation types for massif: 4—pioneer vegetation on sandstone summits; low tepui-summit scrub and meadows on peat and rock; low evergreen, high-tepui forests; and low evergreen, upper-montane forests on slopes and in valleys between tepuis; date of first herpetological collection-exploration: 1985.

Herpetofauna: 5 species, 0 endemics: Amphibia (2)—*Stefania ginesi, Tepuihyla edelcae;* Reptilia (3)—*Arthrosaura* sp. A, *Atractus steyermarki, Thamnodynastes chimanta.*

Pertinent literature: General: Huber 1992. Herpetological: Barreát et al. 1986; Gorzula 1992; Gorzula and Señaris 1998; Señaris et al. 1996.

Remarks: Gorzula and Señaris (1998) listed 05°15' N 62°01' W as coordinates for their collection site. *Arthrosaura* sp. A, *Atractus steyermarki,* and *Thamnodynastes chimanta* are endemic to the Chimantá Massif. A single site on this tepui in the Chimantá Massif was sampled on three different occasions for a total of 11 days (Gorzula and Señaris 1998).

Murey-tepui

Venezuela (Bolívar); latitude × longitude: 05°22′ N 62°05′ W (camp XVIII) from Huber (1992) and Gorzula and Señaris (1998); maximum elevation: 2,650 m; summit area: 51.25 km²; slope area: not available, 915 km² for Chimantá Massif; massif: Chimantá; number of vegetation types for massif: 4—pioneer vegetation on sandstone summits; low tepui-summit scrub and meadows on peat and rock; low evergreen, high-tepui forests; and low evergreen, upper-montane forests on slopes and in valleys between tepuis; date of first herpetological collection-exploration: 1978.

Herpetofauna: 6 species, 1 endemic: Amphibia (4)—*Eleutherodactylus* sp. I*, *Epipedobates rufulus, Stefania ginesi, Tepuihyla edelcae;* Reptilia (2)—*Anadia* sp. A, *Thamnodynastes chimanta.*

Pertinent literature: General: Huber 1992. Herpetological: Barreát et al. 1986; Duellman and Hoogmoed 1984; Gorzula 1988, 1992; Gorzula and Señaris 1998; Señaris et al. 1996.

Remarks: McDiarmid collected on this tepui in 1978 with Charles Brewer-Carías, and they called it Eruoda-tepui (also see Duellman and Hoogmoed 1984); Brewer-Carías (1978b) subsequently adopted Murey-tepui for the site. Others also have used two names for this tepui. Initially, Huber (1992) used Eruoda-tepui but later recommended Murey-tepui (Huber 1995a). USBGN (1961) listed Erueda-tepui as the preferred name (but also listed Eruoda-tepui) and gave coordinates as 05°26′ N 62°03′ W. If these coordinates are accurate, this site is at the northern tip of Murey (= Eruoda) and represents the northernmost locality on the Chimantá Massif (see Huber 1992; Gorzula and Señaris 1998). The following species are endemic to the Chimantá Massif: *Epipedobates rufulus, Stefania ginesi, Anadia* sp. A, and *Thamnodynastes chimanta.*

Tirepón-tepui

Venezuela (Bolívar); latitude × longitude: ~05°23′ N 62°02′ W based on extrapolation from Huber's map (1995a, fig. 1.27) and a description in McDiarmid's field notes; maximum elevation: 2,650 m; summit area: 8.75 km²; slope area: not available, 915 km² for Chimantá Massif; massif: Chimantá; number of vegetation types for massif: 4—pioneer vegetation on sandstone summits; low tepui-summit scrub and meadows on peat and rock; low evergreen, high-tepui forests; and low evergreen, upper-montane forests on slopes and in valleys between tepuis; date of first herpetological collection-exploration: 1978.

Herpetofauna: 4 species, 1 endemic: Amphibia (3)—*Epipedobates rufulus, Stefania ginesi, Tepuihyla edelcae;* Reptilia (1)—Gymnophthalmid sp. B*.

Pertinent literature: General: Huber 1992. Herpetological: Duellman and Hoogmoed 1984.

Remarks: McDiarmid collected here in 1978 with Charles Brewer-Carías. In the field they referred to this site as Toroná-tepui (different from Toronó); later they realized that Tirepón was the more appropriate name and made pertinent corrections to the field catalog. Unfortunately, specimens of *Stefania ginesi* that had been collected by McDiarmid from this tepui were loaned to Duellman with the original field locality information. This has been subsequently corrected so that the specimens of *Stefania ginesi* (USNM 212041-43) reported by Duellman and Hoogmoed (1984, 17, 37) as coming from a locality at 2,450 m on Toroná-tepui are actually from Tirepón-tepui. *Stefania ginesi* and *Epipedobates rufulus* are endemic to the Chimantá Massif. Gymnophthalmid sp. B is known from a shed skin. This isolated, eastern tepui in the Chimantá Massif has been sampled only for a few hours on one day.

Toronó-tepui

Venezuela (Bolívar); latitude × longitude: 05°12′ N 62°10′ W based on an approximation from Huber's map (1995a, fig. 1.27); maximum elevation: 2,500 m; summit area: 59.38 km²; slope area: not available, 915 km² for Chimantá Massif; massif: Chimantá; number of vegetation types for massif: 4—pioneer vegetation on sandstone summits; low tepui-summit scrub and meadows on peat and rock; low evergreen, high-tepui forests; and low evergreen, upper-montane forests on slopes and in valleys between tepuis; date of first herpetological collection-exploration: 1955.

Herpetofauna: 1 species, 0 endemics: Reptilia (1)—*Anolis nitens* ssp.

Pertinent literature: General: Huber 1992. Herpetological: Roze 1958b.

Remarks: Apparently the only herpetological material known from this tepui is the type specimen of *Anolis eewi* Roze 1958, that was collected by Julian Steyermark and John Wurdack during their botanical explorations of the Chimantá Massif in the period of January through March 1955. We use the name *Anolis nitens* ssp., rather than *Anolis eewi* (contra Gorzula and Señaris 1998; see appendix 18.3). The coordinates listed for Toronó-tepui in USBGN (1961) are 05°24′ N 62°00′ W; these are incorrect and contributed to the confusion with an early name for Tirepón-tepui (see earlier remarks).

Angasima-tepui

Venezuela (Bolívar); latitude × longitude: 05°05′ N 62°03′ W (USBGN 1961); maximum elevation: 2,250 m; summit area: 2.00 km²; slope area: 32 km²; massif: not part of a massif; number of vegetation types: 1—low tepui-summit scrub and meadows on peat and rock; date of first herpetological collection-exploration: 1986.

Herpetofauna: 1 species, 0 endemics: Amphibia (1)—*Stefania satelles*.

Pertinent literature: Gorzula and Señaris 1998; Señaris et al. 1996.

Remarks: Gorzula and Señaris (1998) included two localities, 05°03′ N 62°04′ W and 05°03′ N 62°07′ W, that were sampled for two days in August and one day in March.

Upuigma-tepui

Venezuela (Bolívar); latitude × longitude: 05°07′ N 61°56′ W (Paynter 1982; USBGN 1961); maximum elevation: 2,100 m; summit area: 0.63 km²; slope area: 13 km²; massif: not part of a massif; number of vegetation types for massif: 1— low tepui-summit scrub and meadows on peat and rock; date of first herpetological collection-exploration: 1986.

Herpetofauna: 1 species, 0 endemics: Amphibia (1)—*Stefania satelles*.

Pertinent literature: General: Mayr and Phelps 1967. Herpetological: Gorzula and Señaris 1998; Señaris et al. 1996.

Remarks: The collection site described by Gorzula and Señaris (1998) was listed at 05°05′ N 61°57′ W and sampled once in March. Mayr and Phelps (1967) estimated the summit elevation at 2,200 m. This site is also called El Castillo.

Western Pantepui District

MAIGUALIDA SUBDISTRICT

Isolates

Sierra de Maigualida

Venezuela (Bolívar); latitude × longitude: 05°30′ N 65°10′ W (USBGN 1961); maximum elevation: 2,400 m; summit area: 440 km²; slope area: not available; massif: none; number of vegetation types: 2—broad-leaved, shrubby upland and highland meadows on rock and peat and low to medium upper-montane evergreen forest; date of first herpetological collection-exploration: 1988.

Herpetofauna: 2 species, 1 endemic: Amphibia (2)—Hylid sp. A*, *Leptodactylus rugosus*.

Pertinent literature: Gorzula and Señaris 1998.

Remarks: Gorzula and Señaris (1998) listed one set of coordinates (05°43′ N 65°19′ W) for a site in Bolívar state and another (05°33′ N 65°13′ W) for an Amazonas site. As currently known, Maigualida is primarily granitic and has little or none of the Roraima sandstone characteristic of tepuis. This extensive area has received little work (7 days at two sites in March).

YUTAJÉ SUBDISTRICT

Isolates

Cerro Guanay

Venezuela (Amazonas); latitude × longitude: 05°51′ N 66°18′ W (Paynter 1982; USBGN 1961); maximum elevation: 2,080 m; summit area: 165 km²; slope area: 113 km²; massif: none; number of vegetation types: 1—low to tall tepui scrub on rock; date of first herpetological collection-exploration: 1985.

Herpetofauna: 8 species, 3 endemics: Amphibia (5)—*Colostethus guanayensis**, *Eleutherodactylus* sp. A*, *Hyalinobatrachium orientale, Hyla sibleszi, Leptodactylus rugosus;* Reptilia (3)—*Anolis nitens* ssp., *Plica pansticta, Prionodactylus nigroventris**.

Pertinent literature: General: Mayr and Phelps 1967. Herpetological: Gorzula and Señaris 1998; La Marca 1996.

Remarks: Gorzula and Señaris (1998) reported on collections made at and near a camp with the following coordinates: 05°55′ N 66°23′ W at 1,650 m. La Marca (1996) reported an elevation of 1,800 m for a topoparatype of *Colostethus guanayensis* and referred to the tepui as Serranía de Guanay. Myers and Donnelly (2001, 75) reported *Plica pansticta* (as *Tropidurus panstictus*) from Guanay and gave an indication of work in progress. We use the name *Anolis nitens* ssp. rather than *Anolis eewi* (contra Gorzula and Señaris 1998; see appendix 18.3).

Cerro Yaví

Venezuela (Amazonas); latitude × longitude: 05°32′ N 65°59′ W (Paynter 1982; USBGN 1961); maximum elevation: 2,300 m; summit area: 5.62 km²; slope area: 70 km²; massif: none; number of vegetation types: 2—broad-leaved, shrubby upland and highland meadows on rock and peat, and low evergreen, uppermontane forest; date of first herpetological collection-exploration: 1986.

Herpetofauna: 7 species, 5 endemics: Amphibia (3)—*Eleutherodactylus cantitans, Eleutherodactylus pruinatus**, *Eleutherodactylus yaviensis;* Reptilia (4)—Gymnophthalmid sp. A*, *Phenacosaurus bellipeniculus**, *Prionodactylus goeleti**, *Thamnodynastes yavi**.

Pertinent literature: General: Hitchcock 1947; Mayr and Phelps 1967. Herpetological: Gorzula and Señaris 1998; Myers and Donnelly 1996.

Remarks: The collection of 63 specimens described by Myers and Donnelly (1996) was made during five days in February 1995 in the vicinity of a summit camp at 2,150 m and coordinates of 05°43′ N 65°54′ W. Gorzula and Señaris (1998) reported eight specimens collected over four days in October at a locality at 2,100 m elevation and coordinates of 05°42′ N 65°53′ W. Given this level of sampling, we think the herpetofauna of Yaví is reasonably well known compared with that of many tepuis in the Western Pantepui District, some of which (e.g., Cuao-Sipapo Massif) have not been sampled for amphibians and reptiles. The record for Gymnophthalmid sp. A is based on the tail of a lizard removed from the stomach of a specimen of *Thamnodynastes yavi*. Myers and Donnelly (1996) compared the tail to those of *Prionodactylus goeleti* and found that they were different; they suggested that the tail might belong to a species of *Arthrosaura*. Four habitat pictures, one showing a lake, were published in Myers and Donnelly (1996).

Yutajé Massif

Cerro Corocoro

Venezuela (Amazonas); latitude × longitude: 05°46' N 66°11' W (Gorzula and Señaris 1998); maximum elevation: 2,400 m; summit area: 179.38 km²; slope area: not available, 143 km² for Yutajé Massif; massif: Yutajé; number of vegetation types for massif: 3—low evergreen, upper-montane forest; broad-leaved, shrubby upland and highland meadows on rock and peat; and low to tall tepui scrub on rock; date of first herpetological collection-exploration: 1987.

Herpetofauna: 5 species, 1 endemic: Amphibia (2)—*Eleutherodactylus yaviensis, Pseudopaludicola llanera;* Reptilia (3)—*Plica pansticta, Prionodactylus goeleti, Thamnodynastes corocoroensis**.

Pertinent literature: Gorzula and Ayarzagüena 1995; Gorzula and Señaris 1998; Myers and Donnelly 2001.

Remarks: This tepui was called Cerro Coro Coro by Huber and other authors in various chapters in *Flora of the Venezuelan Guayana* (Steyermark et al. 1995a). For purposes of utility we have used the orthography (Corocoro) found on some older maps and recently adopted by Gorzula and Señaris (1998) and Myers and Donnelly (2001). Myers and Donnelly (2001) noted that a single specimen from Corocoro called Eleutherodactylinae series b by Gorzula and Señaris (1998, 55) likely represented *Eleutherodactylus yaviensis,* and we have followed that suggestion. Myers and Donnelly (2001) also reported the frog *Pseudopaludicola llanera* and a new lizard they called *Tropidurus panstictus* (= *Plica pansticta,* following Frost et al. 2001) that were taken along a small rocky stream bordered by a narrow strip of gallery forest on a small plateau at 1,220 m (1,400 m on some maps) and coordinates of 05°42' N 66°10' W on the south end of Cerro Corocoro. Although they were taken from a relatively low site, we have included these two species in the Pantepui list (appendix 18.1) but do so cautiously because of the relatively low elevation and some concern about the lower elevational limits of Cerro Corocoro. For additional comments on *Thamnodynastes corocoroensis,* see remarks under Cerro Yaví.

Serranía Yutajé

Venezuela (Amazonas); latitude × longitude: 05°45' N 66°03' W (Gorzula and Señaris 1998); maximum elevation: 2,140 m; summit area: 95.63 km²; slope area:

143 km² for Yutajé Massif; massif: Yutajé; number of vegetation types for massif: 3—low evergreen, upper-montane forest; broad-leaved, shrubby upland and highland meadows on rock and peat; and low to tall tepui scrub on rock; date of first herpetological collection-exploration: 1988.

Herpetofauna: 7 species, 3 endemics: Amphibia (4)—*Colostethus undulatus*, Eleutherodactylus cantitans, Eleutherodactylus yaviensis, Hyalinobatrachium eccentricum**; Reptilia (3)—*Adercosaurus vixadnexus*, Mabuya bistriata, Prionodactylus goeleti.*

Pertinent literature: Gorzula and Señaris 1998; Myers and Donnelly 2001.

Remarks: The collecting site listed by Myers and Donnelly was on a branch of the upper Río Corocoro at about 1,700 m (1,920 m on some maps) on the north end of Serrania Yutajé (see map in Myers and Donnelly 2001, 5) at 05°46′ N 66°08′ W. The area has a mosaic of dense scrub on the rocky, exposed western side of the river and low, wet mossy forest along the eastern side. These forests are extensions of the medium, evergreen, lower-montane to montane forests that surround the uplands in the Yaví-Corocoro-Yutajé-Guanay mountain complex and reach into the highlands in wetter places (Huber 1995c).

Central Pantepui District

GUAIQUINIMA SUBDISTRICT

Isolates

Cerro Guaiquinima

Venezuela (Bolívar); latitude × longitude: 05°49′ N 63°40′ W (Paynter 1982; USBGN 1961); maximum elevation: 1,650 m; summit area: 1,096.26 km²; slope area: 410 km²; massif: none; number of vegetation types: 3—broad-leaved, shrubby upland meadows on peat; tall, upland scrub on rock; and medium, evergreen montane forest; date of first herpetological collection-exploration: 1984.

Herpetofauna: 22 species, 5 endemics: Amphibia (9)—*Eleutherodactylus* sp. S*, *Hyalinobatrachium* sp. A*, *Hyla benitezi, Hyla boans, Hyla crepitans, Hyla minuta, Leptodactylus longirostris, Leptodactylus rugosus, Osteocephalus taurinus*; Reptilia (13)—*Ameiva ameiva, Anolis nitens* ssp., *Anolis fuscoauratus, Neusticurus racenisi, Neusticurus* sp. A*, *Plica lumaria*, Tropidurus hispidus, Imantodes lentiferus, Leptotyphlops albifrons, Liophis torrenicola, Mastigodryas boddaerti, Philodryas cordata*, Thamnodynastes pallidus.*

Pertinent literature: Donnelly and Myers 1991; Gorzula and Señaris 1998; Mägdefrau et al. 1991; Schlüter 1994.

Remarks: Huber (1995a) treated Guaiquinima as an upland rather than a highland tepui, presumably because most of the cerro is below 1,500 m. Huber's definition of highlands (above 1,500 m) results in a somewhat subjective decision to treat Guaiquinima separately from most other tepuis (see text for discussion). We place Guaiquinima in the Central District primarily because of its geographic position. Donnelly and Myers (1991) listed approximate coordinates for this extensive tepui as 05°46′ N 63°36′ W. Gorzula and Señaris (1998) listed four localities at which collections of amphibians and reptiles were made on this tepui: 05°53′ N 63°41′ W, 05°58′ N 63°29′ W, 05°54′ N 63°26′ W, and 05°47′ N 63°48′ W. Mägdefrau et al. (1991) and Schlüter (1994) gave coordinates and elevations for four other sites at which they collected: 05°54′50″ N 63°28′ W, 1,180 m; 05°45′50″ N 63°33′40″ W, 780 m; 05°44′10″ N 63°38′ W, 980 m; and 05°47′10″ N 63°50′30″ W, 1,520 m. Mägdefrau et al. (1991) used *Neusticurus rudis* for the taxon referred to here as *Neusticurus* sp. A, but the species from Guaiquinima is distinct from *N. rudis* and an undescribed form from Marutaní. Preliminary reports seem to suggest that more than a single species is represented by *Eleutherodactylus* sp. S (= *Eleutherodactylus* spp. of Mägdefrau et al. 1991 and Schlüter 1994); we prefer to list only one pending publication of descriptions or additional information.

JAUA SARISARIÑAMA SUBDISTRICT

Isolates

Cerro Guanacoco
Venezuela (Bolívar); latitude × longitude: 04°40′ N 63°51′ W (Paynter 1982; US-BGN 1961); maximum elevation: 1,500 m; summit area: 526.25 km²; slope area: 400 km²; massif: none; number of vegetation types: 2—tubiform, shrubby highland meadows on rock and peat and medium, evergreen upper-montane forest; date of first herpetological collection-exploration: 1974.

Herpetofauna: 1 species, 0 endemics: Amphibia (1)—*Hyla minuta*.

Pertinent literature: General: Orejas-Miranda and Quesada 1976.

Remarks: Braulio Orejas Miranda visited this tepui for one day in 1974 with Charles Brewer-Carías during the Jaua-Sarisariñama Expedition.

Sierra Marutaní
Venezuela (Bolívar); latitude × longitude: 03°46′ N 63°03′ W (Paynter 1982 as Cerro Urutaní; USBGN 1961 as Sierra Marutani); maximum elevation: 1,500 m; summit area: not available; slope area: 740 km²; massif: none; number of vegetation types: 2—tall, upland scrub on rock and medium, evergreen montane forest; date of first herpetological collection-exploration: 1981.

Herpetofauna: 2 species, 1 endemic: Reptilia (2)—*Anolis nitens* ssp., *Neusticurus* sp. B*.

Pertinent literature: Gorzula and Señaris 1998.

Remarks: This tepui is included in the Jaua Sarisariñama subdistrict primarily because of geography, but as noted by Huber (1995a), this serranía is so poorly known that placement is preliminary. Julian Steyermark gave McDiarmid a specimen of *Neusticurus* that he had collected on Marutaní in 1981. Huber (1995a) noted that the toponymy for this area is unreliable; other names for the mountain range (cerro or serranía) include Urutaní, Pia-Zoi, Piazoi, and Pia Soi. We use the name *Anolis nitens* ssp. rather than *Anolis eewi* (contra Gorzula and Señaris 1998; see appendix 18.3). As currently known, Marutaní is primarily granitic, with little or none of the Roraima sandstone characteristic of tepuis.

Jaua Massif

Cerro Jaua

Venezuela (Bolívar); latitude $\times$ longitude: 04°48′ N 64°26′ W (Paynter 1982; USBGN 1961); maximum elevation: 2,250 m; summit area: 625.62 km²; slope area: 482 km²; massif: Jaua; number of vegetation types for massif: 4—low, high-tepui scrub on peat and rock; tubiform, shrubby highland meadows on rock and peat; medium, evergreen montane forest; and medium, evergreen upper-montane forest; date of first herpetological collection-exploration: 1967.

Herpetofauna: 14 species, 8 endemics: Amphibia (10)—*Colostethus ayarzaguenai**, *Colostethus* sp. A*, *Eleutherodactylus* sp. F*, *Hyalinobatrachium* sp. B*, *Hyla benitezi, Hyla minuta,* Hylid sp. G*, *Otophryne steyermarki, Stefania oculosa*, Stefania percristata**; Reptilia (4)—*Anolis nitens* ssp., *Arthrosaura tyleri, Neusticurus* sp. C*, *Prionodactylus phelpsorum.*

Pertinent literature: General: Orejas-Miranda and Quesada 1976. Herpetological: Barrio 1998, 1999; Donnelly et al. 1992; La Marca 1996; Lancini 1968; Señaris et al. 1996.

Remarks: Braulio Orejas-Miranda visited this tepui in 1974 as part of the Charles Brewer-Carías Expedition and collected several species, including *Prionodactylus phelpsorum.* Unfortunately, only a few species from this collection have been described. Señaris and Ayarzagüena collected at two sites in the central part of the tepui at 1,600 m during 10–12 June 1994. Coordinates listed for the type locality of *Colostethus ayarzaguenai* and *Stefania oculosa* are 04°49′55″ N 64°25′54″ W; those for that of *Stefania percristata* are 04°49′55″ N 64°25′59″· W. La Marca (1996) did not give an elevation for the type locality of *Colostethus ayarzaguenai;* however, he did list the collectors (Ayarzagüena and Señaris), the date (10 June 1994), and a set of coordinates. The latter match exactly those given by Señaris et al. (1996) in the description of *Stefania oculosa,* the type of which was collected at 1,600 m on 11 June 1994. Accordingly, we list 1,600 m as the elevation at the type locality of *Colostethus ayarzaguenai.* Other specimens of amphibians and reptiles have been collected incidental to botanical and ornithological interests on Jaua. Two major streams that make up the Río Marajano drain Jaua from its high point (~2,250 m) in the south to about 1,400 m in the north; gallery forests alternate with tepui meadows along the river on the south half of the tepui (Huber 1995a).

Cerro Sarisariñama

Venezuela (Bolívar); latitude × longitude: 04°30′ N 64°14′ W (Paynter 1982; USBGN 1961); maximum elevation: 2,350 m; summit area: 546.88 km²; slope area: 482 km²; massif: Jaua; number of vegetation types for massif: 4—tall, upland scrub on rock; tubiform, shrubby highland meadows on rock and peat; medium, evergreen upper-montane forest; and medium, evergreen montane forest; date of first herpetological collection-exploration: 1974.

Herpetofauna: 5 species, 2 endemics: Amphibia (2)—*Eleutherodactylus* sp. G*, *Stefania riae**; Reptilia (3)—*Prionodactylus phelpsorum, Tropidurus hispidus, Chironius* sp.

Pertinent literature: General: Brewer-Carías 1976, 1983; Mayr and Phelps 1967. Herpetological: Duellman and Hoogmoed 1984; Orejas-Miranda and Quesada 1976.

Remarks: Braulio Orejas-Miranda visited this tepui in 1974 with Charles Brewer-Carías and collected a few species of amphibians and reptiles. Although Lancini (1968) listed the type locality for *Prionodactylus phelpsorum* as Cerro Jaua, Gorzula and Señaris (1998) corrected the type locality to Sarisariñama. In this action, they accepted the correction provided by Steyermark (Steyermark and Brewer-Carías 1976, 180) for all material collected during the Phelps Expedition of 1967, which incidentally was the first expedition to use helicopters in support of scientific collecting on tepui summits (J. Steyermark, pers. comm., as cited in Huber 1995b). According to Steyermark, who was the botanist on the Phelps Expedition, all botanical specimens were collected from the western summit of Sarisariñama, not Jaua as originally reported; apparently the expedition never reached the real Meseta Jaua. Consequently, one assumes that the holotype of *Prionodactylus phelpsorum* also was from Sarisariñama. As pointed out by Myers and Donnelly (2001), this was confirmed by Phelps (1977), who changed the Jaua locality to Cumbre Occidental, Meseta de Sarisariñama, at approximately 04°45′ N 64°26′ W (new coordinates seem unlikely according to a map in Huber 1995a, fig. 1.29). The *Chironius* sp. record is based on a piece of a shed skin.

DUIDA-MARAHUAKA SUBDISTRICT

Duida-Marahuaka Massif

Cerro Duida

Venezuela (Amazonas); latitude × longitude: 03°25′ N 65°40′ W (Paynter 1982; USBGN 1961); maximum elevation: 2,358 m; summit area: 1,089.00 km²; slope area: 715 km²; massif: Duida-Marahuaka; number of vegetation types for massif: 7—broad-leaved, shrubby highland meadows on peat; broad-leaved, shrubby high-tepui meadows on peat and rock; low to tall upland and high-tepui scrub; low evergreen high-tepui forest; low evergreen upper-montane forest; medium evergreen montane forest; tall evergreen lower-montane forest; date of first herpetological collection-exploration: 1928.

Herpetofauna: 15 species, 10 endemics: Amphibia (7)—*Cochranella duidaeana**, *Colostethus shrevei, Dischidiodactylus duidensis**, *Eleutherodactylus* sp. K*, Hylid sp. F*, *Stefania goini, Tepuihyla aecii**; Reptilia (8)—*Anolis nitens* ssp., *Arthrosaura tyleri, Neusticurus tatei**, *Riolama* sp. C*, *Atractus duidensis**, *Atractus riveroi**, *Chironius exoletus, Thamnodynastes duida**.

Pertinent literature: General: Mayr and Phelps 1967; Tate and Hitchcock 1930. Herpetological: Ayarzagüena 1992; Ayarzagüena et al. 1992a, 1992b; Burt and Burt 1931; Duellman and Hoogmoed 1984; Myers and Donnelly 1996; Rivero 1966, 1968; Roze 1961; Señaris et al. 1996; Vanzolini and Williams 1970; Van Devender 1969.

Remarks: *Stefania goini* is endemic to the Duida Massif. We follow La Marca (1996), who considered *Colostethus shrevei* to be endemic to the Duida Massif and surrounding region, in contrast to the earlier treatment of Hoogmoed (1979a), who assessed the species as peripheral amazonian. The record for *Riolama* sp. C is based on a partially digested lizard removed from the stomach of the holotype of *Thamnodynastes duida* (Myers and Donnelly 1996). *Stefania marahuaquensis* and *Neusticurus racenisi* are known from intermediate elevations (near 600 m) on Cerro Duida and at higher elevations on Marahuaka but are not included in this analysis pending their collection from elevations above 1,500 m. Duida and Marahuaka are joined by upland, but Huachamacari is separated from them by a considerable expanse (10–20 km) of lowland (< 500 m) tree and shrub savanna that apparently does not flood. A map and information about the geography and habitats on the southern portion of the summit are available in Tate and Hitchcock (1930). Myers and Donnelly (1996) described the area near the type locality of *Thamnodynastes duida* as a broad ravine at 2,015 m that lies between peaks 7 and 16 on the southeastern side of the south end of the massif at approximate coordinates of 03°20′ N 65°35′ W.

Cerro Huachamacari

Venezuela (Amazonas); latitude × longitude: 03°48′ N 65°46′ W (Paynter 1982; USBGN 1961); maximum elevation: 1,900 m; summit area: 8.75 km²; slope area: 60 km²; massif: Duida-Marahuaka; number of vegetation types for massif: 7—broad-leaved, shrubby highland meadows on peat; broad-leaved, shrubby high-tepui meadows on peat and rock; low to tall upland and high-tepui scrub; low evergreen high-tepui forest; low evergreen upper-montane forest; medium evergreen montane forest; and tall evergreen lower-montane forest; date for first herpetological collection-exploration: 1992.

Herpetofauna: 2 species, 1 endemic: Amphibia (2)—*Hyla aromatica**, *Stefania goini*.

Pertinent literature: General: Mayr and Phelps 1967. Herpetological: Ayarzagüena and Señaris 1993; Rivero 1966; Señaris et al. 1996.

Remarks: *Stefania goini* is endemic to the Duida-Marahuaka Massif. Huachamacari is the smallest, lowest, and least studied of the three tepuis in the massif.

Cerro Marahuaka

Venezuela (Amazonas); latitude × longitude: 03°34' N 65°27' W (Paynter 1982; USBGN 1961); maximum elevation: 2,800 m; summit area: 121.00 km²; slope area: 325 km²; massif: Duida-Marahuaka; number of vegetation types for massif: 7—broad-leaved, shrubby highland meadows on peat; broad-leaved, shrubby high-tepui meadows on peat and rock; low to tall upland and high-tepui scrub; low evergreen high-tepui forest; low evergreen upper-montane forest; medium evergreen montane forest; and tall evergreen lower-montane forest; date for first herpetological collection-exploration: 1950.

Herpetofauna: 8 species, 6 endemics: Amphibia (7)—*Colostethus brunneus, Colostethus shrevei, Dischidiodactylus colonnelloi**, *Eleutherodactylus* sp. T*, *Hyla inparquesi**, *Metaphryniscus sosai**, *Tepuihyla luteolabris**; Reptilia (1)—*Thamnodynastes marahuaquensis**.

Pertinent literature: Ayarzagüena 1983; Ayarzagüena and Señaris 1993; Ayarzagüena et al. 1992a, 1992b; Duellman and Hoogmoed 1984; Gorzula and Ayarzagüena 1995; Rivero 1961, 1966; Señaris et al. 1994, 1996.

Remarks: A common, alternate spelling for this tepui is Marahuaca. The frogs *Eleutherodactylus marmoratus* and *Stefania marahuaquensis,* and the lizard *Neusticurus racenisi* have been reported from elevations near 1,200 m on Cerro Marahuaka; they are excluded from our analysis pending their collection from higher elevations. Also, the snake *Atractus riveroi* has been collected at Temiche, a camp on the southern slope of Marahuaka at 1,300 m (Roze 1961); also given as 4,050 ft. (1,234 m) by Rivero (1961). We did not include the *Atractus* in the Marahuaka fauna because of the elevation; it also is known from Cerro Duida. Broad-leaved, shrubby high-tepui meadows on peat and rock are known only from Marahuaka, and low evergreen high-tepui forest is restricted to the summit. Molina and Señaris (2001) recently described *Riolama uzzelli* from two sites on Cerro Marahuaka (1,850 and 2,600 m). We do not know if *R. uzzelli* is the same as *Riolama* sp. C from Duida, but we treat them as the same species in this chapter.

Cerro Yapacana

Venezuela (Amazonas); latitude × longitude: approximately 03°42' N 66°45' W (Paynter 1982); maximum elevation: 1,300 m; summit area: 10.50 km²; slope area: 38 km²; massif: none; number of vegetation types: 1—low evergreen montane forest; date of first herpetological collection-exploration: 1970.

Herpetofauna: 1 species, 1 endemic: Amphibia (1)—*Minyobates steyermarki*.

Pertinent literature: General: Mayr and Phelps 1967. Herpetological: Gorzula and Señaris 1998; Rivero 1971.

Remarks: Huber (1995a) considered Yapacana as an upland rather than highland tepui because of its low maximum elevation (given as 1,345 m in Mayr and Phelps 1967, map 1). Although we understand the need for specific definitions and elevational limits for physiographic units, we include Yapacana here because it is a sandstone mountain, was considered a low-elevation tepui by Huber (1995a), and has had some herpetological attention. We provisionally place it in the Duida-Marahuaka Subdistrict strictly on the basis of its geographic position. Additional samples of the herpetofauna should provide some guidance relative to the future treatment of Cerro Yapacana and its fauna. Paynter (1982) listed an approximate locality for this mountain. Gorzula and Señaris (1998) listed 03°42′ N 66°46′ W as the coordinates for the dendrobatid collection.

Southern Pantepui District

SIERRA TAPIRAPECÓ

Cerro Tamacuari

Venezuela (Amazonas); latitude × longitude: 01°13′ N 64°42′ W (Myers and Donnelly 1997); maximum elevation: 2,340 m; summit area: < 0.01 km²; slope area: not available; massif: none; number of vegetation types: 1—low evergreen upper-montane forest; date of first herpetological collection-exploration: 1989.

Herpetofauna: 11 species, 6 endemics: Amphibia (7)—*Colostethus tamacuarensis*, *Eleutherodactylus avius*, *Eleutherodactylus cavernibardus*, *Eleutherodactylus memorans*, *Hyalinobatrachium crurifasciatum*, *Hyla benitezi*, *Stefania tamacuarina*; Reptilia (4)—*Anolis nitens* ssp., *Arthrosaura synaptolepis*, *Neusticurus racenisi*, *Thamnodynastes pallidus*.

Pertinent literature: Myers and Donnelly 1997.

Remarks: This locality, also known as Pico Tamacuari (Mayr and Phelps 1967, map 1), is a conical granite mountain in the poorly explored Sierra Tapirapecó (Huber 1995a), which, as currently known, has little or none of the Roraima sandstone characteristic of tepuis. The immediate region near Cerro Tamacuari has received reasonable attention (Myers and Donnelly 1997), but the Sierra Tapirapecó, which extends for about 100 km along the Venezuelan–Brazilian border, is unknown from a herpetological perspective.

NEBLINA-ARACAMUNI MASSIF

Cerro Aracamuni-Avispa

Venezuela (Amazonas); latitude × longitude: 01°14′ N 65°26′ W (USBGN 1961); maximum elevation: 1,600 m; summit area: 238 km²; slope area: 658 km²; massif: Neblina-Aracamuni; number of vegetation types for massif: 5—low, high-tepui scrub on peat; broad-leaved, shrubby upland and high-tepui meadows on peat; broad-leaved, shrubby upland meadows on peat; medium to tall evergreen montane forest; and low evergreen upper-montane forest; date of first herpetological collection-exploration: 1987.

Herpetofauna: 1 species, 1 endemic: Amphibia (1)—*Cochranella riveroi**.

Pertinent literature: Ayarzagüena 1992.

Remarks: Cerro Avispa is the larger, more southern section of this largely unexplored system. This tepui complex is virtually unknown and compared with its neighbors (Duida to the north and Neblina to the south) deserves much more herpetological work.

Cerro de la Neblina

Venezuela (Amazonas) and Brazil (Amazonas); latitude × longitude: 00°48′ N 65°59′ W (Brewer-Carías 1978b); maximum elevation: 3,014 m; summit area: 235 km²; slope area: 857 km²; massif: Neblina-Aracamuni; number of vegetation types for massif: 5—low, high-tepui scrub on peat; broad-leaved, shrubby upland and high-tepui meadows on peat; broad-leaved, shrubby upland meadows on peat; medium to tall evergreen montane forest; and low evergreen upper-montane forest; date of first herpetological collection-exploration: 1984.

Herpetofauna: 26 species, 20 endemics: Amphibia (17)—*Adelophryne* sp. A*, *Adelophryne* sp. B*, Bufonid sp.*, *Cochranella* sp.*, *Colostethus* sp. B*, *Eleutherodactylus* sp. L*, *Eleutherodactylus* sp. M*, *Eleutherodactylus* sp. N*, *Eleutherodactylus* sp. O*, *Eleutherodactylus* sp. P*, *Eleutherodactylus* sp. Q*, *Eleutherodactylus* sp. R*, Hylid sp. C*, Hylid sp. D*, Hylid sp. E*, Leptodactylid sp.*, *Leptodactylus* sp.*; Reptilia (9)—*Arthrosaura synaptolepis, Mabuya* sp., *Phenacosaurus neblininus*, Riolama* sp. A*, *Riolama* sp. B*, *Dipsas* cf. *indica, Dipsas* sp., *Leptophis ahaetulla, Leptophis* sp.

Pertinent literature: General: Brewer-Carías 1978b, 1988; Maguire 1955; Mayr and Phelps 1967; Ort 1965; Givnish et al. 1986; Willard et al. 1991. Herpetological: Donnelly et al. 1992; McDiarmid and Paolillo 1988; Myers et al. 1993.

Remarks: Also known as the Sierra de la Neblina (Huber 1995a); contains Pico da Neblina (3,014 m), the highest mountain in Brazil and the Guayana Highlands. Maguire and Wurdack (1959) listed 00°59′26″ N 65°58′05″ W for their summit camp at 1,800 m on the northeastern part of the complex. Coordinates and elevations for the camps established during the Neblina Expedition (Brewer-Carías 1988) are as follows: camp 1—00°52'10" N 66°05'25" W, 1,820 – 1,880 m elevation; camp 2—00°50'00" N 66°58'48" W, 2,085–2,100 m; camp 3—00°54'10" N 66°03'50" W, 1,820 m; camp 4—00°56'20" N 65°56'50" W, 2,085–2,770 m; camp 5—00°49'10" N 66°00'02" W, 1,250 m; camp 6— 00°52'20" N 65°56'20" W, 2,000 m; camp 7—00°50'40" N 65°58'10" W, 1,730 – 1,850 m; camp 8—00°50'00" N 65°57'30" W, 2,300–2,400 m; camp 9— 01°00'00" N 65°53'00" W, 1,780–1,830 m; camp 10—00°54'40" N 66°02'30" W, 1,690 m; camp 11—00°51'45" N 65°58'52" W, 1,390–1,515 m; camp 12— 00°47'50" N 66°00'10" W, 1,950–2,000 m.

Appendix 18.5   History of Exploration

The isolated nature of the Guayana Highlands has made exploration of the tepuis and sampling of its herpetofauna a difficult and sporadic endeavor. General accounts of some nineteenth-century explorations serve to introduce the region, and when combined with reports resulting from the significant, though primarily botanical and ornithological, expeditions made during the mid-twentieth century and those from recent interdisciplinary field efforts when the collection of amphibians and reptiles was a primary goal, they provide ample material for our review. We briefly review the history of exploration to augment understanding of the region and to provide a historical context for our treatment of the Pantepui herpetofauna.

Scientific exploration of the interior Guayana region began in the eighteenth century and was concentrated primarily in lowland areas. Huber and Wurdack (1984) and Huber (1995b) provided interesting accounts of the history of botanical exploration in the Venezuelan Guayana. Prominent among the early explorers were the German naturalist Alexander von Humboldt and the French botanist Aimé Bonpland, who, as near as we can tell, may have been the first scientists to actually see a tepui. In May 1800, during their epic explorations of the upper Orinoco and Cassiquiare rivers, Humboldt (1907) described the "solemn grandeur" of Cerro Duida as seen from the small mission settlement of Esmeralda (fig. 18.1). Another remarkable explorer was Robert H. Schomburgk, who reached Esmeralda, which by then had dwindled to a single family, 39 years after Humboldt. His travels (Schomburgk 1841) from Roraima to Duida (approximately 600 km of straight-line distance) took him through parts of the Serra Pacaraima (in northern Roraima, Brazil), along the southern reaches of Sarisariñama, through the northern portions of the Sierra Parima, and along the eastern side of Cerro Marahuaka (fig. 18.1). Schomburgk also was impressed by the view of Duida and noted that Nature had remained the same. Schomburgk (1841, 245) wrote, "Duida still raises its lofty summit to the clouds, and flat savannahs, interspersed with tufts of trees and the majestic Mauritia palm stretch from the banks of the Orinoco to the foot of the mountains beyond, giving to the landscape that grand and animated appearance which so much delighted Humboldt." More than 11 years later, Richard Spruce reached Esmeralda by boat from the Cassiquiare. In December 1853, he described Duida from about the same spot as had Humboldt and Schomburgk; Spruce provided an illustration of Cerro Duida (Spruce 1908). As far as we know, none of these explorers actually collected material on Duida.

Exploration of specific tepuis became the goal of explorers and scientists in the mid- to late 1800s. At that time, travel in this part of the world and especially to the Guayana Highlands was arduous and time-consuming. Field logistical support for these early expeditions usually was nonexistent; nearly everything had to be carried, initially by river and then on foot, sometimes over considerable distances. Food frequently was a major limiting factor, and success often depended on weather, the knowledge and hunting skills of local guides, and luck. Accounts of these early explorations make for exciting reading (e.g., Schomburgk 1841; Quelch 1921). Not surprisingly, the first tepuis to be explored were located close to major rivers, and mostly appeared on some general maps. Specimens were collected in the surrounding lowlands and on the slopes of a few tepuis, but access to summits often was limited or impossible. Conditions on tepuis whose summits were reachable were harsh, and relatively little time was available for collecting. Even so, scientific discoveries were frequent and important enough to convince a few scientists and explorers that the tepuis deserved a lot more attention.

An intensive period of exploration of the tepuis, driven primarily by a few prominent scientists/explorers (e.g., George Tate, William H. Phelps Sr., William H. Phelps Jr., Julian Steyermark, Bassett Maguire, John Wurdack) associated with certain museums and botanical gardens, began around 1930. In addition to basic exploration and description of the geography and geology of this area, the goal of these individuals and their institutions was to amass collections of specific biotic groups adequate to serve as a baseline for future studies on the ecology and evolution of the flora and fauna of the tepuis. Most of these expeditions were major undertakings and sometimes involved weeks to months in the field. Although birds and plants were the target organisms for much of the early work, an occasional herpetological specimen found its way back to a museum. The resulting collections and publications set the stage for the next phase of scientific exploration, which began in the late 1960s. The goals of earlier investigators (i.e., to understand the biota of the region and elucidate the factors that played prominent roles in its evolution) continue to drive much of the science in the region.

Exploration since the early 1970s has focused on tepuis whose summits had never been explored or were poorly known. Fieldwork often involves multiple visits to specific tepuis, usually with helicopter support. In contrast to most previous work, these expeditions frequently involve botanists and zoologists, often with a major contingent of Venezuelan scientists. Recent expeditions have been sponsored by scientific societies, government agencies, private foundations, and other entities. Helicopters have enabled scientists to access formerly unexplored areas and have facilitated visits to multiple sites on each tepui, leading to broader coverage than was previously possible. Also, the significant decrease in travel time and improved logistical support meant that more sites could be visited and more time spent searching for organisms in field situations. Even though helicopters require considerable logistical support and fuel, are sometimes plagued by mechanical problems, and are especially susceptible to changes in local weather conditions, helicopter access allows scientists to commence fieldwork almost immediately upon arriving at a site. As a result, the numbers of scientific visits made and samples collected on tepuis in the past 20 years far exceed those made in all the previous explorations combined, and the number of tepuis visited has increased dramatically. This increased activity is clearly reflected in the dramatic rise in the species accumulation curve for amphibians and reptiles from Pantepui (fig. 18.6). Even so, most tepuis have been sampled relatively few times, and several are completely unknown herpetologically (appendix 18.1).

We review the exploration of Roraima in more detail not only because of its long history but also because its patterns of investigation reflect those characteristic of most of the highlands of Pantepui. In writing about explorations of interior British Guiana, Sir Walter Raleigh (1848) referred to a mountain of Christall (Crystal) with a large waterfall plummeting from its summit. He indicated that he had seen this from a great distance and described the noise from the falls. We suspect that Raleigh may have been reporting what someone told him rather than describing what he actually saw. Warren (1973) suggested that if true, Raleigh's account almost certainly represented the first sighting of Roraima by someone other than an Amerindian.

The first known description of the environs of Roraima was by the German geographer Robert H. Schomburgk, who explored the interior of Guayana for the British Crown. He reached the southern flank of Roraima in November 1838 and for the next 25 days explored the surrounding region. On his approach from the south, Schomburgk reported visiting a small hillock (about 35 miles south of Roraima), which his guides called the Crystal Mountain. Here, a layer of short and greatly weathered crystals, an intrusive characteristic of Roraima Sandstone, covered the surface. According to the local Arecunas, some Portuguese had previously carried away crystals that were clear as water and four to five inches long. Those sorts of exposed strata, several of which occur on the summit, account for Roraima being known as the Crystal Mountain. In

1842 Robert Schomburgk and his brother Richard returned to Roraima and made extensive plant collections in the region. The results of that work were published in *Reisen in Britisch-Guiana* (Schomburgk 1847–48) and included some of the first descriptions of plants and animals from that part of the world. The only reptile from the region specifically mentioned by Schomburgk (1841) was a rattlesnake (*Crotalus durissus*) that bit one of his men. Vivid accounts by other explorers and sporadic descriptions of new taxa based on samples supplied by purveyors of natural history specimens sparked the curiosity of key individuals in the scientific community, especially in British Guiana. Over the next 40 years, a number of explorers and collectors tried to get to Roraima with the goal of reaching the top. Many of them approached Roraima from the north and east and concentrated their efforts along the eastern and southern slopes of the mountain. Nearly all pronounced the summit inaccessible (Warren 1973), and a few even suggested that a balloon might provide the only access to the top (Whitley 1884). Almost all of those explorers and scientists mentioned that a lack of provisions cut short the time available to explore Roraima's vertical faces, and some indicated that dense forests along the north side made that approach especially challenging. Most of them also commented that poor weather and dense cloud cover often made visual inspections difficult. In September 1883, Henry Whitley, who had spent considerable time over the past three years collecting birds in the environs of Roraima, spied a spot on the southwest side where the vertical cliff seemed to have broken away and possibly offered a route to the top. He learned from some of the locals who had accompanied Schomburgk some years earlier that Schomburgk also had seen the route but had not explored it, thinking the precipice halfway up would be insuperable. The ledge depicted in Whitley's (1884) drawing (fig. 18A.1) ascends to the left at a moderate slope across part of the southwest face of Roraima's cliffs; it also seems to have been depicted on the face of the mountain shown to the right in Schomburgk's (1841, 207) drawing (fig. 18A.1). Whitley's assistants spent nearly six days cutting a trail through dense undergrowth and around large boulders to the base of the southwest cliff face near where he thought passage might begin. Unfortunately, dense mist and a misjudgment relative to the point of access prevented further exploration.

Word about a possible route to the summit apparently spread quickly through British Guiana and, after Whitley's account of his explorations were published in the *Proceedings of the Royal Geographical Society* in August 1884, through London as well. The year after Whitley's explorations, a German orchid collector named Seidel, who was on his second trip to Roraima, met Everard Im Thurn, curator at the Georgetown Museum, and Harry I. Perkins, assistant Crown surveyor, on the southwest side of Mount Roraima. Both groups had departed from the mouth of the Mazaruni River on 17 October; Seidel followed the northern route, while Im Thurn and Perkins headed south on the

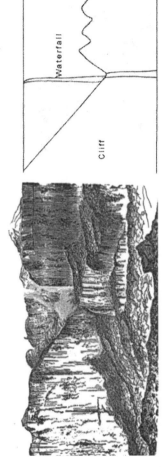

Figure 18A.1 (*Top*) Reproduction of Whitley's (1884) drawing of the southwestern face of Mount Roraima, showing the ledge that Im Thurn subsequently used to reach the top. (*Lower left*) Sketch (from Warren 1973) of the route along the ledge to the summit; this image is similar to a photograph by McConnell (1916). (*Lower right*) Line drawing (adapted from Clementi 1920) depicting the path to the top.

Essequibo River (fig. 18A.2). The primary goal of the Im Thurn and Perkins Expedition, which had been supported in part by the Royal Geographical Society, was to reach Roraima's summit. Im Thurn and Perkins traveled up the Essequibo and Potaro rivers to Chinebowie (= Chenapou) and then overland through forest and savanna, crossing the Ireng, Cotinga, and Arapu rivers to Roraima. They arrived at the small village of Teruta, 3–4 miles south of the Roraima and Kukenán massifs on 4 December, one day after Seidel. On 7 December, they moved closer to Roraima and built a hut near Seidel's. For the next week they collected plants and explored the region. On 14 December, with extensive cutting of new trail, they intercepted Whitley's trail near the base of the cliff and had a good view of the ledge (fig. 18A.1). After cutting additional trail to the ledge and waiting a few days for the weather to clear, they attempted on the bright morning of 18 December 1884 to reach the top (Perkins 1885; Im Thurn 1885). The trail passes over three rounded spurs and then descends to a point where the waterfall from above contacts the ledge. The first part of their ascent (about two-thirds of the distance) was slippery because of the recent rains and reportedly especially treacherous in many areas because the cut and previously trodden plants were slippery. In many places they described climbing on all fours over masses of vegetation dense enough to bear their weight and over and under high-piled rocks and tree stumps covered with moss and filmy ferns. After walking about 150 yards beneath the waterfall, whose flow was relatively light on that particular morning, they made their way up a relatively steep (about 30°) rocky slope to the ledge and then through dense vegetation to the final rise, which was easier than the previous parts, and gradually merged onto the relatively barren summit.

The following excerpt from Im Thurn's account of 18 December (1885, 517) more than adequately captures the excitement and emotion of the moment:

> Up this part of the slope we made our way with comparative ease till we reached a point where one step would bring our eyes on a level with the top—and we should see what never had been seen since the world began; should see that of which, if it cannot be said all the world has wondered, at least many people have long and earnestly wondered; should see that of which all the few, white men or red, whose eyes have ever rested on the mountain had declared would never be seen while the world lasts—should learn what is on top of Roraima.
>
> Then the step was taken—and we saw surely as strange a sight, regarded simply as a product of nature, as may be seen in this world; nay, it would probably not be rash to assert that very few sights even as strange can be seen. The first impression was one of inability mentally to grasp such surroundings; the next that one was entering on some strange country of nightmares for which an appropriate and wildly fantastic landscape had been formed, some dreadful and stormy day, when, in their mid ca-

reer, the broken and chaotic clouds had been stiffened in a single instant into stone. For all around were rocks and pinnacles of rocks of seemingly impossible fantastic forms, standing in apparently impossible fantastic ways—nay, placed one on or next to the other in positions seeming to defy every law of gravity—rocks in groups, rocks standing singly, rocks in terraces, rocks as columns, rocks as walls and rocks as pyramids, rocks ridiculous at every point with countless apparent caricatures of the faces and forms of men and animals, apparent caricatures of umbrellas, tortoises, churches, cannons, and of innumerable other incongruous and unexpected objects. And between the rocks were level spaces, never of great extent, of pure yellow sand, with streamlets and little waterfalls and pools and shallow lakelets of pure water; and in some places there were little marshes filled with low scanty and bristling vegetation. And here and there, alike on level space and jutting from some crevice in the rock, were small shrubs, in form like miniature trees, but all apparently of one species. Not a tree was there; no animal life was visible, or it seemed, so intensely quiet and undisturbed did the place look, ever had been there. Look where one would, on every side it was the same; and climb what high rock one liked, in every direction as far as the eye could see was the same wildly extraordinary scenery.

Other than describing small fleeting masses of clouds, Im Thurn made no mention of the weather on the summit, but Perkins (1885, 532) noted that shortly after arriving on top, the mist closed in and prevented them from exploring much of the summit. They returned to the hut that afternoon, where they remained preparing specimens and collecting on the slopes. On 24 December they returned to Teruta, where they spent Christmas day, departing for home on the following day. They arrived at the junction of the Mazaruni and Essequibo rivers on 28 January. Mr. Seidel left Roraima on 28 December and, following the northern route, arrived at the same river junction two days later than Im Thurn and Perkins.

The accounts of this and other explorations in the Roraima region that appeared in the publications of the Royal Geographic Society and elsewhere served to feed the imagination of many Londoners. The following editorial appeared in *The Spectator* in April 1877 and perhaps reflects some of the attitudes at the time: "Will no one explore Roraima and bring back the tidings which it has been waiting these thousands of years to give us? One of the great marvels of the mysteries of the Earth lies on the outskirts of one of our colonies—British Guiana—and we leave the mystery unsolved, the marvel uncared for." As Warren (1973) pointed out, many people at the time were attracted by the idea that some place on earth might still harbor remnants of prehistoric life. This notion caught on when someone suggested that Roraima had been cut off from the rest of the world for millions of years and that life on its

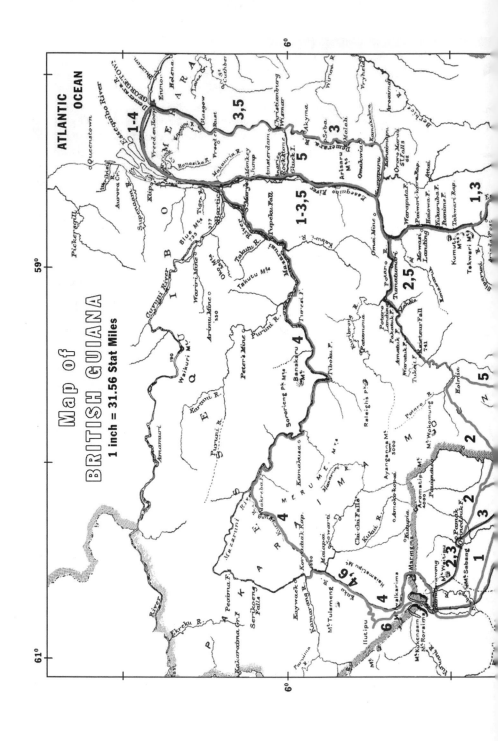

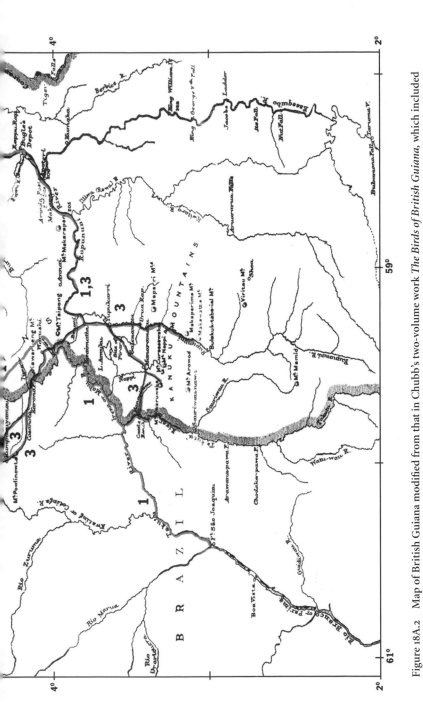

Figure 18A.2   Map of British Guiana modified from that in Chubb's two-volume work *The Birds of British Guiana*, which included McConnell's (1916) and Quelch's (1921) accounts of their expeditions. The primary routes used by explorers in their approaches to Roraima are indicated by number, as follows: 1, Schomburgk (1838); 2, Im Thurn and Perkins (1884); 3, McConnell and Quelch (1894); 4, McConnell and Quelch (1898); 5, Clementi party (1916); and 6, Warren group (1971). The latter has become known as the Waruma route.

summit might be suspended in evolutionary time. Sir Joseph Hooker's comment in 1884 (Warren 1973) that a detailed examination of the top would provide some interesting results and probably would show a flora that was similar to that which had existed in the old times, only raised expectations in some people's minds; perhaps dinosaurs and pterodactyls lived on Roraima's summit! Some say that publications and public presentations at Royal Geographical Society meetings around this time provided ample material for Arthur Conan Doyle to write *The Lost World* (Doyle 1912).

Publications resulting from ornithological and botanical collections from the area served as an incentive to others to get to Roraima and collect their favorite group. This scientific inquisitiveness, coupled with a desire to learn more about the native populations in the area, or to establish a political and economic presence in a region where country boundaries were not well established, resulted in a considerable increase of exploration of the region. From a herpetological perspective, the most important expeditions were those led by John J. Quelch, a curator of zoology at the Georgetown Museum, and Frederick V. McConnell, a collector of birds in British Guiana. The itineraries of their two expeditions were published in the preface of volume 1 (McConnell 1916) and the introduction of volume 2 (Quelch 1921) of Chubb's *The Birds of British Guiana*. The first expedition was sponsored by the Royal Agricultural Society of British Guiana and led by Quelch; it departed Georgetown on 7 July 1894. The group followed the southern route to Roraima (fig. 18A.2), traveling south up the Essequibo and Rupununi rivers, and after a few side trips reached Kwaimatta on 6 September. That village of about 200 inhabitants was located at the eastern edge of extensive savannas and served as the base of operations for the next six weeks. On 16 October they started north on foot into Brazil and then Venezuela on trails that generally tracked the courses of the Ireng, Cotinga, and Arabato rivers. On 3 November they reached Kamaiwawong, a little village at the end of the valley between Roraima and Kukenaam. After two days of exploration and clearing trails through the dense bush and forest on Roraima's slope, they established a rough camp near the base of the cliff. The next morning they reached the cliff and began moving up the ledge. The descent into the gully with the waterfall was made in a sitting position, and the climb out was oftentimes on all fours. From this point the climb to the summit, though initially steep, was not arduous. Once on top, Quelch and one worker decided to stay the night and maximize the time for collecting. McConnell returned to the camp for blankets, food, and photographic equipment and returned the next day. All three spent the second night on the summit and returned to Kamaiwawong the following day. Shortly thereafter, they returned to Georgetown. Although time on top was short, Quelch and McConnell made some collections, among which were several specimens of a tepui toad that Boulenger (1895a) described as a new genus and species, *Oreophryne quelchii*, later changed to *Oreophrynella quelchii*, for reasons of preoccupation (Boulenger 1895b).

The second expedition to Roraima was led by McConnell and was organized specifically to explore the mountain and spend more time on the summit. For various reasons, McConnell and Quelch decided to follow the northern route going west up the Mazaruni River (fig. 18A.2). They left Bartika at the junction of the Mazaruni and Essequibo rivers on 20 August 1898 and, making reasonable time, reached the headwaters of the Kurubung River in slightly less that three weeks. From this point they began a seven-day trek on bad trail over the Pacaraima Mountains. After descending again to the Mazaruni River, they obtained small woodskins (bark-covered canoes) and continued upstream via the Kako River and then a smaller creek, the Aruparu, for three days to a point where the overland trek began. After several (3–5) days moving east around Marima Mountain and then south and west primarily through forest and small savannas, they approached Roraima's eastern flank between "other smaller Roraimas" (likely Wei-assipu, Appokailang, and Yakontipu) to the north and Mount Weitipu (Uei) to the south. From there they headed southwest toward the village of Kamaiwawong at the base of Roraima, which they reached early in October, 40 days after their departure.

Preparations for the summit ascent began immediately. Old trails were overgrown and had to be cleared. Some new trails were cut, and a temporary camp was established at the cliff base. On the fourth day they reached the top. The plan was for Quelch to spend as much time as possible collecting and exploring on Roraima, while McConnell would take a series of photographs on the summit and then concentrate his efforts on the forested slopes and savannas below, making the best collection possible, chiefly of birds. On this trip they brought a small tent and were able to establish a camp near their previous camp. Over the next two days McConnell took photographs of the terrain and vegetation (fig. 18A.3) and collected several specimens of the only bird seen on the summit (subsequently named *Zonotrichia macconnelli*), after which he went down off the mountain. Quelch on the other hand spent the next nine days on top, completely caught up in making collections and exploring the mountain. When discovery of new materials slowed, Quelch decided to join McConnell below and work the sunny slopes of Kukenaam, the agricultural clearings, and possibly the unexplored gorge between the two mountains. Unfortunately, the cold, wet climate on top had taken its toll. Quelch was suffering from stiffness in the back and joints and had a hard time descending the steep slippery trails. On top of this, he had to cover the final two miles back to the village in a cold, drenching rain. He spent the next week nearly helpless in his hammock and a second week regaining strength for the return trip. Finally, at the end of October they departed Roraima, retracing their steps to Bartica.

Among the herpetological material collected by Quelch on Roraima's top were additional specimens of *Oreophrynella quelchii* and a single lizard later described as *Prionodactylus* (= *Riolama*) *leucostictus*. McConnell and others collected herpetological material from the base of Roraima at approximately

Figure 18A.3   Two photographs taken by McConnell (1916) on the summit of Roraima in 1898. McConnell near sculptured rocks (*left*) and with Quelch and assistant at their primitive camp (*right*).

3,500 ft. (~1,067 m). Among these were unique specimens of the following taxa: *Neusticurus rudis, Oreophrynella macconnelli, Hylodes* (= *Eleutherodactylus*) *marmoratus,* and *Otophryne robusta* (Boulenger 1900a, 1900b; see fig. 18A.4). All are recognized species today. Quelch and McConnell were especially successful in adding to the herpetological knowledge of the region. In the 100 years following their expeditions, only one other species of amphibian (*Eleutherodactylus* sp.) has been collected on the summit and two others (*Colostethus praderioi* and *Colostethus roraima*) taken from the southwestern slopes of Roraima (appendix 18.4).

During the next twenty years, several other expeditions reached the summit of Roraima, including three to document national borders (see Warren 1973 and Huber 1995b for brief reviews). Although of minimal herpetological interest, one of the better-documented trips during this time was that undertaken by Cecil Clementi, his wife, and John C. Menzies, a prospector and rancher. Their goal was to reach the interior tablelands of British Guiana and to see the forests and savannas of the interior and perhaps even to make their way to "Mount Roraima, of which the residents in British Guiana hear so much and see so little" (Clementi 1920). The Clementi party traveled by steamer up the Demerara River south from Georgetown, crossed to the Essequibo by railroad, and continued up the Essequibo River by launches to a major cataract on the Potaro River. Here they met Menzies and continued up the Potaro; after several portages they reached Holmia, a place above Kaieteur Falls. From there they walked southwest through the forests, then across the tablelands and savannas of the eastern part of the country through the village of Puwa to the Ireng River, where they turned to the northwest and continued toward Roraima (fig. 18A.2). On the day after their arrival at Kamaiwawong, the same village that Quelch and McConnell had used as their base of operation at Roraima, they climbed to the summit and spent that afternoon, night, and part of the next day on top. On the day after their descent to the village, they departed for the Ireng River, following a slightly different route, and from there retraced their steps along the previously traversed Potaro route. The entire trip from Georgetown to Roraima and back following the Potaro route was accomplished in 46 days, which was quite an undertaking and, as pointed out by Quelch (1921, lxxviii), "to say the least of it, is an exceedingly fine feat for a lady." The account of this exploration was effectively captured by Mrs. Clementi's (1920) book titled *Through British Guiana to the Summit of Roraima.*

The next major exploration of Roraima probably was that accomplished by the Lee Garnett Day Roraima Expedition of the American Museum of Natural History, which was led by the American mammalogist G. H. H. Tate (Tate 1928). Tate and his party reached Roraima on 21 October 1927, traveling by boat up the Río Branco from Brazil. Tate (1930) lamented the common practice of the Arekuna Indians of setting fire to grasslands in their territory; he noted that

Figure 18A.4   Reproduction of the illustration that accompanied Boulenger's (1900a, 1900b) descriptions of specimens collected by Quelch and McConnell on their second Roraima expedition in 1898. *Oreophrynella quelchii* is shown at lower left, and *Riolama leucosticta,* the smaller of the two lizards, is shown on the outside left and right.

during a period of extreme drought some two years before their arrival, a fire started on the savanna had burned onto the slopes of the mountain, destroying the greater part of the forests therein (Means 1995 discussed the role of fire in the region). Tate's party was able to establish three camps in the upper savannas and in patches of forest on the slopes and one on top of Roraima. In total, they spent about 80 days in the region, 13 of which were on the summit. In the history of herpetology at the American Museum of Natural History, Myers (2000) included three previously unpublished pictures from the Tate Expedition, two of which were taken on top. The Tate party returned through British Guiana by following a trail that started on the southeastern side of Roraima near the village of Arabopo, headed south around the western side of Weitipu (Uei), and from there continued east through open savannas and across the upper reaches of the Cotinga and Ireng rivers. Soon they crossed the high sandstone divide (southern part of the Pacaraima Mountains) that separates the upper drainages of the Ireng and Potaro rivers north of Mount Kowatipu and descended to the Chenapowu (= Chenapou) River, from whence they traveled by canoe down to the Potaro River and the Kaieteur Falls area.

In the 1970s, several groups explored Roraima and its surroundings and made important collections of amphibian and reptiles. The British Roraima Expedition, led by Adrian Warren (1973), explored the northern slopes of Roraima, following what has come to be known as the Waruma route (fig. 18A.2). They reached Kamarang, where the Kako and Kamarang rivers join the Mazaruni, by plane and established their main camp. From there they moved up the Kako by longboat to a previously established base camp on a savanna at Makuripai. From that camp they moved farther up the Kako to the Waruma River and then headed due south toward Roraima, establishing a series of camps along the Waruma River and onto the northern slopes. They reportedly reached a high point of 7,700 ft. (2,346 m) on the north slope before being forced to turn back. They collected more than 2,500 scientific specimens of plants and animals, including more than 130 amphibians and 28 reptiles. At camps above 3,000 ft. (914 m), they collected specimens of most of the taxa first collected by McConnell and Quelch some 75 years earlier from the southwestern side of Roraima, including the frogs *Oreophrynella macconnelli, Otophryne robusta, Eleutherodactylus marmoratus,* and the lizard *Neusticurus rudis,* all described as new by Boulenger (1900a, 1900b). At localities later reported as lying between 1,430 and 1,480 m, Warren and his group collected two specimens each of hylid frogs that were described as *Hyla roraima* and *Hyla warreni* by Duellman and Hoogmoed (1992).

In the three decades since the Warren Expedition, other groups have explored the slopes of Roraima from the north; some of those expeditions were reviewed briefly by Gradstein (1985). Two expeditions are important from a herpetological perspective. Two years after the Warren explorations, a British climbing group following the Waruma route successfully reached the north side of Roraima and climbed to the summit (MacInnes 1976). Michael Tamessar collected anurans on that trip, including specimens of what were subsequently described as *Stefania roraimae* (Duellman and Hoogmoed 1984) and *Hyla roraima* (Duellman and Hoogmoed 1992). Early in 2001, scientists from the Smithsonian Institution and the University of Kansas conducted an ornithological survey of the north slope of Roraima, with funding from the National Geographic Society (Smithsonian Guyana Expedition, 2001). That group followed the same path and used some of the same camps established by Warren's party. In addition to birds, they collected several interesting specimens of amphibians and reptiles that remain to be studied. Thus, our knowledge of the slope fauna is gradually improving, but much work remains.

As was the case with other tepuis, the summit of Roraima received additional attention with the increase in the availability of helicopters for scientific exploration. In 1977 (January) and 1978 (February) Charles Brewer-Carías (1978a) led expeditions sponsored by the Ministerio del Ambiente y de los Recursos Naturales Renovables (MARNR) that visited Roraima. McDiarmid was the herpetologist on those expeditions and collected specimens of *Oreophrynella quelchii* and *Riolama leucosticta* on the summit. Field observations of the former species were reported by McDiarmid and Gorzula (1989). On a subsequent trip to Roraima, Brewer-Carías was stranded on the mountain for a day and night; during that episode he collected an undescribed species of *Eleutherodactylus* (table 18A.1).

In the last 20 years, several individuals and groups have visited Roraima, either by helicopter or on foot. Carlos Galán, a biologist with the Electrificación del Caroní project of the Corporación Venezolana de Guayana, reportedly has visited Roraima several times (Galán 2000); in April 1985 he collected a series of 33 *Oreophrynella quelchii* from the summit (Gorzula and Señaris 1998). The report of *Tepuihyla edelcae* from Roraima from the Galán collection (Galán 2000; Gorzula and Señaris 1998, 255) is probably an error; no other records of this species from Roraima are known, and Gorzula and Señaris (1998, 49) made no further mention of a Roraima specimen in their account of the species. Scientists working with the Fundación Terramar visited the summits of Roraima, Kukenán, and Ilú-tepui in 1989 (Huber 1995b), and it seems likely that herpetological specimens, if encountered, would have been collected. In August 1996, two species of dendrobatid frogs were collected along the southwestern cliff face on the trail to the summit. La Marca (1996) described two specimens, one from a quebrada at 1,950 m and another from 1,800 m, as *Colostethus praderioi*, and a single specimen taken in the trail at about 2,700 m (60–70 m before the summit) as *Colostethus roraima*. We suspect that other specimens from the slopes and summit of Roraima may reside in collections in Venezuela and elsewhere.

The expeditions and collectors who have contributed to our understanding of the herpetofauna of Roraima are summarized in the following chronological list of principal explorers and expeditions to Mount Roraima and vicinity, the year the expedition took place, and the nature of the exploration. Similar chronologies and patterns of investigation could be compiled for certain of the other, better-known tepuis (e.g., Auyán, Duida), and details can be garnered from references presented for specific sites in appendix 18.4. Data are derived from Huber (1995a), Perkins (1885), Warren (1973), and other sources.

Robert H. Schomburgk; 1838; discovery and exploration, some plant collecting on southern flank; 25 days in November.

Robert H. Schomburgk and M. Richard Schomburgk; 1842; extensive collections of plants and of some animals from the region.

Karl Ferdinand Appun; 1864; German botanist worked the eastern and southern flanks for a month; declared summit inaccessible.

Charles Barrington Brown; 1869 and 1872; geological survey of the area along the southern route; lack of food forced the expedition's return; viewed Mount Roraima on his second trip, this one from the northeast; concluded that a balloon was needed to reach the summit.

Messrs. Flint and Edgington; 1877; 18 days on foot from the Rupununi Savanna; lack of provisions hampered exploration.

Messers. Boddam-Wetham and McTurk; 1878; approached from Mazaruni and explored the north, west, and south sides; ascent was impractical.

Henry Whitley; 1881–83; several trips collecting birds near Roraima; cut a path to the cliff base and observed a ledge in September 1883; published description and sketches in 1884.

Mr. Siedel; 1884; orchid collector; two trips to same area as Whitley; joined by Im Thurn's party in early December.

Everard F. Im Thurn and Harry I. Perkins; Royal Geographical Society; 1884; traveled overland from Kaieteur Falls; first summit ascent; 1 day on top; plant collections; published descriptions and maps.

John J. Quelch and Frederick V. McConnell; first Roraima expedition; 1894; southern route to the southwestern slopes; 2 nights and 3 days on summit (November); first collection of *Oreophrynella;* detailed account of approach.

Frederick V. McConnell and John J. Quelch; second Roraima expedition; 1898; northern route via Mazaruni, along the eastern side to southwest ledge; 9 days on summit (October); major collections from summit and slopes, with several new amphibians and reptiles.

Mr. and Mrs. Cecil Clementi; 1913; no collections; first woman on summit; wrote a book about her experiences.

G. H. H. Tate and others; American Museum of Natural History Lee Garnett Day Cerro Roraima Expedition; 1927–28; well-documented expedition of

80 days in the region and 13 on the summit; made extensive collections of birds and mammals.

Julian A. Steyermark; Field Museum of Natural History Roraima Expedition; 1944; climbed and collected botanical material from the summit in September; often collected amphibians and reptiles from tepui summits (e.g., Chimantá).

Adrian Warren; British expedition to Mount Roraima; 1971; approached north side along the Waruma route; extensive herpetological collections.

Michael Tamessar; British Climbing Group; 1973; used Waruma route; several new frog species; first successful ascent of Roraima from the north.

Charles Brewer-Carías and Roy McDiarmid; Ministerio del Ambiente y de las Recursos Naturales Renovables, Caracas Expedition I; 1977; general collections, including herpetological material for 3 days in January; three other tepuis visited on this expedition.

Charles Brewer-Carías and Roy McDiarmid; Ministerio del Ambiente y de las Recursos Naturales Renovables, Caracas Expedition II; 1978; herpetological specimens collected over 5 days in February; general collections by several scientists on Roraima and four other tepuis.

Carlos Galán; 1985; 33 *Oreophrynella quelchii* collected in April.

Fundación Terramar Eastern Tepui Expedition; 1989; some herpetological materials collected here and on other eastern tepuis.

Smithsonian Guyana Expedition; 2001; Waruma route; ornithologists from Smithsonian and University of Kansas also collected amphibians and reptiles.

Acknowledgments

We have experienced the excitement of tepui exploration with many people and relish the collective memories of good times together in the field. Notable among the many with whom RWM has shared field time are Bill Buck, Rex Cocroft, Mercedes Foster, Leopoldo Garcia, Al Gardner, and Alfredo Paolillo. RWM especially thanks Charles Brewer-Carías, who introduced him to the Pantepui and made possible much of his work in the Guayana Highlands. We also thank Lee-Ann Hayek (Smithsonian Institution) for guidance with the analytical approaches and Joseph J. O'Brien (Florida International University) and Grace Wyngaard (James Madison University) for statistical consultation, assistance with analysis, and fruitful discussion. Mercedes Foster (USGS National Museum of Natural History) commented on sections of the manuscript, helped with the analyses, and provided needed encouragement to RWM throughout the writing phase of the project. Bruce Means (Coastal Plains Institute) discussed his field experiences in Pantepui with RWM and provided insight about the impact of fire in the region. Kevin de Queiroz (Smithsonian Institution) offered suggestions and discussed ideas with RWM at various stages during manuscript preparation. Craig Guyer (Auburn University) and an unknown reviewer read an early draft of the chapter and sharpened our thinking on several issues. Amy Lathrop and Ross MacCulloch (Royal Ontario Museum) graciously provided us with information on their recent collections from Mount Ayanganna in Guyana, and César Barrio (Fundación Andigen, Mérida) did the same for material from Venezuela. Tom Hollowell (Smithsonian Institution) helped with information on localities in Guyana, and Vicki Funk (Smithsonian Institution) loaned us her copies of the *Flora of the Venezuelan Guayana*. Greg Schneider (University of Michigan), Jose Rosado (Harvard University), and Linda Ford (American Museum of Natural History) provided access to information on amphibians and reptiles in their respective institutions. Fiona Wilkinson (USGS National Museum of Natural History) effectively helped with the preparation of the maps (figs. 18.1 and 18A.2) and figures. Librarians (Leslie Overstreet and Claire Catron) at the Museum of Natural History (Smithsonian Institution) were especially helpful in locating copies of older publications on the Roraima region. Fieldwork by RWM in the Guayana Highlands was supported at various times by the Ministerio del Ambiente y de los Recursos Naturales Renovables, Fuerza Aérea Venezolana, Fundación para el Desarrollo de las Ciencias Físicas, Matemáticas y Naturales, Gustavo Cisneros, National Science Foundation (grants BSR83-17687 and 17561), and the Scholarly Studies Program of the Smithsonian Institution.

Charles W. Myers, curator emeritus of herpetology at the American Museum of Natural History (AMNH), made it possible for MAD to visit several tepuis during her tenure as a Boschenstein Postdoctoral Fellow (1989), a research assistant in herpetology (1990), and a research associate in herpetology (1994, 1995). Myers shared his knowledge of the tepuis with us, helped with estimating the number of collecting days for Cerro Guaiquinima and Auyán-tepui, and offered guidance on other issues. The American Museum's participation in the 1989 Tapirapecó FUDECI (Fundación para el Desarrollo de las Ciencias Físicas, Matemáticas y Naturales) Expedition was funded by the Howard Phipps Foundation and Robert G. Goelet. MAD thanks Anne Sidamon-Eristoff, Howard Phipps Jr., and Robert G. Goelet for their financial support. The late Kathleen de Phelps of Caracas provided funds to support American Museum participation in the 1990 FUDECI Expedition to Cerro Guaiquinima. Kathy Phelps supported AMNH tepui exploration and was a generous hostess during our visits to Caracas. Her stories of early expeditions with Billy Phelps were captivating and inspiring. Gracias para todo Missia Katy! Robert G. Goelet provided funding for the 1994 expedition to Auyán-tepui and the 1995 expedition to the western tepuis (Guanay, Yaví, and Yutajé). He has been extremely generous with his support of AMNH field and laboratory activities. Many companions provided support in the field, but MAD especially thanks Chuck Myers, John Daly, Paul Sweet, and George Barrowclough for memorable times exploring Venezuelan tepuis. MAD is extremely grateful to Chuck Myers and the AMNH for the opportunity to explore several tepuis of the Lost World. This chapter is contribution number 79 to the program in tropical biology at Florida International University.

References

Alford, R. A. 1999. Ecology: Resource use, competition, and predation. Pp. 240–78 in R. Altig and R. W. McDiarmid, eds., *Tadpoles*. University of Chicago Press, Chicago.

Alford, R. A., and S. J. Richards. 1999. Global amphibian declines: A problem in applied ecology. *Annual Review of Ecology and Systematics* 30: 133–65.

Allan, D. M. 1973. Some relationships of vocalizations to behavior in the Pacific treefrog, *Hyla regilla. Herpetologica* 29: 366–71.

Allen, P. H. 1956. *The rain forests of Golfo Dulce*. University of Florida Press, Gainesville.

Anastasi, A., G. Bertaccini, J. M. Cei, G. De Caro, V. Erspamer, and M. Impicciatore. 1969. Structure and pharmacological actions of phyllocaerulein, a caerulein-like nonapeptide: Its occurrence in extracts of the skin of *Phyllomedusa sauvagei* and related *Phyllomedusa* species. *British Journal of Pharmacology* 37: 198–206.

Anastasi, A., G. Bertaccini, and V. Erspamer. 1966. Pharmacological data on phyllokinin (bradykinin-isoleucyl-tyrosine O-sulfate) and bradykinin-isoleucyl-tyrosine. *British Journal of Pharmacology* 27: 479–85.

Anderson, K. 1986. A cytotaxonomic analysis of the Holarctic treefrogs in the genus *Hyla*. Ph.D. dissertation, New York University.

Andersson, M. 1994. *Sexual selection*. Princeton University Press, Princeton, NJ.

Andren, C., and G. Nilson. 1981. Reproductive success and risk of predation in normal and melanistic morphs of the adder, *Vipera berus. Biological Journal of the Linnean Society* 15: 235–46.

Andrews, R. M. 1971. Structural habitat and time budget of a tropical *Anolis* lizard. *Ecology* 52: 262–70.

———. 1979. The lizard *Corytophanes cristatus:* An extreme "sit-and-wait" predator. *Biotropica* 11: 136–39.

Angeles, P. A. 1992. *The Harper Collins dictionary of philosophy*. 2nd ed. Harper Perennial, New York.

Appolonio, M., M. Festa-Bianchet, and F. Mari. 1989. Correlates of copulatory success in a fallow deer lek. *Behavioral Ecology and Sociobiology* 25: 89–97.

Arak, A. 1988. Female mate selection in the natterjack toad: Active choice or passive attraction? *Behavioral Ecology and Sociobiology* 22: 317–27.

Arévalo, E., S. K. Davis, and J. W. Sites Jr. 1994. Mitochondrial DNA sequence divergence and phylogenetic relationships among eight races of the *Sceloporus grammicus* complex (Phrynosomatidae) in Central Mexico. *Systematic Biology* 43: 387–418.

Arnold, A., and S. Breedlove. 1985. Organizational and activational effects of sex steroids on brain and behavior: A reanalysis. *Hormones and Behavior* 19: 469–98.

Arnold, S. J., and M. J. Wade. 1984a. On the measurement of natural and sexual selection: Applications. *Evolution* 38: 720–34.

———. 1984b. On the measurement of natural and sexual selection: Theory. *Evolution* 38: 709–19.

Arrieta, M. I., B. Martinez, N. Lobato, B. Criado, T. Nuñez, A. Echarri, and C. M. Lostao. 1993. Cytogenetic study of fragile sites and sister chromatid intercrossing. *Cytologia* 58: 351–54.

Arruga, M. V., and L. V. Monteagudo. 1989. Evidence of Mendelian inheritance of the nucleolar organizer regions in the Spanish common rabbit. *Journal of Heredity* 80: 85–86.

Atchley, W. R. 1972. The chromosome karyotype in estimation of lineage relationships. *Systematic Zoology* 21: 199–209.

Avila-Pires, T. C. S. 1995. Lizards of Brazilian Amazonia (Reptilia: Squamata). *Zoologische Verhandelingen* (Leiden) 299: 1–706.

Avise, J. C., J. Arnold, R. M. Ball, E. Bermingham, T. Lamb, J. E. Neigel, C. A. Reeb, and N. C. Saunders. 1987. Intraspecific phylogeography: The mitochondrial DNA bridge between population genetics and systematics. *Annual Review of Ecology and Systematics* 18: 489–522.

Ax, P. 1987. *The phylogenetic system: The systematization of organisms on the basis of their phylogenesis.* John Wiley, Chichester, U.K.

Ayarzagüena, J. 1983. Una nueva especie de *Dischidodactylus* Lynch (Amphibia, Leptodactylidae) en la cumbre del tepui Marahuaka, Territorio Federal Amazonas—Venezuela. *Memoria Sociedad Ciencias Naturales La Salle* 43: 215–20.

———. 1992. Los centrolenidos de la Guayana Venezolana. *Publicaciones de la Asociación de Amigos de Doñana* 1: 1–48.

Ayarzagüena, J., and J. C. Señaris. 1993. Dos nuevas especies de *Hyla* (Anura; Hylidae) para las cumbres tepuyanas del Estado Amazonas, Venezuela. *Memoria Sociedad Ciencias Naturales La Salle* 53: 127–46.

———. 1996. Dos nuevas especies de *Cochranella* (Anura; Centrolenidae) para Venezuela. *Publicaciones de la Asociación de Amigos de Doñana* 8: 1–16.

Ayarzagüena, J., J. C. Señaris, and S. Gorzula. 1992a. El grupo *Osteocephalus rodriguezi* de las tierras altas de la Guayana Venezolana: Descripción de cinco nuevas especies. *Memoria Sociedad Ciencias Naturales La Salle* 52: 113–42.

———. 1992b. Un nuevo genero para las especies del "grupo *Osteocephalus rodriguezi*" (Anura: Hylidae). *Memoria Sociedad Ciencias Naturales La Salle* 52: 213–21.

Bardin, C., and J. Catterall. 1981. Testosterone: A major determinant of extragenital sexual dimorphism. *Science* 211: 1285–94.

Barreát, F., A. Barreto, H. Briceño, S. Gorzula, O. Huber, G. Medina-Cuervo, and C. Schubert. 1986. Reconocimiento preliminar del Macizo del Chimantá, Estado Bolívar (Venezuela). *Acta Científica Venezolana* 37: 25–42.

Barrett, M., M. J. Donoghue, and E. Sober. 1991. Against consensus. *Systematic Zoology* 40: 486–93.

Barrio, A. C. L. 1998. Sistematica y biogeografia de los anfibios (Amphibia) de Venezuela. *Acta Biologica Venezuelica* 18: 1–93.

———. 1999. Geographic distribution: *Otophryne steyermarki. Herpetological Review* 30: 173.

Barrowclough, G. F., M. Lentino R., and P. R. Sweet. 1997. New records of birds from Auyán-tepui, Estado Bolívar, Venezuela. *Bulletin of the British Ornithological Club* 117: 194–98.

Barthalmus, G. T. 1994. Biological roles of amphibian skin secretions. Pp. 382–410 in H. Heatwole and G. T. Barthalmus, eds., *Amphibian biology*, vol. 1: *The integument.* Surrey Beatty and Sons, Chipping Norton, New South Wales, Australia.

Bates, H. W. 1862. Contributions to an insect fauna of the Amazon Valley, Lepidoptera, Heliconidae. *Transactions of the Linnean Society of London* 23: 495–566.

Bauer, A. M. 1998. Twentieth-century amphibian and reptiles discoveries. *Cryptozoology* 13: 1–17.

Bauer, A. M., and K. D. DeVaney. 1987. Convergence and mimicry in sea snakes and other New Caledonian reef flat organisms. Pp. 43–48 in J. J. Van Gelder, H. Stribosch, and P. J. M. Bergers, eds., *Proceedings of the Fourth Ordinary General Meeting of the European Society of Herpetology.* University of Nijmegen, Nijmegen, Netherlands.

Beckers, G. J. L., T. A. A. M. Leenders, and H. Strijbosch. 1996. Coral snake mimicry: Live snakes not avoided by a mammalian predator. *Oecologia* 106: 461–63.

Beebe, W. 1944. Field notes on the lizards of Kartabo, British Guiana, and Caripito, Venezuela. II. Iguanidae. *Zoologica* 29: 195–216.

Bellairs, A., and G. Underwood. 1951. The origin of snakes. *Biological Review* 26: 193–237.

Bennett, C. F. 1963. A phytophysiognomic reconnaissance of Barro Colorado Island, Canal Zone. *Smithsonian Miscellaneous Collections* 145: 1–8.

Benson, K. R. 1978. Herpetology on the Lewis and Clark expedition, 1804–1806. *Herpetological Review* 9: 87–91.

Berger, L., R. Speare, P. Daszak, E. D. Green, A. A. Cunningham, L. C. Goggin, R. Slocombe, M. A. Ragan, A. D. Hyatt, K. R. McDonald, H. B. Hines, K. R. Lips, G. Marantelli, and H. Parkes. 1998. Chytridiomycosis causes amphibian mortality associated with population declines in the rain forests of Australia and Central America. *Proceedings of the National Academy of Sciences* (USA) 95: 9031–36.

Berry, P. E., B. K. Holst, and K. Yatskievych. 1995. Introduction. Pp. xv–xx in J. A. Steyermark, P. E. Berry, and B. K. Holst, eds., *Flora of the Venezuelan Guayana*, vol. 1: *Introduction.* Missouri Botanical Garden and Timber Press, St. Louis, MO, and Portland, OR.

Berry, P. E., O. Huber, and B. K. Holst. 1995. Floristic analysis and phytogeography. Pp. 161–91 in J. A. Steyermark, P. E. Berry, and B. K. Holst, eds., *Flora of the Venezuelan Guayana*, vol. 1: *Introduction.* Missouri Botanical Garden and Timber Press, St. Louis, MO, and Portland, OR.

Beu, A. G. 2001. Gradual Miocene to Pleistocene uplift of the Central American isthmus: Evidence from tropical American tonnoidean gastropods. *Journal of Paleontology* 75: 706–20.

Bevill, R. L., and S. M. Louda. 1999. Comparisons of related rare and common species in the study of plant rarity. *Conservation Biology* 13: 493–98.

Bigarella, J. J., and D. de Andrade-Lima. 1982. Paleoenvironmental changes in Brazil. Pp. 27–59 in G. T Prance, ed., *Biological diversification in the tropics*. Columbia University Press, New York.

Blackburn, D., and V. Bernardo. 1998. Sexual dimorphism and testosterone responsiveness in hypaxial muscles of the northern leopard frog, *Rana pipiens*. *Amphibia-Reptilia* 19: 269–79.

Blackburn, D., R. Darrell, K. Lonergan, R. Mancini, and C. Sidor. 1995. Differential testosterone sensitivity of forelimb muscles of male leopard frogs, *Rana pipiens:* Test of a model system. *Amphibia-Reptilia* 16: 351–56.

Blackburn, S., ed. 1994. *The Oxford dictionary of philosophy*. Oxford University Press, Oxford.

Blair, W. F. 1979. Nichos ecológicos y la evolución paralela y convergente de los anfibios del Chaco y del Mesquital norteamericano. *Acta Zoológica Lilloana* 27: 261–67.

Blaustein, A. R., and D. B. Wake. 1990. Declining amphibian populations: A global phenomenon? *Trends in Ecology and Evolution* 5: 203–4.

Blaustein, A. R., D. B. Wake, and W. P. Sousa. 1994. Amphibian declines: Judging stability, persistence, and susceptibility of populations to local and global extinctions. *Conservation Biology* 8: 60–71.

Bogart, J. P. 1970. Los cromosomas de anfibios anuros del genero *Eleutherodactylus*. *Acta IV Congreso Latinamericano de Zoologia* 1: 65–78. Although this article is listed as published in 1970, it was not available until 1973.

———. 1972. Karyotypes. Pp. 171–95 in W. F. Blair, ed., *Evolution in the genus Bufo*. University of Texas Press, Austin.

———. 1973. Evolution of anuran karyotypes. Pp. 337–49 in J. L. Vial, ed., *Evolutionary biology of the anurans*. University of Missouri Press, Columbia.

———. 1981. Chromosome studies in *Sminthillus* from Cuba and *Eleutherodactylus* from Cuba and Puerto Rico (Anura, Leptodactylidae). *Life Sciences Contributions, Royal Ontario Museum* 129: 1–22.

Bogart, J. P., and S. B. Hedges. 1995. Rapid chromosome evolution in Jamaican frogs of the genus *Eleutherodactylus* (Leptodactylidae). *Journal of Zoology* (London) 235: 9–31.

Boos, H. E. A. 2001. *The snakes of Trinidad and Tobago*. Texas A&M University Press, College Station.

Boulenger, G. A. 1895a. Description of a new Batrachian (*Oreophryne Quelchii*) discovered by Messers. J. J. Quelch and F. McConnell on the summit of Mount Roraima. *Annals and Magazine of Natural History* (ser. 6) 15: 521–22.

———. 1895b. Correction to p. 521 (*Annals*, June 1895). *Annals and Magazine of Natural History* (ser. 6) 16: 125.

———. 1900a. Reptiles. Pp. 53–54 and pl. 5 in E. R. Lankester, communicator, Report on a collection made by Messrs. F. V. McConnell and J. J. Quelch at Mount Roraima in British Guiana. *Transactions of the Linnean Society of London* (2nd ser.), *Zoology* 8: 51–76.

———. 1900b. Batrachians. Pp. 55–56 and pl. 5 in E. R. Lankester, communicator, Report on a collection made by Messrs. F. V. McConnell and J. J. Quelch at Mount Roraima in British Guiana. *Transactions of the Linnean Society of London* (2nd ser.), *Zoology* 8: 51–76.

Boyd, B., and R. M. Pile. 2000. *Nabokov's butterflies: Unpublished and uncollected writings.* Beacon Press, Boston.

Boyd, R. 1999. Homeostasis, species, and higher taxa. Pp. 141–85 in R. A. Wilson, ed., *Species: New interdisciplinary essays.* MIT Press, Cambridge.

Boza, M. A. 1978. *Los Parques Nacionales de Costa Rica.* INCAFO, Madrid, Spain.

Boza, M. A., and J. H. Sevo. 1998. *The national parks and other protected areas of Costa Rica.* INCAFO, Madrid, Spain.

Brame, A. H., Jr. 1968. Systematics and evolution of the Mesoamerican salamander genus *Oedipina. Journal of Herpetology* 2: 1–64.

Brame, A. H., Jr., J. M. Savage, D. B. Wake, and J. Hanken. 2001. New species of large black salamander, genus *Bolitoglossa* (Plethodontidae) from western Panamá. *Copeia* 2000: 700–704.

Brattstrom, B. H. 1955. The coral snake "mimic" problem and protective coloration. *Evolution* 9: 217–19.

Bremer, K. 1994. Branch support and tree stability. *Cladistics* 10: 295–304.

Brewer-Carías, C. 1976. Las simas de Sarisariñama. *Boletín de la Sociedad Venezolana de Ciencias Naturales* 32: 549–624.

———. 1978a. *Roraima: La Montaña de Cristal.* Editorial Arte, Caracas, Venezuela.

———. 1978b. *La vegetación del mundo perdido.* Cromotip, Caracas, Venezuela.

———. 1983. *Sarisariñama.* Editorial Arte, Caracas, Venezuela.

———, ed. 1988. *Cerro de la Neblina: Resultados de la expedición, 1983–1987.* Fundación para el Desarrollo de las Ciencias Físicas Matemáticas y Naturales, Caracas, Venezuela.

Brodie, E. D., Jr. 1976. Additional observations on the Batesian mimicry of *Notopthalmus viridescens* efts by *Pseudotriton ruber. Herpetologica* 32: 68–70.

Brodie, E. D., III. 1992. Correlational selection for color pattern and antipredator behavior in the garter snake *Thamnophis ordinoides. Evolution* 46: 1284–98.

———. 1993. Differential avoidance of coral snake banded patterns by free-ranging avian predators in Costa Rica. *Evolution* 47: 227–35.

Brodie, E. D., III, and E. D. Brodie Jr. 1999. Predator–prey arms races. *BioScience* 49: 557–68.

Brodie, E. D., III, and F. J. Janzen. 1995. Experimental studies of coral snake mimicry: Generalized avoidance of ringed snake patterns by free-ranging avian predators. *Functional Ecology* 9: 186–90.

Brodie, E. D., III, and A. J. Moore. 1995. Experimental studies of coral snake mimicry: Do snakes mimic millipedes? *Animal Behaviour* 49: 534–36.

Brookfield, H. C., and P. Brown. 1963. Struggle for land: Agriculture and group territories among the Chimbu of the New Guinea highlands. Oxford University Press, Melbourne, Australia.

Brookfield, H. C., and D. Hart. 1971. *Melanesia: A geographical interpretation of an island world.* Methuen, London.

Brooks, D. R. 1981. Hennig's parasitological method: A proposed solution. *Systematic Zoology* 30: 229–49.

———. 1990. Parsimony analysis in historical biogeography and coevolution: Methodological and theoretical update. *Systematic Zoology* 39: 14–30.

———. 2001. Diversity, organism level. Pp. 191–203 in *Encyclopedia of biodiversity,* vol. 2. Academic Press, San Diego.

Brower, A. V. Z. 1996. Parallel race formation and the evolution of mimicry in *Heliconius* butterflies: A phylogenetic hypothesis from mitochondrial DNA sequences. *Evolution* 50: 195–221.

Brown, K. S., Jr. 1975. Geographical patterns of evolution in Neotropical Lepidoptera: Systematics and derivation of known and new Heliconiini (Nymphalidae: Nymphalinae). *Journal of Entomology* 44: 201–42.

———. 1987. Biogeography and evolution of Neotropical butterflies. Pp. 66–104 in T. C. Whitmore and G. T. Prance, eds., *Biogeography and Quaternary history in Tropical America.* Clarendon Press, Oxford.

Brugger, K. E. 1989. Red-tailed hawk dies with coral snake in talons. *Copeia* 1989: 508–10.

Brumfield, R. T., and A. P. Capparella. 1996. Historical diversification of birds in northwestern South America: A molecular perspective on the role of vicariant events. *Evolution* 50: 1607–24.

Bryant, H. N. 1994. Comments on the phylogenetic definition of taxon names and conventions regarding the name of crown clades. *Systematic Biology* 43: 124–30.

———. 1996. Explicitness, stability and universality in the phylogenetic definition and usage of taxon names: A case study of the phylogenetic taxonomy of the Carnivora (Mammalia). *Systematic Biology* 45: 174–89.

———. 1997. Cladistic information in phylogenetic definitions and designated phylogenetic contexts for the use of taxon names. *Biological Journal of the Linnaean Society* 62: 495–503.

Bucher, E. H. 1980. Ecología de la fauna chaqueña: Una revisión. *ECOSUR* (Argentina) 7: 111–59.

Bucher, T. L., M. J. Ryan, and G. A. Bartholomew. 1982. Oxygen consumption during resting, calling, and nest building in the frog, *Physalaemus pustulosus. Physiological Zoology* 55: 10–22.

Buchsbaum, R. 1969. On top of Auyantepui. *Carnegie Magazine,* May 1969, 169–77.

Budgett, J. S. 1899. Notes on the batrachians of the Paraguayan Chaco, with observations upon their breeding habits and development, especially with regard to *Phyllomedusa hypochondrialis,* (Cope). Also a description of a new genus. *Quarterly Journal of Microscopical Science* 42: 305–33.

Bulmer, R. N. H., and M. J. Tyler. 1968. Karam classification of frogs. *Journal of the Polynesian Society* 77: 333–85.

Burbrink, F. T., R. Lawson, and J. B. Slowinski. 2000. MtDNA phylogeography of the North American rat snake (*Elaphe obsoleta*): A critique of the subspecies concept. *Evolution* 54: 2107–18.

Burger, L. W. 1971. Genera of pitvipers (*Serpentes: Crotalidae*). Ph.D. dissertation, University of Kansas, Lawrence.

Burke, K. 1988. Tectonic evolution of the Caribbean. *Annual Review of Earth and Planetary Sciences* 16: 201–30.

Burnell, K. L., and S. B. Hedges. 1990. Relationships of West Indian *Anolis* (Sauria: Iguanidae): An approach using slow-evolving protein loci. *Caribbean Journal of Science* 26: 7–30.

Burnham, R. J., and A. Graham. 1999. The history of Neotropical vegetation: New developments and status. *Annals of the Missouri Botanical Garden* 86: 546–89.

Burrowes, P. 2000. Parental care and sexual selection in the Puerto Rican cave-dwelling frog, *Eleutherodactylus cooki*. *Herpetologica* 56: 375–86.

Burt, C. E., and M. D. Burt. 1931. South American lizards in the collection of the American Museum of Natural History. *Bulletin American Museum of Natural History* 61: 227–395.

Bush, M. B. 1994. Amazonian speciation: A necessarily complex model. *Journal of Biogeography* 21: 5–17.

Cabot, E. L., and A. T. Beckenback. 1989. Simultaneous editing of multiple nucleic acid and protein sequences with ESEE. *Computer Applications in the Biosciences* 5: 233–34.

Cadle, J. E. 1988. Phylogenetic relationships among advanced snakes: A molecular perspective. *University of California Publications in Zoology* 119: 1–77.

———. 1995. A new species of *Boophis* (Anura: Rhacophoridae) with unusual skin glands from Madagascar, and a discussion of variation and sexual dimorphism in *Boophis albilabris* (Boulenger). *Zoological Journal of the Linnean Society* 115: 313–45.

———. 2003. Colubridae. Pp. 997–1004 in S. M. Goodman and J. P. Benstead, eds., *The natural history of Madagascar*. University of Chicago Press, Chicago.

Cadle, J. E., and H. W. Greene. 1993. Phylogenetic patterns, biogeography, and the ecological structure of Neotropical snake assemblages. Pp. 281–93 in R. E. Ricklefs and D. Schluter, eds., *Species diversity in ecological communities: Historical and geographical perspectives*. University of Chicago Press, Chicago.

Cadle, J. E., and C. W. Myers. 2003. Systematics of snakes referred to *Dipsas variegata* in Panama and western South America, with revalidation of two species and notes on defensive behaviors in the Dipsadini (Colubridae). *American Museum Novitates* 3409: 1–47.

Caldwell, M. W., and M. S. Y. Lee. 1997. A snake with legs from the marine Cretaceous of the Middle East. *Nature* 386: 705–9.

Campagna, C. B., and J. LeBoeuf. 1988. Reproductive behavior of southern sea lions. *Behaviour* 104: 233–61.

Campbell, J. A. 1985. A new species of highland pitviper of the genus *Bothrops* from southern Mexico. *Journal of Herpetology* 19: 48–54.

———. 1999. Distribution patterns of amphibians in Middle America. Pp. 111–210 in W. E. Duellman, ed., *Patterns of distribution of Amphibians: A global perspective*. Johns Hopkins University Press, Baltimore.

Campbell, J. A., and B. T. Clarke. 1998. A review of frogs of the genus *Otophryne* (Microhylidae) with the description of a new species. *Herpetologica* 54: 301–17.

Campbell, J. A., and W. W. Lamar. 1989. *The venomous reptiles of Latin America.* Cornell University Press, Ithaca, NY.

————. 1992. Taxonomic status of miscellaneous Neotropical viperids, with the description of a new genus. *Occasional Papers of the Museum of Texas Tech University* 153: 1–32.

Campbell, J. A., and J. M. Savage. 2000. Taxonomic reconsideration of Middle American frogs of the *Eleutherodactylus rugulosus* group (Anura: Leptodactylidae): A reconnaissance of subtle nuances among frogs. *Herpetological Monographs* 14: 186–292.

Campbell, J. A., and E. N. Smith. 2000. A new species of arboreal pitviper from the Atlantic versant of northern Central America. *Revista Biologia Tropical* 48: 1001–13.

Campbell, J. A., and A. Solórzano. 1992. The distribution, variation, and natural history of the Middle American montane pitviper, *Porthidium godmani.* Pp. 223–50 in J. A. Campbell and E. D. Brodie Jr., eds., *Biology of pitvipers.* Selva, Tyler, TX.

Cannatella, D. C., and K. de Queiroz. 1989. Phylogenetic systematics of the anoles: Is a new taxonomy warranted? *Systematic Zoology* 38: 57–68.

Cantino, P. D. 1998. Binomials, hyphenated uninomials, and phylogenetic nomenclature. *Taxon* 47: 425–29.

Cantino, P. D., H. N. Bryant, K. de Queiroz, M. J. Donoghue, T. Eriksson, D. M. Hillis, and M. S. Y. Lee. 1999. Species names in phylogenetic nomenclature. *Systematic Biology* 48: 790–807.

Cantino, P. D., and K. de Queiroz. 2000. PhyloCode: A phylogenetic code of biological nomenclature. http://www.ohiou.edu/phylocode/. Revised 8 April 2000.

Cantino, P. D., R. G. Olmstead, and S. J. Wagstaff. 1997. A comparison of phylogenetic nomenclature with the current system: A botanical case study. *Systematic Biology* 46: 313–31.

Cantino, P. D., S. J. Wagstaff, and R. G. Olmstead. 1999. *Caryopteris* (Lamiaceae) and the conflict between phylogenetic and pragmatic considerations in botanical nomenclature. *Systematic Botany* 23: 369–86.

Cardoso, A. J., and I. Sazima. 1977. Batracofagia na fase adulta e larvária da rã pimenta, *Leptodactylus labyrinthicus* (Spix, 1824)—Anura, Leptodactylidae. *Ciência e Cultura* 29: 1130–32.

Carpenter, J. M. 1992. Random cladistics. *Cladistics* 8: 147–53.

Case, S. M., and M. H. Wake. 1977. Immunological comparisons of caecilian albumins (Amphibia: Gymnophiona). *Herpetologica* 33: 94–98.

Catchpole, C. K., J. Dittami, and B. Leiser. 1984. Differential responses to male song repertoires in male songbirds implanted with oestrodiol. *Nature* 312: 563–64.

Catz, D., L. Fischer, and D. Kelley. 1995. Androgen regulation of a laryngeal-specific myosin heavy chain mRNA isoform whose expression is sexually differentiated. *Developmental Biology* 171: 448–57.

Catz, D., L. Fischer, M. Moschella, M. Tobias, and D. Kelley. 1992. Sexually dimorphic expression of a laryngeal-specific, androgen-regulated myosin heavy chain gene during *Xenopus laevis* development. *Developmental Biology* 154: 366–76.

Cei, J. M. 1955. Chaco batrachians in central Argentina. *Copeia* 1955: 291–93.

————. 1956. Nueva lista sistemática de los batracios de Argentina y breves notas sobre su biología y ecología. *Investigaciones Zoológicas Chilenas* 3: 36–68.

————. 1980. Amphibians of Argentina. *Monitore Zoologico Italiano Monografia*, n.s., no. 2: 1–609.

Cei, J. M., and V. Erspamer. 1966. Biochemical taxonomy of South American amphibians by means of skin amines and polypeptides. *Copeia* 1966: 74–78.

Cei, J. M., V. Erspamer, and M. Roseghini. 1967. Taxonomic and evolutionary significance of biogenic amines and polypeptides in amphibian skin. I. Neotropical leptodactylid frogs. *Systematic Zoology* 16: 328–42.

————. 1968. Taxonomic and evolutionary significance of biogenic amines and polypeptides in amphibian skin. II. Toads of the genera *Bufo* and *Melanophryniscus*. *Systematic Zoology* 17: 232–45.

Chan-ard, T., W. Grossmann, A. Gumprecht, and K.-D. Schultz. 1999. *Amphibians and reptiles of peninsular Malaysia and Thailand: An illustrated checklist*. Bushmaster Publications, Wuerselen, Germany.

Chapman, F. M. 1931. The upper zonal bird-life of Mts. Roraima and Duida. *Bulletin of the American Museum of Natural History* 63: 1–135.

Chen, M. H. 1998. The herpetofauna of Iwokrama Reserve: A comparison of sampling methods. M.S. thesis, Florida International University, Miami.

Chen, S.-H. 2001. Cytogenetic study of the Lower Central American frogs of the subgenus *Craugastor* (Anura: Leptodactylidae: *Eleutherodactylus*). Ph.D. dissertation, University of Miami, Coral Gables, FL.

Chippaux, J.-P. 1986. *Les serpents de la Guyane française*. Institut Français de Recherche Scientifique, Paris.

Clark, D. B. 1990. La Selva Biological Station: A blueprint for stimulating tropical research. Pp. 9–27 in A. H. Gentry, ed., *Four Neotropical rainforests*. Yale University Press, New Haven, CT.

Clarke, D., V. Funk, T. Hollowell, and C. Kelloff. 1999. *A preliminary analysis of the plant diversity of the Iwokrama tropical moist forest, Guyana*. Report produced for the Iwokrama International Centre for Rain Forest Conservation and Development, Biological Diversity of the Guianas Program, Smithsonian Institute, Washington, DC. Produced under the auspices of the Centre for the Study of Biological Diversity, University of Guyana.

Clementi, C., Mrs. 1920. *Through British Guiana to the summit of Roraima*. T. Fisher Unwin, London.

Clutton-Brock, T. H., ed. 1988. *Reproductive success*. University of Chicago Press, Chicago.

Clutton-Brock, T. H., M. Hiraiwa-Hasegawa, and A. Robertson. 1989. Mate choice on fallow deer leks. *Nature* 340: 463–65.

Coates, A. G. 1997. The forging of Central America. Pp. 1–37 in A. G. Coates, ed., *Central America: A natural and cultural history*. Yale University Press, New Haven, CT.

Coates, A. G., and J. A. Obando. 1996. The geologic evolution of the Central American isthmus. Pp. 21–56 in J. B. C. Jackson, A. F. Budd, and A. G. Coates, eds., *Evolution and environment in tropical America*. University of Chicago Press, Chicago.

Coddington, J. A. 1988. Cladistic tests of adaptational hypotheses. *Cladistics* 4: 3–22.

Cogger, H. C. 1987. Classification and nomenclature. Pp. 266–86 in G. R. Dyne and D. W. Walton, eds., *Fauna of Australia*, vol. 1A, General articles. Australian Government Publication Service, Canberra.

Colinvaux, P. A. 1996. Quaternary environmental history and forest diversity in the Neotropics. Pp. 359–405 in J. B. C. Jackson, A. F. Budd and A. G. Coates, eds., *Evolution and environment in tropical America.* University of Chicago Press, Chicago.

Colwell, R. K., and D. C. Lees. 2000. The mid-domain effect: Geometric constraints on the geometry of species richness. *Trends in Ecology and Evolution* 15: 70–76.

Conner, J. 1988. Field measurements of natural and sexual selection in the fungus beetle, *Bolitotherus cornutus. Evolution* 42: 736–49.

Conniff, R., and D. A. Murawski. 2001. Spider webs: Deadly silk. *National Geographic* 200: 30–45.

Cook, R. E. 1974. Origin of the highland avifauna of southern Venezuela. *Systematic Zoology* 23: 257–64.

Cooke, B., G. Tabibniai, and S. M. Breedlove. 1999. A brain sexual dimorphism controlled by adult circulating androgens. *Proceedings of the National Academy of Sciences* (USA) 96: 7538–40.

Cooper, J. G. 1859. Report upon the reptiles collected on the survey. Pp. 292–306 in *Reports of explorations and surveys, to ascertain the most practicable and economical route for a railroad from the Mississippi River to the Pacific Ocean,* vol. 10. Washington, DC.

Cope, E. D. 1860. Catalog of the Colubridae in the Museum of the Academy of Natural Sciences of Philadelphia, with notes and descriptions of new species. Part 2. *Proceedings of the Academy of Natural Science of Philadelphia* 12: 241–66.

———. 1900. The crocodilians, lizards, and snakes of North America. *Report of the United States National Museum* 1898: 153–1270.

Corn, P. S. 1994. Straight-line drift fences and pitfall traps. Pp. 109–17 in W. R. Heyer, M. A. Donnelly, R. W. McDiarmid, L. C. Hayek, and M. S. Foster, eds., *Measuring and monitoring biological diversity: Standard methods for amphibians.* Smithsonian Institution Press, Washington, DC.

Cornelius, S. E. 1982. The status of sea turtles along the Pacific Coast of Middle America. Pp. 211–19 in K. A. Bjorndal, ed., *Biology and conservation of sea turtles.* Smithsonian Institution Press, Washington, DC.

Cox, M. J. 1991. *The snakes of Thailand and their husbandry.* Krieger Publications, Malabar, FL.

Cracraft, J. 1983. Species concepts and speciation analysis. Pp. 159–87 in R. F. Johnston, ed., *Current ornithology,* vol. 1. Plenum Press, New York.

———. 1992. The species of the birds-of-paradise (Paradisaeidae): Applying the phylogenetic species concept to a complex pattern of diversification. *Cladistics* 8: 1–43.

Cracraft, J., and R. O. Prum. 1988. Patterns and processes of diversification: Speciation and historical congruence in some Neotropical birds. *Evolution* 42: 603–20.

Crane, P. R., and P. Kenrick. 1997. Problems in cladistic classification: Higher-level relationships in land plants. *Aliso* 15: 87–104.

Crankshaw, O. S. 1979. Female choice in relation to calling and courtship songs in *Acheta domesticus. Animal Behaviour* 27: 1274–75.

Croat, T. B. 1978. *The flora of Barro Colorado Island.* Stanford University Press, Stanford, CA.

Crocombe, R., and R. Hide. 1987. New Guinea: Unity in diversity. Pp. 324–67 in R. Crocombe, ed., *Land tenure in the Pacific.* University of the South Pacific, Suva, Fiji.

Crother, B. I. 1999. Phylogenetic relationships among West Indian Xenodontine snakes (Serpentes: Colubridae) with comments on the phylogeny of some mainland xenodontines. *Contemporary Herpetology* 2. http://eagle.cc.ukans.edu/~cnaar/CH/ch/1999/2/index.htm.

Crother, B. I., J. A. Campbell, and D. M. Hillis. 1992. Phylogeny and historical biogeography of the palm pitvipers, genus *Bothreichis:* Biochemical and morphological evidence. Pp. 1–19 in J. A. Campbell and E. D. Brodie Jr., eds., *Biology of the pitvipers.* Selva, Tyler, TX.

Crother, B. I., and D. M. Hillis. 1995. Nuclear ribosomal DNA restriction sites, phylogenetic information, and the phylogeny of some xenodontine (Colubridae) snakes. *Journal of Herpetology* 29: 316–20.

Crump, M. L. 1972. Territoriality and mating behavior in *Dendrobates granuliferus* (Anura: Dendrobatidae). *Herpetologica* 28: 195–98.

———. 1982. Amphibian reproductive ecology on the community level. Pp. 21–36 in N. J. Scott Jr., ed., *Herpetological communities.* Wildlife Research Report No. 13. U.S. Fish and Wildlife Service, Washington, DC.

———. 1983. *Dendrobates granuliferus* and *Dendrobates pumilio* (Ranita Roja, Rana Venenosa, poison dart frogs). Pp. 396–98 in D. H. Janzen, ed., *Costa Rican natural history.* University of Chicago Press, Chicago.

———. 1992. Cannibalism in amphibians. Pp. 256–76 in M. A. Elgar and J. B. Crespi, eds., *Cannibalism: Ecology and evolution among diverse taxa.* Oxford University Press, Oxford.

Crump, M. L., and N. J. Scott Jr. 1994. Visual encounter surveys. Pp. 84–92 in W. R. Heyer, M. A. Donnelly, R. W. McDiarmid, L. C. Hayek, and M. S. Foster, eds., *Measuring and monitoring biological diversity: Standard methods for amphibians.* Smithsonian Institution Press, Washington, DC.

Cuadrado, M. 1999. Mating asynchrony favors no assortative mating by size and serial-type polygyny in common chamaeleons, *Chamaeleo chamaeleon. Herpetologica* 55: 523–30.

Cundall, D., and H. W. Greene. 2000. Feeding in snakes. Pp. 293–333 in K. Schwenk, ed., *Feeding: Form, function, and evolution in tetrapod vertebrates.* Academic Press, San Diego.

Cundall, D., V. Wallach, and D. A. Rossman. 1993. The systematic relationships of the snake genus *Anomochilus. Zoological Journal of the Linnean Society* 109: 275–99.

Cunningham, C. W., H. Zhu, and D. M. Hillis. 1998. Best-fit maximum-likelihood models for phylogenetic inference: Empirical tests with known phylogenies. *Evolution* 52: 978–87.

Daly, J. W. 1998. The nature and origin of amphibian alkaloids. Pp. 141–69 in G. A. Cordell, ed., *The alkaloids: Chemistry and pharmacology,* vol. 50: *The alkaloids: Chemistry and biology.* Academic Press, San Diego.

Daly, J. W., C. W. Myers, and N. Whittaker. 1987. Further classification of skin alkaloids from Neotropical poison frogs (Dendrobatidae), with a general survey of toxic/noxious substances in the Amphibia. *Toxicon* 25: 1023–95.

Darwin, C. 1859. *The origin of species by means of natural selection, or the preservation of favoured races in the struggle for life.* John Murray, London. 1964 facsimile of the 1st ed.; Harvard University Press, Cambridge.

Das Munshi, M., and R. Marsh. 1996. Seasonal changes in contractile properties of the trunk muscles in *Hyla chrysoscelis. American Zoologist* 36: 16A.

David, P., and I. Ineich. 1999. Les serpents venimeux du monde: Systématique et répartition. *Dumerilia* 3: 3–499.

Davies, N. B., and T. R. Halliday. 1977. Optimal mate selection in the toad *Bufo bufo. Nature* 269: 56–58.

Davis, T. W. A., and P. W. Richards. 1933. The vegetation of Moraballi creek, British Guiana: An ecological study of a limited area of tropical rain forest. I. *Journal of Ecology* 21: 350–84.

———. 1934. The vegetation of Moraballi creek, British Guiana: An ecological study of a limited area of tropical rain forest. II. *Journal of Ecology* 22: 106–55.

Daws, G., and M. Fujita. 1999. *Archipelago: The islands of Indonesia.* University of California Press, Berkeley.

Delgado, M., P. Gutierrez, and M. Alonso-Bedate. 1989. Seasonal cycles in testicular activity in the frog, *Rana perezi. General and Comparative Endocrinology* 73: 1–11.

Dengo, G., and J. Case, eds. 1990. *The geology of North America: The Caribbean region.* Vol. H. Geological Society of America, Boulder, CO.

de Queiroz, A., M. J. Donoghue, and J. Kim. 1995. Separate versus combined analysis of phylogenetic evidence. *Annual Review of Ecology and Systematics* 26: 657–82.

de Queiroz, A., R. Lawson, and J. A. Lemos Espinal. 2002. Phylogenetic relationships of North American garter snakes (*Thamnophis*) based on four mitochondrial genes: How much DNA sequence is enough? *Molecular Phylogeny and Evolution* 22: 315–29.

de Queiroz, K. 1987. Phylogenetic systematics of iguanine lizards: A comparative osteological study. *University of California Publications in Zoology* 118: 1–203.

———. 1988. Systematics and the Darwinian revolution. *Philosophy of Science* 55: 238–59.

———. 1992. Phylogenetic definitions and taxonomic philosophy. *Biology and Philosophy* 7: 295–313.

———. 1994. Replacement of an essentialistic perspective on taxonomic definitions as exemplified by the definition of Mammalia. *Systematic Biology* 43: 497–510.

———. 1995a. The definitions of species and clade names: A reply to Ghiselin. *Biology and Philosophy* 10: 223–28.

———. 1995b. Phylogenetic approaches to classification and nomenclature, and the history of taxonomy (an alternative interpretation). *Herpetological Review* 26: 79–81.

———. 1997a. Misunderstandings about the phylogenetic approach to biological nomenclature: A reply to Lidén and Oxelman. *Zoologica Scripta* 26: 67–70.

———. 1997b. The Linnaean hierarchy and the evolutionization of taxonomy, with emphasis on the problem of nomenclature. *Aliso* 15: 125–44.

———. 1998. The general lineage concept of species, species criteria, and the process of speciation. Pp. 57–75 in D. J. Howard and S. H. Berlocher, eds., *Endless forms: Species and speciation.* Oxford University Press, Oxford.

———. 1999. The general lineage concept of species and the defining properties of the species category. Pp. 49–89 in R. A. Wilson, ed., *Species: New interdisciplinary essays.* MIT Press, Cambridge.

de Queiroz, K., and M. J. Donoghue. 1988. Phylogenetic systematics and the species problem. *Cladistics* 4: 317–438.

———. 1990a. Phylogenetic systematics and species revisited. *Cladistics* 6: 83–90.

————. 1990b. Phylogenetic systematics or Nelson's version of cladistics. *Cladistics* 6: 61–75.

de Queiroz, K., and J. A. Gauthier. 1990. Phylogeny as a central principle in taxonomy: Phylogenetic definitions of taxon names. *Systematic Zoology* 39: 307–22.

————. 1992. Phylogenetic taxonomy. *Annual Review of Ecology and Systematics* 23: 449–80.

————. 1994. Toward a phylogenetic system of biological nomenclature. *Trends in Ecology and Evolution* 9: 27–31.

————. 1995. Translating phylogenies into classification: Basic principles. *American Journal of Botany* 82 (suppl.): 108.

Dessauer, H. H., J. E. Cadle, and R. Lawson. 1987. Patterns of snake evolution as suggested by their proteins. *Fieldiana Zoology,* n.s., 34: 1–34.

DeVries, P. J., R. Lande, and D. Murray. 1999. Associations of co-mimetic ithomiine butterflies on small spatial and temporal scales in a Neotropical rainforest. *Biological Journal of the Linnean Society* 67: 73–85.

DeWeese, J. E. 1975. Chromosomes in *Eleutherodactylus* (Anura: Leptodactylidae). *Mammalian Chromosomes Newsletter* 16: 121–23.

————. 1976. The karyotypes of Middle American frogs of the Genus *Eleutherodactylus* (Anura: Leptodactylidae): A case study of the significance of the karyologic method. Ph.D. dissertation, University of Southern California, Los Angeles.

Di Bernardo, M. 1998. História natural de uma comunidade de serpentes da borda oriental do Planalto das Araucárias, Rio Grande do Sul, Brazil. Doctoral dissertation, Universidade Estadual Paulista, Rio Claro, Brazil.

Diego-Aransay, A., and S. Gorzula. 1987. Una nueva especie de *Oreophrynella* (Anura: Bufonidae) de la Guayana Venezolana. *Memoria Sociedad Ciencias Naturales La Salle* 473: 233–37.

Disi, A. M., D. Modry, P. Necas, and L. Rifai. 2001. *Amphibians and reptiles of the Hashemite kingdom of Jordan: An atlas and field guide.* Edition Chimaira, Frankfurt, Germany.

Dixon, J. R. 1979. Origin and distribution of reptiles in lowland tropical rainforests of South America. Pp. 214–40 in W. E. Duellman, ed., *The South American herpetofauna: Its origin, evolution, and dispersal.* University of Kansas Museum of Natural History Monograph, no. 7, Lawrence.

Donnellan, S. C., M. N. Hutchinson, and K. M. Saint. 1999. Molecular evidence for the phylogeny of Australian gekkonoid lizards. *Biological Journal of the Linnean Society* 67: 97–118.

Donnelly, M. A. 1989. Demographic effects of reproductive resource supplementation in a territorial frog, *Dendrobates pumilio. Ecological Monographs* 59: 207–21.

————. 1994a. Amphibian diversity and natural history. Pp. 199–209 in L. A. McDade, K. S. Bawa, H. A. Hespenheide, and G. S. Hartshorn, eds., *La Selva: Ecology and natural history of a Neotropical rain forest.* University of Chicago Press, Chicago.

————. 1994b. Appendix 5: Amphibians. Pp. 380–81 in L. A. McDade, K. S. Bawa, H. A. Hespenheide, and G. S. Hartshorn, eds., *La Selva: Ecology and natural history of a Neotropical rain forest.* University of Chicago Press, Chicago.

————. 1999. Reproductive phenology of *Eleutherodactylus bransfordii* (Anura: Leptodactylidae) in northeastern Costa Rica. *Journal of Herpetology* 33: 624–31.

Donnelly, M. A., R. O. de Sá, and C. Guyer. 1990. Description of the tadpoles of *Gas-

trophryne pictiventris and *Nelsonophryne aterrima* (Anura: Microhylidae), with a review of morphological variation in free-swimming microhylid larvae. *American Museum Novitates* 2976: 1–19.

Donnelly, M. A., and C. Guyer. 1994. Patterns of reproduction and habitat use in an assemblage of Neotropical hylid frogs. *Oecologia* 98: 291–302.

Donnelly, M. A., R. W. McDiarmid, and C. W. Myers. 1992. A new lizard of the genus *Arthrosaura* (Teiidae) from southern Venezuela. *Proceedings of the Biological Society of Washington* 105: 821–33.

Donnelly, M. A., and C. W. Myers. 1991. Herpetological results of the 1990 Venezuelan expedition to the summit of Cerro Guaiquinima, with new tepui reptiles. *American Museum Novitates* 3017: 1–54.

Donnelly, T. W. 1989. Geologic history of the Caribbean and Central America. Pp. 299–321 in A. W. Bally and A. R. Palmer, eds., *The geology of North America—An overview.* Geological Society of America, Boulder, CO.

Donnelly, T. W., G. S. Horne, R. C. Finch, and E. López-Ramos. 1990. Northern Central America: The Maya and Chortis blocks. Pp. 37–76 in G. Dengo and J. E. Case, eds., *The geology of North America: The Caribbean region,* vol. H. Geological Society of America, Boulder, CO.

Donoghue, M. J. 1985. A critique of the biological species concept and recommendations for a phylogenetic alternative. *Bryologist* 88: 172–81.

———. 1995. Phylogeny and phylogenetic taxonomy of Dipsacales. *American Journal of Botany* 82 (suppl.): 108.

———. 2001. A wish list for systematic biology. *Systematic Biology* 50: 755–57.

Donoghue, M. J., and P. D. Cantino. 1988. Paraphyly, ancestors, and the goals of taxonomy: A botanical defense of cladism. *Botanical Review* 54: 107–28.

Dorlochter, M., S. Astrow, and A. Herrera. 1994. Effects of testosterone on a sexually dimorphic frog muscle: Repeated in vivo observations and androgen receptor distribution. *Journal of Neurobiology* 25: 897–916.

Dowling, H. G., C. A. Hass, S. B Hedges, and R. Highton. 1996. Snake relationships revealed by slow-evolving proteins: A preliminary survey. *Journal of Zoology* (London) 240: 1–28.

Doyle, A. C. 1912. *The lost world.* Hodder and Stoughton, London.

Drewry, G. E., and K. L. Jones. 1976. A new ovoviviparous frog, *Eleutherodactylus jasperi* (Amphibia, Anura, Leptodactylidae), from Puerto Rico. *Journal of Herpetology* 10: 161–65.

Duellman, W. E. 1966. The Central American herpetofauna: An ecological perspective. *Copeia* 1966: 700–719.

———. 1967. Courtship isolating mechanisms in Costa Rican frogs. *Herpetologica* 23: 169–83.

———. 1970. *The hylid frogs of Middle America.* University of Kansas Press, Lawrence.

———. 1978. The biology of an equatorial herpetofauna in Amazonian Ecuador. *Miscellaneous Publication University of Kansas Museum of Natural History* 65: 1–352.

———. 1979a. *The South American herpetofauna: Its origin, evolution, and dispersal.* University of Kansas Museum of Natural History Monograph, no. 7, Lawrence.

———. 1979b. The South American herpetofauna: A panoramic view. Pp. 1–28 in W. E. Duellman, ed., *The South American herpetofauna: Its origin, evolution, and dispersal.* University of Kansas Museum of Natural History Monograph, no. 7, Lawrence.

————. 1979c. The herpetofauna of the Andes: Patterns of distribution, origin, differentiation, and present communities. Pp. 371–459 in W. E. Duellman, ed., *The South American herpetofauna: Its origin, evolution, and dispersal.* University of Kansas Museum of Natural History Monograph, no. 7, Lawrence.

————. 1982. Quaternary climatic–ecological fluctuations in the lowland tropics: Frogs and forests. Pp. 389–402 in G. T. Prance, ed., *Biological diversification in the tropics.* Columbia University Press, New York.

————. 1990. Herpetofaunas in Neotropical rainforests: Comparative composition, history, and resource use. Pp. 455–505 in A. H. Gentry, ed., *Four Neotropical rainforests.* Yale University Press, New Haven, CT.

————. 1993. Amphibian species of the world: Additions and corrections. *University of Kansas Museum of Natural History Special Publications* 21: 1–372.

————. 1997. Amphibians of the La Escalera region, southeastern Venezuela: Taxonomy, ecology, and biogeography. *Scientific Papers Natural History Museum University of Kansas* 2: 1–52.

————. 1999. Distribution patterns of amphibians in South America. Pp. 255–328 in W. E. Duellman, ed., *Patterns of distribution of amphibians: A global perspective.* Johns Hopkins University Press, Baltimore, MD.

————. 2001. *The hylid frogs of Middle America.* Society for the Study of Amphibians and Reptiles, St. Louis, MO.

Duellman, W. E., and M. S. Hoogmoed. 1984. The taxonomy and phylogenetic relationships of the hylid genus *Stefania. Miscellaneous Publication University of Kansas Natural History Museum* 75: 1–39.

————. 1992. Some hylid frogs from the Guiana Highlands, northeastern South America: New species, distributional records, and a generic reallocation. *Occasional Papers of the Natural History Museum University of Kansas* 147: 1–21.

Duellman, W. E., and J. R. Mendelson III. 1995. Amphibians and reptiles from northern Departamento Loreto, Peru: Taxonomy and biogeography. *University of Kansas Science Bulletin* 55: 329–76.

Duellman, W. E., and R. A. Pyles. 1983. Acoustic resource partitioning in anuran communities. *Copeia* 1983: 639–49.

Duellman, W. E., and L. Trueb. 1986. *Biology of amphibians.* McGraw-Hill, New York.

Dunn, E. R. 1926. *The salamanders of the family Plethodontidae.* Smith College Fiftieth Anniversary Publications, Northampton, MA.

————. 1954. The coral snake "mimic" problem in Panama. *Evolution* 8: 97–102.

Dure, M. I. 1999. Natural history notes. Anura. *Leptodactylus chaquensis* (NCN). Diet. *Herpetological Review* 30: 92.

Dure, M. I., and A. I. Kehr. 1996. Natural history notes. Anura. *Bufo paracnemis* (Kururu guazu, sapo buey, sapo rococo). Diet. *Herpetological Review* 27: 138.

Dyson, M. L., N. I. Passmore, P. J. Bishop, and S. P. Henzi. 1992. Male behavior and correlates of mating success in a natural population of African painted reed frogs (*Hyperolius marmoratus*). *Herpetologica* 48: 236–46.

Edman, K., C. Reggiani, S. Schiaffino, and G. deKronnie. 1988. Maximum velocity of shortening related to myosin isoform composition in frog skeletal muscle fibres. *Journal of Physiology* 395: 679–94.

Edmunds, M. 2000. Why are there good and poor mimics? *Biological Journal of the Linnean Society* 70: 459–66.

Eens, M., R. Pinxten, and R. F. Verheyen. 1991. Male song as a cue for mate choice in the European starling. *Behaviour* 16: 210–38.

Eens, M., E. Van Duyse, and L. Berghman. 2000. Shield characteristics are testosterone dependent in both male and female moorhens. *Hormones and Behavior* 35: 126–34.

Eisenberg, J. 1979. Habitat, economy, and society: Some correlations and hypotheses for the Neotropical primates. Pp. 215–62 in I. S. Bernstein and E. O. Smith, eds., *Primate ecology and human origins*. Garland STPM Press, New York.

Ek, R. C. 1997. Botanical diversity in the tropical rain forest of Guyana. Tropenbos Guyana Series 4. Tropenbos Foundation, Wageningen, The Netherlands.

Eldredge, N., and J. Cracraft. 1980. *Phylogenetic patterns and the evolutionary process*. Columbia University Press, New York.

Emerson, S. B. 1985. Skull shape in frogs: Correlations with diet. *Herpetologica* 41: 177–88.

———. 1994. Testing pattern predictions of sexual selection: A frog example. *American Naturalist* 143: 848–69.

———. 1997. Testis size variation in frogs: Testing the alternatives. *Behavioral Ecology and Sociobiology* 41: 227–35.

———. 2000. Vertebrate secondary characteristics—Physiological mechanisms and evolutionary patterns. *American Naturalist* 156: 84–91.

Emerson, S., L. Carroll, and D. Hess. 1997. Hormonal induction of thumb pads and the evolution of secondary sexual characteristics of the Southeast Asian fanged frog, *Rana blythii*. *Journal of Experimental Zoology* 279: 587–96.

Emerson, S., A. Greig, L. Carroll, and G. Prins. 1999. Androgen receptors in two androgen-mediated, sexually dimorphic characters of frogs. *General and Comparative Endocrinology* 114: 173–80.

Emerson, S., and D. Hess. 1996. The role of androgens in opportunistic breeding, tropical frogs. *General and Comparative Endocrinology* 103: 220–30.

———. 2001. Glucocorticoids, androgens, testis mass and the energetics of vocalization in breeding male frogs. *Hormones and Behavior* 39: 59–69.

Emerson, S., and H. Voris. 1992. Competing explanations for sexual dimorphism in a voiceless Bornean frog. *Functional Ecology* 6: 654–60.

Emlen, S. T., and L. W. Oring. 1977. Ecology, sexual selection and the evolution of mating systems. *Science* 197: 215–23.

Emmons, L. 1984. Geographic variation in densities and diversities of non-flying mammals in Amazonia. *Biotropica* 16: 210–22.

Emsley, M. G. 1966. The mimetic significance of *Erythrolamprus aesculapii ocellatus* Peters from Tobago. *Evolution* 20: 663–64.

Endler, J. A. 1977. *Geographic variation, speciation, and clines*. Princeton University Press, Princeton, NJ.

———. 1982. Pleistocene forest refuges: Fact or fancy. Pp. 179–200 in G. T. Prance, ed., *Biological diversification in the tropics*. Columbia University Press, New York.

Ereshefsky, M. 2001. *The poverty of the Linnaean hierarchy: A philosophical study of biological taxonomy*. Cambridge University Press, New York.

Eriksson, T., M. J. Donoghue, and M. S. Hibbs. 1998. Phylogenetic analysis of *Potentilla* using DNA sequences of nuclear ribosomal internal transcribed spacers (ITS), and implications for the classification of *Rosoideae* (Rosaceae). *Plant Systematics and Evolution* 211: 155–79.

Erspamer, V. 1994. Bioactive secretions of the Amphibian integument. Pp. 178–350 in H. Heatwole and G. T. Barthalmus, eds., *Amphibian biology*, vol. 1: *The integument*. Surrey Beatty and Sons, Chipping Norton, New South Wales, Australia.

Escalante, G. 1990. The geology of southern Central America and western Colombia. Pp. 201–30 in G. Dengo and J. E. Case, eds., *The geology of North America: The Caribbean region*, vol. H. Geological Society of America, Boulder, CO.

Estes, R., and A. Báez. 1985. Herpetofaunas of North and South America during the late Cretaceous and Cenozoic: Evidence for interchange? Pp. 139–97 in F. G. Stehli and S. D. Webb, eds., *The great American biotic interchange*. Plenum Press, New York.

Estes, R., K. de Queiroz, and J. A. Gauthier. 1988. Phylogenetic relationships within Squamata. Pp. 119–281 in R. Estes and G. K. Pregill, eds., *Relationships of the lizard families: Essays commemorating Charles L. Camp*. Stanford University Press, Stanford, CA.

Etheridge, R. E. 1960. The relationships of the anoles (Reptilia: Sauria: Iguanidae): An interpretation based on skeletal morphology. Ph.D. dissertation, University of Michigan, Ann Arbor.

Fairbairn, D. J., and R. F. Preziosi. 1996. Sexual selection and the evolution of sexual size dimorphism in the water strider, *Aquarius remigis*. *Evolution* 50: 1549–59.

Faith, D. P., and P. S. Cranston. 1991. Could a cladogram this short have risen by chance alone? On permutation tests for cladistic structure. *Cladistics* 7: 1–28.

Farris, J. S. 1973. On comparing the shapes of taxonomic trees. *Systematic Zoology* 22: 50–54.

———. 1976a. Phylogenetic classification of fossils with Recent species. *Systematic Zoology* 25: 271–82.

———. 1976b. On the phenetic approach to vertebrate classification. Pp. 823–50 in M. K. Hecht, P. C. Goody, and B. M. Hecht, eds., *Major patterns in vertebrate evolution*. Plenum Press, New York.

———. 1979. The information content of the phylogenetic system. *Systematic Zoology* 28: 483–519.

———. 1980. The efficient diagnoses of the phylogenetic system. *Systematic Zoology* 29: 386–401.

———. 1983. The logical basis of phylogenetic analysis. Pp. 7–36 in N. I. Platnick and V. A. Funk, eds., *Advances in cladistics*, vol. 2. Columbia University Press, New York.

———. 1989. Entropy and fruit flies. *Cladistics* 5: 103–8.

———. 1999 Likelihood and inconsistency. *Cladistics* 15: 199–204.

———. 2000. Diagnostic efficiency of three-taxon analysis. *Cladistics* 16: 403–10.

Farris, J. S, V. A. Albert, M. Källersjö, D. Lipscomb, and A. G. Kluge. 1996. Parsimony jackknifing outperforms neighbor-joining. *Cladistics* 12: 99–124.

Farris, J. S., M. Källersjö, A. G. Kluge, and C. Bult. 1994. Permutations. *Cladistics* 10: 65–76.

Farris, J. S., A. G. Kluge, and J. M. Carpenter. 2001. Popper and likelihood versus "Popper*." *Systematic Biology* 50: 438–44.

Feder, M. E. 1983. Integrating the ecology and physiology of plethodontid salamanders. *Herpetologica* 39: 291–300.

Fellers, G. M. 1979a. Aggression, territoriality, and mating behavior in North American treefrogs. *Animal Behaviour* 27: 107–19.

————. 1979b. Mate selection in the gray treefrog, *Hyla versicolor. Copeia* 1979: 286–90.

Felsenstein, J. 1981. Evolutionary trees from DNA sequences: A maximum likelihood approach. *Journal of Molecular Evolution* 17: 368–76.

————. 1985. Confidence limits on phylogenies: An approach using the bootstrap. *Evolution* 39: 783–91.

————. 1993. *PHYLIP 3.5—Phylogeny inference package.* Version 3.5. Distributed by the author, Department of Genetics, University of Washington, Seattle.

Felsenstein, J., and H. Kishino. 1993. Is there something wrong with the bootstrap on phylogenies? A reply to Hillis and Bull. *Systematic Biology* 42: 193–200.

Feng, A., and P. Narins. 1991. Unusual mating behavior of Malaysian treefrogs, *Polypedates leucomystax. Naturwissenschaften* 78: 362–65.

Fernandez de Oviedo y Valdes, G. 1526. *La historia general y natural de las Indies.* Vol. 9. Toledo, Spain.

Ferrarezzi, H., and E. M. X. Freire. 2001. New species of *Bothrops* Wagler, 1824 from the Atlantic forest of Northeastern Brazil (Serpentes, Viperidae, Crotalinae). *Boletim do Museu Nacional* (Rio de Janeiro, Brazil), n.s., 440: 1–10.

Ferrusquia-Villafranca, I. 1993. Geology of Mexico: A synopsis. Pp. 3–108 in T. P. Ramamoorthy, R. Bye, A. Lot, and J. Fa, eds., *Biological diversity of Mexico: Origins and distributions.* Oxford University Press, New York.

Fincke, O. M., and H. Hadrys. 2001. Unpredictable offspring survivorship in the damselfly, *Megaloprepus coerulatus,* shapes parental behavior, constrains sexual selection and challenges traditional fitness estimates. *Evolution* 55: 762–72.

Findley, J. S., and D. E. Wilson. 1974. Observations on the Neotropical disc-winged bat, *Thyroptera tricolor* Spix. *Journal of Mammalogy* 55: 562–71.

Fischer, L., D. Catz, and D. Kelley. 1995. Androgen-directed development of the *Xenopus laevis* larynx: Control of androgen receptor expression and tissue differentiation. *Developmental Biology* 170: 115–26.

————. 1998. An androgen receptor mRNA isoform associated with hormone-induced cell proliferation. *Proceedings of the National Academy of Sciences* (USA) 90: 8254–58.

Fiske, P., P. T. Rintamaki, and E. Karionen. 1998. Mating success in lekking males: A meta-analysis. *Behavioral Ecology* 9: 328–38.

Fitch, H. S. 1949. Study of snake populations in central California. *American Midland Naturalist* 41: 1–150.

Fjeldså, J., and J. C. Lovett. 1997. Geographical patterns of old and young species in African forest biota: The significance of specific montane areas as evolutionary centres. *Biodiversity Conservation* 6: 325–46.

Fleming, I. A., and M. R. Gross. 1994. Breeding competition in a Pacific salmon (Coho: *Oncorhynchus kisutch*): Measures of natural and sexual selection. *Evolution* 48: 637–57.

Fogden, M., and P. Fogden. 1974. Animals and their colors: Camouflage, warning coloration, courtship and territorial display, mimicry. Crown Publications, New York.

Folstad, I., and A. Karter. 1992. Parasites, bright males and the immunocompetence handicap. *American Naturalist* 139: 603–22.

Ford, L. S., and J. M. Savage. 1984. A new frog of the genus *Eleutherodactylus* (Leptodactylidae) from Guatemala. *Occasional Papers of the Museum of Natural History University of Kansas* 110: 1–9.

Forey, P. L. 2001. The PhyloCode: Description and commentary. *Bulletin of Zoological Nomenclature* 58: 81–96.

Forstner, M. R. J., S. K. Davis, and E. Arévalo. 1995. Support for the hypothesis of anguimorph ancestry for the suborder Serpentes from phylogenetic analysis of mitochondrial DNA sequences. *Molecular Phylogenetics and Evolution* 4: 93–102.

Frazier, J., and S. Salas. 1983. Tortugas marinas del Pacifico oriental: ¿El Recurso que nunca acabará? Pp. 87–98 in *Symposio Conservacion y Manejo Fauna Silvestre Neotropical.* Congreso Latinamericano de Zoologia, Arequipa, Peru.

Frolich, L. M. 1991. Osteological conservatism and developmental constraint in the polymorphic "ring species" *Ensatina eschscholtzii* (Amphibia: Plethodontidae). *Biological Journal of the Linnean Society* 43: 81–100.

Frost, D. R. 1985. *Amphibian species of the world: A taxonomic and geographical reference.* Allen Press and Association of Systematics Collections, Lawrence, KS.

———, ed. 2001. *Amphibian species of the world: An online reference.* V 2.20 (http://research.amnh.org/herpetology/amphibia/index.html).

Frost, D. R., and A. G. Kluge. 1994. A consideration of epistemology in systematic biology, with special reference to species. *Cladistics* 10: 259–94.

Frost, D. R., M. T. Rodrigues, T. Grant, and T. A. Titus. 2001. Phylogenetics of the lizard genus *Tropidurus* (Squamata: Tropiduridae: Tropidurinae): Direct optimization, descriptive efficiency, and sensitivity analysis of congruence between molecular data and morphology. *Molecular Phylogenetics and Evolution* 21: 352–71.

Fuller, R. C. 1999. Costs of group spawning to guarding males in the rainbow darter, *Etheostoma caeruleum. Copeia* 1999: 1084–88.

Gadow, H. 1911. Isotely and coralsnakes. *Zoologische Jahrbücher, Abteilung für Systematik, Ökologie und Geographie der Tiere* 31: 1–24.

Galán, C. 2000. Herpetofauna colectada en expediciones a cavidades en tepuyes de la Guayana Venezolana. *Boletín de la Sociedad Venezolana de Espeleología* 34: 11–19.

Gallardo, J. M. 1957. Las subespecies argentinas de *Bufo granulosus* Spix. *Revista del Museo Argentino de Ciencias Naturales "Bernardino Rivadavia"* 3: 337–74.

———. 1966. Zoogeografía de los anfibios chaqueños. *Physis* 26: 67–81.

———. 1979. Composición, distribución y origen de la herpetofauna chaqueña. Pp. 299–307 in W. E. Duellman, ed., *The South American herpetofauna: Its origin, evolution, and dispersal.* University of Kansas Museum of Natural History Monograph, no. 7, Lawrence.

Gans, C. 1961a. The first record of egg-laying in *Siphonops paulensis* Boettger. *Copeia* 1961: 490–91.

———. 1961b. Mimicry in procryptically colored snakes of the genus *Dasypeltis. Evolution* 15: 72–91.

———. 1961c. A bullfrog and its prey: A look at the bio-mechanics of jumping. *Natural History,* 26–37.

———. 1973. Uropeltids: Survivors in a changing world. *Endeavour* 32: 60–65.

Gans, C., and G. C. Gorniak. 1982. Functional morphology of lingual protrusion in marine toads (*Bufo marinus*). *American Journal of Anatomy* 163: 195–222.

Gansser, A. 1974. The Roraima problem (South America). *Verhandlungen der Naturforschenden Gesellschaft in Basel* 84: 80–97.

García-París, M., D. A. Good, G. Parra-Olea, and D. B. Wake. 2000. Biodiversity of

Costa Rican salamanders: Implications of high levels of genetic differentiation and phylogeographic structure for species formation. *Proceedings of the National Academy of Sciences* (USA) 97: 1640–47.

García-París, M., G. Parra-Olea, and D. B. Wake. 2000. Phylogenetic relationships within the lowland tropical salamanders of the *Bolitoglossa mexicana* complex (Amphibia: Plethodontidae). Pp. 199–214 in R. Bruce, R. G. Jaeger, and L. D. Houck, eds., *The biology of plethodontid salamanders.* Kluwer Academic/Plenum Publishers, New York.

García-París, M., and D. B. Wake. 2000. Molecular phylogenetic analysis of relationships of the tropical salamanders genera *Oedipina* and *Nototriton,* with descriptions of a new genus and three new species. *Copeia* 2000: 42–70.

Gardiner, P. 1952. *The nature of historical explanation.* Oxford University Press, Oxford.

Garton, J. D., and H. R. Mushinsky. 1979. Integumentary toxicity and unpalatability as an antipredator mechanism in the narrow mouthed toad, *Gastrophryne carolinensis. Canadian Journal of Zoology* 57: 1965–73.

Gasc, J. P., and M. T. Rodrigues. 1980. Liste préliminaire des serpents de la Guyane française. *Bulletin Museum Natural Historie Paris* (4th ser.) 2, sec. A, no. 2: 559–98.

Gascon, C. 1991. Population- and community-level analysis of species occurrences of central Amazonian rainforest tadpoles. *Ecology* 72: 1731–46.

———. 1995. Tropical larval anuran fitness in the absence of direct effects of predation and competition. *Ecology* 76: 2222–29.

Gatz, A. J., Jr. 1981a. Size selective mating in *Hyla versicolor* and *Hyla crucifer. Journal of Herpetology* 15: 114–16.

———. 1981b. Non-random mating by size in American toads, *Bufo americanus. Animal Behaviour* 29: 1004–12.

Gauthier, J. A., R. Estes, and K. de Queiroz. 1988. A phylogenetic analysis of Lepidosauromorpha. Pp. 15–98 in R. Estes and G. K. Pregill, eds., *Relationships of the lizard families: Essays commemorating Charles L. Camp.* Stanford University Press, Stanford, CA.

Gauthier, J. A., A. G. Kluge, and T. Rowe. 1988. Amniote phylogeny and the importance of fossils. *Cladistics* 4: 105–208.

Gehlbach, F. R. 1972. Coral snake mimicry reconsidered: The strategy of self mimicry. *Forma et Functio* 5: 311–320.

Gene Codes Corporation. 1999. *Sequencher.* Version 4.0. Ann Arbor, MI.

Gerhardt, H. 1991. Female choice in treefrogs: Static and dynamic acoustic criteria. *Animal Behaviour* 42: 615–35.

———. 1994. The evolution of vocalization in frogs and toads. *Annual Review of Ecology and Systematics* 25: 293–324.

Gerhardt, H. C., R. E. Daniel, S. A. Perrill, and S. Schramm. 1987. Mating behavior and male mating success in the green treefrog. *Animal Behaviour* 35: 1490–1503.

Gerhardt, H. C., and J. J. Schwartz. 1995. Interspecific interactions in anuran courtship. Pp. 603–32 in H. Heatwole, ed., *Amphibian biology,* vol. 2: *Social behavior.* Chipping Norton, New South Wales, Australia.

Gerhardt, H., and G. Watson. 1995. Within-male variability in call properties and female preference in the grey treefrog. *Animal Behaviour* 50: 1187–91.

Ghiselin, M. T. 1966. An application of the theory of definitions to systematic principles. *Systematic Zoology* 15: 127–30.

————. 1995. Ostensive definitions of the names of species and clades. *Biology and Philosophy* 10: 219–22.

————. 1997. *Metaphysics and the origin of species.* State University of New York Press, Albany.

Ghosh, S. K. 1977. Geologia del grupo Roraima en Territorio Federal Amazonas, Venezuela. Pp. 167–93 in *Memorias, 5th Congreso Geologico Venezolano,* vol. 1.

————. 1985. Geology of the Roraima Group and its implications. Pp. 33–50 in M. I. Muñoz, ed., *Memoria i Simposium Amazónico (Puerto Ayacucho 1981),* Boletín de Geología, Publicación Especial no. 10. Ministerio de Energia y Minas, Dirección de Geología, Caracas, Venezuela.

Gibbs, R., W. Kingston, R. Jozefowick, B. Herr, G. Forbes, and D. Halliday. 1989. Effect of testosterone on muscle mass and muscle protein synthesis. *Journal of Applied Physiology* 66: 498–503.

Gibson, R. M. 1987. Bivariate versus multivariate analyses of sexual selection in red deer. *Animal Behaviour* 35: 292–305.

————. 1989. Field playback of male display attracts females in lek breeding sage grouse. *Behavioral Ecology and Sociobiology* 24: 439–43.

Gibson, R. M., and J. W. Bradbury. 1985. Sexual selection in lekking sage grouse: Phenotypic correlates of male mating success. *Behavioral Ecology and Sociobiology* 18: 117–23.

Gillison, G. 1993. *Between culture and fantasy: A New Guinea highlands mythology.* University of Chicago Press, Chicago.

Girgenrath, M., and R. Marsh. 1997. In vivo performance of trunk muscles in tree frogs during calling. *Journal of Experimental Biology* 200: 3101–8.

Given, M. F. 1988. Territoriality and aggressive interactions of male carpenter frogs, *Rana virgatipes. Copeia* 1988: 411–21.

Givnish, T. J., R. W. McDiarmid, and W. R. Buck. 1986. Fire adaptation in *Neblinaria celiae* (Theaceae), a high-elevation rosette shrub endemic to a wet equatorial tepui. *Oecologia* 70: 481–85.

Glaw, F., and J. Köhler. 1998. Amphibian species diversity exceeds that of mammals. *Herpetological Review* 29: 11–12.

Gloyd, H. K., and R. Conant. 1990. *Snakes of the Agkistrodon complex: A monographic review.* Contributions in Herpetology, no. 6. Society for the Study of Amphibians and Reptiles, St. Louis, MO.

Goeldi, E. 1899. Ueber die Entwicklung von *Siphonops annulatus. Zoologische Jahrbuch Abteilung Systematik* 12: 170–73.

Goin, C. J. 1971. A synopsis of the tree frogs of Suriname. *Annals of the Carnegie Museum* 43: 1–23.

Golay, P., H. M. Smith, D. G. Broadley, J. R. Dixon, C. McCarthy, J.-C. Rage, B. Schatti, and M. Toriba. 1993. *Endoglyphs and other major venomous snakes of the world: A checklist.* Azemiops, S.A., Geneva, Switzerland.

Goldstein, P. Z., and R. DeSalle. 2000. Phylogenetic species, nested hierarchies, and character fixation. *Cladistics* 16: 364–84.

Goloboff, P. A. 1993. Estimating character weights during tree search. *Cladistics* 9: 83–91.

Good, D. A., and D. B. Wake. 1993. Systematic studies of the Costa Rican moss salamanders, genus *Nototriton,* with descriptions of three new species. *Herpetological Monographs* 7: 131–59.

————. 1997. Phylogenetic and taxonomic implications of protein variation in the Mesoamerican salamander genus *Oedipina* (Caudata: Plethodontidae). *Revista de Biologia Tropical* 45: 1185–1208.

Goodman, D. E. 1971. Territorial behavior in a Neotropical frog, *Dendrobates granuliferus. Copeia* 1971: 365–70.

Goodpasture, C., and S. E. Bloom. 1975. Visualization of nucleolar organizer regions in mammalian chromosomes using silver staining. *Chromosoma* (Berlin) 53: 37–50.

Gorman, G. C., and L. Atkins. 1969. New karyotypic data for 16 species of *Anolis* (Sauria: Iguanidae) from Cuba, Jamaica, and the Cayman Islands. *Herpetologica* 24: 13–21.

Gorman, G. C., D. G. Buth, M. Soulé, and S. Y. Yang. 1980. The relationships of the *Anolis cristatellus* species group: Electrophoretic analysis. *Journal of Herpetology* 14: 269–78.

————. 1983. The relationships of the Puerto Rican *Anolis:* Electrophoretic and karyotypic studies. Pp. 626–42 in G. J. Rhodin and K. Miyata, eds., *Advances in herpetology and evolutionary biology.* Museum of Comparative Zoology, Harvard University, Cambridge.

Gorman, G. C., D. G. Buth, and J. S. Wyles. 1980. *Anolis* lizards of the eastern Caribbean: A case study in evolution. III. A cladistic analysis of albumin immunological data, and the definition of species groups. *Systematic Zoology* 29: 143–58.

Gorman, G. C., R. B. Huey, and E. E. Williams. 1969. Cytotaxonomic studies on some unusual iguanid lizards assigned to the genera *Chamaeleolis, Polychrus, Polychroides,* and *Phenacosaurus. Breviora* 316: 1–17.

Gorman, G. C., and Y. S. Kim. 1976. *Anolis* lizards of the eastern Caribbean: A case study in evolution. II. Genetic relationships and genetic variation of the *bimaculatus* group. *Systematic Zoology* 25: 62–77.

Gorman, G. C., C. S. Lieb, and R. H. Harwood. 1984. The relationships of *Anolis gadovii:* Albumin immunological evidence. *Caribbean Journal of Science* 20: 145–52.

Gorman, G. C., and B. Stamm. 1975. The *Anolis* lizards of Mona, Redonda, and La Blanquilla: Chromosomes, relationships, and natural history notes. *Journal of Herpetology* 9: 197–205.

Gorzula, S. 1985. Field notes on *Otophryne robusta steyermarki. Herpetological Review* 16: 102–3.

————. 1988. Una nueva especie de *Dendrobates* (Amphibia, Dendrobatidae) del Macizo del Chimantá, Estado Bolívar, Venezuela. *Memoria Sociedad Ciencias Naturales La Salle* 48: 143–49.

————. 1992. La herpetofauna del macizo del Chimantá. Pp. 267–80 and 304–10 in O. Huber, ed., *El macizo del Chimantá.* Oscar Todtmann Editores, Caracas, Venezuela.

Gorzula, S., and J. Ayarzagüena. 1995. Dos nuevas especies del género *Thamnodynastes* (Serpentes; Colubridae) de los tepuyes de la Guayana Venezolana. *Publicaciones de la Asociación de Amigos de Doñana* 6: 1–17.

Gorzula, S., and J. C. Señaris. 1998. Contribution to the herpetofauna of the Venezuelan Guayana. I. A data base. *Scientia Guaianae* 8: 1–269.

Gosner, K. L. 1960. A simplified key for staging anuran embryos and larvae with notes on identification. *Herpetologica* 16: 183–90.

Gradstein, S. R. 1985. The 1985 expedition to Mount Roraima (Guyana). *Botanical Explorations in Guyana: Appendix I. Flora of Guianas,* newsletter no. 2: 1–17.

Graff, L. von. 1899. *Monographie der Turbellarian II. Tricladia terricola.* Engelmann, Leipzig.

Grant, T. 2002. Testing methods: The evaluation of discovery operations in evolutionary biology. *Cladistics* 18: 94–111.

Gray, J. E. 1868. Notice of two species of salamandra from Central America. *Annals and Magazine of Natural History* (ser. 4) 2: 7–28.

Graybeal, A. 1997. Phylogenetic relationships of bufonid frogs and tests of alternate macroevolutionary hypotheses characterizing their radiation. *Zoological Journal of the Linnean Society* 119: 297–338.

Green, A. J. 1990. Determinants of chorus participation and the effects of size, weight, and competition on advertisement calling in the tungara frog, *Physalaemus pustulosus* (Leptodactylidae). *Animal Behaviour* 39: 620–38.

Green, D. M. 1988. Cytogenetics of the endemic New Zealand frog, *Leiopelma hochstetteri:* Extraordinary supernumerary chromosome variation and a unique sex-chromosome system. *Chromosoma* (Berlin) 97: 55–70.

———. 1990. Muller's Ratchet and the evolution of supernumerary chromosomes. *Genome* 33: 818–24.

———. 1991. Supernumerary chromosomes in amphibians. Pp. 333–57 in D. M. Green and S. K. Sessions, eds., *Amphibian cytogenetics and evolution.* Academic Press, New York.

Greenberg, B. 1942. Some effects of testosterone on the sexual pigmentation and other sex characters of the cricket frog (*Acris gryllus*). *Journal of Experimental Zoology* 91: 435–51.

Greene, H. W. 1986. Natural history and evolutionary biology. Pp. 99–108 in M. E. Feder and G. V. Lauder, eds., *Predator–prey relationships: Perspectives and approaches from the study of lower vertebrates.* University of Chicago Press, Chicago.

———. 1988. Antipredator mechanisms in reptiles. Pp. 1–152 in C. Gans and R. B. Huey, eds., *Biology of the Reptilia,* vol. 16: *Ecology B: Defense and life history.* John Wiley, New York.

———. 1989. Defensive behavior and feeding biology of the Asian Mock Viper, *Psammodynastes pulverulentus* (Colubridae), a specialized predator on scincid lizards. *Chinese Herpetological Research* 2: 21–32.

———. 1992. The ecological and behavioral context of pitviper evolution. Pp. 107–17 in J. A. Campbell and E. D. Brodie Jr., eds., *Biology of pitvipers.* Selva, Tyler, TX.

———. 1997. *Snakes: The evolution of mystery in nature.* University of California Press, Berkeley.

Greene, H. W., and D. Cundall. 2000. Limbless tetrapods and snakes with legs. *Science* 287: 1939–41.

Greene, H. W., and R. W. McDiarmid. 1981. Coral snake mimicry: Does it occur? *Science* 213: 1207–12.

Griffiths, G. C. D. 1973. Some fundamental problems in biological classification. *Systematic Zoology* 22: 338–43.

———. 1974. On the biological foundations of biological systematics. *Acta Biotheoretica* 23: 85–131.

———. 1976. The future of Linnaean nomenclature. *Systematic Zoology* 25: 168–73.

Griffiths, P. E. 1999. Squaring the circle: Natural kinds with historical essences. Pp. 209–28 in R. A. Wilson, ed., *Species: New interdisciplinary essays.* MIT Press, Cambridge.

Grismer, L. L. 1983. A reevaluation of the North American gekkonid genus *Anarbylus* Murphy and its cladistic relationships to *Coleonyx* Gray. *Herpetologica* 39: 394–99.

———. 1988. The phylogeny, taxonomy, classification, and biogeography of eublepharid geckos (Reptilia: Squamata). Pp. 369–469 in R. Estes and G. K. Pregill, eds., *Relationships of the lizard families: Essays commemorating Charles L. Camp*. Stanford University Press, Stanford, CA.

Grismer, L. L., H. Ota, and S. Tanaka. 1994. Phylogeny, classification, and biogeography of *Goniurosaurus kuroiwae* (Squamata: Eublepharidae) from the Ryukyu Archipelago, Japan, with description of a new subspecies. *Zoological Science* 11: 319–35.

Grismer, L. L., B. E. Viets, and L. J. Boyle. 1999. Two new continental species of *Goniurosaurus* (Squamata: Eublepharidae) with a phylogeny and evolutionary classification of the genus. *Journal of Herpetology* 33: 382–93.

Gu, X., Y.-X. Fu, and W.-H. Li. 1995. Maximum likelihood estimation of the heterogeneity of substitution rate among nucleotide sites. *Molecular Biology and Evolution* 12: 546–57.

Guix, J. C. 1993. Hábitat y alimentación de *Bufo paracnemis* en una región semiárida del nordeste de Brasil, durante el período de reproducción. *Revista Española de Herpetología* 7: 65–73.

Gurrola-Hidalgo, M. A., and N. Chavez C. 1996. *Lampropeltis triangulum nelsoni* (milk snake). Predation. *Herpetological Review* 27: 83.

Gutberlet, R. L., Jr. 1998. The phylogenetic position of the Mexican black-tailed pitviper (Squamata: Viperidae: Crotalinae). *Herpetologica* 54: 184–206.

Gutberlet, R. L., Jr., and J. A. Campbell. 2001. Generic recognition for a neglected lineage of South American pitvipers (Squamata: Viperidae: Crotalinae), with the description of a new species from the Colombian Chocó. *American Museum Novitates* 3316: 1–15.

Gutberlet, R. L., Jr., and M. B. Harvey. 2002. Phylogenetic relationships of New World pitvipers as inferred from anatomical evidence. Pp. 51–68 in G. W. Schuett, M. Hoggren, M. E. Douglas, and H. W. Greene, eds. *Biology of the vipers*. Eagle Mountain Publishing, Eagle Mountain, UT.

Guyer, C. 1990. The herpetofauna of La Selva, Costa Rica. Pp. 371–85 in A. H. Gentry, ed., *Four Neotropical rainforests*. Yale University Press, New Haven, CT.

———. 1992. A review of estimates of nonreciprocity in immunological studies. *Systematic Biology* 41: 85–88.

———. 1994a. The reptile fauna: Diversity and ecology. Pp. 210–16 in L. A. McDade, K. S. Bawa, H. A. Hespenheide, and G. S. Hartshorn, eds., *La Selva: Ecology and natural history of a Neotropical rain forest*. University of Chicago Press, Chicago.

———. 1994b. Appendix 6: Reptiles. Pp. 382–83 in L. A. McDade, K. S. Bawa, H. A. Hespenheide, and G. S. Hartshorn, eds., *La Selva: Ecology and natural history of a Neotropical rain forest*. University of Chicago Press, Chicago.

Guyer, C., and J. M. Savage. 1986. Cladistic relationships among anoles (Sauria: Iguanidae). *Systematic Zoology* 35: 509–31.

———. 1992. Anole systematics revisited. *Systematic Biology* 41: 89–110.

Hackett, S. J. 1993. Phylogenetic and biogeographic relationships in the Neotropical genus *Gymnopithys* (Formicariidae). *Wilson Bulletin* 105: 301–15.

Haffer, J. 1985. Avian zoogeography of the Neotropical lowlands. *Ornithological Monographs* 36: 113–46.

———. 1987. Quaternary history of tropical America. Pp. 1–18 in T. C. Whitmore and G. T. Prance, eds., *Biogeography and Quaternary history in tropical America*. Oxford Monographs on Biogeography, no. 3. Oxford University Press, Oxford.

———. 1997. Alternative models of vertebrate speciation in Amazonia: An overview. *Biodiversity and Conservation* 6: 451–76.

Hailman, J. P. 1977. *Optical signals: Animal communication and light*. Indiana University Press, Bloomington.

Hamilton, W., and M. Zuk. 1982. Heritable true fitness and bright birds: A role for parasites? *Science* 218: 384–87.

Hanlon, R. T., J. W. Forsythe, and D. E. Joneschild. 1999. Crypsis, conspicuousness, mimicry and polyphenism as antipredator defences of foraging octopuses of Indo-Pacific coral reefs, with a method of quantifying crypsis from video tapes. *Biological Journal of the Linnean Society* 66: 1–22.

Hanski, I., J. Kouki, and A. Halkka. 1993. Three explanations of the positive relationship between distribution and abundance of species. Pp. 108–16 in R. E. Ricklefs and D. Schluter, eds., *Species diversity in ecological communities*. University of Chicago Press, Chicago.

Härlin, M. 1998. Taxonomic names and phylogenetic trees. *Zoologica Scripta* 27: 381–90.

Härlin, M., and P. Sundberg. 1998. Taxonomy and philosophy of names. *Biology and Philosophy* 13: 233–44.

Harris, D. J., E. A. Sinclair, N. L. Mercader, J. C. Marshall, and K. A. Crandall. 1999. Squamate relationships based on c-*mos* nuclear DNA sequences. *Herpetological Journal* 9: 147–51.

Harshman, J. 1994. The effect of irrelevant characters on bootstrap values. *Systematic Biology* 43: 419–24.

Hartmann, P. A. 2001. Hábito alimentar e utilização do ambiente em duas espécies simpátricas de *Philodryas* (Serpentes, Colubridae), no sul do Brasil. Master's thesis, Universidade Estadual Paulista, Rio Claro, Brazil.

Hartshorn, G. S., and B. E. Hammel. 1994. Vegetation types and floritic patterns. Pp. 73–89 in L. A. McDade, K. S. Bawa, H. A. Hespenheide, and G. S. Hartshorn, eds., *La Selva: Ecology and natural history of a Neotropical rain forest*. University of Chicago Press, Chicago.

Hasegawa, M., H. Kishino, and T. Yano. 1985. Dating of the human–ape split by a molecular clock of mitochondrial DNA. *Journal of Molecular Evolution* 22: 160–74.

Hass, C. A., S. B. Hedges, and L. R. Maxson. 1993. Molecular insights into the relationships and biogeography of West Indian anoline lizards. *Biochemical Systematics and Ecology* 21: 97–114.

Hass, C. A., R. A. Nussbaum, and L. R. Maxson. 1993. Immunological insights into the evolutionary history of caecilians (Amphibia: Gymnophiona): Relationships of the Seychellean caecilians and a preliminary report on family-level relationships. *Herpetological Monographs* 6: 56–63.

Hay, O. P. 1892. *Batrachians and reptiles of Indiana*. Annual Report of the State Geologist of Indiana, no. 10.

Hayes, M. P. 1983. A technique for partitioning hatching and mortality estimates in leaf-breeding frogs. *Herpetological Review* 14: 115–16.

Hayes, M., and D. Krempels. 1986. Vocal sac variation among frogs of the genus *Rana* from western North America. *Copeia* 1986: 927–36.

Hayes, T., and K. Menendez. 1999. The effect of sex steroids on primary and secondary sex differentiation in the sexually dichromatic reedfrog (*Hyperolius argus:* Hyperoliidae) from the Arabuko Sokoke Forest of Kenya. *General and Comparative Endocrinology* 115: 188–99.

Healy, K. 1994. An international travel map—Venezuela, no. 596. International Travel Maps, Vancouver, B.C., Canada.

Hecht, M., and J. Edwards. 1976. The determination of parallel or monophyletic relationships: The proteid salamanders—A test case. *American Naturalist* 110: 653–77.

Heck, K. L., Jr., and M. P. Weinstein. 1978. Mimetic relationships between tropical burrfishes and opisthobranchs. *Biotropica* 10: 78–79.

Hedges, S. B. 1989. Evolution and biogeography of West Indian frogs of the genus *Eleutherodactylus:* Slow-evolving loci and the major groups. Pp. 305–70 in C. Woods, ed., *Biogeography of the West Indies: Past, present, and future.* Sandhill Crane Press, Gainesville, FL.

Hedges, S. B., and K. L. Burnell. 1990. The Jamaican radiations of *Anolis* (Sauria: Iguanidae): An analysis of relationships and biogeography using sequential electrophoresis. *Caribbean Journal of Science* 26: 31–44.

Hedges, S. B., R. A. Nussbaum, and L. R. Maxson. 1993. Caecilian phylogeny and biogeography inferred from the mitochondrial DNA sequences of the 12S rRNA and 16S rRNA genes (Amphibia: Gymnophiona). *Herpetological Monographs* 6: 64–76.

Hedrick, A. V. 1986. Female preferences for male calling bout duration in a field cricket. *Behavioral Ecology and Sociobiology* 19: 73–77.

Heinrich, G. 1996. *Micrurus fulvius fulvius* (Eastern Coral Snake): Diet. *Herpetological Review* 27: 25.

Hcise, P. J., L. R. Maxson, H. G. Dowling, and S. B. Hedges. 1995. Higher-level snake phylogeny inferred from mitochondrial DNA sequences of 12S and 16S rRNA genes. *Molecular Biology and Evolution* 12: 259–65.

Henderson, L. M., and A. N. Bruère. 1980. Inheritance of Ag-stainability of the nucleolus organizer regions in domestic sheep, *Ovis aries. Cytogenetics and Cell Genetics* 26: 1–6.

Henderson, R. W. 1997. A taxonomic review of the *Corallus hortulanus* complex of Neotropical tree boas. *Caribbean Journal of Science* 33: 198–221.

Hendrickson, D. A. 1986. Congruence of bolitoglossine biogeography and phylogeny with geologic history: Paleotransport on displaced suspect terranes? *Cladistics* 2: 113–29.

Hendy, M. D., M. A. Steel, D. Penny, and T. M. Henderson. 1988. Families of trees and consensus. Pp. 355–62 in H. H. Bock, ed., *Classification and related methods of analysis.* Elsevier Science Publishers B.V., North Holland.

Hennig, W. 1965. Phylogenetic systematics. *Annual Review of Entomology* 10: 97–116.

———. 1966. *Phylogenetic systematics.* University of Illinois Press, Urbana.

———. 1981. *Insect phylogeny.* John Wiley, Chichester, U.K.

Hews, D. K. 1990. Examining hypotheses generated by field measurements of sexual selection on male lizards, *Uta palmeri. Evolution* 44: 1956–66.

Hews, D., and M. Moore. 1995. Influence of androgens on differentiation of secondary sex characteristics in tree lizards, *Urosaurus ornatus. General and Comparative Endocrinology* 97: 86–102.

Heyer, W. R. 1975. A preliminary analysis of the intergeneric relationships of the frog family Leptodactylidae. *Smithsonian Contributions to Zoology* 199: 1–55.

———. 1984. Variation, systematics, and zoogeography of *Eleutherodactylus guentheri* and closely related species (Amphibia: Anura: Leptodactylidae). *Smithsonian Contributions to Zoology* 402: 1–42.

———. 1988. On frog distribution patterns east of the Andes. Pp. 245–73 in P. E. Vanzolini and W. R. Heyer, eds., *Proceedings of a workshop on Neotropical distribution patterns.* Academia Brasileira de Ciencias, Rio de Janeiro, Brazil.

———. 1995. South American rocky habitat *Leptodactylus* (Amphibia: Anura: Leptodactylidae) with description of two new species. *Proceedings of the Biological Society of Washington* 108: 695–716.

———. 1998. The relationships of *Leptodactylus diedrus* (Anura, Leptodactylidae). *Alytes* 16: 1–24.

Heyer, W. R., R. de Sá, J. R. McCranie, and L. D. Wilson. 1996. *Leptodactylus silvanimbus* (Amphibia: Anura: Leptodactylidae): Natural history notes, advertisement call, and relationships. *Herpetological Natural History* 4: 169–74.

Heyer, W. R., M. A. Donnelly, R. W. McDiarmid, L. C. Hayek, and M. S. Foster, eds. 1994. *Measuring and monitoring biological diversity: Standard methods for amphibians.* Smithsonian Institution Press, Washington, DC.

Heyer, W. R., and L. R. Maxson. 1982. Distributions, relationships, and zoogeography of lowland frogs, the *Leptodactylus* complex in South America, with special reference to Amazonia. Pp. 375–88 in G. T. Prance, ed., *Biological diversification in the tropics.* Columbia University Press, New York.

Heyer, W. R., R. W. McDiarmid, and D. L. Weigmann. 1975. Tadpoles, predation and pond habitats in the tropics. *Biotropica* 7: 100–111.

Hibbett, D. S., and M. J. Donoghue. 1998. Integrating phylogenetic analysis and classification in fungi. *Mycologia* 90: 347–56.

Highton, R. 2000. Detecting cryptic species using allozyme data. Pp. 215–41 in R. C. Bruce, R. G. Jaeger, and L. D. Houck, eds., *The biology of plethodontid salamanders.* Kluwer Academic/Plenum Publishers, New York.

Hillis, D. M. 1987. Molecular versus morphological approaches to systematics. *Annual Review of Ecology and Systematics* 18: 23–42.

———. 1991. Discriminating between phylogenetic signal and random noise in DNA sequences. Pp. 278–94 in M. M. Miyamoto and J. Cracraft, eds., *Phylogenetic analysis of DNA sequences.* Oxford University Press, New York.

Hillis, D. M., and J. J. Bull. 1993. An empirical test of bootstrapping as a method for assessing confidence in phylogenetic analysis. *Systematic Biology* 42: 182–92.

Hillis, D. M., and M. T. Dixon. 1991. Ribosomal DNA: Molecular evolution and phylogenetic inference. *Quarterly Review of Biology* 66: 411–53.

Hillis, D. M., and J. P. Huelsenbeck. 1992. Signal, noise, and reliability in molecular phylogenetic analyses. *Journal of Heredity* 83: 189–95.

Hillis, D. M., B. K. Mable, A. Larson, S. K. Davis, and E. A. Zimmer. 1996. Nucleic acids IV: Sequencing and cloning. Pp. 321–81 in D. M. Hillis, C. Moritz, and B. K. Mable, eds., *Molecular systematics,* 2nd ed. Sinauer Associates, Sunderland, MA.

Hillman, P. E. 1969. Habitat specificity in three sympatric species of *Ameiva* (Reptilia: Teiidae). *Ecology* 50: 476–81.

Himstedt, W. 1996. *Die Blindwühlen.* New Brehm Library, vol. 630, Westarp Wissenschaften, Magdeburg. Spektrum Adademischer Verlag, Heidelberg, Germany.

Hitchcock, C. B. 1947. The Orinoco-Ventuari region, Venezuela. *Geographical Review* 37: 525–66.

Högland, J., and J. G. M. Robertson. 1990. Female preferences, male decision rules and the evolution of leks in the great snipe *Gallinago media. Animal Behaviour* 40: 15–22.

Holdridge, L. R. 1967. *Life zone ecology.* 2nd ed. Tropical Science Center, San José, Costa Rica.

———. 1982. *Life zone ecology.* Tropical Science Center, San José, Costa Rica.

Holdridge, L. R., W. C. Grenke, W. H. Hathaway, T. Liang, and J. A. Tosi Jr. 1971. *Forest environments in tropical life zones: A pilot study.* Pergamon Press, Oxford.

———. 1978. *Forest environments in tropical lifezones.* Pergamon Press, New York.

Hoogmoed, M. S. 1973. Notes on the herpetofauna of Surinam. IV. The lizards and amphisbaenians of Surinam. *Biogeographica* 4: 1–419.

———. 1979a. The herpetofauna of the Guianan region. Pp. 241–79 in W. E. Duellman, ed., *The South American herpetofauna: Its origin, evolution, and dispersal.* University of Kansas Museum of Natural History Monograph, no. 7, Lawrence.

———. 1979b. Resurrection of *Hyla ornatissima* Noble (Amphibia, Hylidae) and remarks on related species of green tree frogs from the Guiana area. Notes on the herpetofauna of Surinam VI. *Zoologische Verhandelingen* 172: 1–46.

———. 1980. Revision of the genus *Atractus* in Surinam, with the resurrection of two species (Colubridae, Reptilia). Notes on the herpetofauna of Surinam VII. *Zoologische Verhandelingen* 175: 1–47.

———. 1982. Snakes of the Guianan region. *Memorias do Instituto Butantan* 46: 219–54.

———. 1985. *Xenodon werneri* Eiselt, a poorly known snake from Guiana, with notes on *Waglerophis merremii* (Wagler) (Reptilia: Serpentes: Colubridae): Notes on the herpetofauna of Surinam IX. *Zoologische Mededelingen* 59: 79–88.

Hoogmoed, M. S., and T. C. S. Avila-Pires. 1991. Annotated checklist of the herpetofauna of Petit Saut, Sinnamary River, French Guiana. *Zoologische Mededelingen* (Leiden) 65: 63–88.

Hoogmoed, M. S., and J. Lescure. 1984. A new genus and two new species of minute leptodactylid frogs from northern South America, with comments upon *Phyzelaphryne* (Amphibia: Anura: Leptodactylidae). *Zoologische Mededelingen* 58: 84–115.

Hoorn, C. 1993. Marine incursions and the influence of Andean tectonics on the Miocene depositional history of northwestern Amazonia: Results of a palynostratigraphic study. *Paleoclimatology, Paleogeography, Paleoecology* 105: 267–309.

Hoorn, C., J. Guerrero, G. A. Sarmiento, and M. A. Lorente. 1995. Andean tectonics as a cause for changing drainage patterns in Miocene northern South America. *Geology* 23: 237–40.

Houtman, A. M. 1992. Female zebra finches choose extra pair copulations with genetically attractive males. *Proceedings of the Royal Society of London B* 249: 3–6.

Hovenkamp, P. H. 1997. Vicariance events, not areas, should be used in biogeographic analysis. *Cladistics* 13: 67–79.

Howard, R. D. 1978a. The influence of male-defended oviposition sites on early embryo mortality in bullfrogs. *Ecology* 59: 789–98.

————. 1978b. The evolution of mating strategies in bullfrogs, *Rana catesbeiana*. *Evolution* 32: 850–71.

Howard, R. R., and E. D. Brodie Jr. 1973. A Batesian mimicry complex in salamanders: Responses of avian predators. *Herpetologica* 29: 33–41.

Howell, W. M., and D. A. Black. 1980. Controlled silver-staining of nucleolus organizer regions with a protective colloidal developer: A 1-step method. *Experientia* 36: 1014–15.

Huber, O. 1987. Consideraciones sobre el concepto de Pantepui. *Pantepui* 2: 2–10.

————, ed. 1992. *El macizo del Chimantá*. Oscar Todtmann Editores, Caracas, Venezuela.

————. 1994. Recent advances in the phytogeography of the Guayana region. *Mémoires de la Société de Biogéographie* (Paris) 4: 53–63.

————. 1995a. Geographical and physical features. Pp. 1–61 in J. A. Steyermark, P. E. Berry, and B. K. Holst, eds., *Flora of the Venezuelan Guayana*, vol. 1: *Introduction*. Missouri Botanical Garden and Timber Press, St. Louis, MO, and Portland, OR.

————. 1995b. History of botanical exploration. Pp. 63–95 in J. A. Steyermark, P. E. Berry, and B. K. Holst, eds., *Flora of the Venezuelan Guayana*, vol. 1: *Introduction*. Missouri Botanical Garden and Timber Press, St. Louis, MO, and Portland, OR.

————. 1995c. Vegetation. Pp. 97–160 in J. A. Steyermark, P. E. Berry, and B. K. Holst, eds., *Flora of the Venezuelan Guayana*, vol. 1: *Introduction*. Missouri Botanical Garden and Timber Press, St. Louis, MO, and Portland, OR.

Huber, O., and J. J. Wurdack. 1984. History of botanical exploration in Territorio Federal Amazonas, Venezuela. *Smithsonian Contributions to Botany* 56: 1–83.

Huelsenbeck, J. P. 1991. Tree-length distribution skewness: An indicator of phylogenetic information. *Systematic Zoology* 40: 257–70.

Huheey, J. E. 1988. Mathematical models of mimicry. Pp. 22–41 in L. P. Brower, ed., *Mimicry and the evolutionary process*. University of Chicago Press, Chicago.

Hull, D. L. 1964. Consistency and monophyly. *Systematic Zoology* 13: 1–11.

————. 1974. *Philosophy of biological science*. Prentice-Hall, Englewood Cliffs, NJ.

————. 1978. A matter of individuality. *Philosophy of Science* 45: 335–60.

————. 1980. Individuality and selection. *Annual Review of Ecology and Systematics* 11: 311–32.

————. 1988. *Science as a process: An evolutionary account of the social and conceptual development of science*. University of Chicago Press, Chicago.

Humboldt, A. von. 1907. *Personal narrative of travels to the equinoctial regions of America during the years 1799–1804*. Trans. and ed. Thomasina Ross. 3 vols. George Bell and Sons, London.

Humphries, C. J., and L. R. Parenti. 1986. *Cladistic biogeography*. Clarendon Press, Oxford.

Hunter, J. P. 1998. Key innovations and the ecology of macroevolution. *Trends in Ecology and Evolution* 13: 31–36.

Ibáñez D. R., A. S. Rand, and C. A. Jaramillo A. 1999. *The amphibians of Barro Colorado Nature Monument, Soberania National Park and adjacent areas*. Editorial Mizrachi et Pujol, S.A., Panama City, Panama.

ICBN. 1994. *International Code of Botanical Nomenclature*. Fifteenth International Botanical Congress, Yokohama, Japan.

ICNB. 1992. *International Code of Nomenclature of Bacteria*. International Association of Microbiological Societies, Washington, DC.

ICZN. 1999. *International Code of Zoological Nomenclature*. 4th ed. International Commission on Zoological Nomenclature, Berkeley, CA.

Im Thurn, E. F. 1885. The ascent of Mount Roraima. *Proceedings of the Royal Geographical Society*, August, 497–521.

Inger, R. 1966. The systematics and zoogeography of the Amphibia of Borneo. *Fieldiana Zoology* 52: 1–402.

Iturralde-Vinent, M. A., and R. D. E. MacPhee. 1999. Paleogeography of the Caribbean region: Implications for Cenozoic biogeography. *Bulletin of the American Museum of Natural History* 239: 1–95.

Jackman, T. R., G. Applebaum, and D. B. Wake. 1997. Phylogenetic relationships of bolitoglossine salamanders: A demonstration of the effects of combining morphological and molecular data sets. *Molecular Biology and Evolution* 14: 883–91.

Jackman, T. R., A. Larson, K. de Queiroz, and J. B. Losos. 1999. Phylogenetic relationships and tempo of early diversification in *Anolis* lizards. *Systematic Biology* 48: 254–85.

Jackman, T., J. B. Losos, A. Larson, and K. de Queiroz. 1997. Phylogenetic studies of convergent adaptive radiations in Caribbean *Anolis* lizards. Pp. 535–57 in T. J. Givnish and K. J. Sytsma, eds., *Molecular evolution and adaptive radiation*. Cambridge University Press, Cambridge.

Jacobson, S. K., and M. Fogden. 1984. Frog feats. *International Wildlife* 14: 12–17.

Jaeger, R. G., and R. F. Inger. 1994. Quadrat sampling. Pp. 97–102 in W. R. Heyer, M. A. Donnelly, R. W. McDiarmid, L. C. Hayek, and M. S. Foster, eds., *Measuring and monitoring biological diversity: Standard methods for amphibians*. Smithsonian Institution Press, Washington, DC.

Janzen, D. H., ed. 1983. *Costa Rican natural history*. University of Chicago Press, Chicago.

Jennions, M., and N. Passmore. 1993. Sperm competition in frogs: Testis size and a "sterile male" experiment on *Chiromantis xerampelina* (Rhacophoridae). *Biological Journal of the Linnean Society* 50: 211–20.

Jiggins, C. D., R. E. Naisbit, R. L. Coe, and J. Mallet. 2001. Reproductive isolation caused by colour pattern mimicry. *Nature* 411: 302–5.

Jirak, I. L., L. Wolfe, E. Foldats, and R. Buchsbaum. 1968. *Angel Falls Expedition*. Explorer's Club of Pittsburgh and Pittsburgh Zoological Society, Pittsburgh, PA.

John, B. 1981. Chromosome change and evolutionary change: A critique. Pp. 23–51 in W. R. Atchley and D. S. Woodruff, eds., *Evolution and speciation: Essays in honor of M. J. D. White*. Cambridge University Press, New York.

Johnson, P. T. J., K. B. Lunde, E. G. Ritchie, and A. E. Launer. 1999. The effects of trematode infection on amphibian limb development and survivorship. *Science* 284: 802–4.

Johnstone, R. A. 2002. The evolution of inaccurate mimics. Nature 418: 524–26.

Joron, M., and J. L. B. Mallet. 1998. Diversity in mimicry: Paradox or paradigm? *Trends in Ecology and Evolution* 13: 461–66.

Joubert, Y., and C. Tobin. 1995. Testosterone treatment results in quiescent satellite cells being activated and recruited into cell cycle in rat leavtor ani muscle. *Developmental Biology* 169: 286–94.

Joubert, Y., C. Tobin, and M. C. Lebart. 1994. Testosterone induced masculinization of the rat levator ani muscle during puberty. *Developmental Biology* 162: 104–10.

Jukes, T. H., and C. R. Cantor. 1969. Evolution of protein molecules. Pp. 21–132 in H. N. Munro, ed., *Mammalian protein metabolism.* Academic Press, New York.

Kalisz, S. 1986. Variable selection on the timing of germination in *Collinsia verna* (Scrophulariaceae). *Evolution* 40: 479–91.

Källersjö, M., J. S. Farris, A. G. Kluge, and C. Bult. 1992. Skewness and permutation. *Cladistics* 8: 275–87.

Kardong, K. V. 1980. Gopher snakes and rattlesnakes: Presumptive Batesian mimicry. *Northwest Science* 54: 1–4.

Katsikaros, K., and R. Shine. 1997. Sexual dimorphism in the tusked frog, *Adelotus brevis* (Anura: Myobatrachidae): The roles of natural and sexual selection. *Biological Journal of the Linnean Society* 60: 39–51.

Keferstein, W. 1868. Über einige Batrachier aus Costa Rica. *Archiv für Naturgeschicte* 34: 300.

Kelley, D. 1996. Sexual differentiation in *Xenopus laevis.* Pp. 143–76 in R. Tinsley and H. Kobel, eds., *The biology of Xenopus.* Clarendon Press, Oxford.

Kelley, D., D. Sassoon, N. Segil, and M. Scudder. 1989. Development and hormone regulation of androgen receptor levels in the sexually dimorphic larynx of *Xenopus laevis. Developmental Biology* 131: 111–18.

Keogh, J. S. 1998. Molecular phylogeny of elapid snakes and a consideration of their biogeographic history. *Biological Journal of the Linnean Society* 63: 177–203.

Kerr, B. 1993. Iwokrama: The commonwealth rain forest programme in Guyana. *Commonwealth Forest Review* 72: 305–9.

Kerr, J. G. 1950. *A naturalist in the Gran Chaco.* Cambridge University Press, Cambridge.

Kessing, B., H. Crooms, A. Martin, C. McIntosh, W. O. McMillan, and S. Palumbi. 1989. *The simple fool's guide to PCR.* University of Hawaii Press, Honolulu.

Kezer, J., and S. K. Sessions. 1979. Chromosome variation in the plethodontid salamander, *Aneides ferreus. Chromosoma* (Berlin) 71: 65–80.

Kiesecker, J. M., and A. R. Blaustein. 1997. Influences of egg laying behavior on pathogenic infection of amphibian eggs. *Conservation Biology* 11: 214–20.

Kiesecker, J. M., A. R. Blaustein, and L. K. Belden. 2001. Complex causes of amphibian population declines. *Nature* 410: 681–84.

Kimura, M. 1980. A simple method for estimating evolutionary rates of base substitutions through comparative studies of nucleotide sequences. *Journal of Molecular Evolution* 16: 111–20.

King, M. 1980. C-banding studies on Australian hylid frogs: Secondary constriction structure and the concept of euchromatin transformation. *Chromosoma* (Berlin) 80: 191–217.

———. 1985. The canalization model of chromosomal evolution: A critique. *Systematic Zoology* 34: 69–75.

———. 1990. Amphibia. Pp. 1–241 in B. John, ed., *Animal cytogenetics,* vol. 4: *Chordata* 2. Gebruder Borntraeger, Berlin.

Kitchen, D. W. 1974. Social behavior and ecology of the pronghorn. *Wildlife Monographs* 38: 1–96.

Klauber, L. M. 1956. *Rattlesnakes: Their habits, life histories, and influence on mankind.* University of California Press, Berkeley.

———. 1972. *Rattlesnakes: Their habits, life histories, and influence on mankind.* 2 vols. University of California Press, Berkeley.

Kluge, A. G. 1976a. A reinvestigation of the abdominal musculature of gekkonid lizards and its bearing on their phylogenetic relationships. *Herpetologica* 32: 295–98.

———. 1976b. Phylogenetic relationships in the lizard family Pygopodidae. *Miscellaneous Publications Museum of Zoology University of Michigan* 152: 1–72.

———. 1981. The life history, social organization, and parental behavior of *Hyla rosenbergi* Boulenger, a nest-building gladiator frog. *Miscellaneous Publications Museum of Zoology University of Michigan* 160: 1–170.

———. 1987. Cladistic relationships in the Gekkonoidea (Squamata, Sauria). *Miscellaneous Publications Museum of Zoology University of Michigan* 173: 1–54.

———. 1989a. A concern for evidence and a phylogenetic hypothesis of relationships among *Epicrates* (Boidae, Serpentes). *Systematic Zoology* 37: 7–25.

———. 1989b. Metacladistics. *Cladistics* 5: 291–94.

———. 1990. Species as historical individuals. *Biology and Philosophy* 5: 417–31.

———. 1991. Boine snake phylogeny and research cycles. *Miscellaneous Publications Museum of Zoology University of Michigan* 178: 1–58.

———. 1994. Principles of phylogenetic systematics and the informativeness of the karyotype in documenting Gekkotan lizard relationships. *Herpetologica* 50: 210–21.

———. 1997. Testability and the refutation and corroboration of cladistic hypotheses. *Cladistics* 13: 81–96.

———. 1998a. Sophisticated falsification and research cycles: Consequences for differential character weighting in phylogenetic systematics. *Zoologica Scripta* 26: 349–60.

———. 1998b. Total evidence or taxonomic congruence: Cladistic or consensus classification. *Cladistics* 14: 151–58.

———. 1999. The science of phylogenetic systematics: Explanation, prediction, and test. *Cladistics* 15: 429–36.

———. 2001a. Philosophical conjectures and their refutation. *Systematic Biology* 50: 322–30.

———. 2001b. Gekkotan lizard taxonomy. *Hamadryad* 26: 1–209.

———. 2002. Distinguishing "or" from "and," and the case for historical identification. *Cladistics* 18: 585–93.

———. 2003a. On the deduction of species relationships: A précis. *Cladistics* 19: 233–39.

———. 2003b. The repugnant and the mature in phylogenetic inference: Atemporal similarity and historical identity. *Cladistics* 19: 356–68.

Kluge, A. G., and R. A. Nussbaum. 1995. A review of African-Madagascan gekkonid lizard phylogeny and biogeography (Squamata). *Miscellaneous Publications, Museum of Zoology, University of Michigan* 183: 1–20.

Kluge, A. G., and A. J. Wolf. 1993. Cladistics: What's in a word? *Cladistics* 9: 183–99.

Klump, G. M., and H. C. Gerhardt. 1987. Use of non-arbitrary acoustic criteria in mate choice by female gray treefrogs. *Nature* 326: 286–88.

Knight, A., L. D. Densmore, and E. D. Rael. 1992. Molecular systematics of the *Agkistrodon* complex. Pp. 49–69 in J. A. Campbell and E. D. Brodie Jr., eds., *Biology of the pitvipers.* Selva, Tyler, TX.

Kocher, T. D., W. K. Thomas, A. Meyer, S. V. Edwards, S. Paabo, F. X. Villablanca, and A. C. Wilson. 1989. Dynamics of mitochondrial DNA evolution in animals: Amplification and sequencing with conserved primers. *Proceedings of the National Academy of Sciences* (USA) 86: 6196–200.

Koenig, W. D., and S. S. Albano. 1987. Lifetime reproductive success, selection, and the opportunity for selection in the white-tailed skimmer *Plathemis lydia* (Odonata: Libellulidae). *Evolution* 41: 22–36.

Köhler, G. 2001. *Reptilien und Amphibien Mittelamerikas.* Verlag Herpeton, Offenbach, Germany.

———. 2002. A new species of salamander of the genus *Nototriton* from Nicaragua (Amphibia: Caudata: Plethodontidae). *Herpetologica* 58: 205–10.

Kraus, F., and W. M. Brown. 1998. Phylogenetic relationships of colubroid snakes based on mitochondrial DNA sequences. *Zoological Journal of the Linnean Society* 122: 455–87.

Kraus, F., D. G. Mind, and W. M. Brown. 1996. Crotaline intergeneric relationships based on mitochondrial DNA sequence data. *Copeia* 1996: 763–73.

Kron, K. A. 1997. Exploring alternative systems of classification. *Aliso* 15: 105–12.

Kroonenberg, S. B., J. G. M. Bakker, and A. Marian van der Weil. 1990. Late Cenozoic uplift and paleogeography of the Colombian Andes: Constraints on the development of high-Andean biota. *Geologie en Mijnbouw* 69: 279–90.

Krujit, J. P., and G. J. de Vos. 1988. Individual variation in reproductive success in male black grouse, *Tetrao tetrix* L. Pp. 279–90 in T. H. Clutton-Brock, ed., *Reproductive success.* University of Chicago Press, Chicago.

Kuch, U. 1997a. Mimikry bei Schlangen. *Reptilia* (Munster) 2, 4: 25–32.

———. 1997b. Comment on the proposed conservation of the specific and subspecific names of *Trigonocephalus pulcher* Peters, 1863 [recte 1862] and *Bothrops albocarinatus* Shreve, 1934 (Reptilia, Serpentes) by the designation of a neotype for *T. pulcher. Bulletin of Zoological Nomenclature* 54: 245–49.

Kuch, U., and A. Götzke. 2000. Eine Freilandbeobachtungen des Kinabalu-Kraits, *Bungarus flaviceps baluensis,* Loveridge, 1938 (Serpentes: Elapidae). *Sauria* 22: 19–22.

Kumar, S., K. Tamura, and M. Nei. 1994. MEGA: Molecular evolutionary genetics analysis software for microcomputers. *CABIOS* 10: 189–91.

Kumazawa, Y., H. Ota, M. Nishida, and T. Ozawa. 1996. Gene arrangements in snake mitochondrial genomes: Highly concerted evolution of control-region-like sequences duplicated and inserted into a tRNA gene cluser. *Molecular Biology and Evolution* 13: 1242–54.

———. 1998. The complete nucleotide sequence of a snake (*Dinodon semicarinatus*) mitochondrial genome with two identical control regions. *Genetics* 150: 313–29.

Kunin, W. E., and K. J. Gaston. 1993. The biology of rarity: Patterns, causes and consequences. *Trends in Ecology and Evolution* 8: 298–301.

———. 1997. The biology of rarity: Causes and consequences of rare common differences. Chapman and Hall, London.

Kusano, T., M. Toda, and K. Fukama. 1991. Testis size and breeding systems in Japanese anurans with special reference to large testes size in the treefrog, *Rhacophorus arboreus* (Amphibia, Rhacophoridae). *Behavioral Ecology and Sociobiology* 29: 27–31.

Lahanas, P. N., and J. M. Savage. 1992. A new species of caecilian from the Peninsula de Osa of Costa Rica. *Copeia* 1992: 703–8.

La Marca, E. 1996. Ranas del género *Colostethus* (Amphibia: Anura; Dendrobatidae) de la Guayana Venezolana, con la descripción de siete especies nuevas. *Publicaciones de la Asociación de Amigos de Doñana* 9: 1–64.

Lambrachts, M., and A. A. Dhondt. 1988. The anti-exhaustion hypothesis: A new hypothesis to explain song performance and song switching in the great tit. *Animal Behaviour* 36: 327–34.

Lancini, A. R. 1968. El genero *Euspondylus* (Sauria: Teiidae) en Venezuela. *Publicaciones Ocasionales del Museo de Ciencias Naturales, Zoologia* 12: 1–8.

———. 1978. Un nuevo record de serpiente "Mapanare" para Venezuela. *Natura* (Caracas) 65: 16–17.

Lande, R., and S. J. Arnold. 1983. The measurement of selection on correlated characters. *Evolution* 37: 1210–26.

Lannergren, J. 1987. Contractile properties and myosin isoenzymes of various kinds of *Xenopus* twitch muscle fibres. *Journal of Muscle Research and Cell Motility* 8: 260–73.

Larson, A., and J. B. Losos. 1996. Phylogenetic systematics of adaptation. Pp. 187–220 in M. R. Rose and G. V. Lauder, eds., *Adaptation*. Academic Press, San Diego.

Larson, P. M., and R. O. de Sá. 1998. Chondrocranial morphology of *Leptodactylus* larvae (Leptodactylidae: Leptodactylinae): Its utility in phylogenetic reconstruction. *Journal of Morphology* 238: 287–305.

Lee, M. S. Y. 1996a. The phylogenetic approach to biological taxonomy: Practical aspects. *Zoologica Scripta* 25: 187–90.

———. 1996b. Stability in meaning and content of taxon names: An evaluation of crown–clade definitions. *Proceedings of the Royal Society of London B Biological Science* 263: 1103–9.

———. 1998a. Phylogenetic uncertainty, molecular sequences, and the definition of taxon names. *Systematic Biology* 47: 719–26.

———. 1998b. Convergent evolution and character correlation in burrowing reptiles: Towards a resolution of squamate relationships. *Biological Journal of the Linnean Society* 65: 369–453.

———. 1999. Reference taxa and phylogenetic nomenclature. *Taxon* 48: 31–34.

Lee, M. S. Y., and M. W. Caldwell. 1998. Anatomy and relationships of *Pachyrhachis problematicus,* a primitive snake with hindlimbs. *Philosophical Transactions of the Royal Society of London B* 353: 1521–52.

Leenders, T., B. Beckers, and H. Strijbosch. 1996. *Micrurus mipartitus* (NCN): Polymorphism. *Herpetological Review* 27: 25.

Lees, D. C., C. Kremen, and L. Andriamampianina. 1999. A null model for species richness gradients: Bounded range overlap of butterflies and other rainforest endemics in Madagascar. *Biological Journal of the Linnean Society* 67: 529–54.

Leigh, E. G., Jr. 1999. *Tropical forest ecology: A view from Barro Colorado Island.* Oxford University Press, New York.

Leigh, E. G., Jr., A. S. Rand, and D. M. Windsor, eds. 1982. *The ecology of a tropical forest: Seasonal rhythms and long-term changes.* Smithsonian Institution Press, Washington, DC.

León, P. E. 1970. Report of the chromosome numbers of some Costa Rican anurans. *Revista Biologica Tropical* 17: 119 – 24.

Leonard, W. P., and R. C. Stebbins. 1999. Observations of antipredator tactics of the sharp-tailed snake (*Contia tenuis*). *Northwestern Naturalist* 80: 74 – 77.

Lescure, J. 1976. Contribution à l'étude des amphibiens de Guyane française. VI. Liste préliminaire des Anoures. *Bulletin du Muséum National d'Histoire Naturelle* (3rd ser.), no. 377, *Zoologie* 265: 476 – 524.

Levan, A., K. Fredga, and A. A. Sandberg. 1964. Nomenclature for centromeric position on chromosomes. *Hereditas* 52: 201 – 20.

Lev-Yadun, S., and M. Inbar. 2002. Defensive ant, aphid and caterpillar mimicry in plants? *Biological Journal of the Linnean Society* 77: 393 – 98.

Lidén, M. 1990. Replicators, hierarchy, and the species problem. *Cladistics* 6: 183 – 86.

Lidén, M., and B. Oxelman. 1996. Do we need "phylogenetic taxonomy"? *Zoologica Scripta* 25: 183 – 85.

Lieb, C. S. 1981. Biochemical and karyological systematics of the Mexican lizards of the *Anolis gadovi* and *A. nebulosus* species groups (Reptilia: Iguanidae). Ph.D. dissertation, University of California, Los Angeles.

Lieberman, S. S. 1986. Ecology of the leaf litter herpetofauna of a Neotropical rainforest: La Selva, Costa Rica. *Acta Zoologica Mexicana,* n.s., 15: 1 – 71.

Liem, S. 1970. The morphology, systematics, and evolution of the old world treefrogs (Rhacophoridae and Hyperolidae). *Fieldiana Zoology* 57: 1 – 145.

Lim, F. L. K., and M. T.-M. Lee. 1989. *Fascinating snakes of Southeast Asia — An introduction.* Tropical Press, Kuala Lumpur, Malaysia.

Limeses, C. E. 1964. La musculatura del muslo en los ceratofrínidos y formas afines: Con un análisis crítico sobre la significación de los caracteres miológicos en la sistemática de los anuros superiores. *Universidad de Buenos Aires Facultad de Ciencias Exactas y Naturales Contribuciones Científicas* (zoology series) 1: 188 – 245.

Lincoln, R. J., G. A. Boxshall, and P. F. Clark. 1982. *A dictionary of ecology, evolution, and systematics.* Cambridge University Press, New York.

Linnaeus, C. 1758. *Systema naturae.* 10th ed. Stockholm.

Lips, K. R. 1995. The population biology of *Hyla calypsa,* a stream-breeding treefrog from Lower Central America. Ph.D. dissertation, University of Miami, Coral Gables, FL.

―――. 1996. New treefrog from the Cordillera de Talamanca of Central America with a discussion of systematic relationships in the *Hyla lancasteri* group. *Copeia* 1996: 615 – 26.

―――. 1998. Decline of a tropical montane amphibian fauna. *Conservation Biology* 12: 106 – 17.

―――. 1999. Mass mortality and population declines of anurans at an upland site in western Panama. *Conservation Biology* 13: 117 – 25.

―――. 2001. Reproductive trade-offs and bet-hedging in *Hyla calypsa,* a Neotropical treefrog. *Oecologia* 128: 509 – 18.

List, J. C. 1966. Comparative osteology of the snake families Typhlopidae and Leptotyphlopidae. *Illinois Biological Monographs* 36: 1 – 112.

Locke, J. 1700. An essay concerning human understanding. In *Four Books,* 1975 reprint;

edited with an introduction, critical apparatus and glossary by P. H. Nidditch. Clarendon Press, Oxford.

Lombard, R. E., and D. B. Wake. 1986. Tongue evolution in the lungless salamanders, family Plethodontidae. IV. Phylogeny of plethodontid salamanders and the evolution of feeding dynamics. *Systematic Zoology* 35: 532–51.

Loop, M. S., and D. K. Crossman. 2000. High color-vision sensitivity in macaque and humans. *Visual Neuroscience* 17: 119–25.

López-Luna, M. A., R. C. Vogt, and M. A. de la Torre-Loranca. 1999. A new species of montane pitviper from Veracruz, Mexico. *Herpetologica* 55: 382–89.

Losos, J. B., T. R. Jackman, A. Larson, K. de Queiroz, and L. Rodrigues-Schettino. 1998. Contingency and determinism in replicated adaptive radiations of island lizards. *Science* 279: 2115–18.

Loveridge, A. 1950. History and habits of the East African bullfrog. *Journal of the East African Natural History Society* 19: 253–55.

Lundberg, J. G., L. G. Marshall, J. Guerrero, B. Horton, M. C. S. L. Malabarba, and F. Wesselingh. 1998. The stage for Neotropical fish diversification: A history of tropical South American rivers. Pp. 13–48 in L. R. Malabarba, R. E. Reis, R. P. Vari, Z. M. Lucena, and C. A. S. Lucena, eds., *Phylogeny and classification of Neotropical fishes.* EDIPUCRS, Porto Alegre, Brazil.

Lynch, J. D. 1965. A review of the eleutherodactyline frog genus *Microbatrachylus* (Leptodactylidae). *Natural History Miscellanea* 182: 1–12.

———. 1971. Evolutionary relationships, osteology, and zoogeography of leptodactylid frogs. *Miscellaneous Publication University of Kansas Museum of Natural History* 53: 1–238.

———. 1976. The species groups of the South American frogs of the genus *Eleutherodactylus* (Leptodactylidae). *Occasional Papers of the Museum of Natural History University of Kansas* 61: 1–24.

———. 1979. The amphibians of the lowland tropical forests. Pp. 189–215 in W. E. Duellman, ed., *The South American herpetofauna: Its origin, evolution, and dispersal.* University of Kansas Museum of Natural History Monograph, no. 7, Lawrence.

———. 1980. A new frog of the genus *Eleutherodactylus* from western Panama. *Transactions of the Kansas Academy of Science* 83: 101–5.

———. 1986. The definition of the Middle American clade of *Eleutherodactylus* based on jaw musculature (Amphibia: Leptodactylidae). *Herpetologica* 42: 248–58.

———. 1989. The gauge of speciation: On the frequencies of modes of speciation. Pp. 527–53 in D. Otte and J. A. Endler, eds., *Speciation and its consequences.* Sinauer Associates, Sunderland, MA.

———. 1993. The value of the *m. depressor mandibulae* in phylogenetic hypotheses for *Eleutherodactylus* and its allies (Amphibia: Leptodactylidae). *Herpetologica* 49: 32–41.

———. 2000. The relationships of an ensemble of Guatemalan and Mexican frogs (*Eleutherodactylus:* Leptodactylidae: Amphibia). *Revista de la Academia Colombiana de Ciencias Exactas, Fisicas y Naturales* 24: 67–94.

Macartney, J. M., and P. T. Gregory. 1981. Differential susceptibility of sympatric garter snake species to amphibian skin secretions. *American Midland Naturalist* 106: 271–81.

Macey, J. R., and A. Verma. 1997. Homology in phylogenetic analysis: Alignment of transfer RNA genes and the phylogenetic position of snakes. *Molecular Phylogenetics and Evolution* 7: 272–79.

MacInnes, H. 1976. *Climb to the lost world.* Penguin Books, Harmondsworth, U.K.

Madsen, T., and R. Shine. 1994. Toxicity of a tropical Australian frog, *Litoria dahlii,* to sympatric snakes. *Australian Wildlife Research* 21: 21–25.

Mägdefrau, H., and K. Mägdefrau. 1994. Biologie von Anuren auf Tepuis der Roraima-Gruppe in Venezuela. Pp. 84–102 in H.-J. Herrmann and H. Zimmermann, eds., *Beiträge zur Biologie der Anuren.* Tetra Verlag, Melle, Germany.

Mägdefrau, H., K. Mägdefrau, and A. Schlüter. 1991. Herpetologische daten vom Guaiquinima-Tepui, Venezuela. *Herpetofauna* 13: 13–26.

Magee, B. 1973. *Karl Popper.* Viking Press, New York.

Maguire, B. 1955. Cerro de la Neblina, Amazonas, Venezuela: A newly discovered sandstone mountain. *Geographical Review* 45: 27–51.

———. 1970. On the flora of the Guayana Highland. *Biotropica* 2: 85–100.

———. 1972. Guayana as a floristic province: Its relationship within the Neotropics and to the Paleotropics. Pp. 55–56 in *Resúmenes de los Trabajos I Congreso Latinoamericano.* Sociedad Botánica de México, México.

———. 1979. Guayana, region of the Roraima sandstone formation. Pp. 223–38 in K. Larsen and L. B. Holm-Nielsen, eds., *Tropical botany.* Academic Press, London.

Maguire, B., and J. J. Wurdack. 1959. The position of Cerro de la Neblina, Venezuela. *Geographical Review* 49: 566–68.

Mahner, M., and M. Bunge. 1997. *Foundations of biophilosophy.* Springer Verlag, New York.

Mahony, M. J., and E. S. Robinson. 1986. Nucleolar organizer region (NOR) location in karyotypes of Australian ground frogs (family Myobatrachidae). *Genetica* 68: 119–27.

Maina, G. G., and H. F. Howe. 2000. Inherent rarity in community restoration. *Conservation Biology* 14: 1335–40.

Maiorana, V. C. 1976. Predation, submergent behavior, and tropical diversity. *Evolutionary Theory* 1: 157–77.

Maldonado-Koerdell, M. 1964. Geohistory and paleogeography of Middle America. Pp. 3–32 in R. C. West, ed., *Handbook of Middle American Indians,* vol. 1. University of Texas Press, Austin.

Malfait, B. T., and M. G. Dinkelman. 1972. Circum-Caribbean tectonic and igneous activity and the evolution of the Caribbean plate. *Geological Society of America Bulletin* 83: 251–72.

Mallet, J., and M. Joron. 1999. Evolution of diversity in warning color and mimicry. *Annual Review of Ecology and Systematics* 30: 201–33.

Maniatis, T., E. F. Fritsch, and J. Sambrook. 1982. *Molecular cloning: A laboratory manual.* Cold Spring Harbor, NY.

Mann, P., C. Schubert, and K. Burke. 1990. Review of Caribbean neotectonics. Pp. 307–38 in G. Dengo and J. E. Case, eds., *The geology of North America: The Caribbean region,* vol. H. Geological Society of America, Boulder, CO.

Manthey, U., and W. Grossmann. 1997. *Amphibien und Reptilien Südostasiens.* Natur und Tier-Verlag, Munster, Germany.

Margush, J., and F. R. McMorris. 1981. Consensus n-trees. *Bulletin of Mathematical Biology* 43: 239–44.

Marin, M., M. Tobias, and D. Kelley. 1990. Hormone-sensitive stages in the sexual differentiation of laryngeal muscle fiber number in *Xenopus laevis. Development* 110: 703–12.

Marler, C., and M. Ryan. 1996. Energetic constraints and steroid hormone correlates of male calling behavior in the tungara frog. *Journal of Zoology* 240: 397–409.

Marques, O. A. V. 1999. Defensive behavior of the green snake *Philodryas viridissimus* (Linnaeus) (Colubridae, Reptilia) from the Atlantic forest in northeastern Brazil. *Revista Brasileira de Zoologia* 16: 265–66.

———. 2000. Tail displays of the false coral snake *Simophis rhinostoma* (Colubridae). *Amphibia-Reptilia* 22: 127–29.

———. 2002. Natural history of the coral snake *Micrurus decoratus* (Elapidae) from the Atlantic forest in southeast Brazil, with comments on possible mimicry. *Amphibia-Reptilia* 23: 228–32.

Marques, O. A. V., A. Enterovic, and W. Endo. 2000. Seasonal activity of snakes in the Atlantic forest in southeastern Brazil. *Amphibia-Reptilia* 22: 103–11.

Marques, O. A. V., A. Enterovic, and I. Sazima. 2001. *Serpentes da Mata Atlântica: Guia ilustrado para a Serra do Mar.* Editora Holos, Ribeirão, Brazil.

Marques, O. A. V., and G. Puorto. 1991. Padrões cromáticos, distribuição e possível mimetismo em *Erythrolamprus aesculapii* (Serpentes, Colubridae). *Memórias do Instituto Butantan* 53: 127–34.

Marquez, R. 1993. Male reproductive success in two midwife toads, *Alytes obstetricans* and *A. cisternasii. Behavioral Ecology and Sociobiology* 32: 283–91.

Marquis, R. J., M. A. Donnelly, and C. Guyer. 1986. Aggregations of calling males of *Agalychnis calcarifer* Boulenger (Anura: Hylidae) in a Costa Rican lowland wet forest. *Biotropica* 18: 173–75.

Marshall, C. J., and J. K. Liebherr. 2000. Cladistic biogeography of the Mexican transition zone. *Journal of Biogeography* 27: 203–16.

Martin, A. P., and S. R. Palumbi. 1993. Body size, metabolic rate, generation time, and the molecular clock. *Proceedings of the National Academy of Sciences* (USA) 90: 4087–91.

Martins, M. 1996. Defensive tactics in lizards and snakes: The potential contribution of the Neotropical fauna. Pp. 185–99 in K. Del Claro, ed., *Anais do XIV encontro anual de etologia: Sociedade Brasileira de etologia.* Universidade Federal de Uberlandia, Uberlandia, Brazil.

Martins, M., M. S. Araujo, R. J. Sawaya, and R. Nanes. 2001. Diversity and evolution of macrohabitat use, body size, and morphology in a monophyletic group of Neotropical pitvipers (*Bothrops*). *Journal of Zoology* (London) 254: 529–38.

Martins, M., and M. E. Oliveira. 1993. The snakes of the genus *Atractus* Wagler (Reptilia: Squamata: Colubridae) from the Manaus region, central Amazonia, Brazil. *Zoologische Mededelingen* 67: 21–40.

———. 1998. Natural history of snakes in forests of the Manaus region, central Amazonia, Brazil. *Herpetological Natural History* 6: 78–150.

Mayhew, W. W. 1962. *Scaphiopus couchi* in California's Colorado Desert. *Herpetologica* 18: 153–61.

Mayr, E. 1970. *Populations, species, and evolution.* Harvard University Press, Cambridge.

Mayr, E., and P. D. Ashlock. 1991. *Principles of systematic zoology.* 2nd ed. McGraw-Hill, New York.

Mayr, E., and W. H. Phelps Jr. 1955. Origin of the bird fauna of Pantepui. Pp. 399–400 in A. Portmann and E. Sutter, eds., *Acta XI Congressus Internationalis Ornithologici*. Birkhäuser Verlag, Basel, Switzerland.

———. 1967. The origin of the bird fauna of the south Venezuelan highlands. *Bulletin of the American Museum of Natural History* 136: 269–328.

McCarthy, C. J. 1985. Monophyly of elapid snakes (Serpentes: Elapidae): An assessment of the evidence. *Zoological Journal of the Linnean Society* 83: 79–93.

McConnell, H. M. [Mrs. F. V. McConnell]. 1916. Preface and itinerary [expedition to Mount Roraima in 1894]. Pp. iv–xxxv in C. Chubb, *The birds of British Guiana, based on the collections of Frederick Vavasour McConnell*, vol. 1. Bernard Quaritch, London.

McCrainie, J. R., L. D. Wilson, and L. Porras. 1980. A new species of *Leptodactylus* from the cloud forests of Honduras. *Journal of Herpetology* 14: 361–67.

McCrainie, J. R., L. D. Wilson, and K. L. Williams. 1986. The tadpole of *Leptodactylus silvanimbus*, with comments on the relationships of the species. *Journal of Herpetology* 20: 560–62.

McDade, L. A. 1992. Hybrids and phylogenetic systematics II. The impact of hybrids on cladistic analysis. *Evolution* 46: 1329–46.

McDade, L. A., and G. S. Hartshorn. 1994. La Selva Biological station. Pp. 6–14 in L. A. McDade, K. S. Bawa, H. A. Hespenheide, and G. S. Hartshorn, eds., *La Selva: Ecology and natural history of a Neotropical rain forest*. University of Chicago Press, Chicago.

McDiarmid, R. W. 1971. Comparative morphology and evolution of frogs of the Neotropical genera *Atelopus, Dendrophryniscus, Melanophryniscus,* and *Oreophrynella*. *Natural History Museum of Los Angeles County Science Bulletin* 12: 1–66.

———. 1975. Glass frog romance along a tropical stream. *Terra* 13 (4): 14–18.

———. 1978. Evolution of parental care in frogs. Pp. 127–47 in G. Burghardt and M. Bekoff, eds., *The development of behavior*. Garland Publishing, New York.

———. 1983. *Centrolenella fleischmanni* (Ranita de Vidrio, glass frog). Pp. 389–90 in D. H. Janzen, ed., *Costa Rican natural history*. University of Chicago Press, Chicago.

McDiarmid, R. W., and K. Adler. 1974. Notes on territorial and vocal behavior of Neotropical frogs of the genus *Centrolenella*. *Herpetologica* 30: 75–78.

McDiarmid, R. W., J. A. Campbell, and T. T. Touré. 1999. *Snake species of the world: A taxonomic and geographic reference*. Vol. 1. Herpetologists' League Publications, Washington, DC.

McDiarmid, R. W., and J. DeWeese. 1977. The systematic status of the lizard *Bachia blairi* (Dunn) 1940 (Reptilia: Teiidae) and its occurrence in Costa Rica. *Brenesia* 12/13: 143–53.

McDiarmid, R. W., and M. S. Foster. 1975. Unusual sites for two Neotropical tadpoles. *Journal of Herpetology* 9: 264–65.

McDiarmid, R. W., and S. Gorzula. 1989. Aspects of the reproductive ecology and behavior of the tepui toads, genus *Oreophrynella* (Anura: Bufonidae). *Copeia* 1989: 445–51.

McDiarmid, R. W., and A. Paolillo. 1988. Preliminary field report: Herpetological collections—Cerro de la Neblina. Pp. 667–70 in C. Brewer-Carías, ed., *Cerro de la Neblina: Resultados de la expedición, 1983–1987*. Caracas, Venezuela.

McDonald, D. B. 1989. Correlates of male mating success in a lekking bird with male–male cooperation. *Animal Behaviour* 37: 1007–22.

McDowell, S. B. 1968. Affinities of the snakes usually called *Elaps lacteus* and *E. dorsalis*. *Zoological Journal of the Linnean Society* 47: 561–78.

————. 1987. Systematics. Pp. 3–50 in R. A. Seigel, J. T. Collins, and S. S. Novak, eds., *Snakes: Ecology and evolutionary biology.* Macmillan, New York.

McDowell, S. B., and C. M. Bogert. 1954. The systematic position of *Lanthanotus* and the affinities of the anguimorphan lizards. *Bulletin of the American Museum of Natural History* 105: 1–142.

Meachem, A., and C. W. Myers. 1961. An exceptional pattern variant of the coral snake, *Micrurus fulvius* (Linnaeus). *Quarterly Journal of the Florida Academy of Science* 24: 56–58.

Means, D. B. 1995. Fire ecology of the Guayana region, northeastern South America. Pp. 61–77 in S. I. Cerulean and R. T. Engstrom, eds., *Fire in wetlands: A management perspective.* Proceedings of the Tall Timbers Fire Ecology Conference, no. 19. Tall Timbers Research Station, Tallahassee, FL.

Melichna, J., E. Gutmann, A. Herbrychova, and J. Stichova. 1972. Sexual dimorphism in contraction properties and fibre pattern of the flexor carpi radialis muscle of the frog (*Rana temporaria* L.). *Experientia* 28: 88–91.

Mendeleyev, D. I. 1871. *Osnovy khimúii* [Principles of chemistry]. Tip. V. Demakova, St. Petersburg.

Merriam-Webster's geographical dictionary. 1997. 3rd ed. Merriam-Webster, Springfield, MA.

Mertens, R. 1956. Das Problem der Mimikry bei Korallenschlangen. *Zoologische Jahrbücher, Abteilung für Systematik, Ökologie und Geographie der Tiere* 84: 541–76.

Michelangeli, F. A. 2000. Species composition and species-area relationships in vegetation isolates on the summit of a sandstone mountain in southern Venezuela. *Journal of Tropical Ecology* 16: 69–82.

Michener, C. D. 1963. Some future developments in taxonomy. *Systematic Zoology* 12: 151–72.

Mijares-Urrutia, A. 2000. Taxonomía de algunos microtéidos de Venezuela (Squamata). II. Situación nomenclatural de *Prionodactylus ampuedai* y *Prionodactylus phelpsorum. Revista de Biología Tropical* 48: 681–88.

Miller, D. A., V. G. Dev, R. Tantravahi, and O. J. Miller. 1976. Suppression of human nucleolus organizer activity in mouse–human somatic hybrid cells. *Experimental Cell Research* 101: 235–43.

Miller, O. J., and E. Therman. 2001. *Human chromosomes.* 4th ed. Springer Verlag, New York.

Mindell, D. P., and R. L. Honeycutt. 1990. Ribosomal RNA in vertebrates: Evolution and phylogenetic applications. *Annual Review of Ecology and Systematics* 21: 541–66.

Miranda-Ribeiro, A. de. 1937. Uma salamandra no Baixo-Amazonas. *O Campo* (Rio de Janeiro, Março), pp. 42–46. Reprinted in *Arquivos do Museu Nacional, Rio de Janeiro* 42 (1955): 51–62.

Mishler, B., and R. Brandon. 1987. Individuality, pluralism, and the phylogenetic species concept. *Biology and Philosophy* 2: 397–414.

Mishler, B., and M. J. Donoghue. 1982. Species concepts: A case for pluralism. *Systematic Zoology* 31: 491–503.

Miyamoto, M. M. 1981. Cladistic studies of Costa Rican frogs, genus *Eleutherodactylus:* Phylogenetic relationships based on multiple character sets. Ph.D. dissertation, University of Southern California, Los Angeles.

―――――. 1983a. Biochemical variation in the frog *Eleutherodactylus bransfordii*: Geographic patterns and cryptic species. *Systematic Zoology* 32: 43–51.

―――――. 1983b. Frogs of the *Eleutherodactylus rugulosus* group: A cladistic study of allozyme, morphological, and karyological data. *Systematic Zoology* 32: 109–24.

―――――. 1985. Consensus cladograms and general classifications. *Cladistics* 1: 186–89.

Miyamoto, M. M., and W. M. Fitch. 1995. Testing phylogenies and phylogenetic methods with congruence. *Systematic Biology* 44: 64–76.

Moffett, M. W. 1998. Worm with a warning: "Don't eat me." *National Geographic* 193, 6: 142.

Molina, C., and J. C. Señaris. 2001. Una nueva especie del género *Riolama* (Reptilia: Gymnophthalmidae) de las tierras altas del Estado Amazonas, Venezuela. *Memoria Fundación Ciencias Naturales La Salle* 155: 5–19.

Moore, A. J. 1990. The evolution of sexual dimorphism by sexual selection: The separate effects of intrasexual selection and intersexual selection. *Evolution* 44: 315–31.

Moore, G. 1998. A comparison of traditional and phylogenetic nomenclature. *Taxon* 47: 561–79.

Moore, M. 1991. Application of organization-activation theory to alternative male reproductive strategies: A review. *Hormones and Behavior* 25: 154–79.

Morales, V. R., and R. W. McDiarmid. 1996. Annotated checklist of the amphibians and reptiles of Pakitza, Manu National Park Reserve Zone, with comments on the herpetofauna of Madre de Dios, Perú. Pp. 503–22 in D. E. Wilson and A. Sandoval, eds., *Manu: The biodiversity of southeastern Peru.* Smithsonian Institution Press, Washington, DC.

Moritz, C., J. L. Patton, C. J. Schneider, and E. L. Smith. 2000. Diversification of rainforest faunas: An integrated molecular approach. *Annual Review of Ecology and Systematics* 31: 533–63.

Moritz, C., C. J. Schneider, and D. B. Wake. 1992. Evolutionary relationships within the *Ensatina eschscholtzii* complex confirm the ring species interpretation. *Systematic Biology* 41: 273–91.

Morrone, J. J., and J. V. Crisci. 1995. Historical biogeography: Introduction to methods. *Annual Review of Ecology and Systematics* 26: 373–402.

Muller, E., G. Galavazi, and J. Zirmai. 1969. Effect of castration and testosterone treatment on fiber width of the flexor carpi radialis muscle in the male frog (*Rana temporaria* L.). *General and Comparative Endocrinology* 13: 275–84.

Müller, P. 1973. The dispersal centres of terrestrial vertebrates in the Neotropical realm: A study in the evolution of the Neotropical biota and its native landscapes. *Biogeographica* 2: 1–244.

Murphy, C. G. 1994a. Determinants of chorus tenure in barking treefrogs (*Hyla gratiosa*). *Behavioral Ecology and Sociobiology* 34: 285–95.

―――――. 1994b. Chorus tenure of male barking treefrogs, *Hyla gratiosa. Animal Behaviour* 48: 763–77.

―――――. 1998. Interaction-dependent sexual selection and the mechanisms of sexual selection. *Evolution* 52: 8–18.

Murphy, R. W., J. W. Sites Jr., D. G. Buth, and C. H. Haufler. 1996. Protein isozyme electrophoresis. Pp. 51–120 in D. M. Hillis, C. Moritz, and B. K. Mable, eds., *Molecular systematics.* Sinauer Associates, Sunderland, MA.

Murray, J. D., and M. R. Myerscough. 1991. Pigmentation pattern formation on snakes. *Journal of Theoretical Biology* 149: 339–60.

Myers, C. W. 1974. The systematics of *Rhadinaea* (Colubridae), a genus of New World snakes. *Bulletin of the American Museum of Natural History* 153: 1–262.

———. 1997. Preliminary remarks on the summit herpetofauna of Auyantepui, eastern Venezuela. *Acta Terramaris* 10: 1–8.

———. 2000. A history of herpetology at the American Museum of Natural History. *Bulletin of the American Museum of Natural History* 252: 1–32.

Myers, C. W., J. W. Daly, H. M. Garraffo, A. Wisnieski, and J. F. Cover Jr. 1995. Discovery of the Costa Rican poison frog *Dendrobates granuliferus* in sympatry with *Dendrobates pumilio,* and comments on taxonomic use of skin alkaloids. *American Museum Novitates* 3144: 1–21.

Myers, C. W., and M. A. Donnelly. 1996. A new herpetofauna from Cerro Yaví, Venezuela: First results of the Robert G. Goelet American Museum–Terramar Expedition to the northwestern tepuis. *American Museum Novitates* 3172: 1–56.

———. 1997. A tepui herpetofauna on a granitic mountain (Tamacuari) in the borderland between Venezuela and Brazil: Report from the Phipps Tapirapecó Expedition. *American Museum Novitates* 3213: 1–71.

———. 2001. Herpetofauna of the Yutajé-Corocoro Massif, Venezuela: Second report from the Robert G. Goelet American Museum–Terramar Expedition to the northwestern tepuis. *Bulletin of the American Museum of Natural History* 261: 1–85.

Myers, C. W., and A. S. Rand. 1969. Checklist of amphibians and reptiles of Barro Colorado Island, Panama, with comments on faunal change and sampling. *Smithsonian Contributions to Zoology* 10: 1–11.

Myers, C. W., E. E. Williams, and R. W. McDiarmid. 1993. A new anoline lizard (*Phenacosaurus*) from the highland of Cerro de la Neblina, southern Venezuela. *American Museum Novitates* 3070: 1–15.

Nelson, G. J. 1983. Reticulation in cladograms. Pp. 105–14 in N. I. Platnick and V. A. Funk, eds., *Advances in cladistics II.* Columbia University Press, New York.

Nelson, G. J., and N. I. Platnick. 1981. *Systematics and biogeography: Cladistics and vicariance.* Columbia University Press, New York.

Nicholson, K. E. 2002. Phylogenetic analysis of *Norops* (beta *Anolis*): An historical review of the genus and a test of the current infrageneric classification. *Herpetological Monographs* 16: 93–120.

Nixon, K. C., and J. M. Carpenter. 2000. On the other "phylogenetic systematics." *Cladistics* 16: 298–318.

Noble, G. 1931. *The biology of the Amphibia.* McGraw-Hill, New York.

Noble, G., and S. Pope. 1929. The modification of the cloaca and teeth of the adult salamander, *Desmognathus,* by testicular transplants and by castration. *British Journal of Experimental Biology* 6: 399–411.

Norris, D. 1997. *Vertebrate endocrinology.* 3rd ed. Academic Press, San Diego.

Nussbaum, R. A. 1985. Systematics of caecilians (Amphibia: Gymnophiona) of the family Scolecomorphidae. *Occasional Papers of the Museum of Zoology University of Michigan* 713: 1–49.

———. 1988. On the status of *Copeotyphlinus syntremus, Gymnopis oligozona,* and

Siphonops confucionis Taylor (Gymnophiona, Caeciliidae): A comedy of errors. *Copeia* 1988: 921–28.

Nussbaum, R. A., and H. Hinkel. 1994. Revision of East African caecilians of the genera *Afrocaecilia* Taylor and *Boulengerula* Tornier (Amphibia: Gymnophiona: Caeciliaidae). *Copeia* 1994: 750–60.

Nussbaum, R. A., and M. Wilkinson. 1989. On the classification and phylogeny of caecilians (Amphibia: Gymnophiona): A critical review. *Herpetological Monographs* 2: 1–42.

O'Hara, R. J. 1992. Telling the tree: Narrative representation and the study of evolutionary history. *Biology and Philosophy* 7: 135–60.

———. 1993. Systematic generalization, historical fate, and the species problem. *Systematic Biology* 42: 231–46.

Oka, Y., R. Ohtani, M. Satou, and K. Ueda. 1984. Sexually dimorphic muscles in the forelimb of the Japanese toad, *Bufo japonicus. Journal of Morphology* 180: 297–308.

Olert, S. J., and C. Klett. 1978. Chromosome banding in Amphibia. *Chromosoma* (Berlin) 71: 29–55.

Oliveira, M. E., and R. T. Santori. 1999. Predatory behavior of the opossum *Didelphis albiventris* on the pitviper *Bothrops jararaca. Studies on Neotropical Fauna and Environment* 34: 72–75.

Orejas-Miranda, B., and A. Quesada. 1976. Ecosistemas frágiles. *Ciencia Interamericana* 17: 9–15.

Ort, P. 1965. The expedition of the Brazilian–Venezuelan Boundary Commission to Cerro de la Neblina. *Garden Journal, New York Botanical Garden* 15: 199–203.

O'Shea, M. 1996. *A guide to the snakes of Papua New Guinea.* Port Moresby, Papua New Guinea.

Ota, H., M. Honda, M. Kobayashi, S. Sengoku, and T. Hikida. 1999. Phylogenetic relationships of eublepharid geckos (Reptilia: Squamata): A molecular approach. *Zoological Science* 16: 659–66.

Palumbi, S. R. 1996. Nucleic acids II: The polymerase chain reaction. Pp. 205–320 in D. M. Hillis, C. Moritz, and B. K. Mable, eds., *Molecular systematics.* Sinauer Associates, Sunderland, MA.

Palumbi, S. R., A. P. Matin, S. Romano, W. O. McMillan, L. Stice, and G. Grabowski. 1991. *The simple fool's guide to PCR.* Special Publication, Department of Zoology, University of Hawaii, Honolulu.

Papavero, N., J. Lorente-Bousquets, and J. M. Abe. 1992. Propuesta de un nuevo sistema de nomenclatura para la Sistemática Filogenética I. *Publicaciones especiales del Museo de Zoología, UNAM* 5: 1–20.

———. 2001. Proposal of a new system of nomenclature for phylogenetic systematics. *Arquivos de Zoologia* 36: 1–145.

Papenfuss, T. J., D. B. Wake, and K. Adler. 1983. Salamanders of the genus *Bolitoglossa* from the Sierra Madre del Sur of southern Mexico. *Journal of Herpetology* 17: 295–307.

Parker, G. 1983. Arms races in evolution—An ESS to the opponent-independent costs game. *Journal of Theoretical Biology* 101: 619–48.

Parkinson, C. L. 1999. Molecular systematics and biogeographical history of pitvipers as determined by mitochondrial ribosomal DNA sequences. *Copeia* 1999: 576–86.

Parkinson, C. L., J. A. Campbell, and P. T. Chippindale. 2002. Multigene phylogenetic analysis of pitvipers, with comments on the biogeographical history of the group. Pp. 93–110 in G. W. Schuett, M. Hoggren, M. E. Douglas, and H. W. Greene, eds., *Biology of the vipers*. Eagle Mountain Publishing, Eagle Mountain, UT.

Parkinson, C. L., K. R. Zamudio, and H. W. Greene. 2000. Phylogeography of the pitviper clade *Agkistrodon:* Historical ecology, species status, and conservation of cantils. *Molecular Ecology* 9: 411–20.

Parra-Olea, G. 1999. Molecular evolution and systematics of Neotropical salamanders (Caudata: Plethodontidae: Bolitoglossini). Ph.D. dissertation, Integrative Biology, University of California, Berkeley.

Parra-Olea, G., and D. B. Wake. 2001. Extreme morphological and ecological homoplasy in tropical salamanders. *Proceedings of the National Academy of Sciences* (USA) 98: 7888–91.

Pasteur, G. 1982. A classificatory review of mimicry systems. *Annual Review of Ecology and Systematics* 13: 169–99.

Paterson, H. 1985. The recognition species concept. Pp. 21–29 in E. Vrba, ed., *Species and speciation*. Transvaal Museum, Pretoria.

Patton, J. L., and M. F. Smith. 1992. mtDNA phylogeny of Andean mice: A test of diversification across ecological gradients. *Evolution* 46: 174–83.

Paynter, R. A., Jr. 1982. *Ornithological gazetteer of Venezuela*. Museum of Comparative Zoology, Harvard University, Cambridge.

Paynter, R. A., Jr., and M. A. Traylor Jr. 1991. *Ornithological gazetteer of Brazil*. Museum of Comparative Zoology, Harvard University, Cambridge.

Pearl, M. 1994. Local initiatives and the rewards for biodiversity conservation: Crater Mountain Wildlife Management Area, Papua New Guinea. Pp. 193–214 in D. Western and R. M. Wright, eds., *Natural connections: Perspectives in community-based conservation*. Island Press, Washington, DC.

Pechmann, J. H. K., D. E. Scott, R. D. Semlitsch, J. P. Caldwell, L. J. Vitt, and J. W. Gibbons. 1991. Declining amphibian populations: The problem of separating human impacts from natural fluctuations. *Science* 253: 892–95.

Pechmann, J. H. K., and H. M. Wilbur. 1994. Putting declining amphibian populations into perspective: Natural fluctuations and human impacts. *Herpetologica* 50: 65–84.

Péfaur, J. E., and J. A. Rivero. 2000. Distribution, species-richness, endemism, and conservation of Venezuelan amphibians and reptiles. *Amphibian and Reptile Conservation* 2: 42–70.

Pennisi, E. 1996. Evolutionary and systematic biologists converge. *Science* 273: 181–82.

Perkins, H. I. 1885. Notes on a journey to Mount Roraima, British Guiana. *Proceedings of the Royal Geographical Society* 8: 522–34.

Perry, G. 1996. The evolution of sexual dimorphism in the lizard *Anolis polylepis* (Iguania): Evidence from intraspecific variation in foraging behavior and diet. *Canadian Journal of Zoology* 74: 1238–45.

Peters, J. A. 1960. The snakes of the subfamily Dipsadinae. *Miscellaneous Publications Museum of Zoology University of Michigan* 114: 1–224.

Peters, J. A., and R. Donoso-Barros. 1970. Catalogue of the Neotropical Squamata. II. Lizards and Amphisbaenians. *United States National Museum Bulletin* 297: 1–293.

Peters, J. A., and B. Orejas-Miranda. 1970. Catalogue of the Neotropical Squamata. I. Snakes. *United States National Museum Bulletin* 297: 1–347.

Peters, S., and D. Aulner. 2000. Sexual dimorphism in forelimb muscles of the bullfrog *Rana catesbeiana:* A functional analysis of isometric contractile properties. *Journal of Experimental Biology* 203: 3639–54.

Pfennig, D. W., W. R. Harcombe, and K. S. Pfennig. 2001. Frequency-dependent Batesian mimicry. *Nature* 410: 323.

Phelps, W. H., Jr. 1977. Aves colectadas en las mesetas de Sarisariñama y Jaua durante tres expediciones al macizo, Estado Bolívar: Descripciones de dos nuevas subespecies. *Boletín de la Sociedad Venezolana de Ciencias Naturales* 33: 15–24.

Phoenix, C., R. Goy, A. Gerall, and W. Young. 1959. Organizing actions of prenatally administered testosterone propionate on tissues mediating mating behavior in the female guinea pig. *Endocrinology* 65: 369–82.

Pielou, E. C. 1977. *Mathematical ecology.* John Wiley, Toronto.

Pindell, J. L., and J. F. Dewey. 1982. Permo-Triassic reconstruction of western Pangea and the evolution of the Gulf of Mexico, Bahamas, and proto-Caribbean sea. *Tectonics* 4: 1–39.

Platnick, N. I. 1977. The hypochiloid spiders: A cladistic analysis, with notes on the Atypoidea (Arachnida, Araneae). *American Museum Novitates* 2627: 1–23.

Platnick, N. I., C. E. Griswold, and J. A. Coddington. 1991. On missing entries in cladistic analysis. *Cladistics* 7: 337–43.

Pleijel, F., and G. W. Rouse. 2000. A new taxon, *capricornia* (*Hesionidae, Polychaeta*), illustrating the LITU ("Least-Inclusive Taxonomic Unit") concept. *Zoologica Scripta* 29: 157–68.

Poe, S. 1998. Skull characters and the cladistic relationships of the Hispaniolan dwarf twig *Anolis. Herpetological Monographs* 12: 192–236.

Popper, K. 1957. *The poverty of historicism.* Routledge and Kegan Paul, London.

———. 1959. *The logic of scientific discovery.* Harper and Row, New York.

———. 1976. *Unended quest: An intellectual autobiography.* Open Court, Chicago.

Posada, D., and K. A. Crandall. 1998. MODELTEST: Testing the model of DNA substitution. *Bioinformatics* 14: 817–18.

Pough, F. H. 1988a. Mimicry and related phenomena. Pp. 153–234 in C. Gans and R. B. Huey, eds., *Biology of the Reptilia,* vol. 16: *Ecology B: Defense and life history.* John Wiley, New York.

———. 1988b. Mimicry of vertebrates: Are the rules different? Pp. 67–102 in L. P. Brower, ed., *Mimicry and the evolutionary process.* University of Chicago Press, Chicago.

Pough, F. H., R. M. Andrews, J. E. Cadle, M. L. Crump, A. H. Savitzky, and K. D. Wells. 1998. *Herpetology.* Prentice Hall, Upper Saddle River, NY.

Pough, F. H., W. E. Magnusson, M. J. Ryan, K. D. Wells, and T. L. Taigen. 1992. Behavioral energetics. Pp. 395–436 in M. E. Feder and W. W. Burggren, eds., *Environmental physiology of the amphibians.* University of Chicago Press, Chicago.

Pounds, J. A. 2000. Amphibians and reptiles. Pp. 150–77 in N. A. Nadkarni and N. T.

Wheelwright, eds., *Monteverde: Ecology and conservation of a tropical cloud forest*. Oxford University Press, Oxford.

Pounds, J. A., and M. L. Crump. 1994. Amphibian declines and climate disturbance: The case of the golden toad and the harlequin frog. *Conservation Biology* 8: 72–85.

Pounds, J. A., M. P. L. Fogden, J. M. Savage, and G. C. Gorman. 1996. Tests of null models for amphibian declines on a tropical mountain. *Conservation Biology* 11: 1306–22.

Prance, G. T. 1982. Forest refuges: Evidence from woody angiosperms. Pp. 137–58 in G. T. Prance, ed., *Biological diversification in the tropics*. Columbia University Press, New York.

———. 1987. Biogeography of Neotropical plants. Pp. 46–65 in T. C. Whitmore and G. T. Prance, eds., *Biogeography and Quaternary history in tropical America*. Oxford Monographs on Biogeography, no. 3. Oxford University Press, Oxford.

Prestwich, K. N. 1994. The energetics of acoustic signaling in anurans and insects. *American Zoologist* 34: 625–43.

Price, T. D. 1984. Sexual selection on body size, territory and plumage variables in a population of Darwin's finches. *Evolution* 38: 327–41.

Puorto, G., M. Da Graca Salomão, R. D. G. Theakston, R. S. Thorpe, D. A. Warrell, and W. Wüster. 2001. Combining mitochondrial DNA sequences and morphological data to infer species boundaries: Phylogeography of lanceheaded pitvipers in the Brazilian Atlantic forest, and the status of *Bothrops pradoi* (Squamata: Serpentes: Viperidae). *Journal of Evolutionary Biology* 14: 527–38.

Quammen, D. 1996. *The song of the dodo: Island biogeography in an age of extinctions*. Scribner, New York.

Quelch, J. J. 1921. Itinerary [expedition to Mount Roraima in 1898]. Pp. v–lxxviii in C. Chubb, *The birds of British Guiana, based on the collections of Frederick Vavasour McConnell*, vol. 2. Bernard Quaritch, London.

Quine, W. V. O. 1958. Speaking of objects. *Proceedings and Addresses of the American Philosophical Association* 31: 5–22.

———. 1960. *Word and object*. Technology Press of MIT, Cambridge.

Rabinowitz, D. 1981. Seven forms of rarity. Pp. 205–17 in H. Synge, ed., *The biological aspects of rare plant conservation*. John Wiley, New York.

Rabinowitz, D., S. Cairnes, and T. Dillon. 1986. Seven forms of rarity and their frequency in the flora of the British Isles. Pp. 182–204 in M. E. Soulé, ed., *Conservation biology: The science of scarcity and diversity*. Sinauer Associates, Sunderland, MA.

Raby, P. 2001. *Alfred Russel Wallace: A life*. Princeton University Press, Princeton, NJ.

Rachowiecki, R. 1988. The lost world of Venezuela. *Americas* 40: 45–49, 64–65.

Rage, J. C. 1987. Fossil history. Pp. 51–76 in R. A. Seigel, J. T. Collins, and S. S. Novak, eds., *Snakes: Ecology and evolutionary biology*. Macmillan, New York.

Raleigh, Sir W. 1848. *The discovery of the . . . empire of Guiana . . . in the year 1595, by Sir W. Raleigh Kut*. Reprinted from the 1596 edition; ed. and with a biographical memoir by Sir R. H. Schomburgk.

Rand, A. S., and C. W. Myers. 1990. The herpetofauna of Barro Colorado Island, Panama: An ecological summary. Pp. 386–409 in A. H. Gentry, ed., *Four Neotropical rainforests*. Yale University Press, New Haven, CT.

Rasmussen, J. B., and K. M. Howell. 1998. A review of Barbour's short-headed viper,

Adenorhinos barbouri (Serpentes: Viperidae). *African Journal of Herpetology* 47: 69–75.

Reeder, T. W. 1995. Phylogenetic relationships among phrynosomatid lizards as inferred from mitochondrial ribosomal DNA sequences: Substitutional bias and information content of transitions relative to transversions. *Molecular Phylogenetics and Evolution* 4: 203–22.

Regal, P. J., and C. Gans. 1976. Functional aspects of the evolution of frog tongues. *Evolution* 30: 718–34.

Regnier, M., and A. Herrera. 1993. Changes in contractile properties by androgen hormones in sexually dimorphic muscles of male frogs (*Xenopus laevis*). *Journal of Physiology* 461: 565–81.

Reid, A. R. 1974. Stratigraphy of the type area of the Roraima Group, Venezuela. *Memoria de la Novena Conferencia Geológica Inter-Guayanas, Publicación Especial* 6: 343–53.

Reiskind, J. 1976. *Orsima formica:* A Bornean salticid mimicking an insect in reverse. *Bulletin of the British Arachnological Society* 3: 235–36.

Remsen, J. V., Jr., M. A. Hyde, and A. Chapman. 1993. The diets of Neotropical trogons, motmots, barbets and toucans. *Condor* 95: 178–92.

Richard, J. D., and D. A. Hughes. 1972. Some observations of sea turtle nesting activity in Costa Rica. *Marine Biology* 16: 297–309.

Rieppel, O. 1979. A cladistic classification of primitive snakes based on skull structure. *Zeitschrift für Zoologische Systematik und Evolutionforschung* 17: 140–50.

———. 1988. A review of the origin of snakes. *Evolutionary Biology* 22: 37–130.

Rivero, J. A. 1961. Salientia of Venezuela. *Bulletin of the Museum of Comparative Zoology* 126: 1–207.

———. 1966. Notes on the genus *Cryptobatrachus* (Amphibia, Salientia) with the description of a new race and four new species of a new genus of hylid frogs. *Caribbean Journal of Science* 6: 137–49.

———. 1967. A new race of *Otophryne robusta* Boulenger (Amphibia, Salientia) from the Chimantá-tepui of Venezuela. *Caribbean Journal of Science* 7: 155–58.

———. 1968. A new species of *Elosia* (Amphibia, Salientia) from Mt. Duida, Venezuela. *American Museum Novitates* 2334: 1–9.

———. 1971. Un nuevo e interesante *Dendrobates* (Amphibia, Salientia) del Cerro Yapacana de Venezuela. *Kasmera* 3: 389–396.

Robinson, S. K. 1986. Benefits, costs, and determinants of dominance in a polygynous oriole. *Animal Behaviour* 34: 241–55.

Rodriguez, L. B., and J. E. Cadle. 1990. A preliminary overview of the herpetofauna of Cocha Cashu, Manu National Park, Peru. Pp. 410–25 in A. H. Gentry, ed., *Four Neotropical forests.* Yale University Press, New Haven, CT.

Rodríguez-Robles, J. A., and J. M. de Jesús-Escobar. 2000. Molecular systematics of New World gopher, bull, and pinesnakes (*Pituophis:* Colubridae), a transcontinental species complex. *Molecular Phylogenetics and Evolution* 14: 35–50.

Romer, A. S. 1956. *Osteology of the reptiles.* University of Chicago Press, Chicago.

Rosen, D. E. 1976. A vicariance model of Caribbean biogeography. *Systematic Zoology* 24: 431–64.

Ross, M. I., and C. R. Scotese. 1988. A hierarchical tectonic model of the Gulf of Mexico and Caribbean region. *Tectonophysics* 155: 139–54.

Rossman, D. A., N. B. Ford, and R. A. Seigel. 1996. *The garter snakes: Evolution and ecology.* University of Oklahoma Press, Norman.

Rowe, C., and T. Guilford. 2000. Aposematism: To be red or dead. *Trends in Ecology and Evolution* 15: 261–62.

Rowe, M. P., R. G. Coss, and D. H. Owings. 1986. Rattlesnake rattles and burrowing owl hisses: A case of acoustic Batesian mimicry. *Ethology* 72: 53–71.

Rowe, T. 1987. Definition and diagnosis in the phylogenetic system. *Systematic Zoology* 36: 208–11.

———. 1988. Definition, diagnosis, and origin of Mammalia. *Journal of Vertebrate Paleontology* 8: 241–64.

Rowe, T., and J. A. Gauthier. 1992. Ancestry, paleontology and definition of the name Mammalia. *Systematic Biology* 41: 372–78.

Roze, J. A. 1958a. Resultados zoológicos de la expedición de la Universidad Central de Venezuela a la región del Auyantepui en la Guayana Venezolana, abril de 1956. 5. Los reptiles del Auyantepui, Venezuela, basándose en las colecciones de las expediciones de Phelps-Tate, del American Museum of Natural History, 1937–1938, y de la Universidad Central de Venezuela, 1956. *Acta Biologica Venezuelica* 2: 243–70.

———. 1958b. Los reptiles del Chimantá tepui (Estado Bolívar, Venezuela) colectados por la expedición botanica del Chicago Natural History Museum. *Acta Biologica Venezuelica* 2: 299–314.

———. 1961. El genero *Atractus* (Serpentes: Colubridae) en Venezuela. *Acta Biologica Venezuelica* 3: 103–19.

———. 1996. *Coral snakes of the Americas: Biology, identification, and venoms.* Krieger Publishing, Malabar, FL.

Rudak, E., and H. G. Callan. 1976. Differential staining and chromatin packing of the mitotic chromosomes of the newt *Triturus cristatus. Chromosoma* (Berlin) 56: 349–62.

Runkle, L., K. Wells, C. Robb, and S. Lance. 1994. Individual, nightly, and seasonal variation in calling behavior of the gray tree frog, *Hyla versicolor:* Implications for energy expenditure. *Behavioral Ecology* 5: 318–25.

Russell, F. E. 1967. Bites by the Sonoran coral snake *Micruroides euryxanthus. Toxicon* 5: 39–42.

Ryan, M. J. 1985. *The túngara frog.* University of Chicago Press, Chicago.

Saiki, R. K., D. H. Delfand, S. Stooffel, S. J. Scharf, R. Higuchi, G. T. Horn, K. B. Mullis, and H. A. Erlich. 1988. Primer-directed enzymatic amplification of DNA with a thermostable DNA polymerase. *Science* 239: 487–91.

Saint, K. M., C. C. Austin, S. C. Donnellan, and M. N. Hutchinson. 1998. C-*mos,* a nuclear marker useful for squamate phylogenetic analysis. *Molecular Phylogenetics and Evolution* 10: 259–63.

Salmon, W. C. 1966. *The foundations of scientific inference.* University of Pittsburgh Press, Pittsburgh, PA.

Salomão, M. G., W. Wüster, and R. S Thorpe. 1997. DNA evolution of South American pitvipers of the genus *Bothrops* (Reptilia: Serpentes: Viperidae). Pp. 89–98 in R. S. Thorpe, W. Wüster, and A. Malhotra, eds., *Venomous snakes: Ecology, evolution and snakebite*. Symposia of the Zoological Society of London. Clarendon Press, Oxford.

———. 1999. mtDNA phylogeny of Neotropical pitvipers of the genus *Bothrops* (Squamata: Serpentes: Viperidae). *Kaupia* 8: 127–34.

Salvador, A., and A. K. Green. 1980. Opening of the Caribbean Tethys. Pp. 224–29 in *Geologie des Chaines Alpines: Issues de la Tethys*. Memoires du Bureau des Recherches Geologiques et Minìeres 115.

Sánchez-Herrera, O., H. M. Smith, and D. Chiszar. 1981. Another suggested case of ophidian deceptive mimicry. *Transactions of the Kansas Academy of Sciences* 84: 121–27.

Sanford, R. L., Jr., P. Paaby, J. C. Luvall, and E. Phillips. 1994. Climate, geomorphology, and aquatic systems. Pp. 19–34 in L. A. McDade, K. S. Bawa, H. A. Hespenheide, and G. S. Hartshorn, eds., *La Selva: Ecology and natural history of a Neotropical rain forest*. University of Chicago Press, Chicago.

Santiago, R. R. 2000. Biogeographic relationships of lowland Neotropical rainforest based on raw distributions of vertebrate groups. *Biological Journal of the Linnean Society* 71: 379–402.

SAS Institute. 1989. *SAS*. Version 6.08. SAS Institute, Cary, NC.

———. 2000. *JMP statistics and graphics guide*. Version 4. Cary, NC.

Sasa, M. 1996. Morphological variation in the lancehead snake *Bothrops asper* (Garman) from Middle America. M.S. thesis, University of Texas at Arlington, Arlington.

———. 1997. *Cerrophidion godmani* in Costa Rica: A case of extremely low allozyme variation? *Journal of Herpetology* 31: 569–72.

Sasa, M., and A. Solórzano. 1995. The reptiles and amphibians of Santa Rosa National Park, Costa Rica, with comments about the herpetofauna of xerophytic areas. *Herpetological Natural History* 3: 113–26.

Sassoon, D., N. Segil, and D. Kelley. 1986. Androgen-induced myogenesis and chondrogenesis in the larynx of *Xenopus laevis*. *Developmental Biology* 113: 135–40.

Savage, J. M. 1960. Evolution of a peninsular herpetofauna. *Systematic Zoology* 9: 184–212.

———. 1966. The origins and history of the Central American herpetofauna. *Copeia* 1966: 719–66.

———. 1970. On the trail of the golden frog: With Warszewicz and Gabb in Central America. *Proceedings of the California Academy of Sciences* 28: 273–88.

———. 1973. The geographic distribution of frogs: Patterns and predictions. Pp. 349–445 in J. L. Vial, ed., *Evolutionary biology of the Anura*. University of Missouri Press, Columbia.

———. 1975. Systematics and distribution of the Mexican and Central American stream frogs related to *Eleutherodactylus rugulosus*. *Copeia* 1975: 254–306.

———. 1982. The enigma of the Central American Herpetofauna: Dispersals or vicariance? *Annals of the Missouri Botanical Garden* 69: 464–547.

———. 1987. Systematics and distribution of the Mexican and Central American rainfrogs of the *Eleutherodactylus gollmeri* group (Amphibia: Leptodactylidae). *Fieldiana Zoology* 33: 1–57.

———. 1997. A new species of rainfrog of the *Eleutherodactylus diastema* group from the Alta Talamanca region of Costa Rica. *Amphibia-Reptilia* 18: 241–47.

————. 2002. *The amphibians and reptiles of Costa Rica: A herpetofauna between two continents, between two seas.* University of Chicago Press, Chicago.

Savage, J. M., and B. I. Crother. 1989. The status of *Pliocercus* and *Urotheca* (Serpentes: Colubridae), with a review of included species of coral snake mimics. *Zoological Journal of the Linnean Society* 95: 335–62.

Savage, J. M., and J. E. DeWeese. 1979. A new species of leptodactylid frog, genus *Eleutherodactylus*, from the Cordillera de Talamanca, Costa Rica. *Bulletin of the Southern California Academy of Science* 78: 107–15.

Savage, J. M., and S. B. Emerson. 1970. Central American frogs allied to *Eleutherodactylus bransfordii* (Cope): A problem of polymorphism. *Copeia 1970*: 623–44.

Savage, J. M., and C. Guyer. 1989. Infrageneric classification and species composition of the anole genera *Anolis, Ctenonotus, Dactyloa, Norops* and *Semiurus* (Sauria: Iguanidae). *Amphibia-Reptilia* 10: 105–16.

————. 1991. Nomenclatural notes on anoles (Sauria: Polychridae): Stability over priority. *Journal of Herpetology* 25: 365–66.

Savage, J. M., J. R. McCranie, and M. Espinal. 1996. A new species of *Eleutherodactylus* from Honduras related to *Eleutherodactylus bransfordii* (Anura: Leptodactylidae). *Proceedings of the Biological Society of Washington* 109: 366–72.

Savage, J. M., and J. B. Slowinksi. 1990. A simple consistent terminology for the basic colour patterns of the venomous coral snakes and their mimics. *Herpetological Journal* 1: 530–32.

————. 1992. The colouration of the venomous coral snakes (family Elapidae) and their mimics (families Aniliidae and Colubridae). *Biological Journal of the Linnean Society* 45: 235–54.

————. 1996. Evolution of colouration, urotomy and coral snake mimicry in the snake genus *Scaphiodontophis* (Serpentes: Colubridae). *Biological Journal of the Linnean Society* 57: 129–94.

Savage, J. M., and J. J. Talbot. 1978. The giant anoline lizards of Costa Rica and western Panama. *Copeia* 1978: 105–16.

Savage, J. M., and J. Vial. 1974. The venomous coral snakes (genus *Micrurus*) of Costa Rica. *Revista de Biología Tropical* 21: 295–349.

Savage, J. M., and J. Villa. 1986. *An introduction to the herpetofauna of Costa Rica.* Society for the Study of Amphibians and Reptiles, St. Louis, MO.

Savage, J. M., and M. H. Wake. 1972. Geographic variation and systematics of the Middle American caecilians, genera *Dermophis* and *Gymnopis. Copeia* 1972: 680–95.

————. 2001. A re-evaluation of the status of taxa of Central American caecilians (Amphibia: Gymnophiona), with comments on their origin and evolution. *Copeia* 2001: 52–64.

Sawyer, J. O., and A. A. Lindsey. 1971. *Vegetation of the life zones in Costa Rica.* Indiana Academy of Sciences, Bloomington.

Sazima, I. 1974. Experimental predation on the leaf-frog *Phyllomedusa rohdei* by the water snake *Liophis miliaris. Journal of Herpetology* 8: 376–77.

————. 1992. Natural history of the jararaca pitviper, *Bothrops jararaca.* Pp. 199–216 in J. A. Campbell and E. D. Brodie Jr., eds., *Biology of pitvipers.* Selva, Tyler, TX.

Sazima, I., and A. S. Abe. 1991. Habits of five Brazilian snakes with coral-snake pattern, including a summary of defensive tactics. *Studies on Neotropical Fauna and Environment* 26: 159–64.

Schander, C. 1998a. Types, emendations and names: A reply to Lidén et al. *Taxon* 47: 401–6.

———. 1998b. Mandatory categories and impossible hierarchies: A reply to Sosef. *Taxon* 47: 407–10.

Schander, C., and M. Thollesson. 1995. Phylogenetic taxonomy: Some comments. *Zoologica Scripta* 24: 263–68.

Schemske, D. W., and C. C. Horvitz. 1989. Temporal variation in selection in a floral character. *Evolution* 43: 461–65.

Schleich, H. H., W. Kästle, and K. Kabisch. 1996. *Amphibians and reptiles of North Africa.* Koelz Scientific Publishers, Koenigstein, Germany.

Schlüter, A. 1994. Anuren des Guaiquinima-Tepuis, Venezuela. Pp. 69–83 in H.-J. Herrmann and H. Zimmermann, eds., *Beiträge zur Biologie der Anuren.* Tetra Verlag, Melle, Germany.

Schluter, D., and J. N. M. Smith. 1986. Natural selection on beak and body size in the song sparrow. *Evolution* 40: 221–31.

Schmid, M. 1978a. Chromosome banding in Amphibia. I. Constitutive heterochromatin and nucleolus organizer regions in *Bufo* and *Hyla. Chromosoma* (Berlin) 66: 361–88.

———. 1978b. Chromosome banding in Amphibia. II. Constitutive heterochromatin and nucleolus organizer regions in Ranidae, Microhylidae and Rhacophoridae. *Chromosoma* (Berlin) 68: 131–48.

———. 1982. Chromosome banding in Amphibia. VII. Analysis of the structure and variability of NORs in Anura. *Chromosoma* (Berlin) 87: 327–44.

Schnell, C. E., ed. 1971. *Handbook for tropical biology in Costa Rica.* Organization for Tropical Studies, San José, Costa Rica.

Schomburgk, M. R. 1841. Journey from Fort San Joaquim, on the Rio Branco, to Roraima, and thence by the Rivers Parima and Merewari to Esmeralda, on the Orinoco, in 1838–9. *Journal of the Royal Geographic Society* 10: 191–247.

———. 1847–48. *Reisen in Britisch-Guiana in dem Jahren, 1840–1844.* 3 vols. Leipzig, Verlagsbuchhandlung von J. J. Weber. As seen in Los Schomburgk en el Roraima, *Boletín de la Sociedad Venezolana de Ciencias Naturales* (1944) 9: 285–320.

Schubert, C., and H. O. Briceño. 1987. Origen de la topografía tepuyana: Una hipótesis. *Pantepui* 2: 11–14.

Schuh, R. T. 2000. *Biological systematics: Principles and applications.* Cornell University Press, Ithaca, NY.

Schulte, J. A., II, J. R. Macey, R. E. Espinoza, and A. Larson. 2000. Phylogenetic relationships in the iguanid lizard genus *Liolaemus:* Multiple origins of viviparous reproduction and evidence for recurring Andean vicariance and dispersal. *Biological Journal of the Linnean Society* 69: 75–102.

Schwagmeyer, O. L., and C. H. Brown. 1983. Factors affecting male–male competition in thirteen-lined ground squirrels. *Behavioral Ecology and Sociobiology* 13: 1–6.

Schwartz, J. J. 1987. The function of call alternation in anuran amphibians: A test of three hypotheses. *Evolution* 41: 461–71.

Schwartz, J. J., and K. D. Wells. 1983a. An experimental study of acoustic interference between two species of Neotropical treefrogs. *Animal Behaviour* 31: 181–90.

———. 1983b. The influence of background noise on the behavior of a Neotropical treefrog, *Hyla ebraccata*. *Herpetologica* 39: 121–29.

———. 1985. Intra- and interspecific vocal behavior of the Neotropical treefrog *Hyla microcephala*. *Copeia* 1985: 27–38.

Schwenk, K. 1994. Why snakes have forked tongues. *Science* 263: 1573–77.

Scott, N. J., Jr. 1976. The abundance and diversity of the herpetofauna of tropical forest litter. *Biotropica* 8: 41–58.

———. 1983. *Conophis lineatus* (Guarda Camino). Pp. 392–93 in D. H. Janzen, ed., *Costa Rican natural history*. University of Chicago Press, Chicago.

Scott, N. J., Jr., J. M. Savage, and D. G. Robinson. 1983. Checklist of reptiles and amphibians. Pp. 367–74 in D. H. Janzen, ed., *Costa Rican natural history*. University of Chicago Press, Chicago.

Scott, N. J., Jr., and A. Starrett. 1974. An unusual breeding aggregation of frogs, with notes on the ecology of *Agalychnis spurrelli* (Anura: Hylidae). *Bulletin of the Southern California Academy of Sciences* 73: 86–94.

Searcy, W. A. 1982. The evolutionary effect of mate selection. *Annual Review of Ecology and Systematics* 13: 57–85.

Searcy, W. A., and P. Marler. 1984. Interspecific differences in the response of female birds to song repertoires. *Zoologische Tierpsychologie* 66: 128–42.

Seib, R. L. 1980. Human envenomation from the bite of an aglyphous false coral snake, *Pliocercus elapoides* (Serpentes: Colubridae). *Toxicon* 18: 399–401.

Seigel, J. A., and T. A. Adamson. 1983. Batesian mimicry between a cardinalfish (Apogonidae) and a venomous scorpionfish (Scorpaenidae) from the Philippine Islands. *Pacific Science* 37: 75–79.

Señaris, J. C. 1993. Una nueva especie de *Oreophrynella* (Anura; Bufonidae) de la cima del Auyan-tepui, Estado Bolívar, Venezuela. *Memoria Sociedad Ciencias Naturales La Salle* 53: 177–83.

Señaris, J. C., and J. Ayarzagüena. 1993. Una nueva especie de *Centrolenella* (Anura: Centrolenidae) del Auyán-tepui, Edo. Bolívar, Venezuela. *Memoria Sociedad Ciencias Naturales La Salle* 53: 121–26.

Señaris, J. C., J. Ayarzagüena, and S. Gorzula. 1994. Los sapos de la familia Bufonidae (Amphibia: Anura) de las tierras altas de la Guayana Venezolana: Descripción de un nuevo genero y tres especies. *Publicaciones de la Asociación de Amigos de Doñana* 3: 1–37.

———. 1996. Revisión taxonomica del genero *Stefania* (Anura; Hylidae) en Venezuela con la descripción de cinco nuevas especies. *Publicaciones de la Asociación de Amigos de Doñana* 7: 1–57.

Sereno, P. C. 1999. Definitions in phylogenetic taxonomy: Critique and rationale. *Systematic Biology* 48: 329–51.

Sessions, S. K. 1982. Cytogenetics of diploid and triploid salamanders of the *Ambystoma jeffersonianum* complex. *Chromosoma* (Berlin) 84: 599–621.

Sessions, S. K., and J. Kezer. 1991. Evolutionary cytogenetics of bolitoglossine salamanders (Family Plethodontidae). Pp. 89–130 in D. M. Green and S. K. Sessions, eds., *Amphibian cytogenetics and evolution*. Academic Press, San Diego.

Sessions, S. K., and A. Larson. 1987. Developmental correlates of genome size in plethodontid salamanders and their implications for genome evolution. *Evolution* 41: 1239–51.

Shine, R. 1991. Strangers in a strange land: Ecology of the Australian colubrid snakes. *Copeia* 1991: 120–31.

Shochat, D., and H. C. Dessauer. 1981. Comparative immunological study of albumins of *Anolis* lizards of the Caribbean Islands. *Comparative Biochemical Physiology* 68A: 67–73.

Short, L. L. 1975. A zoogeographic analysis of the South American Chaco avifauna. *Bulletin of the American Museum of Natural History* 154: 163–352.

Shreve, B. 1947. On Venezuelan reptiles and amphibians collected by Dr. H. G. Kugler. *Bulletin of the Museum of Comparative Zoölogy* 99: 519–37.

Siddall, M. E. 2002. Measures of support. Pp. 80–101 in R. DeSalle, G. Giribet, and W. C. Wheeler, eds., *Techniques in molecular systematics and evolution*. Birkhäuser Verlag, Basel, Switzerland.

Siddall, M. E., and A. G. Kluge. 1997. Probabilism and phylogenetic inference. *Cladistics* 13: 313–36.

Simmons, L. W. 1988. The calling song of the field cricket, *Gryllus bimaculatus* (De Gree): Constraints on transmission and its role in intermale competition and female choice. *Animal Behaviour* 36: 380–94.

Simpson, B. B. 1979. Quaternary biogeography of the high montane regions of South America. Pp. 157–88 in W. E. Duellman, ed., *The South American herpetofauna: Its origin, evolution, and dispersal*. University of Kansas Museum of Natural History Monograph, no. 7, Lawrence.

Simpson, B. B., and J. Haffer. 1978. Speciation patterns in the Amazon forest biota. *Annual Review of Ecology and Systematics* 9: 497–518.

Simpson, G. G. 1960. Notes on the measurement of faunal resemblance. *American Journal of Science* 258A: 300–311.

Sites, J. W., Jr., and K. M. Reed. 1994. Chromosomal evolution, speciation, and systematics: Some relevant issues. *Herpetologica* 50: 237–49.

Slowinski, J. B. 1994. A phylogenetic analysis of *Bungarus* (Elapidae) based on morphological characters. *Journal of Herpetology* 28: 440–46.

————. 1995. A phylogenetic analysis of the New World coral snakes (Elapidae: *Leptomicrurus, Micruroides,* and *Micrurus*) based on allozymic and morphological characters. *Journal of Herpetology* 29: 325–38.

Slowinski, J. B., J. Boundy, and R. Lawson. 2001. The phylogenetic relationships of Asian coral snakes (Elapidae: *Calliophis* and *Maticora*) based on morphological and molecular characters. *Herpetologica* 57: 233–45.

Slowinski, J. B., and B. I. Crother. 1998. Is the PTP test useful? *Cladistics* 14: 297–302.

Slowinski, J. B., and C. Guyer. 1993. Testing whether certain traits have caused amplified diversification: An improved method based on a model of random speciation and extinction. *American Naturalist* 142: 1019–24.

Slowinski, J. B., and J. S. Keogh. 2000. Phylogenetic relationships of elapid snakes based on cytochrome *b* mtDNA sequences. *Molecular Phylogenetics and Evolution* 15: 157–64.

Slowinski, J. B., A. Knight, and A. R. Rooney. 1997. Inferring species trees from gene trees: A phylogenetic analysis of the Elapidae (Serpentes) based on the amino acid sequences of venom proteins. *Molecular Phylogenetics and Evolution* 8: 349–62.

Slowinski, J. B., and R. Lawson. 2002. Snake phylogeny: Evidence from nuclear and mitochondrial genes. *Molecular Phylogenetics and Evolution* 24: 194–202.

Slowinski, J. B., and R. D. M. Page. 1999. How should species phylogenies be inferred from sequence data? *Systematic Biology* 48: 814–25.

Smith, A. B., and C. Patterson. 1988. The influence of taxonomic method on the perception of patterns of evolution. *Evolutionary Biology* 23: 127–206.

Smith, G., E. Brenowitz, M. Beecher, and J. Wingfield. 1997. Seasonal changes in testosterone neural attributes of song control nuclei and song structure in wild songbirds. *Journal of Neuroscience* 17: 6001–10.

Smith, H. M., and D. Chiszar. 1996. *Species-group taxa of the false coral snake genus Pliocercus*. Ramus Publications, Pottsville, PA.

Smith, S. M. 1975. Innate recognition of coral snake pattern by a possible avian predator. *Science* 187: 759–60.

———. 1977. Coral-snake pattern recognition and stimulus generalization by naïve great kiskadees (Aves: Tyrannidae). *Nature* 265: 535–36.

Sober, E. 1988. *Reconstructing the past: Parsimony, evolution, and inference*. MIT Press, Cambridge.

———. 1991. Organisms, individuals, and units of selection. Pp 275–96 in A. I. Tauber, ed., *Organism and the origins of self*. Kluwer, London.

Sokal, R. R., and F. J. Rohlf. 1969. *Biometry: The principles and practice of statistics in biological research*. W. H. Freeman, San Francisco.

———. 1981. Taxonomic congruence in the Leptopodomorpha re-examined. *Systematic Zoology* 30: 309–25.

Solbrig, O. T. 1976. The origin and floristic affinities of the South American temperate desert and semidesert regions. Pp. 7–49 in D. W. Goodall, ed., *Evolution of desert biota*. University of Texas Press, Austin.

Solis, R., and M. Penna. 1997. Testosterone levels and evoked vocal responses in a natural population of the frog, *Batrachyla taeniata*. *Hormones and Behavior* 31: 101–9.

Sollins, P., F. Sancho M., R. Mata Ch., and R. L. Sanford Jr. 1994. Soils and soil process research. Pp. 34–53 in L. A. McDade, K. S. Bawa, H. A. Hespenheide, and G. S. Hartshorn, eds., *La Selva: Ecology and natural history of a Neotropical rain forest*. University of Chicago Press, Chicago.

Solórzano, A. 1994. Una nueva especie de serpiente venenosa terrestre del gerero *Porthidium* (Serpentes: Viperidae), del suroeste de Costa Rica. *Revista Biologia Tropical* 42: 695–701.

———. 2001. Una nueva especie de serpiente del genero *Sibon* (Serpentes: Colubridae) de la vertiente del Caribe de Costa Rica. *Revista de Biologia Tropical* 49: 1111–20.

Solórzano, A., L. D. Gómez, J. Monge-Nájera, and B. I. Crother. 1998. Redescription and validation of *Bothriechis supraciliaris* (Serpentes: Viperidae). *Revista Biologia Tropical* 46: 453–62.

Sorenson, M. D. 1999. *TreeRot*. Version 2. Boston University, Boston.

Spaur, R. C., and H. M. Smith. 1971. Adrenal enlargement in the hognose snake, *Heterodon platyrhinos*. *Journal of Herpetology* 5: 197–200.

Spawls, S., and B. Branch. 1995. *The dangerous snakes of Africa*. Ralph Curtis Publishing, Sanibel Island, FL.

Speed, M. P., and J. R. G. Turner. 1999. Learning and memory in mimicry. II. Do we understand the mimicry spectrum? *Biological Journal of the Linnean Society* 67: 281–312.

Spruce, R. 1908. *Notes of a botanist on the Amazon & Andes being records of travel on the Amazon and its tributaries, the Trombetas, Rio Negro, Uaupés, Casiquiari, Pacimoni, Huallaga, and Pastasa; as also to the cataracts of the Orinoco, along the eastern side of the Andes of Peru and Ecuador, and the shores of the Pacific, during the years 1849–1864*. Ed. Alfred Russel Wallace. 2 vols. Macmillan, London.

Srygley, R. B. 1999. Incorporating motion into investigations of mimicry. *Evolutionary Ecology* 13: 691–708.

Stafford, P. J. 1999. *Urotheca (Pliocercus) elapoides* (false coral snake): Reproduction and diel activity. *Herpetological Review* 30: 48.

Starace, F. 1998. *Guide des serpents et amphisbènes de Guyane*. Ibis Rouge Editions, Guadeloupe, Guyana.

Statsoft. 1994. *Statistica for the Macintosh*. Vols. 1 and 2. Statsoft, Inc., Tulsa, OK.

Steel, M. A., and D. Penny. 2000. Parsimony, likelihood, and the role of models in molecular phylogenetics. *Molecular Biology and Evolution* 17: 839–50.

Stephens, L., and M. A. Traylor Jr. 1985. *Ornithological gazetteer of the Guianas*. Museum of Comparative Zoology, Harvard University, Cambridge.

Steyermark, J. A. 1982. Relationships of some Venezuelan forest refuges with lowland tropical floras. Pp. 182–220 in G. T. Prance, ed., *Biological diversification in the Tropics*. Columbia University Press, New York.

———. 1986. Speciation and endemism in the flora of the Venezuelan tepuis. Pp. 317–33 in F. Vuilleumier and M. Monasterio, eds., *High altitude tropical biogeography*. Oxford University Press and the American Museum of Natural History, Oxford and New York.

Steyermark, J. A., P. E. Berry, and B. K. Holst, eds. 1995a. *Flora of the Venezuelan Guayana*, vol. 1: *Introduction*. Missouri Botanical Garden and Timber Press, St. Louis, MO, and Portland, OR.

———, eds. 1995b. *Flora of the Venezuelan Guayana*, vol. 2: *Pteridophytes Spermatophytes; Acanthaceae–Araceae*. Missouri Botanical Garden and Timber Press, St. Louis, MO, and Portland, OR.

———, eds. 1997. *Flora of the Venezuelan Guayana*, vol. 3: *Araliaceae–Cactaceae*. Missouri Botanical, St. Louis, MO.

———, eds. 1998. *Flora of the Venezuelan Guayana*, vol. 4: *Caesalpiniaceae–Ericaceae*. Missouri Botanical, St. Louis, MO.

———, eds. 1999. *Flora of the Venezuelan Guayana*, vol. 5: *Eriocaulaceae–Lentibulariaceae*. Missouri Botanical, St. Louis, MO.

Steyermark, J. A., and C. Brewer-Carías. 1976. La vegetación de la cima del Macizo de Jaua. *Boletín de la Sociedad Venezolana de Ciencias Naturales* 32: 179–405.

Strecker, J. K. 1927. Chapters from the life-histories of Texas reptiles and amphibians. II. *Contributions of the Baylor University Museum* 10: 3–14.

————. 1929. Field notes on the herpetology of Wilbarger County, Texas. *Contributions of the Baylor University Museum* 19: 3–9.

Stuart, L. C. 1966. The environment of the Central American cold-blooded vertebrate fauna. *Copeia.* 1966: 684–99.

Stuebing, R. B., and R. F. Inger. 1999. *A field guide to the snakes of Borneo.* Natural History Publications (Borneo), Kota Kinabalu, Malaysia.

Sullivan, B. K., and S. H. Hinshaw. 1992. Female choice and selection of male calling behavior in the grey treefrog *Hyla versicolor. Animal Behaviour* 44: 733–44.

Sullivan, B. K., M. J. Ryan, and P. Verrell. 1994. Female choice and mating system structure. Pp. 470–515 in H. Heatwole and B. Sullivan, eds., *Amphibian biology,* vol. 2: *Social behavior.* Surrey Beatty and Sons, New South Wales, Australia.

Sumner, A. T. 1972. A simple technique for demonstrating centromeric heterochromatin. *Experimental Cell Research* 75: 304–6.

Sundberg, P., and F. Pleijel. 1994. Phylogenetic classification and the definition of taxon names. *Zoologica Scripta* 23: 19–25.

Sweet, S. S. 1985. Geographic variation, convergent crypsis and mimicry in gopher snakes (*Pituophis melanoleucus*) and Western Rattlesnakes (*Crotalus viridis*). *Journal of Herpetology* 19: 55–67.

Swofford, D. 1998. *PAUP*: Phylogenetic analysis using parsimony (*and other methods).* Version 4. Sinauer Associates, Sunderland, MA.

————. 1999. *PAUP*: Phylogenetic analysis using parsimony (*and other methods).* Version 4.0a for Macintosh. Sinauer Associates, Sunderland, MA.

————. 2000. *PAUP*: Phylogenetic analysis using parsimony (*and other methods).* Version 4.0. Sinauer Associates, Sunderland, MA.

————. 2001. *PAUP*: Phylogenetic analysis using parsimony (*and other methods).* Version 4.0. Sinauer Associates, Sunderland, MA.

Swofford, D., G. J. Olsen, P. J. Waddell, and D. M. Hillis. 1996. Phylogenetic inference. Pp. 407–14 in D. M. Hillis, C. Moritz, and B. K. Mable, eds., *Molecular systematics,* 2nd ed. Sinauer Associates, Sunderland, MA.

Sykes, L. R., W. R. McCann, and A. C. Kafka. 1982. Motion of the Caribbean plates during the last 7 million years and implications for earlier Cenozoic movements. *Journal of Geophysical Research* 87: 10656–76.

Symula, R., R. Schulte, and K. Summers. 2001. Molecular phylogenetic evidence for a mimetic radiation of Peruvian poison frogs supports a Müllerian mimicry hypothesis. *Proceedings of the Royal Society of London B* 268: 2415–21.

Szyndlar, Z., and J.-C. Rage. 1990. West palearctic cobras of the genus *Naja* (Serpentes: Elapidae): Interrelationships among extinct and extant species. *Amphibia-Reptilia* 11: 385–400.

Szyndlar, Z., and G. A. Zerova. 1990. Neogene cobras of the genus *Naja* (Serpentes: Elapidae) of east Europe. *Annalen des Naturhistorischen Museums in Wien* 91: 53–61.

Taggart, T. W., B. I. Crother, and M. E. White. 2001. Palm-pitviper (*Bothriechis*) phylogeny, mtDNA and consilience. *Cladistics* 17: 355–70.

Taigen, T. L., and K. D. Wells. 1985. Energetics of vocalizations by anuran amphibians. *Journal of Comparative Physiology* 155: 163–70.

Tanaka, K., and A. Mori. 2000. Literature survey on predators of snakes in Japan. *Current Herpetology* 19: 97–111.

Tate, G. H. H. 1928. The "lost world" of Mount Roraima. *Natural History* 28: 318–28.

———. 1930. Notes on the Mount Roraima region. *Geographical Review* 20: 53–68.

———. 1932. Life zones of Mount Roraima. *Ecology* 13: 235–57.

———. 1938. Auyantepui: Notes on the Phelps Venezuelan Expedition. *Geographical Review* 28: 452–74.

———. 1939. The mammals of the Guiana region. *Bulletin of the American Museum of Natural History* 76: 151–229.

Tate, G. H. H., and C. B. Hitchcock. 1930. The Cerro Duida region of Venezuela. *Geographical Review* 20: 31.

Taylor, E. H. 1952a. The salamanders and caecilians of Costa Rica. *University of Kansas Science Bulletin* 34 (pt. 2): 695–791.

———. 1952b. A review of the frogs and toads of Costa Rica. *University of Kansas Science Bulletin* 35: 577–942.

———. 1954. Further studies on the serpents of Costa Rica. *University of Kansas Science Bulletin* 36: 637–801.

———. 1955. Additions to the known herpetological fauna of Costa Rica with comments on other species. II. *University of Kansas Science Bulletin* 37: 499–575.

———. 1968. *The caecilians of the world: A taxonomic review.* University of Kansas Press, Lawrence.

———. 1969. A new Panamanian caecilian. *University of Kansas Science Bulletin* 48: 315–23.

———. 1973. A caecilian miscellany. *University of Kansas Science Bulletin* 50: 187–231.

Taylor, R. T., A. Flores, F. Guillermo, and R. Bolaños. 1974. Geographical distribution of Viperidae, Elapidae and Hydrophiidae in Costa Rica. *Revista Biologia Tropical* 21: 383–97.

Tchernov, E., O. Rieppel, H. Zaher, M. J. Polcyn, and L. L. Jacobs. 2000. A fossil snake with limbs. *Science* 287: 2010–12.

Templeton, A. R. 1983. Phylogenetic inference from restriction endonuclease cleavage site maps with particular reference to the humans and apes. *Evolution* 37: 221–44.

ter Steege, H. 1993. *Patterns in tropical rain forest in Guyana.* Tropenbos Series 2. Tropenbos Foundation, Wageningen, The Netherlands.

ter Steege, H., R. G. A. Boot, L. C. Brouwer, J. C. Caesar, R. C. Ek, D. S. Hammond, P. P. Haripersaud, P. van der Hout, V. G. Jetten, A. J. van Kekem, M. A. Kellman, Z. Khan, A. M. Polak, T. L. Pons, J. Pulles, D. Raaimakers, S. A. Rose, J. J. van der Sanden, and R. J. Zagt. 1996. *Ecology and logging in a tropical rain forest in Guyana with recommendations for forest management.* Tropenbos Series 14. Tropenbos Foundation, Wageningen, The Netherlands.

Thiele, K. 1993. The holy grail of the perfect character: The cladistic treatment of morphometric data. *Cladistics* 9: 275–304.

Thompson, J. D., T. J. Gibson, F. Plewniak, F. Jeanmougin, and D. Higgins. 1997. The Clustal X windows interface: Flexible strategies for multiple sequence alignment aided by quality analysis tools. *Nucleic Acids Research* 25: 4876–82.

Times Atlas of the World. 1999. 10th ed. Random House, New York.

Tobin, C., and Y. Joubert. 1991. Testosterone-induced development of the rat levator ani muscle. *Developmental Biology* 146: 131–38.

Tosi, J. A., Jr. 1975. The Corcovado Basin on the Península de Osa. Pp. 11–14 in J. A. Tosi

Jr., ed., *Potential national parks, nature reserves, and wildlife sanctuary areas in Costa Rica: A survey of the priorities.* Tropical Science Center Report, San José, Costa Rica.

Townsend, D., and W. Moger. 1987. Plasma androgen levels during male parental care in a tropical frog (*Eleutherodactylus*). *Hormones and Behavior* 21: 93–99.

Trail, P. W. 1984. The lek mating system of the Guianan cock-of-the-rock: A field study of sexual selection. Ph.D. dissertation, Cornell University, Ithaca, NY.

Trenerry, M. P., W. P. Laurance, and K. R. MacDonald. 1994. Further evidence of the precipitous decline of endemic rainforest frogs in tropical Australia. *Pacific Conservation Biology* 1: 150–54.

Tuffley, C., and M. Steel. 1997. Links between maximum likelihood and maximum parsimony under a simple model of site substitution. *Bulletin of Mathematical Biology* 59: 581–607.

Turner, J. R. G., and J. L. B. Mallet. 1996. Did forest islands drive the diversity of warningly coloured butterflies? Biotic drift and the shifting balance. *Philosophical Transactions of the Royal Society of London B* 351: 835–45.

Uetz, P. 2000. How many reptile species? *Herpetological Review* 31: 13–15.

Underwood, G. 1957. On the lizards of the family Pygopodidae, a contribution to the morphology and phylogeny of the Squamata. *Journal of Morphology* 100: 207–68.

———. 1967. *A contribution to the classification of snakes.* British Museum (Natural History), London.

Underwood, G., and E. Kochva. 1993. On the affinities of the burrowing asps *Atractaspis* (Serpentes: Atractaspididae). *Zoological Journal of the Linnean Society* 107: 3–64.

United States Board on Geographic Names [USBGN]. 1961. *Venezuela.* Official standard names approved by the U.S. Board on Geographic Names. Gazetteer no. 56. U.S. Government Printing Office, Washington, DC.

Uy, J. A., and G. Borgia. 2000. Sexual selection drives rapid divergence in bowerbird display traits. *Evolution* 54: 273–78.

Valerio, C. E. 1971. Ability of some tropical tadpoles to survive without water. *Copeia* 1971: 364–65.

Van Devender, R. W. 1969. Resurrection of *Neusticurus racenisi. Journal of Herpetology* 3: 105–7.

Van Devender, T. R., and R. Conant. 1990. Pleistocene forests and copperheads in the eastern United States, and the historical biogeography of New World *Agkistrodon.* Pp. 601–14 in H. K. Gloyd and R. Conant, eds., *Snakes of the Agkistrodon complex: A monographic review.* Contributions in Herpetology, no. 6. Society for the Study of Amphibians and Reptiles, St. Louis, MO.

Van Heest, R. W., and J. A. Hay. 2000. *Charina bottae* (rubber boa): Antipredator behavior. *Herpetological Review* 31: 177.

van Kekem, A. J., J. H. M. Pulles, and Z. Khan. 1996. *Soils of the rainforest in Central Guyana.* Tropenbos Guyana Series 2. Tropenbos Foundation, Wageningen, The Netherlands.

Van Valen, L. 1976. Ecological species, multispecies, and oaks. *Taxon* 25: 233–39.

Van Wijngaarden, R., and F. Bolaños. 1992. Parental care in *Dendrobates granuliferus* (Anura: Dendrobatidae), with a description of the tadpole. *Journal of Herpetology* 26: 102–5.

Vanzolini, P. E., and E. E. Williams. 1970. South American anoles: The geographic differentiation and evolution of the *Anolis chrysolepis* species group (Sauria, Iguanidae). *Arquivos de Zoologia* (São Paulo) 19: 1–298.

Vellard, J. 1948. Batracios del Chaco argentino. *Acta Zoologica Lilloana* 5: 137–74.

Vial, J. L. 1963. A new plethodontid salamander (*Bolitoglossa sooyorum*) from Costa Rica. *Revista de Biologia Tropical* 11: 89–97.

———. 1966. The taxonomic status of two Costa Rican salamanders of the genus *Bolitoglossa*. *Copeia* 1966: 669–73.

———. 1968. The ecology of the tropical salamander, *Bolitoglossa subpalmata*. *Revista de Biologia Tropical* 15: 13–115.

Vidal, N., S. G. Kindl, A. Wong, and S. B. Hedges. 2000. Phylogenetic relationships of xenodontine snakes inferred from 12S and 16S ribosomal RNA sequences. *Molecular Phylogenetics and Evolution* 14: 389–402.

Vidal, N., and G. Lecointre. 1998. Weighting and congruence: A case study on three mitochondrial genes in pitvipers. *Molecular Phylogenetics and Evolution* 3: 366–74.

Villa, J., L. D. Wilson, and J. D. Johnson. 1988. *Middle American herpetology.* University of Missouri Press, Columbia.

Vitt, L. J. 1992. Lizard mimics millipede. *National Geographic Research and Exploration* 8: 76–95.

Vivó, J. A. 1964. Weather and climate of Mexico and Central America. Pp. 187–215 in R. Wauchope and R. C. West, eds., *Handbook of Middle American Indians,* vol. 1: *Natural environment and early cultures.* University of Texas Press, Austin.

Vogt, R. C. 1997. Comunidades de serpientes. Pp. 503–6 in E. G. Soriano, R. Dirzo, and R. C. Vogt, eds., *Historia natural de Los Tuxtlas.* Universidad Nacional Autonoma de Mexico, D. F., Mexico.

Vrana, P., and W. C. Wheeler. 1992. Individual organisms as terminal entities: Laying the species problem to rest. *Cladistics* 8: 67–72.

Waddell, P. J., and D. Penny. 1996. Evolutionary trees of apes and humans from DNA sequences. Pp. 53–73 in A. Lock and C. R. Peters, eds., *Handbook of human symbolic evolution.* Clarendon Press, Oxford.

Wadge, G., and K. Burke. 1983. Neogene Caribbean plate rotation and associated Central American tectonic evolution. *Tectonics* 2: 633–43.

Wagler, J. 1830. *Natürliches System der Amphibien, mit vorangehender Classification der Säugthiere und Vogel.* J. G. Cotta, Munich, Germany.

Wagner, P. L. 1964. Natural vegetation of Middle America. Pp. 216–64 in R. Wauchope and R. C. West, eds., *Handbook of Middle American Indians,* vol. 1: *Natural environment and early cultures.* University of Texas Press, Austin.

Wagner, W. E., Jr. 1989. Fighting, assessment, and frequency alteration in Blanchard's cricket frog. *Behavioral Ecology and Sociobiology* 25: 429–36.

Wake, D. B. 1966. Comparative osteology and evolution of the lungless salamanders, family Plethodontidae. *Memoirs, Southern California Academy of Sciences* 4: 1–111.

———. 1970. The abundance and diversity of tropical salamanders. *American Naturalist* 104: 211–13.

———. 1987. Adaptive radiation of salamanders in Middle American cloud forests. *Annals of the Missouri Botanical Garden* 74: 242–64.

————. 1991a. Homoplasy: The result of natural selection, or evidence of design limitations? *American Naturalist* 138: 543–67.

————. 1991b. Declining amphibian populations. *Science* 203: 860.

Wake, D. B., and A. H. Brame Jr. 1963a. The status of the plethodontid salamander genera *Bolitoglossa* and *Magnadigita. Copeia* 1963: 382–87.

————. 1963b. A new species of Costa Rican salamander, genus *Bolitoglossa. Revista de Biologia Tropical* 11: 63–73.

Wake, D. B., and J. A. Campbell. 2001. An aquatic plethodontid salamander from Oaxaca, Mexico. *Herpetologica* 57: 508–13.

Wake, D. B., and S. Deban. 2000. Terrestrial feeding in salamanders. Pp. 95–116 in K. Schwenk, ed., *Feeding: Form, function, and evolution in tetrapod vertebrates.* Academic Press, San Diego.

Wake, D. B., and I. Dresner. 1967. Functional morphology and evolution of tail autotomy in salamanders. *Journal of Morphology* 122: 265–306.

Wake, D. B., and P. Elias. 1983. New genera and a new species of Central American salamanders, with a review of the tropical genera (Amphibia, Caudata, Plethodontidae). *Contributions to Sciences, Museum of Natural History, Los Angeles County* 345: 1–19.

Wake, D. B., and J. F. Lynch. 1976. The distribution, ecology, and evolutionary history of plethodontid salamanders in tropical America. *Natural History Museum of Los Angeles County Science Bulletin* 25: 1–65.

Wake, D. B., T. J. Papenfuss, and J. F. Lynch. 1992. Distribution of salamanders along elevational transects in Mexico and Guatemala. *Tulane Studies in Zoology and Botany* (suppl.) 1: 303–19.

Wake, M. H. 1978. The reproductive biology of *Eleutherodactylus jasperi* (Amphibia, Anura, Leptodactylidae), with comments on the evolution of live-bearing system. *Journal of Herpetology* 12: 121–33.

————. 1989. Phylogenesis of direct development and viviparity. Pp. 235–50 in D. B. Wake and G. Roth, eds., *Complex organismal functions: Integration and evolution in vertebrates.* John Wiley, Chichester, U.K.

————. 1992. Biogeography of Mesoamerican caecilians (Amphibia: Gymnophiona). *Tulane Studies in Zoology and Botany* (suppl.) 1: 321–25.

————. 1993. The evolution of oviductal gestation in amphibians. *Journal of Experimental Zoology* 266: 394–413.

Wallace, A. R. 1867. Mimicry and other protective resemblances among animals. *West Minster and Foreign Quarterly Review,* n.s., 32: 1–43.

————. 1870. *Natural selection and tropical nature: Essays on descriptive and theoretical biology.* Macmillan, London.

Wallace, D. R. 1992. *The quetzal and the macaw: The story of Costa Rica's national parks.* Sierra Club Books, San Francisco.

Warkentin, K. M. 1995. Adaptive plasticity in hatching age: A response to predation risk trade-offs. *Proceedings of the National Academy of Sciences* (USA) 92: 3507–10.

————. 1997. Life on the leaf. *Fauna* 1: 8–20.

————. 1999a. Effects of hatching age on development and hatchling morphology in the red-eyed treefrog, *Agalychnis callidryas*. *Biological Journal of the Linnean Society* 68: 443–70.

———. 1999b. The development of behavioral defenses: A mechanistic analysis of vulnerability in red-eyed treefrog hatchlings. *Behavioral Ecology* 10: 251–62.

———. 2000. Wasp predation and wasp-induced hatching of red-eyed treefrog eggs. *Animal Behaviour* 60: 503–10.

Warner, R. R. 1987. Female choice of sites versus mates in a coral reef fish, *Thalassoma bifasciatum*. *Animal Behaviour* 35: 1470–78.

Warner, R. R., and E. T. Schultz. 1992. Sexual selection and male characteristics in the bluehead wrasse, *Thalassoma bifasciatum*: Mating site acquisition, mating site defense, and female choice. *Evolution* 46: 1421–42.

Warrell, D. A. 1991. Animals hazardous to humans: Snakes. Pp. 887–89 in G. T. Hunter, *Tropical medicine*. W. B. Saunders, Philadelphia.

Warren, A. 1973. *Report of the British Expedition to Mt. Roraima in South America*. Privately printed, London.

Warren, B. R., and B. I. Crother. 2001. Métodos en biogeografia cladistica: El ejemplo del Caribe. Pp. 233–43 in J. L. Bousquets and J. J. Morrone, eds., *Introducción a la biogeografía en Latinoamérica: Teroías, concepts, métodos, y aplicaciones*. Prensas de Ciencias, Universidad Nacional Autónoma de México.

Wassersug, R. J. 1971. On the comparative palatability of some dry-season tadpoles from Costa Rica. *American Midland Naturalist* 86: 101–9.

Wassersug, R. J., and W. F. Pyburn. 1987. The biology of the Pe-ret' toad, *Otophryne robusta* (Microhylidae), with special consideration of its fossorial larva and systematic relationships. *Zoological Journal of the Linnean Society* 91: 137–69.

Watkins, J. 1984. *Science and scepticism*. Princeton University Press, Princeton, NJ.

Weatherhead, P. J., and R. G. Clark. 1994. Natural selection and sexual size dimorphism in red-winged blackbirds. *Evolution* 48: 1071–79.

Webb, J. K., M. T. Murphy, V. V. Flambaum, V. A. Dzuba, J. D. Barrow, C. W. Churchill, J. X. Prochaska, and A. M. Wolfe. 2001. Further evidence for cosmological evolution of the fine structure constant. *Physical Review Letters* 87.

Welch, A., R. Semlitsch, and H. Gerhardt. 1998. Call duration as an indicator of genetic quality in male gray treefrogs. *Science* 280: 1928–30.

Wells, K. D. 1977. The social behavior of anuran amphibians. *Animal Behaviour* 25: 666–93.

Werman, S. D. 1984a. Taxonomic comments on the Costa Rican pitviper, *Bothrops picadoi*. *Journal of Herpetology* 18: 207–10.

———. 1984b. Taxonomic status of *Bothrops supraciliaris* Taylor. *Journal of Herpetology* 18: 484–86.

———. 1992. Phylogenetic relationships of Central and South American pitvipers of the genus *Bothrops* (sensu lato): Cladistic analyses of biochemical and anatomical characters. Pp. 21–40 in J. A Campbell and E. D. Brodie Jr., eds., *Biology of the pitvipers*. Selva, Tyler, TX.

———. 1997. Systematic implications of lactate dehydrogenase isozyme phenotypes in Neotropical pitvipers (Viperidae: Crotalidae). Pp. 79–88 in R. S. Thorpe, W. Wüster, and A. Malhotra, eds., *Venomous snakes: Ecology, evolution, and snakebite.* Oxford University Press, Oxford.

————. 1999. Molecular phylogenetics and morphological evolution in Neotropical pitvipers: An evaluation of mitochondrial DNA sequence information and the comparative morphology of the cranium and palatomaxillary arch. *Kaupia* 8: 113–26.

Werman, S. D., B. I. Crother, and M. E. White. 1999. Phylogeny of some Middle American pitvipers based on a cladistic analysis of mitochondrial 12S and 16S DNA sequence information. *Contemporary Herpetology* 3.

West, D. A. 2003. *Fritz Müller: A naturalist in Brazil.* Pocahontas Press, Blacksburg, VA.

West-Eberhard, M. J. 2001. The importance of taxon-centered research in biology. Pp. 3–7 in M. J. Ryan, ed., *Anuran communication.* Smithsonian Institution Press, Washington, DC.

Wettstein, O. 1934. Ergebnisse der österreichischen biologischen Costa Rica—Expedition 1930. Die Amphibien und Reptilien. *Sitzungsberichten der Akademie der Wissenschaften in Wien Mathem.-naturw. Klasse* 143: 1–39.

Whitley, H. 1884. Explorations in the neighbourhood of Mounts Roraima and Kukenam, in British Guiana. *Proceedings of the Royal Geographical Society* 6: 452–63.

Whitmore, T. C., and G. T. Prance. 1987. *Biogeography and Quaternary history of tropical America.* Oxford Monographs on Biogeography, no. 3. Oxford University Press, Oxford.

Wickler, W. 1968. *Mimicry in plants and animals.* McGraw-Hill, New York.

Wiest, J. A., Jr. 1982. Anuran succession at temporary ponds in a post oak-savannah region of Texas. Pp. 39–47 in N. J. Scott Jr., ed., *Herpetological communities.* Research Report No. 13. U.S. Fish and Wildlife Service, Washington, DC.

Wilbur, H. M. 1980. Complex life cycles. *Annual Review of Ecology and Systematics* 11: 67–93.

Wilczynski, W., and E. A. Brenowitz. 1988. Acoustic cues mediate intermale spacing in a Neotropical frog. *Animal Behaviour* 36: 1054–63.

Wild, E. R. 1995. New genus and species of Amazonian microhylid frog with a phylogenetic analysis of New World genera. *Copeia* 1995: 837–49.

Wiley, E. O. 1981a. Convex groups and consistent classification. *Systematic Botany* 6: 346–58.

————. 1981b. *Phylogenetics: The theory and practice of phylogenetic systematics.* Wiley-Interscience, New York.

Wilkinson, M. 1995. Coping with abundant missing entries in phylogenetic inference using parsimony. *Systematic Biology* 44: 501–14.

Willard, D. E., M. S. Foster, G. F. Barrowclough, R. W. Dickerman, P. F. Cannell, S. L. Coats, J. L. Cracraft, and J. P. O'Neill. 1991. The birds of Cerro de la Neblina, Territorio Federal Amazonas, Venezuela. *Fieldiana: Zoology* 65: 1–80.

Williams, E. E. 1969. The ecology of colonization as seen in the zoogeography of anoline lizards on small islands. *Quarterly Review of Biology* 44: 345–89.

————. 1974. A case history in retrograde evolution: The *onca* lineage in anoline lizards. I. *Anolis annectens* new species, intermediate between the genera *Anolis* and *Tropidodactylus. Breviora* 421: 1–21.

————. 1976a. West Indian anoles: A taxonomic and evolutionary summary. I. Introduction and a species list. *Breviora* 440: 1–21.

————. 1976b. South American anoles; the species groups. *Papeis Avulsos Zoologia* 29: 259–68.

————. 1983. Ecomorphs, faunas, island size, and diverse end points in island radiations of *Anolis*. Pp. 326–70 in R. B. Huey, E. R. Pianka, and T. W. Schoener, eds., *Lizard ecology: Studies of a model organism*. Harvard University Press, Cambridge.

————. 1989. A critique of Guyer and Savage (1986): Cladistic relationships among anoles (Sauria: Iguanidae): Are the data available to reclassify the anoles? Pp. 433–78 in C. A. Woods, ed., *Biogeography of the West Indies*. Sandhill Crane Press, Gainesville, FL.

Williams, E. E., M. J. Praderio, and S. Gorzula. 1996. A phenacosaur from Chimantá tepui, Venezuela. *Breviora* 506: 1–15.

Williams, K. L. 1988. *Systematics and natural history of the American milk snake, Lampropeltis triangulum*. Milwaukee Public Museum, Milwaukee, WI.

Willmann, R. 1987. Phylogenetic systematics, classification and the plesion concept. *Verhandlungen des naturwissenschaftlichen Vereins in Hamburg* 29: 221–33.

————. 1989. Paleontology and the systematization of natural taxa. *Abhandlungen des naturwissenschaftlichen Vereins in Hamburg* 28: 267–91.

Wilson, C., and M. McPhaul. 1996. A and B forms of the androgen receptor are expressed in a variety of human tissues. *Molecular and Cellular Endocrinology* 120: 51–57.

Wilson, E. O. 1999. Prologue. Pp. xi–xii in G. Daws and M. Fujita, eds., *Archipelago: The islands of Indonesia*. University of California Press, Berkeley.

Wilson, L. D., and J. R. McCranie. 1997. Publication in non-peer-reviewed outlets: The case of Smith and Chiszar's "Species-group taxa of the false coral snake genus *Pliocercus*." *Herpetological Review* 28: 18–21.

Wilson, L. D., J. R. McCranie, and M. R. Espinal. 1996. Coral snake mimics of the genus *Pliocercus* (family Colubridae) in Honduras and their mimetic relationships with *Micrurus* (family Elapidae). *Herpetological Natural History* 4: 57–63.

Winsor, L. 1983. Vomiting of land planarians (Turbellaria: Tricladia) ingested by cats. *Australian Veterinary Journal* 60: 282–83.

Witschi, E. 1961. Sex and secondary sexual characteristics. Pp. 115–68 in A. Marshall, ed., *Biology and comparative physiology of birds*. Academic Press, London.

Woodward, B. D. 1982. Male persistence and mating success in Woodhouse's toad (*Bufo woodhousei*). *Ecology* 63: 583–85.

Wright, D. D., J. H. Jessen, P. Burke, and H. G. de Silva Garza. 1997. Tree and liana enumeration and diversity on a one-hectare plot in Papua New Guinea. *Biotropica* 29: 250–60.

Wright, R. M. 1976. Rain forest: Conservancy assists in the establishment of Costa Rica's Corcovado Park. *Nature Conservancy News* 26: 17–21.

Wüster, W., M. G. Salomão, G. J. Duckett, and R. S. Thorpe. 1999. Mitochondrial DNA phylogeny of the *Bothrops atrox* species complex (Squamata: Serpentes: Viperidae). *Kaupia* 8: 35–144.

Wüster, W., M. G. Salomão, R. S. Thorpe, G. Puorto, M. F. D. Furtado, S. A. Hoge, R. D. G. Theakston, and D. A. Warrell. 1997. Systematics of the *Bothrops atrox* complex: New insights from multivariate analysis and mitochondrial DNA sequence information. Pp. 99–113 in R. S. Thorpe, W. Wüster, and A. Malhotra, eds., *Venomous snakes: Ecology, evolution and snakebite*. Symposia of the Zoological Society of London. Clarendon Press, Oxford.

Wüster, W., R. S. Thorpe, and G. Puorto. 1996. Systematics of the *Bothrops atrox* complex (Reptilia: Serpentes: Viperidae) in Brazil: A multivariate analysis. *Herpetologica* 52: 263–71.

Wyles, J. S., and G. C. Gorman. 1980a. The albumin immunological and Nei eletrophoretic distance correlation: A calibration for the saurian genus *Anolis* (Iguanidae). *Copeia* 1980: 66–71.

———. 1980b. The classification of *Anolis:* Conflict between genetic and osteological interpretation as exemplified by *Anolis cybotes. Journal of Herpetology* 14: 149–53.

Wyss, A. R., and J. Meng. 1996. Application of phylogenetic taxonomy to poorly resolved crown clades: A stem-modified node-based definition of Rodentia. *Systematic Biology* 45: 559–68.

Yang, S. Y., M. Soulé, and G. C. Gorman. 1974. *Anolis* lizards of the eastern Caribbean: A case study in evolution. I. Genetic relationships, phylogeny, and colonization sequence of the *roquet* group. *Systematic Zoology* 23: 387–99.

Yang, Z. 1994a. Estimating the pattern of nucleotide substitution. *Journal of Molecular Evolution* 39: 105–11.

———. 1994b. Maximum likelihood phylogenetic estimation from DNA sequences with variable rates over sites: Approximate methods. *Journal of Molecular Evolution* 39: 306–14.

———. 1996. Among-site rate variation and its impact on phylogenetic analysis. *Trends in Ecology and Evolution* 11: 367–72.

Yanosky, A. A., and J. M. Chani. 1988. Possible dual mimicry of *Bothrops* and *Micrurus* by the colubrid, *Lystrophis dorbignyi. Journal of Herpetology* 22: 222–24.

Yoder, A. E., J. A. Irwin, and B. A. Payseur. 2001. Failure of the ILD to determine data combinability for slow loris phylogeny. *Systematic Biology* 50 (3): 408–24.

Young, B. A., K. Meltzer, and C. Marsit. 1999. Scratching the surface of mimicry: Sound production through scale abrasion in snakes. *Hamadryad* 24: 29–38.

Young, B. A., S. Sheft, and W. Yost. 1995. Sound production in *Pituophis melanoleucus* (Serpentes: Colubridae) with the first description of a vocal chord in snakes. *Journal of Experimental Zoology* 273: 472–81.

Young, B. A., J. Solomon, and G. Abishahin. 1999. How many ways can a snake growl? The morphology of sound production in *Ptyas mucosus* and its potential mimicry of *Ophiophagus. Herpetological Journal* 9: 89–94.

Young, L., and D. Crews. 1995. Comparative neuroendocrinology of steroid receptor gene expression and regulation: Relationship to physiology and behavior. *Trends in Endocrinology and Metabolism* 6: 317–23.

Young, W., R. Goy, and C. Phoenix. 1964. Hormones and sexual behavior. *Science* 143: 212–18.

Zaher, H. 1994. Les Tropidopheoidea (Serpentes; Alethinophidia) sont-ils réellement monophylétiques? Arguments en favor de leur polyphylétisme. *Comptes Rendus de l'Académie des Sciences Paris* 317: 471–78.

———. 1998. The phylogenetic position of *Pachyrhachis* within snakes (Squamata, Lepidosauria). *Journal of Vertebrate Paleontology* 18: 1–3.

———. 1999. Hemipenial morphology of the South American xenodontine snakes with a proposal for a monophyletic Xenodontinae and a reappraisal of colubroid hemipenes. *Bulletin of the American Museum of Natural History* 240: 1–168.

Zaher, H., and O. Rieppel. 1999a. Tooth implantation and replacement in squamates, with special reference to mosasaur lizards and snakes. *American Museum Novitates* 327: 1–79.

———. 1999b. The phylogenetic relationships of *Pachyrhachis problematicus,* and the evolution of limblessness in snakes (Lepidosauria, Squamata). *Comptes Rendus de l'Académie des Sciences Paris* 329: 831–37.

———. 2000. A brief history of snakes. *Herpetological Review* 31: 73–76.

Zamudio, K. R., and H. W. Greene. 1997. Phylogeography of the bushmaster (*Lachesis muta:* Viperidae): Implications for Neotropical biogeography, systematics, and conservation. *Biological Journal of the Linnean Society* 62: 421–42.

Zar, J. H. 1984. *Biostatistical analysis.* Prentice-Hall, Englewood Cliffs, NJ.

Zimmerman, B. L. 1994. Audio strip transects. Pp. 92–96 in W. R. Heyer, M. A. Donnelly, R. W. McDiarmid, L. C. Hayek, and M. S. Foster, eds., *Measuring and monitoring biological diversity: Standard methods for amphibians.* Smithsonian Institution Press, Washington, DC.

Zimmerman, B. L., and M. T. Rodrigues. 1990. Frogs, snakes, and lizards of the INPA-WWF reserves near Manaus, Brazil. Pp. 426–54 in A. H. Gentry, ed., *Four Neotropical forests.* Yale University Press, New Haven, CT.

Zrzavy, J., and O. Nedved. 1999. Evolution of mimicry in the New World *Dysdercus* (Hemiptera: Pyrrhocoridae). *Journal of Evolutionary Biology* 12: 956–69.

Zug, G. R., L. J. Vitt, and J. P. Caldwell. 2001. *Herpetology: An introductory biology of amphibians and reptiles.* Academic Press, San Diego.

Zuk, M. 1988. Parasite load, body size, and age of wild-caught male field crickets (Orthoptera: Gryllide): Effects on sexual selection. *Evolution* 42: 969–76.

Zweifel, R. 1955. Ecology, distribution, and systematics of frogs of the *Rana boylei* group. *University of California Publications in Zoology* 54: 207–92.

Contributors

A. Luz Aquino
Bosque Atlántico Interior–Paraguay
World Wildlife Fund
Cañada del Carmen No. 2780
e/Cap. José Domingo Jara y. Cap. Arturo
   Scario
Edificio Arami, Planta Bajo-Apto. 8
Asunción
Paraguay
laquino@sce.cnc.una.py

David P. Bickford
Section of Integrative Biology
One University Station
University of Texas
Austin, TX 78712
rokrok@mail.utexas.edu

Megan H. Chen
Parks RFTA
Building 791
AFRC–FMC–ENV
Dublin, CA 94568-5201

Shyh-Hwang Chen
Department of Life Sciences
National Taiwan Normal University
88 Ting-chou Road, Sect 4
Taipei, Taiwan 116
Republic of China

Brian I. Crother
Department of Biology
Southeastern Louisiana University
Hammond, LA 70402
bcrother@selu.edu

Rafael O. de Sá
Department of Biology
University of Richmond
Richmond, VA 23173
rdesa@richmond.edu

Maureen A. Donnelly
Department of Biological Sciences
Florida International University
11200 SW 8th Street, OE 167
Miami, FL 33199
donnelly@fiu.edu

Sharon B. Emerson
University of Utah
Department of Biology
257 South, 1400 East
Salt Lake City, UT 84112
emerson@biology.utah.edu

Harry W. Greene
Department of Ecology and Evolutionary
    Biology
Corson Hall
Cornell University
Ithaca, NY 14853-2701
hwg5@cornell.edu

Craig Guyer
Department of Biological Sciences
Auburn University
Auburn, AL 36849
cguyer@acesag.auburn.edu

W. Ronald Heyer
MRC 162, NHB W-201
Smithsonian Institution
P.O. Box 37012
Washington, DC 20013-7012
heyer.ron@nmnh.si.edu

Maria Kelly-Smith
School of Veterinary Medicine
Skip Bertman Drive
Louisiana State University
Baton Rouge, LA 70803

Arnold G. Kluge
Museum of Zoology
University of Michigan
Ann Arbor, MI 48109-1079
akluge@umich.edu

Robin Lawson
Osher Foundation Laboratory for
    Molecular Systematics
Department of Herpetology
California Academy of Sciences
Golden Gate Park
San Francisco, CA 94118
rlawson@calacademy.org

Karen R. Lips
Department of Zoology
Southern Illinois University
Carbondale, IL 62901-6501
klips@zoology.siu.edu

Roy W. McDiarmid
USGS Patuxent Wildlife Research Center
Smithsonian Institution
P.O. Box 37012
National Museum of Natural History
Room 378, MRC 111
Washington, DC 20013-7012
mcdiarmid.roy@nmnh.si.edu

Sarah Muller
Department of Biology
University of Richmond
Richmond, VA 23173

Kirsten E. Nicholson
Department of Biology
Campus Box 1137
Washington University
St. Louis, MO 63130-4899
knicholson@biology2.wustl.edu

Gabriela Parra-Olea
Instituto de Biologia, UNAM
AP 70-153 Ciudad Universitaria
CP 04510
Mexico, D.F.
gparra@ibunam.ibiologia.unam.mx

Jay M. Savage
Rana Dorada Enterprises, S.A.
3401 Adams Avenue, Suite A
San Diego, CA 92116-2126
savy1@cox.net

Norman J. Scott Jr.
3655 Linndquist Lane
P.O. Box 307
Creston, CA 93432
reptile@tcsn.net

Judy P.-Y. Sheen
U.S. Army Corps of Engineers
San Francisco District
333 Market Street
San Francisco, CA 94105
judypsheen@yahoo.com

Joseph B. Slowinski (deceased)
Department of Herpetology
California Academy of Sciences
Golden Gate Park
San Francisco, CA 94118

David B. Wake
Department of Integrative Biology
Museum of Vertebrate Zoology
University of California
Berkeley, CA 94720-3160
wakelab@uclink4.berkeley.edu

Marvalee H. Wake
Department of Integrative Biology
Museum of Vertebrate Zoology
3060 VLSB
University of California
Berkeley, CA 94720-3140
mhwake@socrates.berkeley.edu

Graham G. Watkins
Iwokrama International Centre
P.O. Box 10630
77 High Street
Kingston, Georgetown
Guyana
gwatkins@iwokrama.org

Steven D. Werman
Department of Biological Sciences
Western Colorado Center for Tropical
    Research
Mesa State College
1100 North Avenue
Grand Junction, CO 81501-3122
swerman@mesastate.edu

Mary E. White
Department of Biology
Southeastern Louisiana University
Hammond, LA 70402
mwhite@selu.edu

Subject Index

Note: The letter *f* following a page reference denotes a figure; *t* denotes a table.

Taxonomic Index

Note: The letter *f* following a page reference denotes a figure; *t* denotes a table.